REVERSE OSMOSIS SYSTEMS

REVERSE OSMOSIS SYSTEMS

Design, Optimization, and Troubleshooting Guide

SYED JAVAID ZAIDI

HALEEMA SALEEM

ELSEVIER

Elsevier
Radarweg 29, PO Box 211, 1000 AE Amsterdam, Netherlands
The Boulevard, Langford Lane, Kidlington, Oxford OX5 1GB, United Kingdom
50 Hampshire Street, 5th Floor, Cambridge, MA 02139, United States

Notices

Knowledge and best practice in this field are constantly changing. As new research and experience broaden our understanding, changes in research methods, professional practices, or medical treatment may become necessary.

Practitioners and researchers must always rely on their own experience and knowledge in evaluating and using any information, methods, compounds, or experiments described herein. In using such information or methods they should be mindful of their own safety and the safety of others, including parties for whom they have a professional responsibility.

To the fullest extent of the law, neither the Publisher nor the authors, contributors, or editors, assume any liability for any injury and/or damage to persons or property as a matter of products liability, negligence or otherwise, or from any use or operation of any methods, products, instructions, or ideas contained in the material herein.

British Library Cataloguing-in-Publication Data
A catalogue record for this book is available from the British Library

Library of Congress Cataloging-in-Publication Data
A catalog record for this book is available from the Library of Congress

ISBN: 978-0-12-823965-0

For Information on all Elsevier publications
visit our website at https://www.elsevier.com/books-and-journals

Publisher: Susan Dennis
Acquisitions Editor: Anita Koch
Editorial Project Manager: Allison Hill
Production Project Manager: Bharatwaj Varatharajan
Cover Designer: Greg Harris

Typeset by MPS Limited, Chennai, India

Working together
to grow libraries in
developing countries

www.elsevier.com • www.bookaid.org

Contents

About the Authors

Prof. Syed Javaid Zaidi (Ph.D.), szaidi@qu.edu.qa, smjavaidzaidi@gmail.com

Prof. Syed Javaid Zaidi is presently QAFAC Chair Professor of chemical engineering in the Center for Advanced Materials, Qatar University. He did BSc in chemical engineering from Aligarh Muslim University, India, MS in chemical engineering from King Fahd University of Petroleum & Minerals, and Ph.D. in chemical engineering from Laval University, Canada. He has more than 25 years of teaching and research experience in the area of clean water and clean energy. He has worked in King Fahd University of Petroleum & Minerals, The University of Queensland Australia, Laval University, Canada and had been a visiting scholar at the Massachusetts Institute of Technology (MIT), United States. He has extensive experience in the desalination research and has supervised many graduate students in clean water and clean energy. He has published more than 300 papers in international journals, book chapters, and conferences and has 14 US patents and patent applications. He has conducted joint projects in reverse osmosis desalination in collaboration with the MIT, United States for the development of nanostructured membranes for reverse osmosis desalination. He has collaborated with industry and provided technical support to industry in water technology and clean energy projects. He was a founding member of the Center of Research Excellence in Renewable Energy, a national research center, was its co-Director, and a member of the Board of Directors. He was a member of the Research Advisory Board of University headed by the rector of university, a member of the National Unifying efforts in desalination in KSA, and presently an honorary member of the Advisory Council of Arab Water Desalination (ARWDEX). At Qatar University, he took the initiative for the establishment of the Water Technology unit and coordinated its activities. He has been invited by UNESCO to contribute to the Encyclopaedia of Life Support Systems (EOLSS), which is the world's largest Encyclopedia developed under the auspices of UNESCO and contributed two chapters on "Seawater Desalination Membranes and Experience of SWRO in Qatar" and "Salt and Water Transport in Reverse Osmosis and Biofouling". He is the recipient of many national and international awards including the Excellence in Research Award, Patent Award, Almari Prize for Research Innovation, and Lifetime Achievement Award by Venus International Foundation

Affiliations: Center for Advanced Materials (CAM), Qatar University, Doha, Qatar

Haleema Saleem (M.Sc. Chem Engg), haleema.saleem@qu.edu.qa, haleemasaleem@gmail.com

Haleema Saleem is a Researcher in the Center for Advanced Materials, Qatar University, and she holds an MSc degree in chemical engineering from Khalifa University (Petroleum Institute), UAE. She is also pursuing her Ph.D. in chemical engineering from Universiti Teknologi Malaysia.

Haleema has great interest and knowledge in the reverse osmosis membranes and processes. She has about 13 years of industrial and academic experience. Haleema also has good experience working with several polymers and nanomaterials. She has 45 publications in high impact international peer-reviewed journals and as book chapters. Haleema has extensive work experience in water treatment, reverse osmosis membrane testing and characterization and collaborated on several industrial projects. She has also been an invited speaker at numerous international seminars and conferences on desalination and nanomaterials. She is also the co-author of two chapters on "Seawater Desalination Membranes and Experience of SWRO in Qatar" and "Salt and Water Transport in Reverse Osmosis and Biofouling" in the UNESCO Encyclopaedia of Life Support Systems (EOLSS), which is the world's largest Encyclopedia developed under the auspices of UNESCO. Haleema has also mentored several graduate students on the environmental impact assessment of desalination plants, produced water management, water treatment, and nanofiber membranes. Her research area is mainly focused on desalination, reverse osmosis, membranes, thin-film composites, nanomaterials, etc.

Affiliations: Center for Advanced Materials (CAM), Qatar University, Doha, Qatar

Foreword *by*
Prof. Mariam Ali Al-Maadeed

Water security is one of the major concerns the world is facing today, especially in dry and arid areas such as the Middle East, where water is scarce and limited. This by-product of climate change is perhaps the leading existential crisis in regions with arid/hyperarid climates, where rainfall is rare and erratic and thus lacks permanent sources of surface water. This situation, coupled with the excessive exploitation of the limited amount of underground water, makes seawater desalination the best alternative to produce fresh water and ultimately meet the increasing demand of modern society.

I have known Dr. Syed Zaidi since 2015, when he first joined the Center for Advanced Materials at Qatar University, where he had since contributed to multiple important collaboration initiatives with the local industry in Qatar and international institutes worldwide.

As an expert in Academia/industry relations and with my long experience in technology and education, I consider Dr. Zaidi's research as a valuable addition that can bridge the gap between industries and academia and demonstrate how the basic knowledge that is initiated in universities can be translated efficiently to real-world applications. Most importantly, the information presented by the author is related to innovative advancement in water desalination through the use of the reverse osmosis (RO) technique, an essential and advantageous shift from energy-intensive thermal techniques at a low cost and better quality.

Desalination is a rapidly growing field, inducing more research, teaching, training programs, and broader use. This is an important book for those involved in academic and industrial work in the improvement of design performance and troubleshooting. The book is timely as Qatar desalination industry is shifting from thermal-based desalination to membrane-based RO desalination technology. The book covers latest information on the design of RO system, its operation, and troubleshooting. It provides guidelines for best RO design and operational performance and is a perfect training guide for engineers and technicians who are responsible for RO system design, operation, and maintenance.

The book, as written by a leading expert, provides a high-quality overview on RO desalination from Qatar and demonstrates the progress in this area. It is an essential reference for professionals, researchers, and students in environmental engineering practices and related fields. It is also a helpful tool for managers and official decision-makers that face water scarcity issues worldwide.

Prof. Mariam Ali Al-Maadeed
Vice President for Research and Graduate Studies
Qatar University

Foreword by
Mr. Fahad H. Al-Mohannadi

Considering the freshwater scarcity in different regions of the world, desalination has proved to be the best solution to overcome water scarcity in the Gulf region. The saline water, which is plentiful, could be desalinated for producing potable water appropriate for human consumption. Among various desalination technologies presently available, reverse osmosis (RO) has turned out to be the extensively used process globally for both seawater and brackish water desalination due to its environmental benefits. This membrane-based RO technology is considered to be mature and is available in most of the coastal regions in the world. Because of its low energy consumption and environmental benefits, majority plants are shifting from the energy-intensive thermal-based process to membrane-based technology for seawater desalination.

One of the major grand challenges of Qatar National Vision 2030 (QNV 2030) is producing and providing safe drinking water. As Qatar is among the most arid regions in the world with no natural water resource and very minimal rainfall, the complete water requirement is fulfilled by seawater desalination. The first desalination plant was commissioned in 1953 with a capacity of 680 m3/day. Desalination provides more than 99% of Qatar's municipal water demand. Qatar is investing heavily in new projects and new technologies to further boost the production of desalinated water. Two large RO desalination plants have been set up recently and the third one is in the commissioning stage. One in "Ras Abu Fontas 3" with a capacity of 36 MIGD and the second in "Umm Al Houl" plant with a total capacity of 136.5 MIGD out of which 60 MIGD is based on RO technology, which is the largest RO plant in the Gulf region and other 76.5 MIGD is based on thermal plant integrated with power generation technology.

I have been involved in the commissioning and setting-up of desalination plants in Qatar and have more than 30 years of experience in the desalination and power industry. I have been the Chairman and Board of Directors of many desalination and power companies, such as Umm Al-Houl Power, QPower, Nebras Power, who are involved in the water and electricity production in Qatar. The present book is coming at the right time when Qatar is building its capacity in the RO-based desalination and needs to develop local know-how and train the plant operators. It addresses important issues facing the RO desalination, such as fouling and scaling control, and membrane degradation. The book is very well organized and comprehensive, and it will be a perfect tool for engineers, operators, and decision makers who are responsible for RO system maintenance. The book also describes the important design parameters and step-by-step procedures about designing a reverse osmosis system.

My special appreciations go to the authors for their great effort in writing this timely and needed book. The main author of the book, Prof. Syed Zaidi is not strange to the desalination community. He is an accomplished scholar and internationally renowned for his research in membrane technology for clean water RO desalination and clean energy. He is also collaborating with QEWC for the improvement of RO system operation and pretreatment processes.

I congratulate the authors and the Elsevier publishing company for their valuable contribution to the RO desalination technology. I am glad to write the foreword of this book on the RO system design, which will be a very useful source of information for the desalination industry especially in Qatar and the GCC region.

Fahad H. Al-Mohannadi
Managing Director & General Manager,
Qatar Electricity & Water Company (QEWC)

Preface

There is nothing more essential to life on earth than water. By 2025, almost 1.8 billion people will live in areas affected by water scarcity, with two-thirds of the global population living in water-stressed regions as a result of increased use, population growth, as well as climatic change. Hence, the countries and regions need to urgently tackle the crucial issues caused by water stress. Saline water could be made into freshwater by a process known as desalination. Desalination is considered to be an increasingly common solution to supply potable water in several regions of the world where the water resources are rare. As drought situations magnify, desalination will turn into a long-term solution for such problems instead of a temporary fix. Developing cost-effective and sustainable systems will enable technology providers to capitalise on this potential technology. There are almost 15,906 operational desalination plants producing around 95 million m3/day of desalinated water for human consumption, of which 48% is produced in the Middle East and North Africa region. The majority of desalinated water is produced either by thermal-based or membrane-based technology. Presently, the pressure-driven membrane-based reverse osmosis (RO) has become an extensively used process globally for both seawater desalination and brackish water desalination. Due to the fact that RO systems operate at ambient temperatures and are more energy efficient compared to thermal-based systems, they are more used in the desalination of seawater to make potable water.

Presently used RO membranes are the superior technology for advanced desalination facilities, and they are applied to various salt water resources using customized pretreatment and membrane system design. Membrane operating pressures and membrane costs have decreased remarkably over the past decade and membrane durability has increased, which has lowered RO process costs relative to thermal-based desalination technologies. Well designed and properly operated systems give a trouble-free performance over long periods of time. On the other hand, mistakes made during the design or operation of RO systems can lead to ongoing problems and reduced membrane performance and membrane useful life. The membrane system, which includes the membrane elements housed in pressure vessels and a high pressure pump constitutes the heart of the RO plant. The membrane scaling caused by the precipitation of salts is a common problem in RO process. Membranes are also susceptible to fouling by fine suspended solids, oil, and microbes. Pretreatment to mitigate scaling and fouling must be carefully evaluated for each case. Most of the RO plant design that have problems in the operation and maintenance in the Middle East was due to inadequate and improper pretreatment. To the best of our knowledge, there is no dedicated book for RO systems design, performance and troubleshooting, published from the Gulf region so far, with special focus on RO desalination plants and studies carried out in this region. The most distinguishing feature of this book is that it presents various latest version RO design and simulation software programs (with step-by-step procedure screenshots and video tutorial) used by RO desalination plants for designing the RO system and for performance monitoring. Some of the software programs discussed in the book are also posted in Elsevier website for the readers' reference. The link to the website is as follows:

https://www.elsevier.com/books-and-journals/book-companion/9780128239650

This book summarizes some of the issues faced during the design as well as operation of the RO systems, suggest corrective measures and its troubleshooting. It also provides guidelines for the best RO design and operational performance. It demonstrates the monitoring as well as maintaince of the RO systems with their corrective measures. Although there are books that cover specific aspects of this field, but most of them are not presented as a training guide/handbook mainly for the process engineers and technician working in the RO desalination plant that is very important for the everyday assessment of performance of RO system. The goal of a good RO design for a specific required permeate flow is to limit the membrane costs and feed pressure, thereby maximizing the permeate quality as well as recovery. Therefore, this book will be an asset for the design as well as process engineers working in water treatment and desalination industry using RO systems, RO equipment manufacturing companies, engineering companies, and those who are in-charge of the operation of a reverse osmosis system. Authors believe that proper knowledge about the RO fundamentals is extremely important for analyzing and evaluating the performance of a reverse osmosis desalination process.

In the introductory section, the book covers the history of RO, along with the fundamentals, principles, and equations. Following sections cover the practical areas such as pretreatment, design parameters, design software programs, RO performance monitoring, troubleshooting as well as system engineering. The final section includes the frequently asked questions along with its answers. Moreover, different case studies carried out and recent developments related to RO system performance, membrane fouling, scaling, and degradation studies have been analyzed. The book also has several solved examples, which are detailed in a careful as well as simple manner that help the reader to understand and follow the development. The information used in these case studies are obtained from existing RO desalination plants. The book can also be adopted as the textbook for undergradute and graduate students in Environment Engineering and Chemical Engineering programs.

Chapter 1 is the general introduction to RO process technology. It starts with a short description of the seawater composition and the historical development of RO technology. Further, the essential process steps in a reverse osmosis-based desalination plant, RO technology in GCC countries, the benefits as well as the limitations of RO processes are briefly discussed. Also, different applications of RO process and the future of this technology in desalination are briefly reviewed. At the end of this chapter, explanation of various terms commonly used by industry practitioners in RO process technology is presented.

Chapter 2 discusses about the RO principles and system components. The basics of osmosis as well as RO process are analysed here. Also, the equations for water flux, salt flux, rejection, recovery are also explained. Further, the concentration polarization phenomenon and its associated equations are described. Moreover, the different processes in RO-based desalination plants and various RO system components are presented in this chapter. In the end, the various processes in different seawater desalination plants in the Gulf region are discussed.

Chapter 3 presents the guidelines for quality of water to be used in a reverse osmosis plant. The characteristics of feed water constituents can affect the performance of membrane by generating membrane fouling, scaling, or degradation. It is recommended to follow the limit of each constituent in water for the effective functioning of the RO membrane system. This chapter analyzes different feed water constituents such as suspended solids, hydrogen sulfide, silicon dioxide, microbes, organics, colour, iron, manganese, aluminium, calcium carbonate, barium, strontium, chlorine, magnesium, and hydrocarbons, which influence the RO membrane performance.

Chapter 4 is about the transport models, membrane materials and the basic flow patterns in the RO process technology. This chapter concisely covers the transport models, which describes the flow of water and salt in the RO system, such as non-porous models, porous models, and irreversible thermodynamic models. Also, we discuss in detail the basic membrane materials, their preparation methods, different module configurations, and how they are arranged. Moreover, the the different flow patterns employed in a reverse osmosis system, i.e, arrays, recycle, double pass and multiple trains are analyzed. Also, various seawater reverse osmosis (SWRO) system configurations and hybrid RO desalination systems are described in this chapter.

Chapter 5 discusses about the different pretreatment processes in a reverse osmosis system. For best performance, feed water to the RO system must be properly treated to remove undesirable contaminants and solids, and to prevent fouling of the system. In this chapter, an overview of the conventional, physical, and chemical techniques which are regularly used for pretreatment before the RO process are discussed. Also, different pretreatment techniques for scale prevention and fouling (colloidal, biological, and organic fouling) prevention are analyzed separately in detail. The fouling causes as well as their prevention depend mainly on the feed water being treated, and proper control procedures should be devised for each plant.

Chapter 6 analyses the important parameters to be considered for the design and operation of RO systems. The product water quality and quantity depends on the feed water quality, membrane, system set-up, operation and monitoring techniques, cleaning and maintenance practices. The efficiency of membrane elements operating in a reverse osmosis system is influenced mainly by the feed water source, salinity, composition, feed water flow, pressure, temperature, as well as pH. The key performance parameters of a reverse osmosis system are the permeate flux and its TDS and salt rejection. Also, we discuss the importance of performing a pilot-scale study in RO desalination plant.

Chapter 7 is the designing of a reverse osmosis system. In this chapter, the RO system design guidelines and the main steps for designing a reverse osmosis membrane system are reviewed. The information presented in this chapter can help the plant engineers for designing and optimizing a reverse osmosis system to match their requirements. Also, some designing examples to properly understand the system designing process are analyzed. Some case studies of RO system design and optimization, leading to low-power consumption and highly effective system operation, are also discussed. With the right system design, experienced service support, and maintenance program, the RO system can offer high purity water for several years.

Chapter 8 is about the latest RO design software programs used in desalination plants for properly designing the RO systems. Here, we discuss some commonly available design software programs such as WAVE (DuPont Water Solutions), IMSDesign (Hydranautics),

TORAYDS2 (Toray Industries, Inc), Lewaplus (LANXESS), ROAM Ver. 2.0 (Microdyn-Nadir), Winflows (SUEZ Water Technologies & Solutions), CSMPRO v6.0 (Toray Advanced Materials Korea) and IPSE software (SimTech Simulation Technology). The purpose of this chapter is to discuss these software programs and their applicability under various design conditions and ease of application and versatility. Simplified methods to use these software programs are illustrated and the screenshots of the results, methods etc. are also given here.

Chapter 9 provides a comprehensive overview of RO performance monitoring, process parameters and their control. In this chapter, the initial start-up and shut down procedures of a reverse osmosis system are discussed. Moreover, the instrumentation and controls involved in a reverse osmosis plant such as SCADA system, plant monitoring, and control etc. have been analyzed. Calculations about the system performance, RO automatic system, process parameters as well as equipment performance monitoring in the RO system are also included. Different RO system performance normalization methods and software programs (RODataXL and TorayTrak) are also analyzed in this chapter. Further, the different performance restoration methods such as chemical cleaning and direct osmosis cleaning have been analyzed here.

Chapter 10 presents the RO membrane performance degradation. This chapter covers details on the three main indicators of membrane performance degradation, namely normalized permeate flow (NPF), normalized salt rejection (NSR), and pressure drop. Moreover, the detailed effects of membrane scaling, fouling, and degradation on the pressure drop, normalized salt rejection, and normalized product flow are discussed. Further, the RO data normalization, data interpretation, and the frequent causes as well as the techniques for the prevention of degradation of membrane performance have been examined. This will help to keep the RO system and pretreatment system perform properly.

Chapter 11 deals with RO system troubleshooting. In spite of the entire pretreatment processes as well as special attention to system hydraulics, majority of RO system units will ultimately show some performance degradation. This chapter discusses the techniques helpful in RO system troubleshooting. The commonly observed issues in the RO system, steps to detect the problems, and the probable causes and solutions for these problems are discussed in this chapter. Also, various investigative strategies that could be used for RO system troubleshooting are analyzed. Moreover, the analyses and tests required to be carried out during a membrane autopsy are analyzed in detail.

Finally, Chapter 12 explores the different issues related to RO system engineering such as membrane cleaning, brine disposal, pressure drop tradeoff, pump issues, membrane fouling, scaling, pretreatment problems, chlorine elimination etc. Any compromise in the pretreatment approaches, quality of equipment, monitoring instrumentation, would generally lead to operational issues in the downstream RO system. Moreover, we discuss the frequently asked questions related to RO system dealing with contaminants, chemical injection, pretreatment, silt density index, RO membranes etc. Also, presented here are some surveys carried out for analyzing the challenges faced by the desalination plants based in Qatar.

The main sectors of interests of this book will be desalination and water treatment. The target audience of this book are process engineers, plant operators, researchers, faculty and students. Those who have interest in RO process can benefit from this book in a number of ways. The engineer responsible for the supervision of RO system operation will find this book as a useful source of information and for operator training.

Feedback from practitioners, researchers, faculty and students who use this book are welcomed. We hope you will find it to be beneficial.

Acknowledgments

The authors would like to express gratitude to a number of people who have helped in the preparation of this book, especially Dr. Nasser Alnuami, Director of the Center for Advanced Materials, Qatar University, who have taken personal interest in the RO desalination book project, research, and related activities. The authors also would like to acknowledge the support of Prof. Mariam Ali Al-Maadeed, Vice President for Research and Graduate Studies, Qatar University, who kindly provided the Foreword of the book and supported the book project.

The authors would also like to extend immeasurable appreciation and deepest gratitude for the support and help received from the following persons and organizations, who in one way or another have contributed in making this work possible:

1. Eng. Jamal Alkhalaf, CEO Umm Al-Houl Power, Qatar
2. Eng. Fahad Al-Mohannadi, Qatar Electricity and Water Company
3. Dr. Asem Zino, Veolia Water Technologies, Middle East
4. Metito Overseas Qatar
5. Prof. Datuk Ts. Dr. Ahmad Fauzi Ismail, Vice Chancellor, Universiti Teknologi Malaysia, Malaysia
6. Dr. Pei Sean Goh, Associate Professor, School of Chemical and Energy Engineering, Universiti Teknologi Malaysia, Malaysia
7. Moisés Menéndez, Editor & Community Manager, SAGUENAY, SLU (FuturENERGY & FuturENVIRO), Spain
8. Ellen Moore, Civil Engineering Research Journal (CERJ), Juniper Publishers
9. Guillermo Hijós Gago, Acciona Agua, Qatar
10. Rubén Vivo Fernández, Facility-D SWRO Plant Manager, Acciona Agua, Qatar
11. Coco Wang, APAC Communications, DuPont Water Solutions, China
12. Rita Cheung, DuPont Water Solutions
13. Prof. Miriam Balaban, Balaban Desalination Publications
14. Dr. Mark Wilf, RO Technology
15. Hannah Parsley, Global Marketing Communications Manager, MICRODYN-NADIR, USA
16. Dr. Jens Lipnizki, Head of Technical Marketing Membrane, Business Unit Liquid Purification Technologies, LANXESS Deutschland GmbH
17. Jayesh Shah, Global Marketing & Product Manager, Hydranautics — A Nitto Group Company
18. Deepti Batheja, Global Marketing Communications Specialist, Hydranautics — A Nitto Group Company

19. Dr. Tamás Zsirai, Commercial Engineering Lead, Water Technologies & Solutions, Suez Water Technologies, USA

20. Harish Warsono, RO Membrane Products Department, Toray Industries, Inc.

21. Hammadur Rahman Siddiqui, Qatar University

22. Jasir Jawad, Qatar University

Abbreviations

AC	activated carbon
AOC	assimilable organic carbon
APHA	American Public Health Association
APF	average permeate flux
ASTM	American Society for Testing and Materials
ATD	anti-telescoping device
BOD	biochemical oxygen demand
BW	brackish water
BWRO	brackish water reverse osmosis
C_C	solute conc. in concentrate water
C_F	solute conc. in feed water
C_P	solute conc. in permeate water
CA	cellulose acetate
CDI	capacitive deionization
CF	concentration factor
CFF	cross-flow filtration
CFU	colony-forming units
CFV	cross-flow velocity
CIP	clean in place
COD	chemical oxygen demand
CP	concentration polarization
CTA	cellulose triacetate
DAF	dissolved air flotation
DBPs	disinfection byproducts
DF	disk filter
DI	deionized water
DO	dissolved oxygen
DP	differential pressure
EC	electrocoagulation
ED	electrodialysis
EDI	electro-deionization
EDTA	ethylenediaminetetraacetic acid
EDX	energy dispersive X-ray
EDXRF	energy dispersive X-ray fluorescence
EPS	extracellular polymeric substances
ERD	energy recovery device
FeRB	iron-reducing bacteria
FO	forward osmosis
FR	fouling resistant
FTIR	Fourier transform infrared
GAB	general aerobic bacteria
GAC	granular activated carbon
GC−MS	gas chromatography/mass spectrometry
GFD	gallons per square foot per day

GMF	granular media filtration
GPM	gallons per minute
GRP	glass-reinforced plastic
GSF	green sand filter
HF	hollow fiber
HFF	hollow-fine fiber
HFM	hollow fiber membrane
HMI	human−machine interface
HPP	high-pressure pump
ICP	inductively coupled plasma
ICP−MS	inductively coupled plasma mass spectrometry
IP	interfacial polymerization
IPc	ion product
IR	infrared
ISD	internally staged design
IX	ion exchange
K_{sp}	solubility product
LC−OCD	liquid chromatography with organic carbon detector
LBL	layer-by-layer
LSI	Langlier saturation index
L/m^2/h	Liters per square meter per hour
MF	microfiltration
MFI	modified fouling index
MGD	million gallons per day
MIGD	million imperial gallons per day
MINLP	mixed-integer nonlinear programming
MOM	marine organic matter
MSF	multistage flash
MW	molecular weight
MWCNT	multi-walled carbon nanotubes
NaCl	sodium chloride
NDP	net driving pressure
NF	nanofiltration
NOM	natural organic matter
N_p	normalized productivity
NPD	normalized pressure differential
NPF	normalized permeate flow
NSP	normalized salt passage
NSR	normalized salt rejection
NTU	nephelometric turbidity units
ORP	oxidation−reduction potential
P_{osm}	osmotic pressure
ΔP	pressure drop
PA	polyamide

PAN	polyacrylonitrile		**SP**	salt passage
PD	pressure differential		**SPSP**	split partial second-pass
PE	polyethylene		**SRB**	sulfate-reducing bacteria
PES	polyethersulfone		**STC**	salt transport coefficient
PLC	programmable logic controller		**SWM**	spiral wound membrane
ppm	parts per million		**SWRO**	seawater reverse osmosis
PRO	pressure retarded osmosis		**S&DSI**	Stiff & Davis stability index
PSD	particle-size distribution		**3D−FEEM**	three-dimensional−fluorescence
PSF	polysulfone			excitation emission matrix
PV	pressure vessel		T_{CF}	temperature correction factor
Q_c	concentrate flow		**TBC**	total bacterial count
Q_f	feed flow		**TDS**	total dissolved solids
r	recovery		**TFC**	thin film composite
R	rejection		**TMP**	transmembrane pressure
Re	Reynolds number		**TOC**	total organic carbon
R_s	staging ratio		**TSS**	total suspended solids
RED	reverse electrodialysis		**UF**	ultrafiltration
RIO	remote input/output		**UV**	ultraviolet
RO	reverse osmosis		**VFD**	variable frequency drive
RSM	response surface methodology		**VOC**	volatile organic chemical
Sc	Schmidt number		**WWTP**	wastewater treatment plant
SCADA	supervisory control and data acquisition		**WTC**	water transport coefficient
SDI	silt density index		**WHO**	World Health Organization
SEC	specific energy consumption		**XRD**	X-ray diffraction
SEM	scanning electron microscopy			
SEM−EDS	scanning electron microscopy−energy dispersive X-ray spectroscopy			

Readership

- Design engineeers
- Process engineers
- Plant maintenance engineers
- Plant technicians/operators
- Researchers and scientists
- Faculty and students
- Reverse osmosis desalination practitioners and regulators

1

Introduction to Reverse Osmosis

1.1 Introduction

Desalination is considered as a progressively common solution for supplying potable water in numerous regions of the world where the water resource is rare. Reverse osmosis (RO) technology has evolved into a widely used process for both seawater and brackish water desalination. RO is a separation process, which uses membranes as a physical barrier for removing the dissolved solutes present in water. The membrane used in the RO process is semipermeable, which only allows water to pass through and stops the ions and other solids under applied pressure by preferential diffusion of the separation process. A standard RO membrane is prepared from a polymeric material by the interfacial polymerization process, and consists of three different layers. The top layer ($\sim 0.2 \, \mu m$ thick) performs the function of separation followed by the second layer ($40-50 \, \mu m$), which provides pathways for fluid flow, and the last layer ($120-150 \, \mu m$ thick) provides the mechanical strength. The overall thickness of the membrane is less than 1 mm. With the applied pressure, feed water will be pumped through the membrane surface, triggering a portion of the water to pass across the membrane (Crittenden et al., 2012), as shown in Fig. 1.1. The pure water flowing across the membrane, known as permeate/product water, is comparatively free of targeted dissolved solutes, whereas the unconverted water, termed as reject water (also known as brine, concentrate, or retentate), will be released at the remote end of the pressure vessels. Depending on their size and electric charge, major water constituents will be retained on the feed side of the RO membrane, whereas the pure water will pass across the membrane. These membranes have the ability to reject particulate and dissolved solids of practically any size. Typically, RO membranes can reject solids greater than 1 Å (Angstrom). This confirms that the RO membranes can remove essentially all suspended solids, viruses, bacteria, protozoa, and other human pathogens present in the feed water.

This technology is in an advanced stage, and is available in the majority of the coastal regions in the world where the natural hydrological resources are limited. Primarily, for the aforestated reasons, the research and developments in this technology field are ongoing. Fundamentally, the present researches and innovations concentrate on attempting to further reduce the process energy consumption (Peñate et al., 2012). Properly designed as well as accurately operating systems provide a smooth performance over extended periods

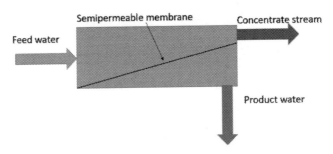

FIGURE 1.1 Diagrammatic representation of membrane-based reverse osmosis separation process.

of time. Conversely, mistakes made in the course of the design or operation of RO systems can result in continuing problems and decreased membrane performance and membrane useful life. The membrane system, which includes the membrane elements housed in pressure vessels and a high-pressure pump, constitutes the heart of the RO plant. The system also includes a pretreatment section to achieve the required feed water quality, a clean in place section to facilitate membrane cleaning, and a post-treatment section to achieve the required product quality.

Generally, the RO membranes can separate monovalent ions, for example, chloride and sodium. Presently, the seawater RO membranes are noted to have salt rejections of more than 99% (Brehant et al., 2003; Reverter et al., 2001); certain membranes, when functioned under standard testing conditions (8% recovery, pH 8, 25°C, 5.5 MPa, and 32,000 parts per million (ppm) NaCl), can accomplish as great as 99.7%−99.8% rejection of salt (Hydranautics, 2007; Reverberi et al., 2007). Pressure used in RO applications vary between 60 and 80 bar in seawater desalination and 15 bar in brackish water desalination. The RO membrane desalination field has quickly developed over the past 40 years to turn into the essential choice for new plant facilities. Membrane technology has permitted remarkable improvements in the production of permeate and cost savings. Progressively strict water quality standards will lead to additional optimization and advancement in RO membrane technology. Specifically, the two important types of feed water, brackish water and seawater, have characteristic features that require system design and specific parameter alteration. The application of membrane desalination has expanded as materials have improved and expenses have reduced. Currently, RO membranes are the superior technology for advanced desalination facilities, and they are applied to various saltwater sources utilizing customized pretreatment and membrane system design.

1.2 Seawater composition

The World Health Organization and the United States Environmental Protection Agency, under the Safe Drinking Water Act, have specified maximal total dissolved solids (TDS) concentration of 300−500 ppm, chloride concentration of 150−240 ppm, and the sodium concentration of 90−180 ppm in the permeate of a reverse osmosis system for drinking water standard. The aforestated level of TDS could be used as a categorization

TABLE 1.1 Average TDS values of different seas and oceans.

Sl. no.	Sea or ocean	TDS (ppm)
1	Baltic Sea	8,000
2	Caspian Sea	13,500
3	Pacific Ocean	33,600
4	Indian Ocean	33,800
5	Atlantic Ocean	37,000
6	Red Sea	43,000
7	Arabian Gulf	50,000
8	Dead Sea	260,000

limit for defining drinking (fresh) water. Usually, water having a concentration of TDS between 1000 ppm and 15,000 ppm is termed as brackish water. Natural water resources such as ocean, bay, and seawaters, which have TDS concentrations greater than 15,000 ppm, are normally categorized as seawater. Seawater with TDS of 35,000 ppm is considered as the standard seawater comprising the largest amount of water around the world. The actual content of TDS might change with geographical location within wide-ranging limits from the Baltic Sea with TDS almost 8,000 ppm to the Arabian Gulf with TDS till 50,000 ppm (Pandey et al., 2012). The average TDS values of different seas and oceans are summarized in Table 1.1. The actual compositions could be relatively assessed from the standard seawater composition. However, the water obtained from seashore wells, contingent upon the soil, influx from inland, and so forth, can usually have salt content and composition totally different from water obtained from the sea itself. Table 1.2 summarizes the standard seawater composition.

The composition of seawater is almost similar everywhere worldwide. The six most plentiful ions present in seawater are chloride (Cl^-), sodium (Na^+), sulfate (SO_4^{2-}), magnesium (Mg^{2+}), calcium (Ca^{2+}), and potassium (K^+). The seawater salt content is illustrated by salinity, which is defined as the salt quantity in grams dissolved in 1 kg of seawater. The seawater viscosity (i.e., internal resistance to flow) is greater than that of freshwater due to its increased salinity. The seawater density increases with the increase of salt content and reduces with increasing temperature. Water quality guidelines are provided in Chapter 3, Guidelines for Water Quality, in a detailed manner.

1.3 History of reverse osmosis technology

The phenomenon of the osmosis process across semipermeable membranes was first observed by Jean Antoine Nollet in the year 1748 (Laidler et al., 1999). Nollet replicated the osmotic process by employing a bladder of a pig as a membrane to demonstrate that solvent molecules can flow from low solute concentration to a high solute concentration through the

TABLE 1.2 Standard seawater composition.

Sl. no.	Ion	Concentration (ppm)
1	Chlorine	19,700
2	Sodium	10,900
3	Sulfate	2,700
4	Magnesium	1,310
5	Calcium	412
6	Potassium	390
7	Bicarbonate	150
8	Bromide	65
9	Carbon	28
10	Nitrogen	11.5
11	Strontium	8 6
12	Oxygen	4.4
13	Boron	2
14	Silicon	1.4
15	Fluoride	1.3

bladder wall. It was confirmed that a solvent can pass selectively across a semipermeable membrane through the process of natural osmotic pressure, and the solvent constantly entered through the cell membrane till dynamic equilibrium was attained on the two sides of the bladder.

The osmosis study almost disappeared for the following 200 years until the late 1940s, when scientists from prestigious American universities started to reevaluate this topic. The aforestated resumed interest was based on an aspiration to identify a solution for filtering or desalinating seawater, which was an objective set by the Kennedy administration for helping to develop solutions for the water shortage issues of their country. The possibility of seawater desalination using semipermeable membranes was primarily explored in 1949 at the University of California at Los Angeles (UCLA), and in 1955 at the University of Florida, with finance contributed by the United States Department of Interior Office of Saline Water (Glater, 1998). In the mid-1950s, researchers from both the University of Florida and UCLA effectively produced freshwater using seawater; however, the water flux was noted to be extremely less to be commercially sustainable. The breakthrough in membrane technology led to the establishment of a reverse osmosis membrane desalination based on cellulose acetate (CA) that contributed to increased salt rejection and higher flux at adequate hydrostatic pressures (Loeb et al., 1962; Reid et al., 1959). The aforestated was a noteworthy development toward the use of RO membranes films as a successful tool for the drinking water production from the sea. In 1960 Loeb and Sourirajan at UCLA succeeded in developing a functional synthetic RO membrane from CA polymer. This membrane had an asymmetric structure with a dense skin at the exterior that controlled the flux as well as selectivity of membrane and exceptionally porous substructure that contributed the mechanical properties. In their tests, the water of high solute concentration was pressurized through an engineered membrane that functioned as a filter by allowing only water molecules to pass through whereas rejecting salt and TDS. Pure water was able to pass across a membrane at a good flow rate for producing potable water and the membrane was in fact durable and can function under normal water pressure and operating conditions. As this innovative technology operated in reverse of the natural osmosis process, it soon became known as the reverse osmosis process. Also, it has been demonstrated that the preparation of asymmetric CA membranes was by a phase inversion technique where a homogeneous polymer solution is transformed into a two-phase system, that is, a solid polymer-rich phase delivering the solid polymer structure and a polymer lean phase forming the liquid-filled pores of the membrane (Kesting, 1971; Strathmann et al., 1975). In 1965 the world's first commercial RO plant was established in Coalinga, California with the assistance and guidance of Sidney Loeb and Joseph W. McCutchan, and its pilot program has drawn the attention of scientists and governments from all over the world.

Subsequently, for the manufacturing of synthetic membranes, other synthetic polymers, for example, polyamide (PA), polysulfone (PSF), polyacrylonitrile (PAN), polyethylene (PE), and so forth were used as basic material. The aforementioned polymers frequently demonstrated increased mechanical properties, thermal stability, and chemical stability relative to the CA. On the other hand, CA continued to be the predominant material for the manufacturing of RO till the evolution of the composite membrane prepared by interfacial polymerization technique (Cadotte et al., 1981; Riley et al., 1967). This type of membrane demonstrated remarkably increased fluxes, superior rejection of salt, and increased mechanical and thermal stability relative to the CA membranes.

During the 1970s, RO applications have been broadened from desalting to softening applications. The primary membranes produced for RO desalination and additional applications were prepared as flat sheets and after that fixed in a spiral wound membrane (SWM) module (Riley et al., 1967; Westmoreland, 1968). An alternate technique to deal with membrane geometry was simply the evolution of self-supporting hollow fiber (HF) membranes, which possessed a wall thickness of just 6–7 μm (Mahon, 1966). DuPont Corporation manufactured the asymmetric HF membranes with the fundamental utilization in seawater as well as brackish water desalination. Subsequent to the establishment of a proficient membrane, proper membrane housing assembly, termed as a module, has been prepared. The major requirements for such module design involved reliability, low cost, concentration polarization control, ease of membrane or module replacement, and high membrane-packing density. Membranes were manufactured in three distinct configurations, such as tubes, as capillaries or HFs, and as flat sheets. Table 1.3 presents the major milestones in the development of RO technology.

With the increasing world population, scarcity of freshwater supplies, coastal area urbanization, larger dependence on oceans and low-quality water supplies (treated wastewater, brackish groundwater), pollution of freshwater supplies, and developments in membrane technology prompted sustained fast growth of RO installations. Thus, the future of this RO technology is very promising. During the continued advancement, innovations and improvements witnessed in the last five decades, the RO technology has been

TABLE 1.3   Major milestones in the development of reverse osmosis technology.

Year	Major milestones
1748	The phenomenon of the osmosis process across semipermeable membranes was observed foremost by Jean Antoine Nollet
1948	In the United States, Hassler studied osmotic properties of cellophane membranes
1959	Reid and Breton demonstrated the desalination ability of CA film
1960	Asymmetric CA membrane developed by Loeb and Sourirajan
1963	General Atomics developed the first spiral wound module
1967	DuPont developed the first efficacious hollow-fiber module
1972	Cadotte prepared the first interfacial composite membrane
1974	The first seawater RO desalination plant came into operation in Bermuda.
2006	At UCLA, a thin-film nanocomposite membrane was developed

RO, reverse osmosis; *UCLA*, University of California at Los Angeles.

shifted from a scientific interest into a rapidly growing, self-supporting industry. Presently, in the RO desalination plants, the spiral wound modules are extensively used.

1.4 Reverse osmosis technology in Gulf cooperation council countries

Presently, the majority of countries face water scarcity problems in different forms depending on the environment, population, and topographical location. In the Gulf cooperation council (GCC) countries, such as Qatar, UAE, Saudi, Bahrain, Oman, and Kuwait, water scarcity is mainly because of the fact that these are arid desert regions with little or no natural clean water. The entire Gulf region has practically no underground water sources. Furthermore, the seawater salinity of the Arabian Gulf is considerably higher, with TDS ranging from 40,000 to 50,000 ppm, as compared to other parts of the world. Because of this distressing level of water scarcity, the authorities of GCC countries having a rich reserve of fossil fuels have moved toward the seawater desalination of seawater for fulfilling their water requirements. Seawater desalination was initially introduced in GCC in 1950 in Kuwait, succeeded by Qatar in 1955. In the early years of seawater desalination, the plants' setup and installation were completely based on the thermal desalination process until 1970. The population of Qatar in the early 2000s was around 0.5 million and by the end of 2012, the population increased more than three times reaching 2 million and had crossed 2.6 million by June 2018 and is predicted to reach 4 million by the year 2030. This exponential increase in population increases the requirement for clean water for domestic and industrial use. Moreover, the massive infrastructure developments in Qatar and higher standards of living are increasing the demand of clean water. Also, environmental concerns of CO_2 emissions as a result of the use of hydrocarbon resources in the thermal desalination process and their cost-effectiveness make the thermal desalination process not an ideal one. Taking these factors into account the GCC governments are shifting to RO seawater desalination. At present, all the GCC countries are in a process of modifying their current plant partially or entirely by RO process technology (Rahman et al., 2018).

In recent years, the majority of the new plants commissioned or planned for seawater desalination in the Gulf countries are based on RO process technology, as the total energy equivalent needed with respect to mechanical energy is equivalent to $5 \, \text{kWh/m}^3$. The expense of which is the same as $0.550 \, \text{dollar/m}^3$ of desalinated water (estimated 1 kWh work equivalent to 0.11 dollar), which is considerably less relative to multistage flash distillation plant. This is due to the fact that seawater RO process does not need heating or condensing, and the major consumption of energy is just by the usage of the pump. Also, the recently advanced increasingly efficient membranes could operate for extended duration before any requirement of cleaning or replacement. Furthermore, the size of a conventional RO plant is approximately 70% smaller as compared to a conventional multistage flash distillation plant of similar capacity. The RO plant will have a modular design in the form of a vertical or horizontal stack of a membrane known as RO trains that are attached in parallel or series, and are generally made up of materials that are noncorrosive

TABLE 1.4 Some of the reverse osmosis desalination plants in Gulf cooperation council countries.

Sl. no.	Plant	Country	Capacity (million imperial gallons per day (MIGD))	Commissioning year
1	Abu Samra	Qatar	0.2	1982
2	Ras Abu Fantas A3	Qatar	36	2016
3	Umm Al Houl	Qatar	60	2016−17
4	Mirfa IWPP	UAE	53	2017
5	Al Taweelah IWP	UAE	200	2022
6	Khor Fakkan Plant	UAE	3	2008
7	Hamriyah	UAE	20	2014
8	Ajman Plant	UAE	10	2011
9	Ras Al Khaimah IWP	UAE	22	2020
10	Ghalilah	UAE	15	2015
11	Shoaiba 3 Expansion II	Saudi Arabia	55	2019
12	Ras Al Khair Desalination Plant	Saudi Arabia	228	2014
13	Shuwaikh	Kuwait	33	2010
14	Barka II power	Oman	26.4	2009
15	Salalah plant	Oman	25	Ongoing
16	Sur plant	Oman	17.7	2014
17	Ghubra IWP	Oman	42	2013
18	Al-Dur	Bahrain	48	2009

FIGURE 1.2 Reverse osmosis desalination facility in Qatar (FuturENVIRO, 2018). *Courtesy: FuturENVIRO.*

to seawater. Hence, it needs very less shutdown, and maintenance operation can be performed when the plant is online. Table 1.4 presents the capacities of different RO desalination plants in GCC countries.

Ras Abu Fontas 3 has a capacity of 164,000 m^3/day (36 MIGD). The construction of the Ras Abu Fontas 3 desalination plant represents a milestone in the desalination world, given that it is the first time RO technology will be used on a large scale in Qatar. Umm Al Houl will produce 284,000 m^3/day (62 MIGD) of seawater desalinated by RO, as part of a large-scale Independent Water & Power Project (IWPP) that will produce around 2500 MW of electric power and will reach 614,000 m^3/day (135 MIGD) after the startup of the new facility (FuturENVIRO, 2018). The different processes in the Umm Al Houl plant are discussed in detail in Chapter 2, Reverse Osmosis Principles and System Components. Fig. 1.2 is the photograph of a reverse osmosis seawater desalination plant in Qatar (FuturENVIRO, 2018).

1.5 Advantages and limiting factors of reverse osmosis technology

The quick development of RO technology is due to the fact that it can produce potable water with reduced cost. The other appealing characteristic of the RO process is that the RO plant design and its operation is simple and is modular in nature. The membrane RO plants are compact, could be scaled up effectively, and set up easily relative to thermal desalination plants. Furthermore, the RO process makes system maintenance simpler (Abbas et al., 2005). An additional benefit of this RO technology is that this process could meet different feed water concentrations and changing product water quantity and quality prerequisite through changing system construction in addition to operating conditions. The manufacturers of RO membranes produced different membrane types for accurately meeting the fluctuating requirement of a wide-ranging scope of drinking water, commercial, municipal as well as industrial applications, for example, low pressure, high salt rejection membranes and fouling resistant membrane, and high-flux, high rejection membrane, and so on (Busch et al., 2004; Redondo et al., 2001). All the aforestated benefits have made the RO process design extra flexible.

The RO process is not just limited by high osmotic pressure due to high salt concentration and concentration polarization along the membrane surface, but also by additional factors, which reduce the separation efficiency significantly, for example, salt rejection. The factors responsible for performance reduction can be identified by their mechanism (Fritzmann et al., 2007).

Fig. 1.3 presents the different limiting factors for membrane desalination by RO technology. Different chemicals can damage the membrane active layer, thereby causing irreversible damage related to decreased salt rejection ability and even membrane destruction. Oxidants used in pretreatment of the RO feed water or as cleaning chemicals are the utmost significant group of chemicals, which results in the degradation of the membrane. In the course of a reverse osmosis plant operation, it has to be made sure that no biological, colloidal, or dissolved matter get accumulated at the surface of the membrane, forming a continual layer, that acts as an additional resistance to mass transfer throughout the membrane and obstructs the flow of permeate. Scaling consistently happens at the surface of the membrane due to the high concentration of salt close to the membrane surface caused by concentration polarization. The different scale control methods are discussed in Chapter 5, Pretreatment: Fouling and Scaling Control, to reduce or avoid scaling. The

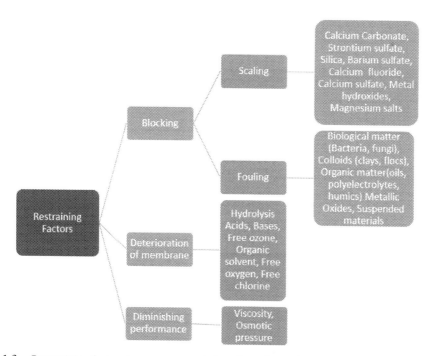

FIGURE 1.3 Constraining factors to reverse osmosis-based membrane desalination.

fouling of membrane is caused by either biological growth (bio-fouling) or diffusive and convective transport of colloidal or suspended matter. Fouling is irreversible and cannot be completely avoided even with upgraded pretreatment. In this manner, periodical cleaning of the membrane must be performed. In recent years, substantial advancement has been made in RO membranes for seawater desalination, which not only have remarkably improved overall performance but also demonstrate increased thermal stability, and chemical stability, and are less sensitive to operational mistakes.

1.6 Essential process steps in reverse osmosis-based desalination plants

The core process of desalination can be based on RO membrane technology, however as an independent unit, it cannot contribute safe potable water, nor does it ensure a proficient plant. Fig. 1.4 is the diagrammatic representation of the basic process steps in RO desalination plants. The pretreatment stage involves all the essential treatment procedures before the RO plant operations. This pretreatment process is essential for an increased plant lifetime and to limit chemical cleaning and replacement of membrane. The pretreatment process directly affects the performance of the plant.

The RO process can be built with one or two passes, on the basis of the seawater salinity, temperature, and requirements of product water. In majority of the cases, one pass is adequate to achieve the EU drinking water standards, particularly with respect to the boron content (1 ppm). For achieving the boron guideline of the World Health Organization (0.5 ppm), a subsequent pass may be required (boron removal process). The device for energy recovery is an important factor, that determines the electrical expenses of the plant. Hence, the device must be selected carefully on the basis of environmental policies and local energy costs.

Post-treatment and/or polishing steps are needed for conditioning the water after the RO procedure to make it appropriate for the specific application. The disposal of brine can be an ecological and economic issue in certain regions where the flora and fauna are sensitive to the salinity increase of local seawater. The disposal of brine must be examined and engineered on a case-by-case basis. The specialty of desalination is to determine and consolidate accessible technologies for optimizing water production and quality. All types of natural seawater sources such as deep seawater, shallow surface seawater, beach well seawater, and brackish river water can be treated in a desalination plant. It is possible to produce drinking water, ultrapure water,

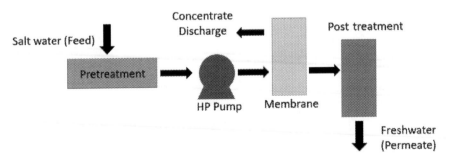

FIGURE 1.4 A conventional reverse osmosis system.

FIGURE 1.5 A typical seawater reverse osmosis desalination plant (FuturENVIRO, 2018). Source: *Courtesy: FuturENVIRO.*

process water, and irrigation water from a desalination plant. Fig. 1.5 is a typical seawater RO desalination plant (Design, construction, operation and maintenance of two desalination plants in Qatar by ACCIONA Agua, 2018). The different RO system components are discussed in Chapter 2, Reverse Osmosis Principles and System Components.

1.7 Manufacturers of membrane

Several types of PA-based composite SWMs are obtainable to satisfy diverse feed water conditions. The seawater membranes are used for treating feed water of increased salinity (TDS from 35,000 to 50,000 ppm). The aforestated membranes could function at pressures till 1500 psi (103.4 bar). Conditions for a standard membrane test are as per the following:

pH of feed water: 6.5−8; Concentration of feed water: 32,000 ppm sodium chloride (NaCl); Temperature: 77 °F (25°C); Operating pressure: 800 psi (55.15 bar); Recovery per module: 8−10.

Test conditions are significant due to the fact that the rated performance will be based on these conditions. Working under various conditions can lead to performance that varies from the rated performance. Also, it is important to note that there is not any single constant test conditions to which entire membrane producers follow. In this manner, as a consequence of the difference in recovery and pH under such membranes are tested, the seawater membrane rated performance from various manufacturers could not be legitimately straightaway compared (Kucera, 2015).

Some of the well-known RO membrane manufacturers are presented in Table 1.5.

Within the seawater membrane classification, there are membrane subsets that are used for various applications. As an illustration, Koch Membrane Systems provide a typical increased rejection seawater membrane module in addition to a superior-flow seawater element membrane module. RO Design software provided by the membrane manufacturers is helpful to a plant designer to decide the configuration of membrane system, which is technically optimized and accurate. It enables a designer to evaluate various combinations rapidly

TABLE 1.5 Some of the well-known reverse osmosis membrane manufacturers.

Parent company	Firm	Location	Membrane material	Membrane configurations
DuPont Water Solutions	DuPont Filmtec	USA	Polyamide	Spiral-wound
Nitto Denko	Hydranautics	USA	Polyamide, piperazine, poly ether sulfone	Spiral-wound
Koch Membrane Systems	Fluid Systems	USA	Polyamide, cellulose acetate	Spiral-wound
Toray Industries, Inc.	Toray Industries, Inc.	Japan	Polyamide	Spiral-wound
SUEZ	SUEZ Water Technologies & Solutions	USA	Polyamide, cellulose acetate	Spiral-wound
Toyobo	Toyobo	Japan	Cellulose triacetate	Hollow-fiber
Mann + Hummel	MICRODYN-NADIR	Germany	Polyamide, cellulose acetate	Spiral-wound and flat-sheet

and furthermore get details of accessories. The currently used RO design software programs are presented in Chapter 8, Reverse Osmosis Design Software Programs.

1.8 Reverse osmosis applications

Currently, spiral-wound RO membrane elements are mainly used for different applications throughout the globe, whereas some manufacturers use hollow-fiber configurations also. The different applications are discussed in the following sections.

1.8.1 Desalination for potable use

With the increasing world population and depletion of natural water resources, the requirement for drinking water has increased sevenfold (Watkins, 2006). In the next 30 years, it is anticipated that the world populace will increase to more than 40% and the demand for industrial, agriculture, and domestic water sources would be expanded, particularly in the developing nations where the requirement of water is higher relative to its population and financial status (Water, U. N., 2007). According to the World Health Organization estimation, 1100 million people need access to pure drinking water (Urgency, U., 2007). Also, it has been reported by the World Water Council by 2030, there will be an incredible chance of 3900 million individuals residing in "water-scarce" regions (WHO, UNICEF, 2010). Just 0.5% of the planet's complete water asset is suitable for the purpose of drinking, while 2.5% is frozen freshwater and the rest of 97% is seawater, which is not fit for human use because of its greater salinity; and as per assessment, there will be an increase of approximately 7–10 billion in total world populace by 2050 (Lalia et al., 2013; Li et al., 2011). To overcome the issue of water scarcity and the need for pure potable water, there is a need for fresh sources of water along with the conservation of existent water assets through an appropriate methodology for the treatment of water (Pendergast et al., 2011). Desalination contributes to the interesting opportunity of extending the natural hydrological cycle by enhancing it with water from brackish reservoirs and oceans. For water desalination, several techniques have been advanced inclusive of membrane-based RO, distillation processes, nanofiltration (NF), and electrodialysis processes (Shenvi et al., 2015).

Municipal water supplies are frequently polluted with several chemicals, which might be dangerous to human health, such as heavy metals (byproducts from consumer waste and industrial processes), chlorine (regularly added to the water supply for killing bacteria), and pesticides. RO elements are very effective and economical for removing TDS, hardness, organics, disinfectant byproducts, harmful heavy metals, and color for meeting the quality requirement of safe drinking water.

1.8.2 Industrial water and wastewater reclamation and its reuse

Because of the rising water demand along with the water scarcity, water recovery as well as wastewater reclamation and its reusability in commercial power plants have turned out to be a developing trend in the previous decade. In actual fact, water treatment inclusive of the utilization of RO for power production is projected at greater than 30% of all commercial water treatment sales worldwide (Water Technology, 2017). Purification of water for steam boilers is an important application of RO desalination. This technology might also be employed for replacing ion exchange, inclusive of sodium softening, for purifying water for medium- and low-pressure boilers. In the case of high-pressure boilers and steam generators, RO technology is generally used in combination with ion exchange and could considerably decrease the periodicity of resin regeneration and significantly reduce the quantity of chemical and regeneration waste.

1.8.3 Superior purity water

Microelectronics and pharmaceutical industries need extremely pure water for the purpose of production. For the microelectronics industries, superior purity water is essential in the course of production phases and is used for rinsing the finished microelectronic components. For pharmaceutical industries, water is considered to be the most commonly used component out of all pharmaceutical drugs, hence it is extremely important that the water should not contain any dissolved solids, organics, as well as bacteria. RO systems with ultraviolet sterilizers are commonly used for treating and disinfecting feed water at commercial levels for producing the extremely pure water needed for several pharmaceutical manufacturing processes. Kim and Hong (2018) proposed an advanced single-pass RO configuration for improving the permeate quality.

1.8.4 Spot-free rinse

In the case of several commercial manufacturing plants, which develop parts for air-crafts or vehicles, and commercial enterprises such as car washing, the cleaning of surfaces and its rinsing could be a significant portion of the production process. Water that is not filtered appropriately might result in the formation of scale and causes gray or white water spots on newly cleaned surfaces. The aforestated spots are due to magnesium as well as calcium salts (hardness) present in the source water. The RO technology has the ability to remove the minerals, which can cause scaling and spotting to occur.

1.8.5 Ingredient water (for beverages and food)

In addition to the fact that a superior quality of water is required for the mineral water industry, water is also an important component in various beverages and food processes. For maintaining the product quality and preventive well-being measures, extremely pure water is needed. RO in combination with ultraviolet sterilizers could fulfill the aforestated require-ments by decreasing the level of dissolved solids and eradicating microorganisms and con-taminants. Moreover, RO systems are beneficial in beverage and food manufacturing, processing, in addition to packaging due to the fact that they can separate several contami-nants, which might change the odor or taste of the food or beverage.

1.8.6 Removal of specific contaminant

An additional application of RO is the removal of specific contaminants. The Environmental Protection Agency has selected RO as the best available technology for the removal of several inorganic contaminants, inclusive of selenium, nitrite, nitrate, fluoride, barium, arsenic, antimony, and radionuclides, inclusive of photon emitters, beta-particle, radium-226, and alpha emitters. Also, RO technology has been proved to be efficient for eliminating higher molecular weight synthetic organics such as pesticides. Utilization of RO for specific contaminants, conversely, is less frequent due to the fact that alternate technologies are often more economical and the reject stream disposal might contribute several challenges.

1.9 Future of reverse osmosis technology in desalination

The desalination of seawater has turned out to be one of the most significant techniques of lessening water scarcities across the globe. Although this RO technology has demon-strated to be very effective so far, the technology continues to progress. Due to the fact that RO is being used for treating a wider category of feed waters, particularly municipal and industrial wastewater effluents and other water resources, additional developments are always essential.

1.9.1 Chlorine-resistant membranes

In order to widen the range of applications appropriate for RO desalination, membranes having superior chemical stability are required. Surface waters as well as wastewater feed sources regularly consist of biological materials along with nutrients that feed bacteria. The aforestated materials might seriously foul RO membranes and are consequently destroyed with the application of chlorine before RO. On the other hand, the majority of the PA membranes have no tolerance toward chlorine and other oxidizers, hence they should be separated in advance of the RO process. Oxidizer exposure might ultimately destroy the membrane triggering the membrane to be exposed to increased permeate flow as well as greater salt passage (Wiles & Peirtsegaele). Developing chlorine-resistant membranes (also resistance against other oxidizing chemicals and halogens)-is crucial as this RO technology is being used for treating ever more challenging feed waters. Improving the tolerance of RO membranes to oxidizers will further reduce the operating costs by eradicating dechlorination steps in advance of the RO system. In a study performed by Gholami et al. (2018), high salt rejection, and appropriate chlorine resistance, has been achieved through polyethylene glycol diacrylate coating onto thin film composite (TFC) RO membranes.

1.9.2 Less-fouling membranes

Due to the fact that the RO feed water sources are becoming increasingly complex, there is a continuing requirement for membrane elements, that have scaling resistance, fouling resistance and degradation resistance along with concurrently contributing increased permeability and rejection of solute. Decreased fouling from suspended solids and organics will turn the membranes furthermore economical by prolonging their operating lifetime and reducing the energy needs. Several less-fouling membranes are available at present: research studies on these membranes are concentrated on modification of the surface of the membrane. Systems using RO technology should have proper pretreatment processes, such that all the impurities present in the water will be separated before the feed water reaches the RO membranes. Park et al. (2019) examined the surface modification of RO membranes to improve fouling resistance. RO membranes have been modified by means of a sol−gel process utilizing mercaptopropyltrimethoxysilane, chlorotrimethoxysilane, and aminopropyltrimethoxysilane.

1.9.3 Alternate membrane materials

RO membranes have been developed over the years; however, most of these advancements have been accomplished by modifying the surface of the membrane or upgrading the membrane element or module design. The membrane manufacturers are continuing to modify the present, commercial TFC PA reverse osmosis membranes and elements, whereas scientists continue to develop other materials that might contribute an excellent substitute to polymer materials. The application of nanotechnology has contributed to the advancement of nanostructured membranes, or membranes functionalized using distinct nanomaterials (Buonomenna, 2013). The commonly used nonpolymeric RO membranes are inorganic, biomimetic, and membranes of a combination of polymer and inorganic materials (Kucera, 2010). More researches are ongoing for the development of advanced membrane materials for water treatment (Saleem et al., 2020a, 2020b, 2020c; Saleem et al., 2020; Yadav et al., 2020; Zaidi et al., 2019). Saleem et al. (2020c) have reviewed the recent developments in the application of nanomaterials in thin-film composite membranes for desalination. In a study by Xia et al. (2020), the nano-sized sodium alginate/Cu^{2+} hydrogel developed using the layer-by-layer method was initially applied for PA RO membrane modification for solving the problems such as complex fabrication, high-cost, single antifouling mechanism, or reduced flux.

1.9.4 Handling of feed water with increased total dissolved solids

With the increase in the zero liquid discharge applications enhancing worldwide, a cost-effective desalination technology was required to handle feed water with increased levels of TDS relative to the level typical seawater RO membranes can handle (higher than TDS 50,000 ppm). Forward osmosis (FO) is considered to be a developing membrane technology in desalination, which has been developed for handling the feed water of increased TDS. FO membrane processes vary from the typical RO process in the sense that FO processes utilize a draw solution of an increased saline concentration relative to the feed solution for forcing water from the feed into the draw solution. In a separate process, the water is subsequently recovered from the draw solution. The aforestated procedure permits the FO system to treat water until 100,000−200,000 ppm feed TDS. Comparable processes that involve two sets of inlet and outlet flows through the membrane for addressing feed water of greater TDS include pressure retarded osmosis and counter-flow RO. As confirmed by the developing membrane-based desalination technologies, the membranes will remain as a backbone in desalination applications in the upcoming years.

1.9.5 Higher boron rejection membranes

Boron is a nonmetallic element that is present in the form of boric acid and borates in surface water and groundwater. The concentration of boron in seawater varies from 0.50 to 9.60 ppm (Woods, 1994) averaging at a value of 4.5−4.6 ppm. The boron concentration has been reported to be as high as 7 ppm in the Arabian Gulf (Busch et al., 2004). In the fourth edition of the Drinking Water Quality guidelines, published by the World Health Organization in 2009, the boron guideline value was revised from 0.5 to 2.4 ppm (World Health Organization, 2009), because of the absence of toxicity data on human beings. Different modifications in reverse osmosis plant configurations have been examined and currently there are various techniques employed in the desalination of seawater for managing the boron level in the permeate water. One of the most commonly used methods is the usage of advanced membranes having high rejection of boron. All major RO membrane manufacturers have invested considerable efforts in developing advanced grades of desalination membranes with boron rejection as higher as 93%−96%. Table 1.6 presents the boron rejection performance of some of the membranes from well-known RO membrane manufacturers. Another most commonly practiced method is to use a second-pass

TABLE 1.6 Boron rejection performance of some of the membranes from well-known RO membrane manufacturers.

Manufacturer	Product Type	Active area (m^2)	Permeate Flowrate (m^3/d)	Stabilized Boron Rejection(%)	Reference
DuPont Water Solutions	FilmTec SW30XHR-440i Element	41	25	93	Dupont Water Solutions
Nitto Denko Hydranautics	SWC4B	37.1	24.6	95	Nitto Hydranautics
Nitto Denko Hydranautics	SWC4B Max	40.8	27.3	95	Nitto Hydranautics
Toray Membrane	TM820K-440	41	24.2	96	TORAY

reverse osmosis stage added subsequent to the first pass, in which the pH of first pass permeate is adjusted prior to feeding to the second pass. The work by Farhat et al. (2013) reported substantial rejections of boron as higher as 96% using actual seawater from the Arabian Gulf (total dissolved solids: 41,700—47,400 ppm) by employing commercially available advanced generation reverse osmosis membranes under a double-pass configuration and with no pH adjustments. This study confirmed that higher rejection of boron was achieved with lower feed temperature, higher second pass pressure, and higher feed velocity.

1.10 Definitions

Algal Count: It is a measure of the count of algal particles present per unit volume of the resource water. This is expressed in total count of algal cells present per milliliter of water. In general, the total algal count in seawater below 10,000 cells/L will not contribute as a challenge to the performance of a desalination plant. It could be measured in a laboratory setting or by online instrumentation. A high algal count can reduce the filtration capacity of pretreatment system, decrease the efficiency of solids removal, and speed up the fouling of cartridge filters.

Antiscalants: These are chemicals included to enhance the solubility of sparingly soluble salts. In the case of RO systems, calcium carbonate, calcium sulfate, barium sulfate, strontium sulfate, and calcium fluoride are the most frequently observed scales (Ning, 2015). Antiscalants are pretreatment additives for RO system, which is extremely efficient in restraining the membranes from scaling.

Array: It is the physical arrangement of the pressure vessels, for example, a 6:3 array configuration is a two-stage configuration with a total of nine vessels. The first stage will have six pressure vessels and the second stage with three pressure vessels. The concentrate of every single stage would act as the feed for the consecutive stage.

Asymmetric membrane: It is a type of membrane made of the same material (PA or CA) and shows an increment in porosity moving from surface to base. Also, there will be a thick porous support layer and a dense thin barrier skin on the membrane surface.

Biochemical oxygen demand: It represents the oxygen amount used by bacteria as well as other microorganisms when these organisms decompose the organic matter at a particular temperature under aerobic conditions. It is a beneficial parameter for evaluating the biodegradability of dissolved organic matter present in water. Biochemical oxygen demand is often employed in wastewater treatment plants, as an indication of the extent of organic contamination in water.

Brackish water: In general, brackish water is distinguished as the water of TDS levels from 1000 ppm up until 15,000 ppm. It is not possible for humans to consume brackish water directly because of its high salinity.

Brine seal: It is a rubber or plastic device that seals the exterior of one of the ends of an SWM element against the wall of the RO pressure vessel housing. The aforestated device inhibits feed water bypassing around the element and forces the feed water through the element.

Calcium bicarbonate ($Ca(HCO_3)_2$): This is a salt available in the majority natural waters. Water consisting of $Ca(HCO_3)_2$ loses CO_2 while it is evaporated or concentrated by RO process and subsequently $CaCO_3$ will get precipitated. $Ca(HCO_3)_2$ solubility is determined by employing Stiff–Davis Index for seawaters and Langlier Saturation Index (LSI) for brackish waters.

Cellulose acetate membranes: CA is hydrophilic in nature, and the membranes composed of CA are usually of asymmetric construction. Formerly CA membranes have been made-up of cellulose triacetate, cellulose diacetate, or a combination of the aforestated materials. These CA-based membranes require a residual of chlorine for protecting these membranes from microbial attack. The CA membranes also have constricted pH operational needs. These membranes are also considered to be uncharged due to the fact that their functional groups are not polar. Because of this nonpolar nature, the CA membranes will never attract foulants to the surfaces effortlessly. Low fouling is noted because of a smoother surface of the CA membrane. Further, it can be noted that the CA membranes are straightforwardly degraded by bio-fouling.

Chemical oxygen demand: This is a nonspecific test used on wastewater where a chemical oxidizing agent is reacted with certain organic matters present in water. Chemical oxygen demand is noted to be extremely accurate relative to the biochemical oxygen demand test; however, it will not measure the entire organic matters existing in the water.

Coagulation: It is a process where minute particles of suspended matter are united by chemical agents to form bigger-sized particles for permitting very fast settling or improved separation. Alum is considered as the most commonly used coagulant. Additional coagulants employed in this process include magnesium oxide, lime, ferric chloride, and sodium aluminate. Also, polyelectrolytes are commonly employed as coagulant aids.

Concentrate: This is the water rejected during the RO process, and it consists of the majority of the dissolved solids present in the feed in a more concentrated form. Brine is another term used for concentrate.

Concentration factor: It is the degree that the RO feed water dissolved solids are concentrated in the reject water. The concentration factor is related to the recovery of RO system and the equation is important for the design of the system.

$$\text{Concentration Factor} = \frac{1}{1 - \text{Recovery}\%} \tag{1.1}$$

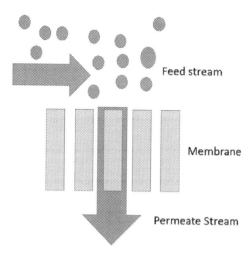

FIGURE 1.6   Cross-flow filtration.

Concentration polarization: It is the salt concentration gradient on the elevated pressure side of the surface of the membrane. The concentration of salt in this boundary layer goes beyond the bulk water concentration. The aforestated phenomenon affects the process efficiency by enhancing the osmotic pressure at the surface of the membrane, rising salt passage, decreasing flux, and widening the possibility of scale formation. Enhancing the concentrate stream velocity or dropping recovery assists to decrease the concentration polarization.

Cross-flow: The running of the feed water stream in parallel to the membrane surface so that it continuously removes contaminants from the membrane surface (Fig. 1.6). The process is termed as cross-flow due to the fact that the direction of feed flow and the direction of filtration flow are at a 90° angle. This is an excellent method for filtering liquids with a high concentration of filterable matter.

Demineralization: This is a type of water purification. It is a process for removing minerals from water, and generally the term is restricted to ion exchange processes. Here, strong acid cation resin in the hydrogen form changes the dissolved salts into their corresponding acids, and strong base anion resin in the hydroxide form eliminates these acids. This process produces water with the same quality as distillation at a lower cost.

Desalination: This is the process of removing salt from seawater or brackish water for producing potable water, utilizing different systems. Desalination techniques are classified into thermal-based processes and membrane-based processes. All these technologies require energy for operation (Soliman et al., 2021).

Dissolved air flotation: Dissolved air floatation is a physio-chemical treatment in which air is closely contacted with an aqueous stream at increased pressure, dissolving the air. This process works on the principle of the transfer of floc to the water surface by the attachment of air bubbles to the floc. This floc accumulated on the surface is termed as the float and is skimmed off as sludge.

Dissolved oxygen: This is the quantity of oxygen dissolved in water at a specific time and it is expressed in terms of mg/L or ppm. Dissolved oxygen level, which is too low or too high, could damage aquatic life and disturb water quality.

Disinfection: This is the process of destroying pathogenic micro-organisms, typically by chemical agents, such as sodium hypochlorite and formaldehyde (Ahuja, 2009). This disinfecting process reduces the count of micro-organisms without essentially destroying all those present. Even though the overall destruction of entire microorganisms is practically difficult, the sterilization process would lessen the count of organisms to a benign predefined value. The sterilization process could typically be accomplished by ethylene oxide, gamma irradiation, heat, and in some instances, special filtration.

Elements: Elements are considered to be physical devices that house the membrane, and are also termed as modules. Spiral wound systems possess up till six elements per pressure vessel. However, the HF RO systems typically possess just one element per pressure vessel.

External fouling: This type of fouling is caused by the accumulation of deposits on the surface of the membrane by three different mechanisms: (1) Mineral deposit accumulation (scale); (2) development of cake of rejected solids, colloids, particulates, and other inorganic or/and organic matter; and (3) development of biofilm.

Feed channel spacers: These spacers offer channels for the feed water flow and are seen in SWM elements. They are considered to be an important part of SWM modules in RO filtration. The aforestated spacers perform a significant role in identifying the hydraulic conditions of the feed channel.

Feed stream: This is the flow into the primary stage of a reverse osmosis system. This stream will be divided to form a product/permeate stream as well as a concentrate stream.

$$\text{Feed Stream} = \text{Permeate Stream} + \text{Concentrate Stream} \tag{1.2}$$

Filtration: This is the process of separation of a liquid and a solid by utilizing a porous substance, that can only allow the liquid to pass through.

Fouling: This is the process of depositing solid substances on the RO membrane surface. This process could be because of the existence of biological growth, sparingly soluble salts, or suspended solids. RO membrane fouling results in a reduction in both the water quality and the amount of water produced. By proper cleaning procedures, the efficiency of fouled membranes can generally be restored. Different types of fouling and its prevention techniques are discussed in Chapter 5, Pretreatment: Fouling and Scaling Control.

Flux or Water Flux: Flux is used to express the rate at which water permeates across a membrane and is normally expressed as volume per area per unit of time. Conventional units are liters per square meter per hour ($L/m^2/h$) or gallons per square foot per day (GFD). The membrane flux is noted to be directly proportional to pressure and temperature. As a general rule, the water flux reduces 1.5% for every 1°F. The salt flux is the quantity of TDS passed across a specific membrane area per unit of time. Also, the salt flux will be a function of the concentration gradient. Hence, with enhancing driving pressure, the permeate salt concentration reduces because of continuous salt leakage and improved water flux. The net effect of increased drive pressure is to dilute a constant quantity of salt with extra pure water.

Note: 1 GFD = $1.660 \text{ L}/(m^2 \cdot h)$

Flocculation: This is the process of aggregation of destabilized particles and micro flakes, and the successive development of sizeable flakes. One should add another chemical known as flocculent for facilitating the formation of flakes termed as flocs. Fig. 1.7 presents the coagulation and flocculation processes.

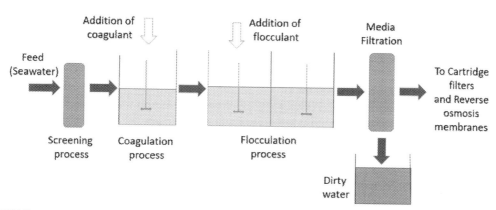

FIGURE 1.7 Coagulation and flocculation processes.

Hard water: It is the water consisting of an excessive number of positive ions. The water hardness is illustrated by the number of magnesium and calcium atoms. Soap typically dissolves poorly in hard water.

Heavy metals: These are metals having a high elemental weight and a density of 5.0 or higher. The majority of heavy metals are considered to be harmful to humans, even in small concentrations.

High-pressure pumps: High-pressure pumps are used to pump the pretreated seawater up to the required pressure by RO membrane to achieve the desired separation of feed water into pure water and concentrated brine streams. For this application, centrifugal pumps are used, which operate in the range of 50–80 bar.

Hollow fiber element: This is one of the four potential membrane configurations (tubular, plate, and frame, and spiral wound are the others) (Berk, 2018). These HF elements are made-up of PA or CA. The pressurized feed stream will pass through the exterior of the fibers. Freshwater gets permeated and will be subsequently accumulated at the element end. HF elements have been considered amongst the first RO systems.

The HF elements do not permit for uniform flow or turbulent flow across the surface of the fiber, thereby turning these elements more prone to scaling and fouling. Once the HF elements are fouled, it will be very difficult for cleaning them because of the incapability to take the cleaning solution to the fouled section. HF elements have been typically used in seawater desalination applications and restricted brackish water applications in which fouling probability is marginal. Generally, the HF module has four main parts, the HF membrane bundle, the tube sheet, the housing, and the end cap. Different membrane module configurations are explained in Chapter 4, Transport Models, Membrane Materials and Basic Flow Patterns.

Hydrolysis: It is a process of chemical breakdown of a membrane when it is exposed to high or low pH, temperature, and bioactivity. The hydrolysis process is generally related to CA membranes in which the acetyl groups are substituted by hydroxyl groups. This process amplifies leakage of salt (i.e., higher permeate conductivity) and a lower feed pressure necessity. Oxidants as well as temperature can lead to hydrolysis in thin-film composite elements.

Internal fouling: This type of fouling is a slow deterioration of membrane efficiency due to variations in the chemical structure of the membrane polymers and activated by chemical degradation or physical compaction.

Langelier Saturation Index: It is considered to be a measure of calcium carbonate scaling probability and is used as a significant efficiency indicator in the RO system management. A positive value of LSI specifies that calcium carbonate can precipitate. A negative value of LSI points out the corrosive nature of the water.

Limiting salt: This is the salt that attains its saturated concentration initially, as the water is concentrated in a reverse osmosis system. The maximum recovery conceivable before any salt starts precipitating is termed as the allowable recovery.

Membrane configuration: It refers to the membrane geometry and its position in space with regard to the feed flow and permeate flow. The two major membrane module configurations employed for RO applications are spiral wound and HF.

Microfiltration: Microfiltration (MF) is a pressure-driven separation process, and is considered to be extensively employed in concentrating, purifying, or separating macromolecules, suspended particles, and colloids from solution. MF can also be used as a pretreatment for RO or NF. Fig. 1.8 is the representation of the contaminants, which can be filtered using the MF technique.

Modified Fouling Index: Modified Fouling Index is proportional to the suspended matter concentration, and is considered to be a very precise index relative to the Silt Density Index (SDI) for foreseeing the water tendency to foul RO/NF membranes.

Nanofiltration: NF is playing a vital role in softening the seawater. An integrated membrane system of NF and ultrafiltration (UF) can be used as the pretreatment for the seawater to increase the overall efficiency of the Seawater Reverse Osmosis (SWRO) plant. Various factors such as operating pressure, cross-flow velocity, and feed temperature can be examined to increase the effectiveness of the process. With the advancement of nanomaterials more efficient quality membranes can be produced to maximize the results.

Nanofiltration membranes: These membranes are the same as RO membranes; however, the NF membranes are not as efficient as RO in separating dissolved solids. NF membranes are generally termed as membrane softeners due to the fact that these membranes will generally reject the twice-positive charged hardness ions (magnesium and calcium)

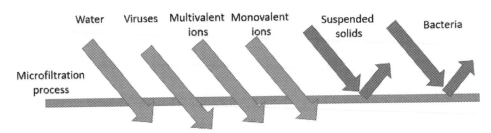

FIGURE 1.8   Representation of the contaminants, which can be filtered using microfiltration technique.

reasonably good, on the other hand, they demonstrate very less rejection for the single-positive and negative charged ions (e.g., chloride, potassium, and sodium). Also, the NF membranes have the ability to reject twice-negative charged ions such as sulfate.

NF membranes are very extensively employed in the potable water industry in which the concentration of dissolved solids should be decreased to less than 500 ppm (Drinking Water Regulations and Contaminants). Also, the NF membranes need lesser pumping pressure relative to RO membranes.

Net driving pressure (NDP): NDP is considered to be the amount of the actual driving pressure accessible for forcing the water across the membrane. It is the difference between the osmotic pressure and feed pressure. As the value of NDP increases, the flux also increases proportionally (provided that all the remaining factors should remain constant).

Normalized permeate flow (NPF): NPF measures the quantity of product water that the RO system is producing. Product flow or permeate flow is a function of membrane condition, temperature, and NDP. By the normalization of the measured product flow for observed NDP and temperature, a measure is achieved, which could be used for comparing the membrane condition to the initial start-up conditions. A decrease in normalized permeate flow of about 10.0%−15.0% specifies that the cleaning of the membrane is needed.

O-Rings: These are used for sealing the product water tube interconnectors of side-by-side elements for preventing the intrusion of increased pressure feed stream (low quality) into the less pressure product stream (high quality). A defective O-ring would lead to a greater saline concentration of the product water in that particular area of the system.

Osmosis: It is a natural process where water undergoes diffusion across a membrane from a low concentrated salt solution to a high concentrated salt solution. The natural osmosis process is shown in Fig. 1.9.

Osmotic pressure: This is the pressure needed for preventing the water flow through a semipermeable membrane separating two solutions with dissimilar ionic strengths (AlZainati et al., 2021). In the case of RO systems, this osmotic pressure needs to be overcome for producing product water. As a general rule, for each 100 ppm of TDS difference between permeate water and feed water, there exists 1 psi of osmotic pressure. RO fundamentals and principles are provided in Chapter 2, Reverse Osmosis Principles and System Components.

The osmotic pressure π of the solution can be determined using van't Hoff's equation:

$$\pi = \vartheta_i R T C_i \tag{1.3}$$

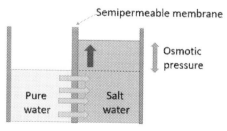

FIGURE 1.9   Representation of an osmosis process.

where π is the osmotic pressure expressed in Pa; ϑ_i is the number of various dissociated ions (e.g., this number is 2 for sodium chloride); R is the gas constant (R = 8.31450 m^3 Pa/(K mol), T is the absolute temperature in Kelvin, and C_i is the concentration of component i (mol/m^3). The concentration should be multiplied by 2, for example, NaCl solute because it dissociates into two ions.

Parts per million: Parts per million (ppm) is considered to be a measurement of concentration, and 1 ppm is one-unit weight of solute per million unit weight of solution. For the analysis of water, the ppm is the same as mg/L.

Permeability: It is the capability of a medium for passing a fluid under pressure.

Permeate: It is the portion of the feed water stream that passes across the membrane and is generally known as "product."

$$Permeate = Feed\ water - Concentrate \tag{1.4}$$

Physical and chemical treatment: These are the processes commonly used in wastewater treatment facilities. In physical treatment, different physical methods are employed for wastewater cleaning. Processes such as sedimentation, screening, flotation, and skimming are used for removing the solids. No chemicals are involved in this physical treatment process. Chemical treatment processes include disinfection, adsorption, neutralization, ion-exchange, and chemical precipitation (coagulation, flocculation).

Pilot test: This is the testing of a cleanup technology under real site conditions in a laboratory for identifying possible difficulties prior to implementation.

Polymers: Polymers are high molecular weight compounds derived by either the addition of several small molecules (such as PE) or the condensation of several small molecules with the removal of alcohol, water (such as nylon). The main polymers that are being utilized for preparing RO membranes are PAs, cellulose diacetate, cellulose triacetate, CA, etc.

Polyamide membranes: Presented at the beginning of the 1970s, the asymmetric PA polymer has been used in the fabrication of the TFC spiral wound RO membrane. PA-based membranes are the most distinctive membrane fabrication material because of their more flexible operating conditions and low-pressure requirements. Also, these PA membranes are oxidant (ozone, bromine, chloramine, chlorine, etc.) intolerant. Different types of membrane materials are discussed in Chapter 4, Transport Models, Membrane Materials and Basic Flow Patterns.

Pretreatment: These are processes used for reducing or eliminating wastewater contaminants before they are discharged. This process is very important while operating nanofiltration and RO membranes because of the feature of their SWM design. The material is designed in a manner that it will permit only a single-way flow across the system. Due to the fact that accumulated material is difficult to be detached from the surface of the membrane, the membranes are extremely vulnerable to fouling (production capacity reduction). Hence, pretreatment is a requirement for any nanofiltration or RO system. Pretreatment typically involves the application of sand filters and fine prefilters. Further, chemical treatment is required in the event that biological fouling or corrosion scaling of the RO membranes is expected.

Post-treatment: In a reverse osmosis system, the post-treatment term normally includes the water treatment processes that happen downstream of the RO plant system. The post-treatment frequently involves disinfection utilizing appropriate bio-cides, probably the inclusion of appropriate corrosion inhibitors, and adjustment of pH. To normalize water for domestic use by stabilizing the alkalinity by adding calcium hydroxide with 30–40 ppm chlorine is added to safeguard water from microorganisms during storage and transportation. Finally, water flows through decarbonization system to maintain the pH of water.

Pressure drop: Consumption of a specific quantity of energy is essential for a fluid to flow by means of channels, such as membrane or pine. The pressure at a specific point is considered to be a measure of the fluid's energy content at the particular point. Due to the fact that part of the energy will be consumed in the flow to a second point downstream, this downstream point pressure will be lower relative to the initial point. Energy consumed, and henceforth the reduction in pressure, depends on the fluid flow rate and fluid viscosity, the shape, and size of the channel.

Pressure vessels: These are tubular devices that comprise the membrane elements. For HF systems, the pressure vessel is normally termed as the "permeator." In the case of SWM elements, the pressure vessels are generally termed as the pressure tubes/housings and could include up till eight membrane elements (generally it is six membrane elements).

Product channel spacer: It is commonly used for preventing the membrane from turning off on itself with the increased operating pressure. During the membrane element fabrication, these types of spacers are located between dual layers of the flat-sheet membranes. It is also known as permeate water carrier or permeate channel spacer. Product water flows in a spiral path through the permeate channel spacers into the permeate collection tubes.

Permeate collection tube: This is also termed as product collection tube, and this tube will collect the product water and directs it to a permeate water header. This tube will be in the middle of an SWM element with the "membrane-product channel spacer—membrane—feed water channel spacer" sandwich wrapped around it.

Permeate staging/Product staging: This is a configuration in which the permeate stream of the first group of RO pressure vessels turns out to be the feed water stream for the second group. This product staging is employed to enhance the product water quality.

Product stream/Permeate stream: This is the share of the feed water stream that passes across the membrane. In a reverse osmosis process, the permeate will have almost 95%–99% of the dissolved salts removed from it.

Pounds per square inch: Pounds per square inch (psi) is a unit of pressure. In SI units, 1 psi is approximately equal to $6895 \, N/m^2$. Different units are utilized to express pressure, and some of these are derived from a unit of force divided by a unit of area.

Recovery: Recovery is the percentage of the feed water that is converted into a product stream (occasionally termed as conversion).

$$\text{Percentage Recovery} = \frac{Q_f - Q_C}{Q_f} \times 100 \qquad (1.5)$$

where Q_f is the feed flow (m^3/h or Gpm) and Q_C is the concentrate flow (m^3/h or Gpm).

By determining the percentage recovery, one will be able to rapidly analyze whether the system is functioning outside of the proposed design.

Reject staging: This is a configuration in which the reject from one group of RO pressure vessels turns out to be the feed stream of a second pressure vessel group. This type of configuration is employed to improve water recovery.

Rejection rate: As hydraulic pressure is applied to water, which is in contact with RO membranes, the water permeates across the membranes and the dissolved solids present in water will get rejected. Rejection rate is defined as the extent to which the dissolved solids are removed. The rate of rejection reduces as the feed water flows across the RO unit due to the fact that the dissolved solids in it are turning out to be increasingly concentrated.

Reverse osmosis: RO desalination is a physical process that works on the movement of water from an area of high level of solute concentration to an area with low solute concentration. It is the reverse of the osmosis process. This movement happens through a special type of barrier called semipermeable membrane, which allows the flow of water and restricts the salts. This flow happens under the influence of a pressure higher than the osmotic pressure, applied on the side with higher salinity. In this process, there is no need for phase separation or heating. The energy required in this process is only used to power the high-pressure centrifugal pumps. The amount of pressure required depends on the salinity of the feed water, temperature (Rahman et al., 2018).

Reverse osmosis membranes: A reverse osmosis membrane is a semipermeable material that will allow the passage of water comparatively fast, whereas other substances are not allowed to pass or will pass moderately slow. These membranes will provide an interface or barrier layer for the cross-flow filtration. The RO membranes are relatively thin and porous materials made of organic polymers (e.g., PSF, PA, and CA). In general, RO membranes would eliminate contaminants having molecular weights higher than 200.

Reverse osmosis membrane compaction: It is the physical membrane compression and will lead to a reduction in water flux (Davenport et al., 2020). The compaction rate is directly proportional to the enhancement in pressure and temperature. This process happens naturally over time demanding increased feed pressure.

Reverse osmosis membrane flushing: RO systems are normally equipped with a permanently piped membrane flushing system for automatically flushing the vessels in the RO trains on shutdown for removing residual concentrate and stop RO membranes from fouling and degradation. This flushing is accomplished utilizing RO system permeate free from disinfectants or other chemicals, or, in the case of shorter RO system shutdowns, the RO vessels can be flushed utilizing nonchlorinated and chemically conditioned filtered water.

Reverse osmosis skids: RO membrane elements are installed in pressure vessels, which typically house 6–8 elements per vessel. Multiple pressure vessels are arranged on support structures (termed as racks or skids). These skids are normally made of plastic, plastic-coated steel, or powder-coated structural steel.

Reverse osmosis trains: The combination of RO feed pump, concentrate, feed and permeate piping, pressure vessels, couplings, valves, and other fittings (instrumentation and controls, and energy-recovery system) installed on a separate support structure (rack/skid), which could operate independently, is termed as RO train. Each RO train is usually designed for producing between 10% and 20% of the whole amount of the membrane desalination permeate water flow.

Salt passage: This is the amount of salt that passes across the membrane into the product stream. The salt passage is expressed in percentage, and is considered to be a function of concentration gradient (brine salt concentration vs the permeate salt concentration), velocity, and temperature.

Note:

$$\text{Salt Passage} = 1 - \text{Salt Rejection} \tag{1.6}$$

$$\text{Percentage salt passage} = \frac{\text{(TDS of product)}}{\text{(TDS of feed)}} \times 100 \tag{1.7}$$

Salt rejection: The term rejection is used for describing the percentage of an influent species a membrane will retain. Salt rejection is the amount of salt separated from the feed water, and is expressed in percentage. As an illustration, 97% salt rejection signifies that the membrane retains 97% of the influent salt. This also indicates that 3% of the influent salt will be passing through the membrane into the permeate, which is referred to as the salt passage.

Note:

$$\text{Salt Rejection} = 1 - \text{Salt Passage} \tag{1.8}$$

$$\text{Percentage salt rejection} = \frac{\text{(TDS of feed)} - \text{(TDS of product)}}{\text{(TDS of feed)}} \times 100 \tag{1.9}$$

Scale: This is precipitate that forms on surfaces in contact with water, as the result of a chemical or physical modification. Scale formation of soluble salts is one of the main factors limiting the efficiency of RO membranes for desalination.

Semipermeable: It is a material that will allow specific size material to pass through whereas rejecting other size material. A reverse osmosis unit utilizes a semipermeable membrane, as shown in Fig. 1.10.

Silt Density Index (SDI): It is a test used for characterizing the fouling probability of a feed stream. SDI is carried out on the basis of determining the rate of plugging a 45-micron filter employing a constant 30 psig feed pressure for a definite time period. SDI_{15} is the SDI test which will run for 15 min.

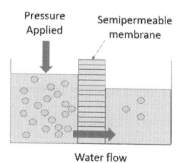

FIGURE 1.10 Reverse osmosis process. Water is forced through a semipermeable membrane via pressure across a concentration gradient.

Generally, the spiral wound systems need an SDI value <5 and HF systems need an SDI value <3. Majority of the deep well waters have an SDI of 3 and common surface water have an SDI higher than 6.

Solution: A solution is referred to as a condition where one or more substances are evenly and uniformly mixed or dissolved. Alternatively stated, a solution is considered to be a homogenous mixture of two or more substances. The solutions can be liquids, solids, or gases, such as seawater, potable water, or air. In this book, we are concentrating mainly with liquid solutions.

The concentration of solution is measured as follows:

$$\text{Percentage Strength} = \frac{\text{Weight of solute}}{\text{Weight of solution}} \times 100 \tag{1.10}$$

$$\text{Percentage Strength} = \frac{\text{Weight of Solute}}{\text{Weight of Solute} + \text{Weight of Solvent}} \times 100 \tag{1.11}$$

Spiral wound element: It is a type of membrane configuration that consists of "flat sheet membrane-permeate channel spacer−flat sheet membrane−feed channel spacer" combinations rolled up around a permeate collection tube.

Telescoping: It is the longitudinal unraveling of SWM elements, which leads to the RO membrane leaves ranging beyond the spacing material separating the leaves. Majority manufacturers establish antitelescoping devices on their membrane elements. Telescoping could be due to the hydraulic surges, excessive differential pressures, or temperature extremes.

Thin-film composite: TFC is a reverse osmosis membrane consisting of and fabricated as three layers joined each other. There will be two base layers (porous structure support) of the asymmetric construction and a thin third skin layer (salt rejecting layer) of PA deposited on the surface, as shown in Fig. 1.11.

Total dissolved solids (TDS): It corresponds to the entire concentration of dissolved substances existing in the water. TDS comprises inorganic salts, along with a modest quantity of organic matter. The commonly found inorganic salts in water are sodium, potassium, magnesium, and calcium, which are all cations, and sulfates, chlorides, bicarbonates, nitrates, and carbonates, which are all anions.

Total organic carbon: It is the carbon mass existing in a sample of water, exclusive of the carbon existing as carbonates or/and carbon dioxide. The value is obtained by catalytically oxidizing the entire dissolved carbon to carbon dioxide. The resultant carbon dioxide

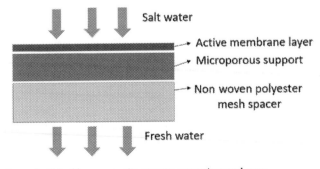

FIGURE 1.11   Structure of a thin-film composite reverse osmosis membrane.

might be determined by infrared absorption, or it might be reduced in a furnace using H_2 to develop CH_4, that could be found using flame ionization detectors.

Total suspended solids (TSS): TSS are the solids present in water, which could be trapped using a filter. For measuring the TSS, the water sample can be filtered by means of a preweighed filter. The residue deposited on the filter is subsequently oven-dried at a temperature of $103°C-105°C$ till the filter weight no longer changes. The rise in the filter weight will represent the TSS.

Turbidity: Turbidity is determined by analyzing the intensity of light scattered by the suspended particles in samples of water. It is generally expressed in terms of nephelometric turbidity units.

Ultrafiltration: UF is a membrane-based separation process used for removing colloidal, extremely fine particles and macromolecules from a water stream. The size of pores in a UF system varies from 0.001 to 0.1 μm. Also, the ultrafiltration membrane systems are characterized by the molecular weight cut-off points. In contrast to RO systems, the ultrafiltration will not eliminate dissolved salts present in water.

Ultrapure water: This is the term employed for characterizing electronic grade process water. Fundamentally, ultrapure water will not have colloids, particles, inorganic, and organic contaminants in it.

1.11 Basic calculations

1.11.1 Example 1.1

If 15.87 kg of a chemical is added to 204.1 kg of water, calculate the percentage strength (by weight) of the solution?

Solution:

$$\text{Percentage strength} = \frac{\text{Weight of solute}}{\text{Weight of solute} + \text{Weight of solvent}} \times 100$$

$$= \frac{15.87 \text{ kg}}{15.87 \text{kg} + 204.1 \text{ kg}} \times 100 = \underline{7.2\%}$$

1.11.2 Example 1.2

What is the system recovery, when feed flow (Q_f) is 63.3 m^3/h and concentrate flow (Q_C) is 13.63 m^3/h?

Solution:

$$\text{Percentage recovery} = \frac{Q_f - Q_C}{Q_f} \times 100$$

$$= \frac{63.3 - 13.63}{63.3} \times 100 = \underline{78.5\%}$$

1.11.3 Example 1.3

Determine the osmotic pressure of 1000 ppm solutions of the solutes (1) $SrSO_4$ and (2) NaCl at 20°C temperature supposing an osmotic coefficient of 0.950. Also, note that $SrSO_4$ and NaCl will dissociate to form two ions while dissolved in water.

Solution:

$$\pi = \Phi CRT$$

where Φ = osmotic coefficient, unitless.

1. Calculate the osmotic pressure for $SrSO_4$, initially by determining the molar concentration of ions and subsequently utilizing the above equation:

$$C = \frac{(2 \text{ mol ion/mol } SrSO_4)\left(1000\frac{mg}{L}\right)}{(10^3 mg/g)(183.60 \text{ g/mol})} = 0.0109 \text{ mol/L}$$

$$\pi = (0.950)\left(0.00556\frac{mol}{L}\right)(0.0831450 \text{ L. bar/K.mol})(293 \text{ K})$$

$$\pi = 0.130 \text{ bar}$$

2. Calculate the osmotic pressure for NaCl:

$$C = \frac{(2 \text{ mol ion/mol NaCl})\left(1000\frac{mg}{L}\right)}{(10^3 mg/g)(58.40 \text{ g/mol})} = 0.0342 \text{ mol/L}$$

$$\pi = (0.950)\left(0.03420\frac{mol}{L}\right)(0.0831450 \text{ Lbar/Kmol})(293 \text{ K})$$

$$\pi = 0.790 \text{ bar}$$

References

Abbas, A., & Al-Bastaki, N. M. (2005). Modeling of a reverse osmosis water desalination unit using neural networks. *Chemical Engineering Journal, 114,* 139–143.

Ahuja, S. (Ed.), (2009). *Handbook of water purity and quality.* Academic Press.

AlZainati, N., Saleem, H., Altaee, A., Zaidi, S. J., Mohsen, M., Hawari, A., & Millar, G. J. (2021). Pressure retarded osmosis: Advancement, challenges and potential. *Journal of Water Process Engineering, 40,* 101950.

Berk, Z. (2018). *Food process engineering and technology.* Academic Press.

Brehant, A., Bonnelye, V., & Perez, M. (2003). Assessment of ultrafiltration as a pretreatment of reverse osmosis membranes for surface seawater desalination. *Water Science and Technology: Water Supply, 3*(5–6), 437–445.

Buonomenna, M. G. (2013). Nano-enhanced reverse osmosis membranes. *Desalination, 314,* 73–88.

Busch, M., & Mickols, W. E. (2004). Reducing energy consumption in seawater desalination. *Desalination, 165,* 299–312.

Busch, M., Mickols, W. E., Jons, S., Redondo, J., & De Witte, J. (2004). Boron removal in sea water desalination. *International Desalination and Water Reuse Quarterly, 13*(4), 25.

Cadotte, J. E., & Petersen, R. I. (1981). Thin film reverse osmosis membranes: Origin, development, and recent ddvances, in synthetic membranes, ACS Symposium Series 153 In A. F. Turbak (Ed.), *Desalination* (Vol. I). American Chemical Society.

Crittenden, J. C., Trussell, R. R., Hand, D. W., Howe, K. J., & Tchobanoglous, G. (2012). Copyright *MWH's water treatment: Principles and design* (Third Edition). John Wiley & Sons, Inc.

Davenport, D. M., Ritt, C. L., Verbeke, R., Dickmann, M., Egger, W., Vankelecom, I. F., & Elimelech, M. (2020). Thin film composite membrane compaction in high-pressure reverse osmosis. *Journal of Membrane Science*, 118268.

Drinking Water Regulations and Contaminants. <https://www.epa.gov/sdwa/drinking-water-regulations-and-contaminants>.

Dupont Water Solutions, FilmTec™ SW30XHR-440i Element https://www.dupont.com/content/dam/dupont/amer/us/en/water-solutions/public/documents/en/45-D00968-en.pdf [Accessed on 10th March 2021].

Farhat, A., Ahmad, F., Hilal, N., & Arafat, H. A. (2013). Boron removal in new generation reverse osmosis (RO) membranes using two-pass RO without pH adjustment. *Desalination, 310,* 50–59.

Fritzmann, C., Löwenberg, J., Wintgens, T., & Melin, T. (2007). State-of-the-art of reverse osmosis desalination. *Desalination, 216*(1–3), 1–76.

FuturEnviro (2018). Design, construction, operation and maintenance of two desalination plants in Qatar ACCIONA Agua, 15915–2013 ISSN: 2340-2628. <http://www.futurenviro.com/pdf/reportajes-especiales/06-2018/FuturENVIRO_Water_June_2018_Desaladoras_Qatar_Acciona.pdf>.

Gholami, S., Rezvani, A., Vatanpour, V., & Cortina, J. L. (2018). Improving the chlorine resistance property of polyamide TFC RO membrane by polyethylene glycol diacrylate (PEGDA) coating. *Desalination, 443,* 245–255.

Glater, J. (1998). The early history of reverse osmosis membrane development. *Desalination, 117*(1–3), 297–309.

Hydranautics. (2007). Press release: Integrated membrane solutions at work in Southern Spain. <http://www.membranes.com/press/Escombreras.Jan%202007.pdf> [Accessed 05 August 2019].

Kesting, R. E. (1971). *Synthetic polymeric membranes.* McGraw-Hill.

Kim, J., & Hong, S. (2018). A novel single-pass reverse osmosis configuration for high-purity water production and low energy consumption in seawater desalination. *Desalination, 429,* 142–154.

Kucera, J. (2010). Reverse osmosis design, processes, and applications for engineers.

Kucera, J. (2015). *Reverse osmosis: Industrial processes and applications.* John Wiley & Sons.

Laidler, K. J., & Meiser, J. H. (1999). *Physical chemistry.* Houghton Mifflin.

Lalia, B. S., Kochkodan, V., Hashaikeh, R., & Hilal, N. (2013). A review on membrane fabrication: Structure, properties and performance relationship. *Desalination, 326,* 77–95.

Li, N. N., Fane, A. G., Winston Ho, W. S., & Matsuura, T. (Eds.), (2011). *Advanced membrane technology and applications.* John Wiley & Sons.

Loeb, S., & Sourirajan, S. (1962). Seawater demineralization by means of a semipermeable membrane. In R. Gould (Ed.), *Advances in chemistry* (pp. 117–132). American Chemical Society, also Loeb S., Sourirajan S., USPatent 3 133 132 (1964).

Mahon,.H.I. (1966). Permeability separatory apparatus, permeability separatory membrane element, method of making the same and process utilizing the same, United States-Patent. 3, 228, 876.

Ning, R. Y. (2015). *Reverse osmosis chemistry—Basics, barriers and breakthroughs. Desalination updates.* IntechOpen.

Nitto Hydranautics, <https://membranes.com/wp-content/uploads/pdf/brochure/RO/PB-116-rev3-SWC-MARCH%202015.pdf> [Accessed on 10th March 2021].

Pandey, S. R., Jegatheesan, V., Baskaran, K., & Shu, L. (2012). Fouling in reverse osmosis (RO) membrane in water recovery from secondary effluent: A review. *Reviews in Environmental Science and Bio/Technology, 11*(2), 125–145.

Park, H. M., Yoo, J., & Lee, Y. T. (2019). Improved fouling resistance for RO membranes by a surface modification method. *Journal of Industrial and Engineering Chemistry, 76,* 344–354.

Peñate, B., & García-Rodríguez, L. (2012). Current trends and future prospects in the design of seawater reverse osmosis desalination technology. *Desalination, 284,* 1–8.

Pendergast, M. T. M., & Hoek, E. M. V. (2011). A review of water treatment membrane nanotechnologies. *Energy & Environmental Science, 4*(6), 1946–1971.

Rahman, H., & Zaidi, S. J. (2018). Desalination in Qatar: Present status and future prospects. *Civil Eng Res J, 6,* 133–138.

Redondo, J. A., & Casanas, A. (2001). Designing seawater RO for clean and fouling RO feed. Desalination experiences with the FilmTec SW30HR-380 and SW30HR-320 elements – technical – economic review. *Desalination, 134,* 83–92.

Reid, C. E., & Breton, E. J. (1959). Water and ion flow across cellulose membranes. *Journal of Applied Polymer Science, 1,* 133.

Reverberi, F., & Gorenflo, A. (2007). Three year operational experience of a spiral-wound SWRO system with a high fouling potential feed water. *Desalination, 203,* 100–106.

Reverter, J. A., Talo, S., & Alday, J. (2001). Las Palmas III – The success story of brine staging. *Desalination, 138,* 207–217.

Riley, R. L., Lonsdale, H. K., Lyons, C. R., & Merten, U. (1967). Preparation of ultrathin reverse osmosis membranes and the attainment of theoretical salt rejection. *Journal of Applied Polymer Science, 11,* 2143.

Saleem, H., Trabzon, L., Kilic, A., & Zaidi, S. J. (2020). Recent advances in nanofibrous membranes: Production and applications in water treatment and desalination. *Desalination, 476,* 114178.

Saleem, H., & Zaidi, S. J. (2020a). Developments in the application of nanomaterials for water treatment and their impact on the environment. *Nanomaterials, 10*(9), 1764.

Saleem, H., & Zaidi, S.J. (2020b). Innovative nanostructured membranes for reverse osmosis water desalination. https://doi.org/10.29117/quarfe.2020.0023

Saleem, H., & Zaidi, S. J. (2020c). Nanoparticles in reverse osmosis membranes for desalination: A state of the art review. *Desalination, 475,* 114171.

Shenvi, S. S., Isloor, A. M., & Ismail, A. F. (2015). A review on RO membrane technology: developments and challenges. *Desalination, 368,* 10–26.

Soliman, M. N., Guen, F. Z., Ahmed, S. A., Saleem, H., Khalil, M. J., & Zaidi, S. J. (2021). Energy consumption and environmental impact assessment of desalination plants and brine disposal strategies. *Process Safety and Environmental Protection.*

Strathmann, H., Kock, K., Amar, P., & Baker, R. W. (1975). The formation mechanism of asymmetric membranes. *Desalination, 16,* 179.

TORAY, Highest rejection SWRO, https://www.toraywater.com/products/ro/pdf/TM800K.pdf [Accessed on 10th March 2021].

Urgency, U. Water caucus summary. World Water Council (WWC), Marseille, France (2007).

Water Technology (20 January 2017). What is reverse osmosis and how is it best used? <http://www.watertechonline.com/what-is-reverse-osmosisand-how-is-it-best-used/>.

Water, U. N. Coping with water scarcity: Challenge of the twenty-first century. Prepared for World Water Day (2007).

Watkins,.K. (2006). Human development report: United Nations development programme.

Westmoreland,.J.C. (1968). Spiral wrapped reverse osmosis membrane cell, United States-Patent 3, 367–504.

WHO, UNICEF. Millennium development goals: Progress on sanitation and drinking-water: 2010 Update Report, Geneva: WHO/UNICEF Joint Monitoring Programme for Water Supply ISBN 978 92 4 156395 6. (2010).

Wiles, L., & Peirtsegaele, E. Reverse osmosis: A history and explanation of the technology and how it became so important for desalination. IWC, 18, 49.

Woods, W. G. (1994). An introduction to boron: history, sources, uses, and chemistry. *Environmental health perspectives, 102*(suppl 7), 5–11.

World Health Organization. (2009). Boron in drinking-water: Background document for development of WHO Guidelines for Drinking-water Quality (No. WHO/HSE/WSH/09.01/2). World Health Organization.

Xia, Y., Wang, Z., Chen, L. Y., Xiong, S. W., Zhang, P., Fu, P. G., & Gai, J. G. (2020). Nanoscale polyelectrolyte/metal ion hydrogel modified RO membrane with dual anti-fouling mechanism and superhigh transport property. *Desalination, 488,* 114510.

Yadav, S., Saleem, H., Ibrar, I., Naji, O., Hawari, A. A., Alanezi, A. A., ... Zhou, J. (2020). Recent developments in forward osmosis membranes using carbon-based nanomaterials. *Desalination, 482,* 114375.

Zaidi, S. J., Fadhillah, F., Saleem, H., Hawari, A., & Benamor, A. (2019). Organically modified nanoclay filled thin-film nanocomposite membranes for reverse osmosis application. *Materials, 12*(22), 3803.

C H A P T E R

2

Reverse Osmosis Principles and System Components

2.1 Osmosis and Reverse osmosis

Osmosis is a naturally occurring phenomenon where a solvent (normally water) flows by means of a semipermeable barrier from the lower solute concentration side to the higher solute concentration side (Baumgarten & Feher, 2012). Osmotic pressure is the force caused by a solution passing through a semipermeable surface by osmosis, which is equal to the force required to resist the solution from passing back through the surface. In other words, it is the hydraulic pressure applied by the water on the membrane in the course of its transfer from the low-concentration solute side to the high-concentration solute side of the membrane. It is considered as a natural force same as gravity and is proportional to the difference in the salt concentrations on the two sides of the membrane, the temperature of source water, and the nature of ions that form the total dissolved solid content of the source water. Also, the osmotic pressure is not related to the membrane type. To remove clean (low-solute concentration) water from a high-solute concentration feed water by employing the membrane separation process, the natural osmosis-powered motion of water should be reversed; in other words, the clean water must be transferred from the high-concentration solute side to the low-concentration solute side of the membrane.

As demonstrated in Fig. 2.1, the flow of water proceeds until the chemical potential equilibrium of the solvent is developed. The pressure difference between both sides of the membrane is the same as the solution osmotic pressure, at the equilibrium condition. In order to reverse the solvent (water) flow from higher concentration to lower concentration, pressure higher than osmotic pressure is applied; subsequently, the water separation from the solution happens because pure water passes from the high concentration side to the low concentration side. This process is termed as reverse osmosis (RO; or hyperfiltration) (Caputo & Giaconia, 2014). RO is a diffusion-controlled pressure-driven membrane process; similar in principle to nanofiltration (NF), a partial membrane separation process having the capability of separating bivalent ions (magnesium, calcium, and so on), dissolved organic matter, and the compounds for odors and tastes in water. The RO process separates the majority of ions, regardless of their valence state (chlorides, sodium, and so forth), primarily based on a solubility—diffusivity mechanism.

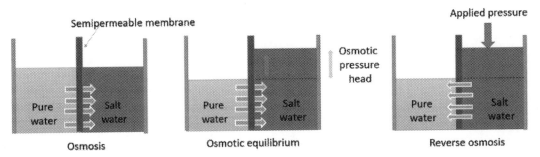

FIGURE 2.1 Reverse osmosis principle. Water is the diluted solution and salt water is the concentrate solution used.

TABLE 2.1 TDS of different categories of water that can be treated by desalination.

Sl. no.	Types of water	TDS range (mg/L or ppm)
1	Seawater	15,000–50,000
2	Brackish water	1,000–15,000
3	River water	500–3,000
4	Waste water (treated domestic)	500–700
5	Waste water (untreated domestic)	250–1000
6	Pure water	Much less than TDS 500

The rate of water transport across the membrane is considered to be several orders of magnitude greater as compared to the rate of salt passage. This difference between salt passage and water passage rates allows the membrane system to produce clean water having extremely low salt concentration. The feed water pressure applied counteracts the osmotic pressure and overpowers the pressure loss that happens while the water flows across the membrane, thus maintaining the fresh water on the low solute concentration side of the membrane until this water leaves the membrane (Nikolay, 2013). The solutes present in the feed water side of the membrane are retained and concentrated; and these salts are eventually discharged from the membrane for removal. Consequently, the RO process results in two streams; the first stream is the fresh water of low salt concentration (known as permeate) and the second stream is the water of high salt concentration (known as retentate, concentrate, or brine).

A reverse osmosis membrane functions as a semipermeable barrier to the flow in the RO process, allowing the water and solvent to pass through and blocking the salts and solutes. The RO desalination process can be used for raw water of different qualities apart from seawater. Waste water, river water, brackish water, and even treated potable water from the municipal supply can be treated by desalination. The total dissolved solids (TDS) of different categories of water that can be treated by desalination are given in Table 2.1.

Thermodynamically, the osmotic pressure can be determined by the following expression:

$$\pi = -\left(\frac{R.T}{V_b}\right)\ln x_W \tag{2.1}$$

where π is the osmotic pressure in bar, R is the ideal gas constant (0.083145 L · bar/mol K), V_b is the molar volume of water in L/mol, and x_W is the mole fraction of water (mol/mol).

In the case of dilute solutions ($x_W \sim 1$), the osmotic pressure is determined by applying van t'Hoff's law as follows:

$$\pi = - \left(\frac{n_s}{V} \right) RT \tag{2.2}$$

$$\text{or } \pi = CRT \tag{2.3}$$

where n_s is the entire solute amount present in solution in mol, C is the total solute concentration in mol/L, and V is the volume of solvent in L.

Eq. (2.3) has been derived under the condition of an infinitely dilute solution, which is normally not possible in a reverse osmosis system. To justify the assumption of diluteness, the nonideal behavior of the concentrated solution, and the compressibility of liquid at elevated pressure, a nonideality coefficient ($\varnothing$) should be included in Eq. (2.3).

$$\pi = \varnothing CRT \tag{2.4}$$

Even though numerous solvents might be used, the fundamental application of RO is water-based systems. Consequently, successive discussions, and examples, will be based on the utilization of water as the solvent.

Also, it can be noted that there are no terms in the osmotic pressure thermodynamic equation [Eq. (2.1)] that recognize the solute (Crittenden et al., 2012). Osmotic pressure is considered to be a function of water concentration in the system. The solutes present will decrease the water mole fraction, and the impact of several solutes will be additive due to the fact that they collectively lessen the mole fraction of water.

2.2 Water flux, solute flux, rejection, and recovery equations

Generally, in the RO process, a polymer membrane is used for treating the solute with a molecular weight below 100 Da (Basile & Nunes, 2011; Jurchevsky & Pervov, 2020). As the RO membranes are very dense (Jornitz, 2020; Kabay et al., 2010), an increased amount of pressure must be applied for diffusing the water to flow across the membrane. Particularly, in the case of seawater, which has high osmotic pressure (almost 25–28 bar) (Sablani et al., 2006; Swarup et al., 1992), if the pressure applied is below this osmotic pressure, then the water will move from the low-concentration solution to the high-concentration solution and vice versa. Therefore, the pressure applied must be always higher as compared to the feed water osmotic pressure. In the case of the RO process, the water and salt will permeate by the solution-diffusion model (Crystal, 1995; Hasmadi et al., 2017) as the phenomena of separation is by diffusion. According to this model, the water molecules will diffuse across the dense membrane. Once the water is dissolved in the membrane, the water molecules will move by random molecular diffusion. However the diffusing fluid concentration will depend on the pressure, temperature, and composition of the fluid on the membrane.

The water flux, J_w, is connected with the pressure and concentration gradient across the membrane as per Eq. (2.5):

$$J_w = K_w(\Delta P - \Delta \pi) \tag{2.5}$$

where K_w is the mass transfer coefficient for water flux, $L/(m^2 \cdot h \cdot bar)$, ΔP is the pressure difference across the membrane, $\Delta \pi$ is the osmotic pressure differential across the membrane, and J_w is the water volumetric flux in $L/(m^2 \cdot h)$. As Eq. (2.5) shows when $\Delta P > \Delta \pi$, then the water flows from the higher concentrated solution to the lower concentrated solution, and when $\Delta P = \Delta \pi$, there will not be any flow happening.

Water flux is usually expressed as a volumetric flux ($gal/ft^2 \cdot d$ or $L/m^2 \cdot h$) and the mass transfer coefficient is normally expressed in $L/m^2 \cdot h \cdot bar$ or $gal/ft^2 \cdot d \cdot atm$ unit.

The solute flux or salt flux could be determined using Eq. (2.6):

$$J_s = K_s(C_f - C_p) \tag{2.6}$$

where J_s is the mass flux of solute, $mg/(m^2 \cdot h)$, K_s is the solute or salt permeability coefficient in $L/(m^2 \cdot h)$ or m/h, C_f is the solute or salt concentration on the feed side of the membrane, and C_p is the solute or salt concentration on the permeate side of the membrane. However the concentration of solute in the permeate side of the membrane will be very low as compared to the concentration of solute in the feed solution. Hence, Eq. (2.6) could be simplified as follows:

$$J_s = K_s(C_f) \tag{2.7}$$

Solute flux will be typically expressed as a mass flux with $mg/m^2 \cdot h$ or $lb/ft^2 \cdot d$ unit. Values of K_w and K_s are experimentally provided by the membrane manufacturers.

From Eq. (2.5), it can be noted that the water flux will increase with applied pressure linearly. On the other hand, the solute flux will be practically not influenced by the applied pressure [Eq. (2.6)], and it is calculated just by the concentration difference across the membrane.

The solute concentration in the product water is determined as the ratio of the solute flux and the water flux, and is presented as

$$C_p = \frac{J_s}{J_w} \tag{2.8}$$

Thus, lower the solute flux or higher the water flux, then better removal of solute will be accomplished and the permeate will contain a low concentration of solute.

Further, the membrane selectivity could be expressed in terms of the rejection "R":

$$R = \left(1 - \frac{C_p}{C_f}\right) \times 100\% \tag{2.9}$$

Therefore, the selectivity of the membrane increases with the applied pressure as the concentration of solute in the permeate decreases. By linking the previous equations, the rejection could be determined as follows:

$$R = \left(\frac{K_w(\Delta P - \Delta \pi)}{K_w(\Delta P - \Delta \pi) + K_s}\right) \times 100\% \tag{2.10}$$

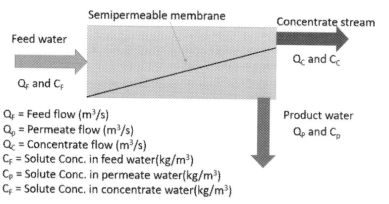

FIGURE 2.2 Separation process using reverse osmosis membrane.

The recovery can be determined as the ratio of permeate flow to feed water flow.

$$r = \frac{Q_p}{Q_f} \tag{2.11}$$

where Q is the flow in m^3/s and r is the recovery (dimensionless).

By employing the principles of flow and mass balance, the concentration of solute in the reject water can be determined from the solute rejection and recovery. The relevant flow along with mass balances employing flow and concentration terminologies are presented in Fig. 2.2:

$$\text{Flow balance: } Q_F = Q_C + Q_P \tag{2.12}$$

$$\text{Mass balance: } C_F Q_F = C_C Q_C + C_P Q_P \tag{2.13}$$

where Q_F = feed flow (m^3/s), Q_p = permeate flow (m^3/s), Q_C = concentrate flow (m^3/s), C_F = solute concentration in feed water (kg/m^3), C_P = solute concentration in permeate water (kg/m^3), and C_C = solute concentration in concentrate water (kg/m^3). Solute concentration is also expressed in mg/L or ppm.

Relating the mass and flow balances with Eq. 2.11 (recovery) and Eq. 2.9 (rejection) gives the succeeding expression for the concentration of solute in the brine:

$$C_C = C_F \left[\frac{1 - (1 - R)r}{1 - r} \right] \tag{2.14}$$

where R = rejection (dimensionless) and r = recovery.

Rejection is often nearly 100%, and thus Eq. (2.14) could be made simpler as below:

$$C_C = C_F \left[\frac{1}{1 - r} \right] \tag{2.15}$$

As presented in Eqs. (2.5) and (2.6), the water flux will depend on the pressure gradient, whereas the solute flux will depend on the concentration gradient. When the concentration of solute in the feed water increases at constant pressure, which will result in the osmotic pressure increase, the water flux will be decreased (due to greater $\Delta\pi$) and the solute flux will be increased (due to greater ΔC), which decreases rejection and leads to a decline in the permeate quality (Jorgensen, 1979). When the pressure of feed water increases, the water flux will be increased; however the solute flux will be constant. Consequently, when the water flux increases, the concentration of solute in the permeate will decrease, and the rejection will increase (Singh, 2014).

Example 2.1

The feed water containing 6000 mg/L of TDS is treated with the RO process for removing the salts. The water flux through RO membranes was found to be 30 L/(m² · h). Calculate the salt concentration in the product water and also the salt passage. Ks value is 2.5 [(mg/m² · h)/(mg/L)].

Solution:

In 1 hour, 30 L water is produced by 1 m² membrane surface area.

In 1 hour, the following amount of salts will pass through 1 m² membrane.

From Eq. 2.6,

$$J_s = K_s(C_f - C_p)$$

$$J_s = 6000 \text{ mg/L} \times 2.5\left((\text{mg/m}^2\text{h})/(\text{mg/L})\right)$$
$$= 15,000 \text{ mg/m}^2 \cdot \text{h}$$

This amount of salt arrives in 30 L/m² · h

From Eq. (2.8),

$$C_p = \frac{J_s}{J_w}$$

$$C_p = \frac{15,000}{30} = 500 \text{ mg/L}$$

$$Pecentage \ Salt \ passage = \frac{(TDS \ of \ Product)}{(TDS \ of \ Feed)} \times 100$$

$$Salt \ passage = \frac{C_p}{C_f} \times 100\% = \frac{500}{6000} \times 100\% = 8.33\%$$

$$Salt \ rejection = 1 - Salt \ passage = 100\% - 8.33\% = 91.67\%$$

Example 2.2

In the above problem, if the water flux is increased from 30 L/m² · h to 60 L/m² · h, what will be:

(1) Salt concentration in the permeate? (2) Salt passage? (3) Salt rejection?

Solution:

From Eqs. (2.8) and (2.7),

$$C_p = \frac{J_s}{J_w}$$

and $J_s = K_s(C_f)$

$$C_p = \frac{K_s(C_f)}{J_w}$$

Here K_s *and* (C_f) are constants, and J_w is $60/30 = 2$ times higher.

As a result, C_p will be 2 times lesser, so 500 mg/L divided by 2 $= 250$ mg/L.
The salt passage will be $250/6000 \times 100\% = 4.16\%$.
The salt rejection will be $100\% - 4.16\% = 95.84\%$.

2.3 Concentration polarization

The accumulation of solutes nearby the surface of the membrane is termed as the concentration polarization, and this will act as a barrier to the water flow across the membrane thereby decreasing the efficiency of the membrane (Sadr & Saroj, 2015). The water flow across the membrane will bring feed water (consisting of solute and water) to the surface of the membrane, and as the pure water passes across the membrane, the solutes will accumulate over the surface of the membrane. During this filtration process, the particles will deposit over the membrane and subsequently develops a cake layer. Due to the fact that the rejection mechanisms for RO are different, these solutes will remain in the solution and leads to the development of a boundary layer of greater concentration at the surface of the membrane. Consequently, the solute concentration in the feed water turns out to be polarized, with the solute concentration at the surface of the membrane greater as compared to the solute concentration in the bulk feed water in the feed stream.

The boundary layer is the layer formed due to the tangential feed water flow in the RO membrane feed/brine spacer and the product water flow in the direction perpendicular to the membrane on both sides of the feed/brine spacer (Saeed, 2012). In Fig. 2.3, C_M is the concentration of salt at the surface of the membrane, C_{FC} is the concentration of salt in the feed−concentrate channel, C_p is the salt concentration on the less salinity (permeate) membrane side, and δ is the boundary layer thickness. Usually, two forms of flow take place in the feed/concentrate spacer boundary layer, that is, (1) fresh water convective flow from the bulk feed water across the membrane and (2) rejected solute diffusion flow from the surface of the

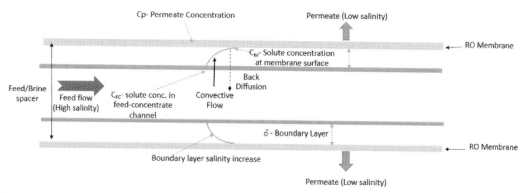

FIGURE 2.3   Reverse osmosis feed spacer boundary layer.

membrane back to the feed flow. As the semipermeable RO membrane is designed in such a way that the water convective flow rate is normally greater relative to the diffusion of salt flow, the membrane-rejected salt will build up in the boundary layer with an utmost concentration of solute happening at the internal surface of the membrane (C_M).

This concentration polarization will have many adverse impacts on the efficiency of the RO process:

1. Low water flux due to the fact that the osmotic pressure gradient is greater because of the increased solute concentration at the surface of the membrane.
2. Low rejection because of an increase in the solute passage through the membrane from an increment in the concentration gradient and a decrease in the flux of water.
3. Solute solubility limit might be surpassed, resulting in precipitation and scaling.

The flux of solute towards the surface of the membrane because of the water convective flow is represented by the following expression (Crittenden et al., 2012):

$$J_s = J_w \times C \tag{2.16}$$

From Fig. 2.4, a mass balance at the surface of the membrane can be found below:

$$\text{Accumulation of mass} = \text{Mass in} - \text{Mass out} \tag{2.17}$$

At steady state with no mass accumulation, the flux of solute towards the surface of the membrane should be balanced by solute fluxes flowing away from the membrane (because of diffusion) and across the membrane (to the product water) as below:

$$\frac{dM}{dt} = 0 = J_w Ca - D_L \frac{dC}{dZ}a - = J_w C_p a \tag{2.18}$$

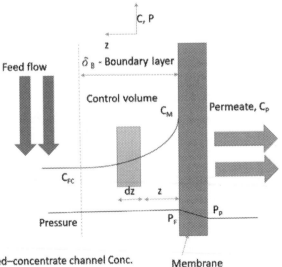

FIGURE 2.4 Diagrammatic representation of concentration polarization.

C_{FC}-Feed–concentrate channel Conc.
C_M–Membrane surface Conc.
C_p–Permeate Conc.

where M = solute mass (g), t = time (s), D_L = solute diffusion coefficient in water (m^2/s), a = membrane surface area (m^2), z = distance perpendicular to the surface of membrane (m), δ_B = boundary layer thickness.

The Eq. 2.18 can be adjusted and integrated across the boundary layer thickness with boundary conditions $C(0)$ = C_M and $C(\delta_B)$ = C_{FC}, where C_M is the solute concentration at the surface of the membrane and C_{FC} is the solute concentration in the feed−concentrate channel:

$$D_L \int_{C_M}^{C_{FC}} \frac{dC}{C - C_p} = -J_w \int_0^{\delta_B} dz \qquad (2.19)$$

Integrating yields

$$\ln\left(\frac{C_M - C_P}{C_{FC} - C_P}\right) = \frac{J_w \delta_B}{D_L} \qquad (2.20)$$

$$\frac{C_M - C_P}{C_{FC} - C_P} = e^{(J_w \delta_B)/D_L} = e^{J_w/k_{CP}} \qquad (2.21)$$

where $k_{CP} = \frac{D_L}{\delta_B}$, mass transfer coefficient of concentration polarization in m/s.

This k_{CP} defines the solute diffusion away from the surface of the membrane.

In general, concentration polarization is the ratio of the membrane surface solute concentration and feed−concentrate channel solute concentration and is expressed as follows:

$$\beta = \left[\frac{C_M}{C_{FC}}\right] \qquad (2.22)$$

Combining Eq. (2.22) with Eqs. (2.9) (rejection equation) and (2.21) results in the subsequent expression:

$$\beta = (1 - R) + R\left(e^{J_w/k_{CP}}\right) \qquad (2.23)$$

When the rejection is greater (>99%), then Eq. (2.23) could be sensibly simplified as below:

$$\beta = e^{J_w/k_{CP}} \qquad (2.24)$$

For predicting the degree of concentration polarization, the k_{CP} value is needed. The k_{CP} can be determined using a relationship between Schmidt (Sc), Reynolds (Re), and Sherwood (Sh) numbers. For promoting turbulent conditions and minimizing the concentration polarization in the RO membranes, the spiral-wound membrane (SWM) element consists of mesh feed channel spacer and maintains an increased velocity flow parallel to the surface of the membrane. Schock and Miquel (1987) studied the spacer-filled feed channel of a spiral-wound membrane membrane element and established that k_{CP} can be predicted by using the subsequent equation (Schock & Miquel, 1987):

$$k_{CP} = 0.023 \frac{D_L}{d_H} (Re)^{0.875} (Sc)^{0.25} \qquad (2.25)$$

$$Re = \frac{\rho \vartheta d_H}{\mu} \qquad (2.26)$$

$$Sc = \frac{\mu}{\rho D_L} \tag{2.27}$$

where ϑ = feed channel velocity (m/s), d_H = hydraulic diameter (m), μ = feed water dynamic viscosity (kg/m · s), ρ = feed water density (kg/m³), Re = Reynolds number (dimensionless), and Sc = Schmidt number (dimensionless).

The Reynolds number (Re) is defined as the ratio of inertial forces to viscous forces. This number is dimensionless and is used for categorizing the fluids systems in which the effect of viscosity is significant in controlling the velocities or the flow pattern of a fluid. Also, Re is used to determine whether fluid is in turbulent or laminar flow. Schmidt number (Sc) is a dimensionless number and is the ratio of momentum diffusivity (kinematic viscosity) and mass diffusivity. Sc is used to characterize fluid flows in which there are simultaneous momentum and mass diffusion convection processes. It physically relates the relative thickness of the hydrodynamic layer and mass-transfer boundary layer.

The hydraulic diameter (d_H) is commonly used when handling flow in noncircular tubes and channels. The need for the hydraulic diameter arises due to the use of a single dimension in case of dimensionless quantity such as Reynolds number. d_H is found using the below equation:

$$d_H = \frac{4(Volume\ of\ flow\ channel)}{wetted\ surface} \tag{2.28}$$

In the case of hollow fiber membranes, the d_H is the same as diameter of the inner fiber. The spiral wound membrane could be approximated by flow across a slit, in which the width $\gg$ feed channel height. For empty channels, the d_H will be two times the height of the feed channel, as presented in the below equation:

$$d_H = \frac{4wh}{2w + 2h} \sim 2h \tag{2.29}$$

where h = height of feed channel in m, and w = width of feed channel in m.

The value of h in standard SWM element ranges from almost 0.40 to 1.20 mm and is controlled by the spacer thickness.

Marinas and Urama established a relationship employing the superficial velocity and the channel height, that eliminates the task of calculating the spacer parameters (Urama & Marinas, 1997). Their correlation is

$$k_{CP} = \lambda \frac{D_L}{d_H} (Re)^{0.50} (Sc)^{\left(\frac{1}{3}\right)} \tag{2.30}$$

The value of β will always be greater than 1, despite the fact that new RO membrane elements are set up and the system is properly configured and functioned (Voutchkov, 2017b). At best, the β can be 1.1−1.2 range, whereas in the worst-case scenario, it can go beyond 2.0.

Eq. (2.31) specifies the effect of concentration polarization on the RO membrane flux:

$$J = A \times \left[F_p - \left(\beta \times O_p + P_p + 0.5 \times P_d\right)\right] \tag{2.31}$$

where A is the membrane water permeability coefficient; J is the membrane permeate flux; O_p is the saline water osmotic pressure; F_p is the feed pressure applied to the RO membranes; P_d is the pressure drop between the concentrate side and feed side of the RO membranes; P_p is the permeate pressure.

From Eq. (2.31), it can be noted that the clean water production (flux) of a membrane reduces as the concentration polarization increases. The above expression demonstrates that in practice, the osmotic pressure needed by the RO system feed pump for producing a similar quantity of water will increase proportionally with the concentration polarization factor.

A method for decreasing the concentration polarization in RO channels using a stable, spatially variant slip velocity profile was studied by Ratnayake and Bao (2017). An approach was well-developed for identifying the utmost efficient wall slip velocity profile for enhancing the diffusive driving force away from the wall, as a result increases mass transfer away from the membrane and decreases concentration polarization. In this work, an approximate solution to the nonlinear system was developed using systems of linearized ordinary differential equations to approximate the behavior of the partial differential equations and determine the steady-state actuation profile that most effectively increases mass transfer at the wall. The developed membranes with the engineered surface feature was demonstrated to decrease membrane fouling and increase water flux. The surface feature pattern could be developed by numerous techniques, like template-based micromolding, thermal embossing using hard stamps, and printing. It has been proposed that the patterns create improved mixing as well as an irregular fluid flow that increases solute mass transfer away from the membrane. Zhou et al. (2020) examined whether improved mixing and increased mass transfer actually does take place for RO membranes operated in laminar flow conditions typical of large-scale application. The methods developed by this team coupled the calculation of fluid flow with mass transport of solute, instead of imposing a flux. A correlation between the mass-transfer coefficient for flat membranes and Sherwood number was used to characterize the hydrodynamic conditions. The results obtained in this study were in perfect agreement with the numerical simulations, offering support for the modeling results.

Example 2.3

Determine the concentration polarization factor and the sodium concentration at the surface of membrane for a spiral-wound membrane element with the following information: water temperature 20°C, the diffusivity of sodium in water 1.350×10^{-9} m²/s, 7000 mg/L sodium concentration, 0.20 m/s velocity of the feed channel, 0.880 mm height of the feed channel, and 30 L/m² · h permeate flux.

Use 0.47 as the coefficient value. It can be assumed that the solute rejection is quite high that the sodium flux impact across the membrane could be neglected. Water viscosity and density at 20°C are 1.002 kg/m · s and 998.21 kg/m³, respectively.

Solution:

1. Determine the Schmidt and Reynolds numbers employing Eqs. (2.27) and (2.26). As the height of feed channel is 0.880 mm, the hydraulic diameter will be 1.760 mm:

$$Re = \frac{\rho \vartheta d_H}{\mu} = \frac{\left(998.21 \text{ kg/m}^3\right)(0.20 \text{ m/s})(1.76 \text{ mm})}{\left(1.002 \times \frac{10^{-3}\text{kg}}{\text{m.s}}\right)\left(\frac{10^3\text{mm}}{\text{m}}\right)} = 350.66$$

$$Sc = \frac{\mu}{\rho D_L} = \frac{\left(1.002 \times \frac{10^{-3} \text{kg}}{\text{m.s}}\right)}{\left(998.21 \frac{\text{kg}}{\text{m}^3}\right)\left(01.35 \times 10^{-9} \frac{\text{m}^2}{\text{s}}\right)} = 743.55$$

2. Calculate k_{CP} using Eq. (2.30):

$$k_{CP} = \lambda \frac{D_L}{d_H}(Re)^{0.50}(Sc)^{\left(\frac{1}{3}\right)}$$

$$= \frac{(0.47)\left(1.35 \times 10^{-9} \frac{\text{m}^2}{\text{s}}\right)(350.66)^{0.5}(743.55)^{\left(\frac{1}{3}\right)}}{(1.76 \text{ mm})\left(\frac{10^{-3} \text{m}}{\text{mm}}\right)} = 6.1 \times 10^{-5} \text{m/s}$$

3. As the rejection is very high, β could be determined from Eq. (2.24)

$$\beta = e^{J_W/k_{CP}} = \exp\left[\frac{J_W}{k_{CP}}\right] = \exp\left[\frac{30 \frac{L}{m^2 h} . 10^{-3} \text{m}^3/L}{\left((6.1 \times 10^{-5}) \frac{\text{m}}{\text{s}}\right).\left(3600 \frac{\text{s}}{\text{h}}\right)}\right] = \exp(0.1366) = 1.14$$

4. Determine the sodium concentration at the surface of the membrane employing Eq. (2.22):

$$\beta = \left[\frac{C_M}{C_{FC}}\right]$$

$$C_M = 1.14 \times 7000 \frac{\text{mg}}{L} = 7980 \text{ mg/L}$$

2.4 Reverse osmosis-based membrane desalination plant processes

The core process of desalination can be based on RO membrane technology; however as an independent unit, it cannot contribute safe potable water, nor does it ensure a proficient plant.

2.4.1 General overview

As in the case of any other natural water resource, seawater consists of solids in two different forms: dissolved and suspended. The dissolved solid exists in a soluble form (mineral ions like sodium, chloride, magnesium, calcium, etc.) whereas the suspended solid exists in the form of the insoluble particles (colloid, silt, marine organism, debris, particulate, etc.) (Nikolay, 2013). Presently, the entire RO desalination plants include three significant treatment steps designed for sequentially removing dissolved solids and suspended solids from the source water. The objective of step-1 (source water pretreatment) is to separate the suspended solid and obstruct certain naturally forming soluble solids from changing into solid form and precipitating on the RO membrane in the course of the salt segregation process. Step-2 (the RO unit) removes the dissolved solids from the water undergone pretreatment, thus producing clean less-salt concentration water appropriate for human usage, for industrial, agricultural application, and other uses. As soon as the desalination operation is finished, the clean water obtained from the RO unit additionally undergoes treatment for health safety and is subjected to disinfection before the distribution for final usage. Step-3 of the water treatment process in a desalination plant is termed as the post-treatment. Post-treatment consists of disinfection and conditioning (blending and remineralization) for decreasing the aggressive nature of the treated water.

Fig. 2.5 is a diagrammatic representation of various processes in a reverse osmosis seawater desalination plant. Generally, brackish water RO desalination plants will also have the same source water treatment stages and processes. The plant typically collects water through open seawater intake, and subsequently conditioned by coagulation and flocculation processes, and then filtered using granular media pretreatment filters for removing the majority colloidal solids, particulates, certain organic foulants, and microbial foulants. This filtered water will be transported using a transfer pump to a micron-sized filter or cartridge filter into the suction header of a high-pressure pump (HPP). This HPP pumps the filtered water to the RO membrane vessel at a net driving pressure suitable for producing the required water quality as well as water flow.

The RO membrane vessels are arranged in distinct sets of individually operating units termed as RO trains or racks. The entire RO trains jointly are referred to as the RO system. This RO system typically has energy recovery equipment that permits it to reutilize the

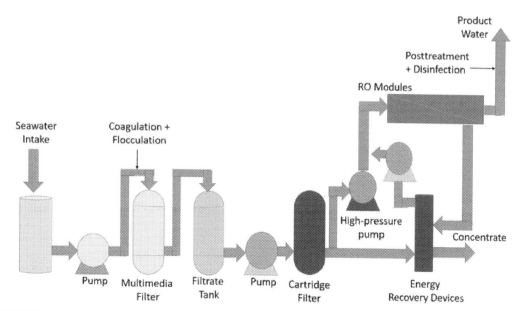

FIGURE 2.5 Diagrammatic representation of seawater reverse osmosis desalination plant processes.

energy present in the reject/brine to pump the feed water into the RO membrane system. In the post-treatment stage, the product water produced by the RO process is stabilized by adding lime or by the addition of carbon dioxide (CO_2) for providing a satisfactory level of alkalinity and hardness, for protecting the permeate distribution system against corrosion. The water conditioned will be stored and subsequently disinfected before its distribution to the ultimate users. The particulate solids separated from the feed water in the pretreatment process, using filters, are brought together in a filter backwash and subsequently concentrated by thickening as well as dewatering for final off-site disposal. Although the aforestated solid management technique is implemented by several newly built desalination plants, some of the old plants mix the brine and backwash water, and subsequently dispose it into the water source body which was used for feed water collection.

2.4.2 Source water intake system

The source water intake system is appropriately designed for collecting adequate feed water to produce the targeted quality and quantity of product water. The major objective of saline water intake is to pull together source feed water of consistent and satisfactory flow in addition to quality during the complete service life of the desalination facility. The configuration and type of intake chosen for a particular desalination plant have a considerable influence on the quantity and nature of foulants present in the feed water and on the complexities of the pretreatment setup required for controlling RO membrane fouling (Voutchkov, 2017c).

The intakes of desalination are categorized as open (surface) intake and groundwater (subsurface) intakes. The open-intake system gets water directly from a saline water source through a submerged inlet framework. This water, which is collected using an open intake system, consists of the same quantity of contaminants, organics, and solids as the surface water source from which the water originates. Up to now, onshore intakes have been noted to have applications mostly for extremely large hybrid or thermal type seawater desalination facilities. These intakes usually have a deep, large intake canal ending in a concrete forebay framework well-equipped with a coarse bar screen succeeded by a fine screen and the intake pump station. The foremost role of the intake pump station is to consistently transport source water brought together by the intake system to the pretreatment units at required flow rates and pressures. The intake pump stations include components such as the source water receiving well, interconnecting piping, pumps, chemical feed systems, service, and auxiliary equipment, and facilities.

2.4.3 Pretreatment

The micro-structured thin-film composite (TFC) membranes currently used for RO-based desalination applications will not allow the passage of particulates present in the feed water or developed in the course of the desalination operation. Hence, if particulates are available in the feed water in substantial quantities, then it might cause the fouling of membranes, which might rapidly lower the productivity of the membrane and lead to plant operation failure. Typically, the membrane foulants are inorganic and organic colloids and particulates, naturally present in the source water or developed on the membrane surface by microorganisms or chemical/physical processes that happen at the time of RO salt separation and concentration (Nikolay, 2013). The main function of the pretreatment unit is to efficiently eliminate foulants from the source water and to safeguard the steady as well as the effective performance of the RO membranes. This pretreatment system is normally positioned downstream of the water intake facility of the desalination plant and upstream of the RO membrane system. On the basis of the quality of source water, the pretreatment processes might consist of one or several water treatments, comprising screening, sedimentation, granular media filtration, cartridge filtration, chemical conditioning, microfiltration or ultrafiltration, and dissolved air flotation (DAF). Filters might be categorized according to the types of media. The different categories include single-media filters (one type of media—typically sand or crushed anthracite coal), dual-media filters (have two types of media, typically crushed anthracite coal and sand), and multimedia filters (three types of media—generally crushed anthracite coal, sand, and garnet).

For reducing the membrane fouling potential, salty source water will be conditioned, before the RO separation process, by different chemicals like flocculants, scale inhibitors, coagulants, oxidants [e.g., chlorine dioxide (ClO_2), chlorine], and oxidant reduction compounds (e.g., sulfuric acid, and sodium bisulfite) (Voutchkov, 2017a). Flocculants and coagulants are introduced for increasing the removal of particulate and colloidal foulant in the source water. Scale inhibitors are added in the salty feed water following the pretreatment filtration for suppressing the crystallization of mineral-scaling foulant on the RO membrane surface. Oxidants are normally introduced to the source water for minimizing the pretreatment, biofouling of membrane, and unnecessary growth of marine organisms on the interior of the water intake piping, structures, as well as equipment. Reducing chemicals such as sodium bisulfite are introduced to the pretreated source water for removing residual chlorine or/and other oxidants prior to the introduction of water into the RO membrane.

The function of sedimentation, sand removal, and DAF pretreatment system is to lower the content of coarse materials like suspended solids, debris, and grit assembled by the water intake system and to shield the downstream filtration units from the overloading of solids. The source water assembled by shallow offshore open intakes and onshore intakes normally will not contain large amounts of sand; however, it can have increased quantities of suspended solids and floating solids. Well-intake usually has a very less quantity of suspended solids; however, on the basis of the subsurface soil conditions and their design, they can produce source water of higher sand content, particularly when they are put into operation after a prolonged shutdown.

Currently, granular media filtration is the most frequently employed source water pretreatment process for RO-based desalination plants, apart from cartridge filtration. The granular media filtration process includes filtration of the source water by means of one or several layers of granular media (e.g., garnet, silica sand, anthracite coal, etc.). The standard filters used for saline water pretreatment are usually rapid single-stage dual-media (sand and anthracite) units. On the other hand, in certain cases where the source water comprises increased levels of suspended solids (monthly average turbidity go above

20 NTU) and organics (concentration of total organic carbon greater than 6 mg/L), two-stage filtration systems are practical. In this configuration, the filtration stage-1 is primarily designed for removing coarse solids as well as organics present in suspended form. The stage-2 filters are configured for retaining fine solids in addition to silt, and to eliminate a portion (20%–40%) of the soluble organics present in the saline water source by biofiltration.

Chapter 5 of this book provides an overview of pretreatment filtration technologies such as chemical pretreatment, conventional pretreatment, and membrane pretreatment. Subsequent to the pretreatment process, the feed water will be pressurized using feed pumps. The feed water pressure range will be 5.0–10.0 bar (73.0–145.0 psi) for NF membranes, 10.0–30.0 bar (145.0–430.0 psi) for brackish water RO, and 55.0–85.0 bar (800.0–1200.0 psi) for seawater RO (Crittenden et al., 2012).

2.4.4 Reverse osmosis separation system

RO technology is a pressure-driven membrane separation process meant for recovering water from a salt solution, pressurized to a point higher than the solution osmotic pressure. The purification process by RO technology comprises setting a semipermeable membrane in contact with a salt solution under a pressure greater than the osmotic pressure of the solution, generally in the range of 50 to 80 bar for seawater. Generally, the membrane separates the salt ions from the feed solution, permitting just the water to flow through. The RO process uses hydraulic pressure for forcing pure water from saline feed water across a semipermeable layer. In general, the membranes used in the RO process are either made from polyamides (PAs) or cellulose acetate (CA). CA membranes, developed in the early 1960s, were the first type of membrane used in commercial RO desalination plants. The CA membranes are available in both hollow fine-fiber and flat sheet configurations. TFC membranes, developed later, demonstrated to exceed the membrane fluxes and salt rejections of the CA membrane. Many research studies are continually being carried out for the advancement of current membrane technology (Saleem & Zaidi, 2020a, 2020b, 2020c; Saleem et al., 2020; Yadav et al., 2020; Zaidi et al., 2019). Unfortunately, many developed membranes fail to meet the fundamental criteria for commercial success.

The main components of a reverse osmosis separation system include filter effluent transfer pump, high pressure pumps, RO trains, energy recovery device (ERD), and membrane flushing/cleaning units. In the RO-based membrane desalination process, the feed is pressurized using an high pressure pump and this feed water is made to pass through the surface of the membrane. A portion of this feed, the permeate, goes through the membrane whereas the remaining emerges from the membrane modules as concentrate/reject stream. The main reverse osmosis system components are discussed in the last section of this chapter. The smallest unit of production capacity in a reverse osmosis membrane plant is termed as a membrane element. This membrane element will be enclosed in PV mounted on a skid, which has piping connection for reject stream, product stream, and feed stream. A certain number of pressure vessels (PVs) arranged in parallel are termed as a stage. The brine from a particular stage could be introduced to a succeeding stage for increasing

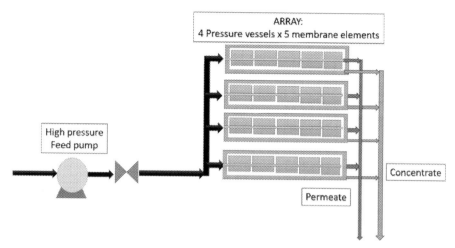

FIGURE 2.6 Schematic representation of an array (4 pressure vessesls × 5 membrane elements).

water recovery (multistage system or concentrate-stage system) or the product water from a specific stage could be introduced to a subsequent stage for increasing the separation of solute (2-pass system or permeate-staged system) (Crittenden et al., 2012). In the case of a multistage system, the number of PVs will decrease in each subsequent stage for maintaining adequate velocity in the feed stream due to the fact that product water is removed from the feed water stream. A unit of production capacity, which might consist of single or several stages, is termed as an array. Diagrammatic representation of an array (4 PVs × 5 membrane elements) is shown in Fig. 2.6. In the case of seawater RO systems, the ratio of product water flow to feed water flow (recovery) will be around 50%, and for low-pressure RO systems, it will be about 90%. A number of factors limit recovery, particularly solubility of sparingly soluble salt, concentration polarization, as well as osmotic pressure.

2.4.5 Post-treatment

Permeate/product water from the RO-based desalination plant will be typically low in mineral content, pH, alkalinity, and hardness. Thus, the water desalinated should be properly conditioned (post-treated) before its final distribution and usage. The post-treatment process of permeate produced by the RO process has two main components: (1) the addition of adequate minerals for protecting human health and for safeguarding the reliability of the water supply system, and (2) disinfection.

The post-treatment process normally involves increased removal of certain minerals like chlorides, sodium, and boron, or/and supplementary addition of some minerals like magnesium and calcium. Typically, the actual application dose of post-treatment chemicals to permeate should be chosen in accordance with the lowest quantities required for achieving all the objectives for which the chemicals are introduced. Usually, the permeate post-treatment includes at least one of the succeeding processes: (1) remineralization for protection against corrosion, (2) disinfection for public health protection as well as biological stability, and (3) improving water quality by increased removal of particular water constituents (e.g., N-nitrosodimethylamine, silica, boron, gases that cause odor and taste, etc.).

2.5 Reverse osmosis system components

This section discusses the RO system components such as filtered water transfer pumps, high-pressure pumps, RO membranes, RO skids (trains), interconnecting piping, PVs, energy-recovery system, instrumentation and controls, membrane cleaning system, and membrane flushing system.

2.5.1 Filtered water transfer pump

The filtered water transfer pump is usually a horizontal centrifugal pump or a vertical turbine pump designed for transporting the filtered water to the RO system. This filtered water from the water intake system (if the quality of water is acceptable) or plant pretreatment system can follow two flow patterns: desalination system with direct flow-through pattern (Fig. 2.7) or interim pumping desalination system (Fig. 2.8).

In the case of a desalination system with a direct flow-through pattern, the intake pump station will be sized to provide the suction pressure required for the effective operating of the HPPs (Nikolay, 2013). Here, the pretreatment unit is configured in such a manner that it will not lose the pressure, which is achieved by the use of either a pressure-driven membrane pretreatment filter or a pressure granular media filter. Thus, the pretreatment unit should be designed for withstanding the extra pressure required for the high-pressure RO pump suction. In the case of seawater RO desalination plants, this suction pressure will be two to six bars, and for brackish water RO desalination plants, this suction pressure is typically less than 1 bar.

For the water transport in an interim pumping desalination system, a separate pump station is a setup for boosting the filtered water to the suction pressure required for the effective functioning of the HPPs. In the latest seawater RO system designs, the filtered water transfer pump is regularly well-equipped with a variable frequency drive (VFD) to

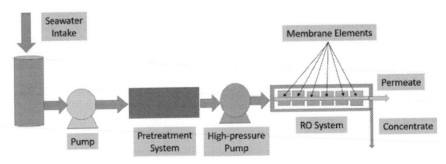

FIGURE 2.7   Reverse osmosis (RO)-based desalination system with direct flow-through pattern.

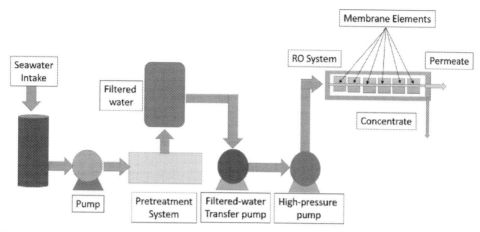

FIGURE 2.8 Reverse osmosis (RO)-based desalination system with interim pumping.

let the feed pressure of the RO unit to be economically regulated by the transfer pump feed pressure.

2.5.2 High-pressure feed pumps

These types of feed pumps are designed for delivering the source feed water to the RO membranes at a pressure that is needed for effective membrane separation of the fresh water from the salt solution or saline water. This pressure is generally in the range of 55—85 bars for seawater RO desalination and 5—25 bars for brackish water RO desalination, depending on feed water temperature and salinity (Voutchkov et al., 2010). The actual feed pressure needed will be dependent on the feed water-quality, and is mostly determined by the temperature, salinity of source water, the RO system configuration, and the target permeate quality. The feed pump will be sized depending on the flow needed and the operating pressure by means of standard performance curves provided by the pump manufacturer. The entire wetted pump materials must be of high-quality stainless steel, which is considered as a function of the salt concentration of water to be pumped. Normally, a low-salinity brackish water RO system uses 316 L or higher quality stainless steel. Super-duplex stainless steel and duplex stainless steel are generally suggested for seawater RO and high salt-concentrated brackish water RO applications, respectively.

In some cases, VFDs are connected to the HPP motors for adjusting the motor speed to retain optimal pump performance with varying feed pressure requirements driven by natural variations in source water temperature and salinity. Furthermore, VFDs enable the pumps to retain the finest performance even when the membranes scale or foul and lose permeability eventually. If the VFD is not fixed on the HPP motor or the filtered water transfer pump, then the feed flow as well as the pressure of the centrifugal HPP will be regulated using a pressure control valve (Fig. 2.9). This valve will be throttled along with the flow control valve connected to the reject pipe for setting the RO unit operations at the

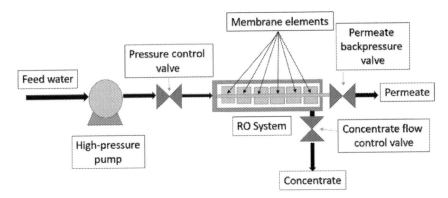

FIGURE 2.9 Main reverse osmosis (RO) system control valves.

targeted pressure, recovery, and feed flow. When the RO membranes age over a period of time, permeability and, consequently, their productivity will be decreased permanently. Usually, a reverse osmosis system drops 8.0%−15.0% of its preliminary productivity over a period of 3−5 years. Permeate backpressure valve is fitted for controlling the Beta factor (in particular, decrease concentration polarization) to increase the permeate quality and decrease the fouling rate of the membrane. The capability to control water flux and membrane fouling by back-pressure is limited because of the possibility for thin-film delamination if the product water backpressure exceeds 0.30 bar above the reject pressure.

2.5.2.1 High-pressure feed pump types

There are two most widely used pumps for RO systems, namely centrifugal pumps and reciprocating (piston) pumps.

Reciprocating HPPs: These pumps are also known as piston pumps or positive displacement pumps. They are normally employed for small-sized desalination plants of capacity of about 4000 m³/day, or lower. For these pumps, the motor's rotating motion is transformed into reciprocating motion that forces the pump piston. The flow of water conveyed by these type of pumps varies with respect to the stroke length of the pump, the number of pistons, and piston surface area. The foremost benefits of the reciprocating pump relative to the centrifugal pump are that this pump usually has greater efficiency (reciprocating pump 90.0%−95.0% and centrifugal pump 80.0%−88.0%) and an extremely flat pump curve. The main drawback of this pump is that it delivers pulsating water flow, which might lead to a pressure surge in the RO vessel and lead to membrane element deterioration.

Centrifugal pumps: These pumps can be used for desalination plants of all sizes. The main drawback of centrifugal pump as compared to the reciprocating HPP is that the pressure delivered varies with the flow and vice versa (no flat pump curve). The most frequently used types of centrifugal pumps for the desalination application are high-speed single-stage pumps, vertical turbine pumps, split-case multistage pumps, and segmental-ring multistage pumps.

1. High-speed single-stage pump: These pumps are normally applied in small-sized as well as medium-sized desalination plants and are normally connected with a particular

kind of energy-recovery device that functions on a common shaft with the RO feed pumps and increase their pressure by using the energy recovered from the RO system brine.

2. Vertical turbine pump: Single-stage and multistage vertical turbine pumps are extensively used in brackish water RO desalination plants.

3. Segmental-Ring Multistage Pumps/ Ring-Section Multistage Pump: This pump is usually employed in high-pressure RO applications in medium-sized and small-sized desalination plants (i.e., plants having capacity in the range of 1500−10,000 m³/day). The main benefit of these pumps is their low equipment costs. These pumps comprise distinct pump stages positioned between the discharge castings and the pump suction.

4. Split-case multistage centrifugal pump: Currently, these pumps are extensively used as HPPs for medium-sized and large-sized seawater RO desalination plants. The split-case multistage centrifugal pumps typically give superior efficiency (almost 80%−88%). The commercial pumps for large installations will have the capacity in the range of 600−3000 m³/h.

2.5.3 Spiral-wound polyamide membrane elements

Common size spiral-wound PA-based RO membrane elements are commercially available from different membrane manufacturers such as DuPont-Filmtec, Hydranuatics-Nitto Denko, Koch Membrane Systems, Toray Membranes, and others. Table 2.2 presents some of the commercially available membrane element models.

Seawater RO membranes could also be categorized into four major groups depending on their performance: (1) low-energy, (2) high-rejection, (3) high-productivity, and (4) low-fouling. High productivity/low-energy seawater RO membrane elements have been designed with features for operating at lower feed pressure or produce extra permeate per membrane element, specifically, greater permeability and increased surface area. Typical rejection membrane elements have been designed for removing up to 99.6% of the salts present in the source water. The high-rejection membrane elements are extensively employed nowadays, and have proven applications in different RO system arrangements. These membrane elements have been designed with close-fitted membrane structure, which permits to increase the mass of rejected ions and to reject small sized ions; for example, boron. Usually, the high-boron rejection membranes will also have a high salt rejection feature. For example, FilmTec Element SW30XHR-400440 has high stabilized salt rejection (99.8%), high stabilized boron rejection (92%), and high active area (41 m²).

Increasing the total active surface area of membrane leaf permits to gain substantial productivity for the membrane element with the same size (diameter). The low-fouling or low-differential pressure or fouling-resistant feature of majority commercial seawater RO membranes presently is achieved by incorporating a wider feed/brine spacer in the membrane element configuration. Real-world experience at seawater RO plants so far demonstrates that the use of wider spacer seawater RO membrane elements can be advantageous for source waters having high fouling tendencies like the Red Sea and the Persian Gulf.

2.5.4 Pressure vessels

RO membrane elements are mounted within the PV (housing) in a series of six to eight membrane elements per vessel. In the case of large systems, six-element vessels are

TABLE 2.2 Some of the commercially available membrane element models.

Element models	Feature	Active area (m²)	Permeate flowrate (m³/d)	Maximum operating pressure (bar)	Stabilized boron rejection (%)	Stabilized salt rejection (%)	Ref
FilmTec Element SW30XHR-400	High-rejection seawater RO elements, very high NaCl and boron rejection	37	23	83	93	99.82	Dupont, 2020a
FilmTec Element SW30HR-380	High salt rejection, highest boron rejection	35	23	69	90	99.7	Lenntech, 2020a
FilmTec Element SW30ULE-400i	high flow rates, high rejection of NaCl and boron	37	41.6	83	89	99.7	Lenntech, 2020b
FilmTec Element SW30HRLE-440i	Lower operating cost through reduced energy consumption, high NaCl and boron rejection	41	30.2	83	92	99.8	Dupont, 2020b
Hydranautics SWC4 +	High Salt rejection	37.1	24.6	82.7	-	99.8	Hydranautics, 2020a
Hydranautics SWC4B MAX	High Salt rejection	40.8	27.3	82.7	-	99.8	Hydranautics, 2020b
Hydranautics SWC6 MAX	High salt rejection, high flow	40.8	50	82.7	91.0	99.7	Hydranautics, 2020c
Toray TM820K-400	High Salt rejection	37	21.9	82.7	96	99.86	Toray Membranes, 2020

standard; however, vessels with up to eight elements are available. The interconnection of membrane elements inside the PVs is normally achieved by using small plastic spool pipe segments having O-rings or by means of specifically framed interlocking devices. All the PVs are enclosed on their sides with close-fitting enclosures termed as end caps. These end caps are configured in such a way that it can withstand the operating pressures of the membrane, and restrict the movement of the RO membrane elements inside the PVs.

A modern design trend in seawater RO plants is to mount eight membrane elements per PV. In addition to the capital cost decline, the utilization of an 8-element PV arrangement can also reduce the overall concentration polarization factor for the RO membrane because of greater feed/concentrate velocity and decreased recovery of the individual membrane elements, which are considered to be advantageous in terms of fouling. On the other hand, the higher the number of RO membrane elements in the PVs, the higher the differential pressure inside the PVs, and the closest the PVs will operate to the maximum limit of pressure drop recommended by the membrane manufacturers of 4 bars above which irreversible damage and compaction of the elements might happen. Further, the usage of eight elements would lead to a marginally higher feed pressure.

For preventing the movement of the membrane elements inside the PV, the end connection of the product water side of the RO PV has to be shimmed. The RO membrane vessel will be coupled with steel pipe sections to the reject line and feed line of the RO train and with a plastic port to the product water line. The recommended maximal feed flow and minimum feed flow per individual 8-inch vessels are 17 and 10 m^3/h, respectively. The minimum possible flow of reject per vessel is recommended to be 2.7 m^3/h.

2.5.4.1 Classification of membrane vessels

Membrane vessels vary by the location of their feed port, the material from which they are produced, by their diameters, and by their pressure class (i.e., the maximal operating pressure). Generally, the PV could be designed to house from one element to up to eight elements connected in series.

By the location of feed ports: Depending on the feed port location, the PVs can be categorized as multiple-port vessels, side-port vessels, and end-port vessels. Normal designs generally have end-port vessels.

By materials: The most extensively used material for the PV is fiberglass-reinforced plastic (Towler & Sinnott, 2012). In the case of definite industrial applications, in which the PVs should be sanitized or/and functioned at increased temperatures (i.e., temperature of 65°C or greater), then stainless-steel PVs will be more appropriate. The stainless-steel PV could be employed for the treatment of municipal water; however this vessel will be heavier, very difficult to handle, and more expensive, and these will limit its extensive application.

By diameters: The RO PVs can be designed to house definite standard diameter membranes. Hence, they are designed in typical membrane diameter sizes of 400, 200, 102, and 63 mm.

By pressure class: When the RO systems are operated, the PVs will be totally enclosed as well as pressurized at the system operating pressure. Depending on the maximum pressure rating, the PVs are categorized into three types: (1) seawater RO PV having operating pressure of 42.0−105.0 bars, (2) brackish water RO PV for handling operating pressure of 10.5−42.0 bars, and (3) water softening PV designed for handling operating pressure of 3.5−10.5 bars.

2.5.5 Reverse osmosis system piping

Superior quality stainless steel can be normally used for high-pressure feed and reject piping of RO systems (American Water Works Association, 2007). As the salinity of the feed water as well as reject concentration increases, superior quality stainless steel will be needed for preventing the RO system piping from corrosion and to sustain its durability. In addition to the stainless steel, copper−nickel alloys are also noted to have applications in seawater and brackish water intake screens and other facilities. Fiberglass-reinforced plastic and high-density polyethylene piping are employed for low-pressure applications.

Schedule 80 polyvinyl chloride material is most extensively used for low-pressure product water piping and valves. The permeate port connections to the end caps of the RO PV are regularly made up of inexpensive and low-pressure flexible tubing for simplifying the RO membrane inspection, maintenance, and decrease total equipment costs. In case of using flexible tubing, this tubing must be enclosed with ultraviolet-resistant coating for RO systems which are fixed outside due to the fact that the sunlight exposure might degrade the piping, and consequently, these types of piping must be replaced every 2—4 years. It has to be noted that the recommended RO distribution pipe velocities change with the material and flow rate.

2.5.6 Reverse osmosis skids and trains

The RO membrane elements are set up in PVs that generally house six to eight elements per vessel. Multiple PVs will be configured on support frameworks (termed as racks or skids). These skids are usually madeup of plastic, plastic-coated steel, or powder-coated structural steel. The grouping of the RO feed pump, feed piping, permeate piping, concentrate piping, PVs, couplings, valves, and other fittings (instrumentation and controls, energy-recovery system, etc.) set up on a distinct support structure (rack/skid), which can work individually, is termed as RO train. Each RO train is normally designed for producing between 10.0 and 20.0 percentage of the entire quantity of the membrane desalination permeate flow. The RO trains are arranged as well as designed in such a way that each individual train has the capability of separately controlling total product water flows and reject flows (Nikolay, 2013). RO train consisting of an array of interconnected PVs is shown in Fig. 2.10.

FIGURE 2.10 Reverse osmosis (RO) train consisting of an array of interconnected pressure vessels (PVs) (A). The PVs contain the RO membranes (B). Panel (C) shows a cross section of a membrane and its spiral wound configuration. Source: *Reproduced from Kress, 2019. Desalination technologies. Marine impacts of seawater desalination, pp. 11—34. 10.1016/b978-0-12-811953-2.00002-5.*

2.5.7 Systems for energy recovery

A great portion (about 40.0%−50.0%) of the energy applied for the seawater desalination is contained in the reject generated by the RO system. This maximal energy amount that could be recovered from the reject stream, expressed in terms of percent of the whole amount of energy provided with the RO feed flow, could be determined as follows:

$$ER_{max} = \frac{[(F_p - P_d) \times (1 - R)]}{F_p} \tag{2.32}$$

where ER_{max} is the maximum energy recovered from brine, and expressed in terms of % of the energy introduced to the RO with the feed flow (%), P_d is the pressure drop across the membrane, expressed in bars, F_p is the applied RO feed pressure, expressed in bars; and R is the RO system recovery, expressed in %.

Example 2.4

In the case of a seawater RO system operating at 64 bars feed pressure, the recovery is 42% and the differential pressure is 1.5 bars. What will be the maximal energy that can be recovered from the reject stream?
Solution:

$$ER_{max} = \frac{[(F_p - P_d) \times (1 - R)]}{F_p}$$

$$ER_{max} = \frac{[(64\ bars - 1.5\ bars) \times (1 - 0.42)]}{64} = 56.6\%$$

This confirms that if the efficiency of the energy recovery equipment is 100%, then it has the ability to recover 56.6% of the energy provided to the RO system.

This type of energy could be recovered and reused to pump new saltwater source by equipment explicitly designed for this objective, termed as an ERD. As energy used for seawater desalination is about 50.0%−70.0% of the overall plant yearly operation and maintenance expenses and about 25.0%−35.0% of the entire expenses of potable water production, reutilization of this energy is considered to be advantageous as well as economical. The settlement of equipment expenses for setting up of energy recovery systems in seawater RO plants by means of energy saving is typically below 5 years. Developments in the technology and equipment letting the recovery and reuse of the energy used for seawater desalination have led to a decrease of 80.0% of the energy used for the production of water over the last 20 years.

2.5.8 Membrane flushing and cleaning systems

RO membrane flushing system: RO system will be normally well-equipped with a perpetually piped membrane flushing unit for automatically flushing vessel in the RO train on shutdowns to eliminate residual brine and prevent the fouling and deterioration of RO membrane. The flushing is achieved by means of RO unit product water free from disinfectants or any other chemicals. For shorter RO unit shutdowns, the RO vessels can be flushed by using nonchlorinated and chemically conditioned filtered water. The flush water can be stored in an on-site tank of adequate capacity for flushing the entire installed trains with no simultaneous refill.

RO membrane cleaning system: All RO membranes build up the foulants in the feed/concentrate spacer cavity over a period of time, which subsequently leads to an increase in the differential pressure and, hence, it has to be cleaned periodically for maintaining their efficiency as well as service life. The objective of membrane cleaning is to dissolve and remove inorganic scales, displace and eliminate particulate and colloidal foulants, and break down as well as get rid of biological film collected in the feed/concentrate spacers. In actual practice, typical criteria for membrane cleaning include: (1) 10%−15% rise in normalized differential pressure, (2) 10%−15% reduction in normalized permeate flow, (3) 10%−15% rise in normalized product water TDS concentration, and (4) cleaning after and before long-standing RO train shutdown. Based on the real membrane fouling rate, RO trains must be cleaned once a month for plants treating the water having a high-fouling tendency, and cleaned once in every year for plants treating water having extremely less fouling tendency. In majority of the desalination plants, the RO membrane train will be cleaned once in every 4−6 months.

Fig. 2.11 presents the diagrammatic representation of a clean-in-place (CIP) unit for the cleaning of RO membranes. The CIP system comprises at least one clean-in-place tanks, cartridge filters, cleaning pumps, feed piping, recirculation piping, control equipment, instrumentation, and power supply. The clean-in-place unit is designed for mixing and recirculating a range of different cleaning chemicals made up with dechlorinated potable water or RO permeate. The components of the clean-in-place system are designed depending on the PV number that would be cleaned in individual stage. In the case of multistage system, the vessel in each stage will be cleaned in a distinct step to prevent forcing the foulants into the succeeding stages.

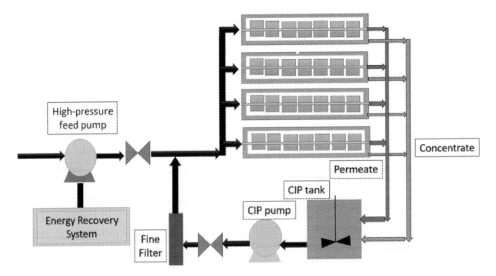

FIGURE 2.11   Diagrammatic representation of a reverse osmosis membrane cleaning system.

2.5.9 Instrumentation and controls

Instrumentation and controls could be as complex as a Supervisory Control And Data Acquisition (SCADA) system or as simple as a manual control having automatic shutdown features for pumps and membrane safety. This SCADA system is generally based on programmable logic controller (Wiles et al., 2008) in addition to distant telemetry units which are managed by a host computer situated in a control room nearby the high-pressure pumps and membrane skids. Currently, the use of private computers or human—machine interface computers is very common. The systems which are designed for automated control could perform chemical feed system monitoring, and these systems possess alarm and report-generating capabilities. In many plants, private computers are utilized for the purpose of calculating the performance normalizations of membrane train and for the graph generation, that enable proper monitoring of plant performance and determine the time for membrane cleaning. The SCADA system, plant control system, and basic instrumentation needed for monitoring and controlling any RO system are discussed in detail in Chapter 9.

2.5.10 Reverse osmosis system types

On the basis of the number of sequential RO systems for product water treatment and brine treatment, the RO system configurations can be categorized into two types: (1) single-pass and multipass RO systems, and (2) single-stage and multistage RO systems. For all desalination plant types, the multipass and multistage RO systems can also be combined into configurations that permit to accomplish target RO system recovery and permeate quality at optimal lifetime expense of water production. The single-pass, multipass, single-stage, and multistage RO systems are discussed in the subsequent section.

Single-pass and multipass reverse osmosis system: A reverse osmosis system where the saline water source is desalinated just a single time is termed as a single-pass RO system (Gude, 2018). The RO systems which are designed for retreating the product water multiple times are referred to as multipass RO systems. The single-pass and two-pass RO systems are shown in Fig. 2.12. Since each RO pass offers extra treatment of product water

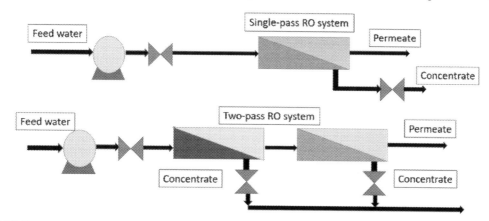

FIGURE 2.12   Single-pass and multipass reverse osmosis systems.

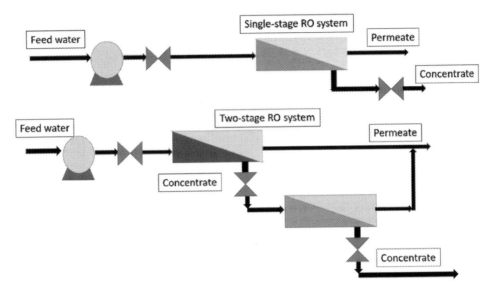

FIGURE 2.13   Single-stage and two-stage reverse osmosis(RO) systems.

produced by the preceding RO pass, the overall system product water quality improves with each pass. Consequently, multipass RO systems are applied when source water salt concentration is comparatively high and the targeted permeate quality could not be accomplished by treating the salty source water by RO just one time.

Single-stage and multiple-stage reverse osmosis systems: A significant challenge related to the application of multipass RO system is that the complete recovery of this kind of system reduces with the number of the installed RO passes due to the fact that some of the feed water is converted into brine at each pass. Thus, to decrease the total brine volume from the same volume of feed water, the brine produced by individual RO passes could be undergone treatment by a distinct RO system, termed as a stage. Fig. 2.13 presents the single-stage and two-stage RO systems.

2.6   Processes in reverse osmosis plants based in Gulf Cooperation Council (GCC)

2.6.1   Ras Abu Fontas 3 Reverse osmosis desalination plant (Qatar)

This plant is divided into two sub-RO desalination plants SP1 and SP2. The SP1 is designed for producing 22.0 million imperial gallons per day, whereas SP2 is designed for 14.0 million imperial gallons per day. The remineralization process and the DAF system are designed to be common for both plants (Menéndez, 2018).

The plant has the areas such as (a) DAF, (b) disk filters, (c) ultrafiltration, (d) RO high-pressure pumps, (e) RO racks, (f) remineralization, (g) auxiliary equipment, and (h) chemical dosing system.

The total capacity of the plant is 36 MIGD (163,660 m^3/day). To this total production, an extra 2270 m^3/day must be added for the internal water consumption of the plant (chemical preparation, carrier water, CIP for RO membranes, and external auxiliary water). The pretreatment package, the RO package, and the post-treatment package are designed to operate for at least 8,760 hours per year with a minimum uptime of 97%. Sufficient spare capacity is installed to allow for shutdown due to maintenance or plant failure.

2.6.2 Umm Al Houl plant (Qatar)

Umm Al Houl plant (Fig. 2.14) will produce 284,000 m^3/day of seawater desalinated by RO process (Menéndez, 2018).

1. Seawater feed

The seawater pumping station is sized to supply the required seawater flow to RO pretreatment. Two pipes of 1800 mm take the seawater into the RO plant.

2. Pretreatment

It includes pH correction (sulfuric acid), coagulant dosing, mixing and flocculation, DAF, and intermediate water pumping station (Fig. 2.15).

- Filtration stage through disk filters.
- Filtration stage through ultrafiltration membranes.

This pretreatment system produces a reverse osmosis feed water with a Silt Density Index ≤ 3.5. These parameters are required by the membranes for ensuring the correct operation of the RO process. The implementation of a DAF system is necessary for improving the removal of the light contaminants, algae, and microorganisms that are a feature of the Gulf seawater in ordinary conditions. The DAF system also protects against exceptional conditions of black or red tides.

Coagulation: Coagulant dosing in the pipe is completed with ferric chloride and sulfuric acid dosing prior to DAF. The Ultrafiltration (UF) ferric chloride dosing is also done at this point.

FIGURE 2.14   Umm Al Houl plant (Menéndez, 2018). *Courtesy: FuturENVIRO.*

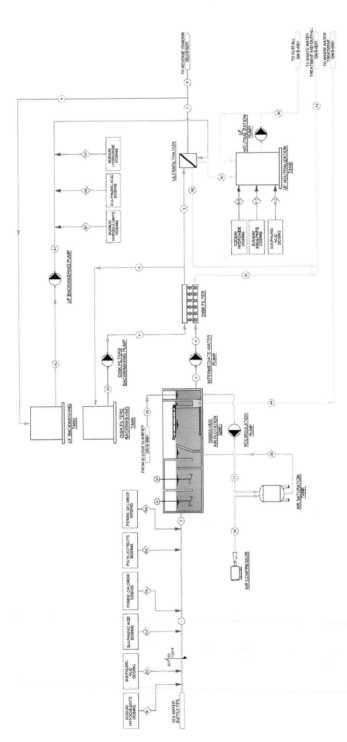

FIGURE 2.15 Umm Al Houl plant pretreatment process flow diagram (Acciona, 2020). *Courtesy: Acciona Agua.*

Flocculation: This stage comprises two stages, each of which is fitted with two axial propeller type mixers.

Sludge removal: The sludge on the surface of the water is removed by a rotary scraper designed to operate either continuously or intermittently.

Intermediate pumping station: The intermediate water pumps send the seawater through the disk filters and ultrafiltration membranes to the HPPs. The intermediate pumps feeding the disk filters and ultrafiltration membranes adjust the flow and pressure to the needs of the process by means of a variable speed drive.

Disk filter (DF): These filters are needed to remove coarse solids and restrict them from reaching the ultrafiltration membranes. The filters installed consist of thin polypropylene disks, with a filtration grade of 100 microns. The DF system has sufficient hydraulic capacity to treat the raw water flows needed for maximum capacity with one-DF battery out of service for backwashing or maintenance operations.

Ultrafiltration: Ultrafiltration racks are installed as part of the pretreatment for the RO plant. A total of 20 ultrafiltration trains are installed with 192 modules per rack. Each membrane has a surface area of 55 m^2, giving a total membrane surface area of 211,200 m^2.

3. RO system

The RO system is designed in accordance with high energy efficiency specifications and features ERDs with very high recovery rates (Fig. 2.16). The low-pressure pumps (11 + 1 standby) send the seawater to the HPPs. The HPP (11 + 1 standby) then sends the seawater to the first RO pass. The filtered water is transferred to the RO system. The pretreated seawater flow enters each RO rack in two streams. One of the streams, with a flow rate of slightly less than the product water flow, is pumped to the membranes by the high-pressure pumps. The booster pump then increases the pressure to overcome head loss in the reject pipe and reject membranes, in order to achieve the pressure required in the RO rack inlet. The model PX-Q300 ERDs selected are manufactured by ERI. The installed membranes, which have a high salt rejection and large membrane surface, are made of aromatic PA and feature a spiral configuration.

4. Wastewater treatment

The main process line generates secondary flows, some of which are sent directly to the outfall, due to the fact that they have similar characteristics to seawater and do not have any environmental impact. Others, however, require specific treatment to remove solids prior to discharge into the sea.

The secondary flows at the plant are as follows:

• Floating sludge from the DAF system.
• Disk filter and ultrafiltration backwash water.

These flows are treated in the wastewater treatment building. The design of the sludge treatment system is based on the removal of suspended solids by means of two sludge clarifiers and three dewatering centrifuges (Fig. 2.17). The floating sludge from the sludge settling tanks is pumped to the floating sludge tank. Same as the Ras Abu Fontas A3 facility, the Umm Al Houl facility is designed for operation with an uptime of over 97%. For this reason, it features great redundancy in the design and installation of process, control, and electrical equipment.

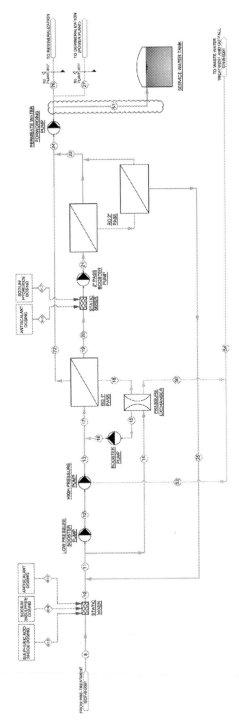

FIGURE 2.16 Umm Al Houl plant reverse osmosis process flow diagram (Acciona, 2020). *Courtesy: Acciona Agua.*

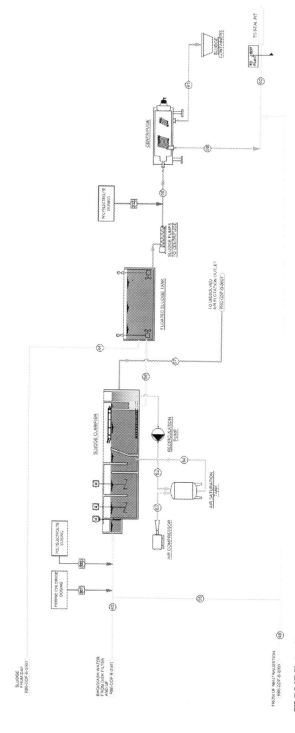

FIGURE 2.17 Umm Al Houl plant waste water treatment process flow diagram (Acciona, 2020). *Courtesy: Acciona Agua.*

2.6.3 Barka II seawater reverse osmosis desalination plant (Oman)

1. Pretreatment: two-stage filtration
2. Two stages of dual-media filter in series:
   a. 23 horizontal pressure filters
      Filtering media: 1.65 anthracite and 0.55 mm sand
      Filtration rate: 11.0–14.0 m/h
   b. 16 horizontal pressure filters
      Filtering media: 1.95 anthracite and 0.30 sand
      Filtration rate: 17.0–19.0 m/h
3. Safety filtration by means of 14 cartridges filters (5 μm)
4. RO: 2-pass configuration
   a. 14 first-pass RO trains (six modules each)
   b. 14 HPPs
   c. 3724 brackish water membranes
   d. 12,250 seawater membranes
   e. Seven second-pass RO trains (four modules each)
5. Post-treatment
   a. Water remineralization carried out for distribution as potable water. The potabilization treatment involves a remineralization process, with the addition of water lime and CO_2, and fluorination.
   b. Ultimate disinfection: This is carried out using sodium hypochlorite (Suez Water Handbook, 2020a).

2.6.4 Al-Dur seawater reverse osmosis desalination plant (Bahrain)

This seawater RO desalination facility, with a capacity of 218,000 m^3/day, distributes superior quality drinking water and fulfills the increased necessities with respect to pretreatment and post-treatment considering the Persian Gulf water quality (Suez Water Handbook, 2020b).

1. Pretreatment
   The process is divided in two independent lines, housing each the subsequent facilities and equipment:
   a. Chemical dosing system.
   b. 7 flocculation/flotation units, that is the very appropriate and established technology for facing the threat of probable algal blooms and the increased fouling probability of the Persian seawater.
   c. 22 pressurized dual-media filters.
   d. Polishing step comprising 12 cartridge filters with 360 cartridges of a five-micron each.
   e. Chemical dosing system.
2. RO
   Double-pass RO membrane treatment for reducing the concentration of boron to 1 mg/L:

a. First pass: 26 trains with 42% recovery rate. The Toray membranes and the Pelton turbines as ERD are used.

b. Second pass: 12 two-stage trains and 90% recovery rate (Toray membranes).

3. Post-treatment

Remineralization (two separate lines).

2.6.5 Fujairah seawater reverse osmosis desalination plant (United Arab Emirates)

1. Pretreatment unit

a. A dosing system which has the ability to dose up polymer, sulfuric acid, and $FeCl_3$ (pure) are mixed with the seawater in two coagulation tanks and two static mixers.

b. Two lines of seven open dual-media filters fed by gravity and handling up till 1,500 m^3/h of seawater.

c. Two lines of nine cartridge filters each having 360 cartridges (5 microns).

2. RO process

To accomplish a salt concentration lesser than 180 mg/L, the RO design has a two-pass system:

- First pass: Seventeen + one trains, each one with 136 PVs × seven elements SWC3 hydranautics high-rejection modules with 43% recovery rate
- Second pass: eight trains, each one with 70 PVs × seven elements with 90% recovery rate.

A pressure regulation is provided for compensating automatically for the variation in membrane permeability and temperature in a manner to lessen the energy necessity of the entire system (Suez Water Handbook, 2020c).

References

American Water Works Association. (2007). *Reverse osmosis and nanofiltration* (pp. 117–118). American Water Works Association, M46.

Basile, A., & Nunes, S. P. (Eds.), (2011). *Advanced membrane science and technology for sustainable energy and environmental applications*. Elsevier.

Baumgarten, C. M., & Feher, J. J. (2012). *Osmosis and regulation of cell volume. Cell Physiology Source Book* (Fourth Edition). Academic Press.

Caputo, G., & Giaconia, A. (2014). *Membrane technologies for solar-desalination plants. Membranes for clean and renewable power applications* (pp. 347–364). Woodhead Publishing.

Crittenden, J. C., Trussell, R. R., Hand, D. W., Howe, K. J., & Tchobanoglous, G. (2012). *Reverse Osmosis. MWH's water treatment: Principles and design* (pp. 1335–1414). Available from http://doi.org/10.1002/9781118131473.ch17.

Crystal, H. T. L. (1995). The solution-diffusion model: A review. *Journal of Membrane Science, 107*, 1–21.

Dupont 2020a FilmTec™ SW30 XHR-400 Element seawater reverse osmosis element <https://www.dupont.com/content/dam/dupont/amer/us/en/water-solutions/public/documents/en/45-D00973-en.pdf>. [Accessed 24 June 2020].

Dupont 2020b FilmTec™ SW30HRLE-440i Element <https://www.dupont.com/content/dam/dupont/amer/us/en/water-solutions/public/documents/en/45-D00965-en.pdf>. [Accessed 24 June 2020].

Gago, G.H., O&M desalination Middle East Director, Acciona Agua, personal communication, November 8th, 2020. <https://www.acciona.com/projects/middle-east/>.

Gude, G. (Ed.), (2018). *Sustainable desalination handbook: Plant selection, design and implementation*. Butterworth-Heinemann.

Hasmadi, N. S., Jullok, N., & Fadzilah, M. H. H. (2017). Solution-diffusion model for a small scale reverse osmosis system. *Jurnal Teknologi, 79*(1-2).

Hydranautics, 2020a. Membrane Element SWC4+ <https://membranes.com/docs/8inch/SWC4+.pdf>. [Accessed 24 June 2020].

Hydranautics, 2020b. Membrane Element SWC4B MAX, <https://membranes.com/docs/8inch/SWC4B%20MAX.pdf>. [Accessed 24 June 2020].

Hydranautics, 2020c. Membrane Element SWC6 MAX <https://membranes.com/docs/8inch/SWC6MAX.pdf>. [Accessed 24 June 2020].

Jorgensen, S. E. (1979). *Industrial waste water management*. Elsevier.

Jornitz, M. W. (2020). *Sterile filtration: A practical approach*. CRC Press.

Jurchevsky, E. B., & Pervov, A. G. (2020). Potentialities of membrane water treatment for removing organic pollutants from natural water. *Thermal Engineering, 67*, 484−491. Available from https://doi.org/10.1134/S0040601520070095.

Kabay, N., Bundschuh, J., Hendry, B., Bryjak, M., Yoshizuka, K., Bhattacharya, P., & Anac, S. (Eds.), (2010). *The global arsenic problem: Challenges for safe water production*. CRC Press.

Kress, N. (2019). Desalination technologies. *Marine impacts of seawater desalination*, 11−34. Available from https://doi.org/10.1016/b978-0-12-811953-2.00002-5.

Lenntech 2020a. FILMTEC SW30HR-380 High Rejection Seawater RO Element <https://www.lenntech.com/Data-sheets/Dow-Filmtec-SW30HR-380.pdf>. [Accessed 24 June 2020].

Lenntech 2020b FILMTEC™ SW30ULE-400i Seawater reverse osmosis element <https://www.lenntech.com/Data-sheets/Dow-Filmtec-SW30ULE-400i.pdf>. [Accessed 24 June 2020].

Menéndez, M., Editor & Community Manager, SAGUENAY, SLU (FuturENERGY & FuturENVIRO) Design, construction, operation and maintenance of two desalination plants in Qatar by ACCIONA Agua, FuturEnviro, 15915-2013 ISSN: 2340-2628, June 2018 <http://www.futurenviro.com/pdf/reportajes-especiales/06-2018/FuturENVIRO_Water_June_2018_Desaladoras_Qatar_Acciona.pdf>. [Accessed 24 June 2020].

Nikolay, V. (2013). *Desalination engineering: planning and design*. McGraw-Hill Professional.

Ratnayake, P., & Bao, J. (2017). Actuation of spatially-varying boundary conditions for reduction of concentration polarisation in reverse osmosis channels. *Computers & Chemical Engineering, 98*, 31−49.

Sablani, S. S., Datta, A. K., Rahman, M. S., & Mujumdar, A. S. (Eds.), (2006). *Handbook of food and bioprocess modeling techniques*. CRC Press.

Sadr, S. M., & Saroj, D. P. (2015). *Membrane technologies for municipal wastewater treatment. Advances in membrane technologies for water treatment* (pp. 443−463). Woodhead Publishing.

Saeed, A. (2012). Effect of feed channel spacer geometry on hydrodynamics and mass transport in membrane modules (Doctoral dissertation, Curtin University).

Saleem, H., Trabzon, L., Kilic, A., & Zaidi, S. J. (2020). Recent advances in nanofibrous membranes: Production and applications in water treatment and desalination. *Desalination, 476*, 114178.

Saleem, H., & Zaidi, S. J. (2020a). Developments in the application of nanomaterials for water treatment and their impact on the environment. *Nanomaterials, 10*(9), 1764.

Saleem, H., & Zaidi, S.J. (2020b). Innovative nanostructured membranes for reverse osmosis water desalination. Qatar University Annual Research Forum and Exhibition (QUARFE 2020), Doha. Available from https://doi.org/10.29117/quarfe.2020.0023.

Saleem, H., & Zaidi, S. J. (2020c). Nanoparticles in reverse osmosis membranes for desalination: A state of the art review. *Desalination, 475*, 114171.

Schock, G., & Miquel, A. (1987). Mass transfer and pressure loss in spiral wound modules. *Desalination, 64*, 339−352.

Singh, R. (2014). *Membrane technology and engineering for water purification: Application, systems design and operation*. Butterworth-Heinemann.

Suez Water Handbook, 2020a. Al-Dur seawater reverse osmosis desalination plant (Bahrain) <https://www.suezwaterhandbook.com/case-studies/desalination/Al-Dur-seawater-reverse-osmosis-desalination-plant-Bahrain>. [Accessed 24 June 2020].

Suez Water Handbook, 2020b. Barka II seawater reverse osmosis desalination plant (Oman) < https://www.suez-waterhandbook.com/case-studies/desalination/Barka-II-seawater-reverse-osmosis-desalination-plant-Oman > . [Accessed 24 June 2020].

Suez Water Handbook. 2020c. Fujairah seawater reverse osmosis desalination plant (United Arab Emirates). < https://www.suezwaterhandbook.com/case-studies/desalination/Fujairah-seawater-reverse-osmosis-desalination-plant-United-Arab-Emirates > . [Accessed 24 June 2020].

Swarup, R., Mishra, S. N., & Jauhari, V. P. (Eds.), (1992). *Environmental water pollution and its control* (Vol. 15). Mittal Publications.

Toray Membranes, 2020. Toray TM820K 400 Sea Water RO Element L < https://www.lenntech.com/Datasheets/Toray-TM820K-400-Sea-Water-RO-Element-L.pdf?h = 1&language_id = 1 > . [Accessed 24 June 2020].

Towler, G., & Sinnott, R. (2012). Chemical engineering design: Principles, practice and economics of plant and process design. Elsevier.

Urama, R. I., & Marinas, B. J. (1997). Mechanistic interpretation of solute permeation through a fully aromatic polyamide reverse osmosis membrane. *Journal of Membrane Science, 123,* 267−280.

Voutchkov, N. (2017a). Conditioning of saline water. *Pretreatment for Reverse Osmosis Desalination,* 113−135. Available from https://doi.org/10.1016/b978-0-12-809953-7.00006-1.

Voutchkov, N. (2017b). Introduction to saline water pretreatment. *Pretreatment for Reverse Osmosis Desalination,* 1−10. Available from https://doi.org/10.1016/b978-0-12-809953-7.00001-2.

Voutchkov, N. (2017c). Saline water intakes and pretreatment. *Pretreatment for Reverse Osmosis Desalination,* 65−94. Available from https://doi.org/10.1016/b978-0-12-809953-7.00004-8.

Voutchkov, N., Sommariva, C., Pankratz, T., & Tonner, J. (2010). Desalination process technology. In J. Cotruvo, N. Voutchkov, J. Fawell, P. Payment, D. Cunliffe, & S. Lattemann (Eds.), Desalination technology−Health and environmental impacts (pp. 21−90). Boca Raton, FL, USA: Taylor & Francis Group.

Wiles, J., Claypoole, T., Drake, P., Henry, P. A., Johnson, L. J., Jr, Lowther, S., & Windle, J. H. (2008). Techno security's guide to securing SCADA: *A comprehensive handbook on protecting the critical infrastructure.* Syngress.

Yadav, S., Saleem, H., Ibrar, I., Naji, O., Hawari, A. A., Alanezi, A. A., ... Zhou, J. (2020). Recent developments in forward osmosis membranes using carbon-based nanomaterials. *Desalination, 482,* 114375.

Zaidi, S. J., Fadhillah, F., Saleem, H., Hawari, A., & Benamor, A. (2019). Organically modified nanoclay filled thin-film nanocomposite membranes for reverse osmosis application. *Materials, 12*(22), 3803.

Zhou, Z., Ling, B., Battiato, I., Husson, S. M., & Ladner, D. A. (2020). Concentration polarization over reverse osmosis membranes with engineered surface features. *Journal of Membrane Science,* 118199.

CHAPTER

3

Guidelines for Water Quality

3.1 Source water and its possible risks

The source water for desalination application could be seawater, brackish water, or extremely mineralized groundwater (Bundschuh & Hoinkis, 2012). By its nature, this water will have a substantial amount of naturally occurring inorganic ions, and the purpose of water treatment is to remove or reduce the concentration of these substances (El-Dessouky & Ettouney, 2002; Micale et al. 2009). Some of these naturally occurring substances might be of possible concern if they exist in substantial concentrations following the water treatment. Proper knowledge about the risks that are probably present in the source water is a significant factor for the appropriate design of the desalination operation. It stresses the requirement of necessary pretreatment steps and the elimination of contaminants during the treatment process. With regard to the possible problems due to the presence of contaminants, either microbial or chemical, the primary step for decreasing the related hazards is to try to prevent or decrease the inputs at the source. Adequate knowledge of possible contaminants is essential in developing plans for protecting the quality of the source water (World Health Organization, 2011).

The source water for the desalination plant is collected from a water body employing either open surface intakes or subsurface intakes. The open seawater intake collects water straightaway from a saline surface water source and has facilities (fine traveling screen, bar rack, strainer, or/and microscreen) for prescreening coarse sand, large aquatic organisms, floating materials, large debris, and stringy materials present in the source water. The subsurface intake will be naturally prescreening and prefiltering the salt water collected and thus separate coarse debris and the majority of the particulates and sand from the saline water.

Thus subsequent to the initial screening by the intake units, the source water usually consists of the succeeding five types of compounds that might lead to the fouling of reverse osmosis (RO) membranes, and hence should be separated using the pretreatment processes (Voutchkov, 2017). Some of the main factors that affect the level of membrane fouling are presented in Fig. 3.1. The different conventional, chemical, and membrane-based pretreatments are discussed in Chapter 5.

1. Microbial foulants: These are marine organisms and organic compounds discharged by these organisms (could function as food to the microbes that live in the source water).

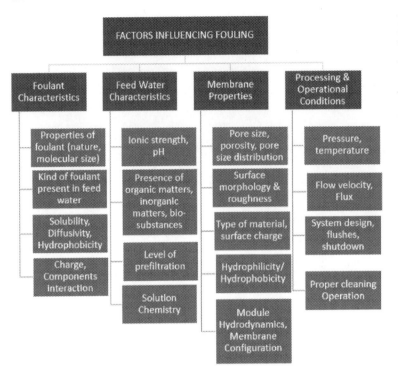

FIGURE 3.1 Main factors that affect the level of membrane fouling. *Source: Adapted from Saleh, T.A., & Gupta, V.K. (2016). Membrane fouling and strategies for cleaning and fouling control. In: Nanomaterial and polymer membranes. Amsterdam: Elsevier, pp. 25–53.*

These could develop biofilm, thereby dropping the product water transportation across the membranes.

2. Natural organic foulants: These include natural organic matter (NOM) that has the ability to attach with the membranes and cause membrane fouling.

3. Mineral-scaling foulants: These foulants include inorganic compounds such as calcium, magnesium, strontium, and barium salts, which might precipitate in the course of the salt separation process and causes scaling on the surface of the membrane (e.g., calcium sulfate, calcium carbonate, barium sulfate, strontium sulfate, and magnesium hydroxide) or might cause blocking of the membrane separation layer (e.g., manganese and iron).

4. Colloidal foulants: These are compounds of moderately small size (almost 0.20–1.00 μm) and are not in completely dissolved condition, which while concentrated at the time of the membrane separation process might unite and subsequently get precipitated on the surface of the membrane.

5. Particulate foulants: These are mostly suspended solids and silt.

Presently, numerous standards are available for water analysis techniques. It is recommended to use the American Society for Testing and Materials (ASTM) International Standards (http://www.astm.org) or the latest edition of "Standard Methods for the Examination of Water and Wastewater." ASTM D 4195 covers the analyses that should be performed on any given water sample if reverse osmosis or nanofiltration (NF) application is being considered. Standard methods for the examination of water and wastewater is the result of a joint effort by three technical societies: American Public Health Association (APHA), American Water Works Association (AWWA), and Water Environment Federation. Standard methods are also offered online at standardmethods.org.

Table 3.1 provides a proper listing of the appropriate ASTM procedures and standard methods for the analysis of water and wastewater. A guidance for water examination for the reverse osmosis application is provided in ASTM D 4195/2/.

TABLE 3.1 Standard procedures associated with water analysis for reverse osmosis applications.

Sl. no.		ASTM	Standard methods
1	Chloride	D 512	4500-Chloride
2	Calcium, magnesium	D 511	3500-Ca, Mg
3	Carbon dioxide, carbonate, bicarbonate	D 513	4500-Carbon dioxide, 2320
4	Phosphorus	D 515	4500-P
5	Aluminum	D 857	3500-Al
6	Sulfate	D 516	4500-Sulfate
7	Manganese	D 858	3500-Mn
8	Silica	D 859	4500-Silica
9	Fluoride	D 1179	4500-Fluoride
10	Iron	D 1068	3500-Fe
11	Dissolved oxygen	D 888	4500-O
12	Chemical oxygen demand	D 6697, D 1252	5220
13	pH	D 1293	4500-pH value
14	Residual chlorine	D 1253	4500-Cl
15	Sodium, potassium, lithium	D 1428, D 3561	3500-Na, K, Li
16	Turbidity	D 1889	2130
17	Particulate and dissolved matter	D 1888	2560
18	Ammonia nitrogen	D 1426	45-NH$_3$
19	Total organic carbon	D 2579, D 4129, D 4839, D 5904	5310
20	Strontium	D 3352	3500-Sr
21	Boron	D 3082	4500-B
22	Arsenic	D 2972	3500-As
23	Practices for water sampling	D 3370	1060
24	Barium	D 4382	3500-Ba
25	Silt density index	D 4189	—

(Continued)

TABLE 3.1   (Continued)

Sl. no.		ASTM	Standard methods
26	Nitrite—nitrate	D 3867	4500-Nitrogen
27	Oxidation—reduction potential (ORP)	D 1498	2580
28	Microbiological contaminants in water	F 60	—
29	Biochemical oxygen demand (BOD)	—	5210
30	Assimilable organic carbon (AOC)	—	9217

Courtesy: DuPont Water Solution. Adapted from DuPont Water Solution's FilmTec™ reverse osmosis technical manual; Version 3; Form No. 45-D01504-en, Rev. 3, April 2020.

3.2 Different feed water constituents affecting the reverse osmosis performance

In this section, we discuss the different feed water constituents such as suspended solids, hydrogen sulfide, silicon dioxide, microbes, organics, color, iron, manganese, aluminum, calcium carbonate, barium, strontium, chlorine, magnesium, and hydrocarbons, and their effect on the reverse osmosis membrane performance. Different studies are carried out for improving the efficiency of membranes for water treatment (Saleem & Zaidi, 2020a, 2020b, 2020c; Saleem et al., 2020; Yadav et al., 2020; Zaidi et al., 2019). Table 3.2 details the guidelines for water quality against which the reverse osmosis feed water (and reject stream) must be compared, to decide if membrane scaling, fouling, or degradation is conceivable. The reverse osmosis membrane scaling might happen when the sparingly soluble salts get concentrated inside the element more than its solubility limits. In order to minimize the precipitation and membrane scaling, it is important to develop properly designed scaling control methods and prevent exceeding the solubility limit of sparingly soluble salts. For a reverse osmosis system, the commonly observed sparingly soluble salts are calcium sulfate, calcium carbonate, and silica (Ning, 2015). Other salts leading to probable scaling issues are calcium fluoride, strontium sulfate, barium sulfate, and calcium phosphate. Solubility products of different sparingly soluble inorganic compounds are given in Table 3.3. The different scale control techniques used in the reverse osmosis process are discussed in Chapter 5. A picture of a scaled reverse osmosis membrane is shown in Fig. 3.2.

3.2.1 Suspended solids

In reverse osmosis, the suspended solids and colloidal materials are considered as one of the major problems that influence the performance of the reverse osmosis process. Total suspended solids (TSS) are characterized as solids present in water that could be trapped using a filter. For measuring the TSSs, the sample of water is filtered by means of a preweighed filter. Despite the fact that majority systems have pretreatments inclusive of five-micron prefilter, these fine particles can lead to the reverse osmosis membrane fouling. TSS concentration is usually expressed in mg/L or ppm. TSSs are measured by the

TABLE 3.2 Commonly acceptable water quality guidelines for reverse osmosis feed water and reject stream.

Sl. no.	Species	Guideline value/range	Units
1	Chemical oxygen demand (COD)	<10	ppm
2	Color	<3	APHA
3	Organics (TOC)	<3	ppm
4	Silica (soluble)	$140-200$[a]	ppm
5	Microbes	<1000[b]	CFU/mL
6	Hydrogen sulfide	<0.1	ppm
7	Barium, strontium	<0.05	ppm
8	Calcium carbonate	<0[c]	LSI
9	Suspended solids	<1	NTU
10	Silt density index—Colloids	<5	1
11	Modified fouling index (0.45)	<4	1
12	Total organic carbon	<3	ppm
13	Oil and grease	<0.1	ppm
14	Assimilable organic carbon	<10	µg/L Ac-C
15	Manganese	<0.05	ppm
16	Ferric iron	<0.05	ppm
17	Ferrous iron	<4	ppm
18	Aluminum	<0.05	ppm
19	Temperature—PA membranes	<45	°C
20	Temperature—CA membranes	<30	°C
21	Chlorine, free-PA membranes	<0.02	ppm
22	Chlorine, free-CA membrane	<1	ppm
23	pH—PA membranes	$2-12$	pH units
24	pH—CA membranes	$4-6$	pH units

[a]*In reverse osmosis concentrate stream, changes as functions of temperature and pH.*
[b]*In reverse osmosis concentrate stream.*
[c]*In reverse osmosis concentrate stream, and could be till 2.0–2.5 on the basis of the type of antiscalants used.*

filtration of a specific volume of water (usually 1.0 L) using a preweighed glass fiber filter, and then drying the filter with the solids held on it at temperature 103°C, and subsequently taking the weight of the filter once more subsequent to drying. The difference between the dried filter weight and the fresh filter weight, divided by the filtered sample volume, gives the total quantity of suspended solids present in the source water (Voutchkov, 2017).

TABLE 3.3 Solubility product of different sparingly soluble inorganic compounds.

Sl. no.	Substance	Formula	Temp. (°C)	Solubility product	Negative log K_{sp}
1	Aluminum hydroxide	$Al(OH)_3$	25	3×10^{-34}	33.5
2	Aluminum phosphate	$AlPO_4$	25	9.84×10^{-21}	20
3	Barium carbonate	$BaCO_3$	25	2.58×10^{-9}	8.6
4	Barium sulfate	$BaSO_4$	25	1.1×10^{-10}	10
5	Calcium carbonate	$CaCO_3$	25	Calcite: 3.36×10^{-9}	8.5
				Aragonite: 6×10^{-9}	8.2
6	Calcium sulfate	$CaSO_4$	25	4.93×10^{-5}	4.3
7	Calcium phosphate	$Ca_3(PO_4)_2$	25	2.07×10^{-33}	32.70
8	Calcium fluoride	CaF_2	25	3.45×10^{-11}	10.50
9	Iron(II) hydroxide	$Fe(OH)_2$	25	4.87×10^{-17}	16.3
10	Iron(II) sulfide	FeS	25	$25\ 8 \times 10^{-19}$	18.1
11	Iron(III) phosphate dihydrate	$FePO_4 \times 2H_2O$	25	9.91×10^{-16}	15
12	Iron(III) hydroxide	$Fe(OH)_3$	25	2.79×10^{-39}	38.6
13	Lead sulfate	$PbSO_4$	25	2.53×10^{-8}	7.6
14	Lead fluoride	PbF_2	25	3.3×10^{-8}	7.5
15	Lead carbonate	$PbCO_3$	25	7.4×10^{-14}	13.1
16	Magnesium ammonium phosphate	$MgNH_4PO_4$	25	2.5×10^{-13}	12.6
17	Magnesium carbonate	$MgCO_3$	12	2.6×10^{-5}	4.58
			25	6.82×10^{-6}	5.17
18	Magnesium hydroxide	$Mg(OH)_2$	18	1.2×10^{-11}	10.9
			25	5.61×10^{-12}	11.25
19	Magnesium fluoride	MgF_2	18	7.10×10^{-9}	8.15
			25	5.16×10^{-11}	10.3
20	Magnesium phosphate	$Mg_3(PO_4)_2$	25	1.04×10^{-24}	24
21	Manganese hydroxide	$Mn(OH)_2$	18	4.0×10^{-14}	13.4
			25	2×10^{-13}	12.7
22	Strontium carbonate	$SrCO_3$	25	5.6×10^{-10}	9.25
23	Strontium sulfate	$SrSO_4$	17.4	3.80×10^{-7}	6.42
24	Zinc carbonate	$ZnCO_3$	25	1.46×10^{-10}	9.84

Courtesy: DuPont Water Solution. Reproduced from DuPont Water Solution's FilmTec™ reverse osmosis technical manual; Version 3; Form No. 45-D01504-en, Rev. 3, April 2020.

Suspended solids are usually estimated by turbidity. Turbidity is referred to the extent to which light can be scattered by particles remaining suspended in a liquid. As per the water quality guidelines, there is a requirement for influent turbidity of below 1 nephelometric turbidity units (NTU), which additionally turns out to be a guarantee prerequisite of manufacturers of the membranes. If the turbidity is above 1 NTU, then the warranty of

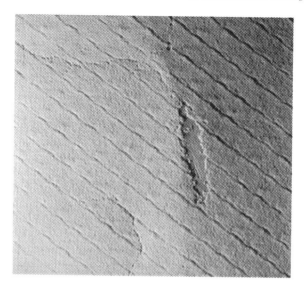

FIGURE 3.2 Photograph of the scaled surface of a membrane with imprints from the feed spacer. *Source: Courtesy: DuPont Water Solution. Reproduced from DuPont Water Solution's FilmTec™ reverse osmosis technical manual; Version 3; Form No. 45-D01504-en, Rev. 3, April 2020.*

the membrane is voided. The higher the turbidity, the membranes will be more susceptible to foul by suspended solids. The finest practices for reverse osmosis demand turbidity of feed water under 0.50 NTU. particle-size distribution is considered to be another measure of suspended solids.

For measuring the fouling potential of suspended solids, a method called silt density index (SDI) is employed. The SDI measures the suspended solids, especially colloids; for example, clay, microbes, iron corrosion products, and alumina- or iron silicates, which have an incredible potential for reverse osmosis membrane fouling. In the SDI empirical test, the feed water is filtered by means of a 0.450-micron filter membrane at consistent pressure in dead-end filtration, as shown in Fig. 3.3, and the filtration rates are calculated. The SDI must be very low in order to limit membrane fouling; however it should be lesser than 5 for meeting warranty requirements set by the manufacturers of membranes (in reverse osmosis feed water, good practices call for SDI to be under 3).

Calculation of SDI

$$\text{SDI} = \frac{P_{30}}{T_t} = 100 \times \frac{\left(1 - \frac{T_i}{T_f}\right)}{T_t} \tag{3.1}$$

where SDI = silt density index, P_{30} = percentage pluggage at feed pressure 30 psig, T_i = time required to collect the first 500 mL filtrate, T_t = total elapsed test time (5, 10, or 15 min), T_f = time needed for collecting final 500 mL of the filtrate.

Table 3.4 presents the recommendation of maximum SDI values by reverse osmosis manufacturers. No direct correlation of turbidity to the SDI exists, except that the increased turbidity generally implies increased SDI. The membranes fouled with suspended solids would have low productivity and an increment in pressure drop. Suspended solids could be reduced or removed in reverse osmosis feed water using clarification, coagulation, and filtration. The details are given in Chapter 5.

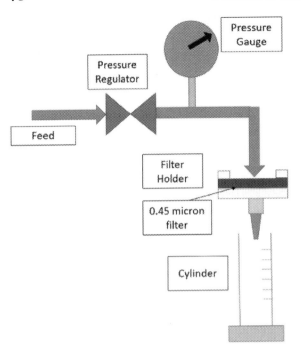

FIGURE 3.3 Apparatus for measuring Silt density index.

TABLE 3.4 Recommendation of maximum SDI values by reverse osmosis manufacturers.

Membrane manufacturer	Recommended maximum SDI$_{15}$	Reference
DuPont Filmtec	5	FilmTec™ Seawater RO Elements for Marine Systems
Toyobo	4	HR5255PIElement Configuration: Hollow Fiber
Hydranautics	4–5	Hydranautics SWC4 + Membrane
Toray membrane	5	Toray TM820C-370 Membrane
Koch membrane	5	FLUID SYSTEMS TFC-SS 8

3.2.2 Hydrogen sulfide

Hydrogen sulfide (H$_2$S) is a gas that leads to a recognizable smell of "rotten egg "in feed waters, with 3–5 ppm noticeable offensive odor level and 0.1 ppm threshold odor level. H$_2$S is promptly oxidized to elemental sulfur by oxidants (e.g., potassium permanganate, chlorine, or air). Sulfur functions as a colloidal foulant and possesses a record of not being expelled properly by typical multimedia filtration. The preferable design of reverse osmosis system proposes leaving the H$_2$S in its gas state, and allows it to go through the reverse osmosis membrane into the product stream, and subsequently treat the product water for the H$_2$S removal (Kucera, 2015).

H_2S is considered as a weak acid, and its dissociation will depend on the pH value.

$$H_2S + H_2O = H_3O^+ + HS^- \quad pK_1 = 7 \tag{3.2}$$

$$HS^- + H_2O = H_3O^+ + S^{2-} \quad pK_2 = 14.0 \tag{3.3}$$

H_2S levels as less as 0.10 ppm could unfavorably influence the reverse osmosis system performance.

3.2.2.1 Preventing the possible issues on the feed/reject side of the membranes

The existence of H_2S in feed water exposed to oxidants can lead to a metallic sulfide or elemental sulfur precipitation. The deposit will be a dark pasty residue that will clog the filter cartridge and get deposited on the feed water piping. This precipitated solid will lead to a higher filter cartridge replacement rate. Moreover, since the particle size of colloidal sulfur and metallic sulfides is in the submicrometer range, a substantial amount of precipitants would move across the standard 5.0-μm cartridge filter. These solids would accumulate in the feed/reject channel spacer of the reverse osmosis membrane element, resulting in an increase in the operational differential pressure. Also, an additional buildup of metallic sulfide and sulfur on the surface of the membrane would lead to an increased salt passage along with a reduction in water flux, thereby leading to a decrease in the efficiency of the reverse osmosis system. Also, air could be fed into the feed/reject zone in reverse osmosis element due to siphoning in the reject piping. The abovementioned process is most probable when an extended run of pipe is used for the reject line. A siphon breaker must be used for preventing the development of the partial vacuum that shows a tendency to draw water from the feed/reject side of the reverse osmosis membranes again developing voids that could present air. The pipe arrangements must be designed for keeping the reverse osmosis membrane skid assembly "flooded" and without any air at the time of the resting periods. The colloidal sulfur might be hard to separate. A sodium hydroxide solution with a chelating agent, like ethylenediamine-tetraacetic acid, is a suitable cleaner. In the case that foulants are not profoundly consisting of elemental sulfur, then a solution of phosphoric acid might be used to dissolve the sulfide components (Dupont Water Solution, 2020).

3.2.2.2 Preventing the possible issues on permeate side of the membranes

Due to the fact that hydrogen sulfide is a gas, it can pass across the barrier layer of the membrane, and under specific conditions, it would get precipitated as elemental sulfur in the membrane polyester support webs, microporous polysulfone (PSF) substrates, and product water channel spacers. A precipitate will be developed on the backside of the membrane composite while hydrogen sulfide is subjected to an oxidizing environment. Due to the natural osmosis phenomenon, there will be a trend for a water reverse flow from the permeate membrane side to the feed/reject side. The aforementioned phenomenon is predominantly important in the case of highly saline water consisting of TDS > 6000 ppm. This back-flow could cause the introduction of air into the product side of the elements. A pure water flush can be suggested, particularly for feed consisting of hydrogen sulfide, for displacing the concentrated solution as part of any shutdown sequencing. This would eliminate all osmotic driving forces for the backflow (Dupont Water Solution,

2020). For preventing a negative transmembrane pressure higher than 5.0 psi, a dump valve can be used for relieving the pressure on the product line upon system shutdowns. In the case of systems with feed water consisting of hydrogen sulfide, this should be carried out in such a way so as not to permit air to be brought into the system. The piping arrangement must be designed for keeping the reverse osmosis membrane skid assembly "flooded" and with no air in the course of resting periods.

The elemental sulfur precipitation on the product water side will not lead to a sudden deterioration in efficiency; however after a while, a progressive rise in feed pressure will be observed that could ultimately result in a substantial loss in efficiency (specific flux reduction). It is practically not possible to clean the previously mentioned precipitate from the membrane backside and the product channel spacer.

3.2.2.3 Pretreatment

The finest pretreatment technique for hydrogen sulfide is maintaining the system in anaerobic conditions. Water should not be exposed to chlorine, air, or any additional oxidizers from the well till the water leaves the membrane system. Hydrogen sulfide is usually removed from the permeate. The aforesaid rule is applicable to both seawater and brackish water; and is of particular relevance when iron is existing in groundwater. In practical terms, where it could be generally acceptable to employ oxidation or media filtration with greensand, the existence of hydrogen sulfide turns out to be the principal factor that eradicates this technique of iron removal from consideration.

3.2.2.4 Post-treatment

As H_2S normally passes through the reverse osmosis membranes, it is essential to separate this intolerable contaminant from the product stream as a post-treatment process. The technique used in majority membrane systems is air stripping that uses forced draft degasifiers. The previously stated degasifiers utilize a packed tower having counter-current airflow for stripping the H_2S from the water.

3.2.3 Silicon dioxide

Silicon is considered to be the second most plentiful element in the earth's crust by mass. Hence, natural water sources typically consist of a specific level of silicon dioxide/silica (SiO_2), in a concentration from 1.0 to 40.0 ppm. Silica is present in either amorphous or crystalline form. The amorphous silica could be categorized as particulate, colloidal, polymeric, and dissolved. Silicon dioxide in natural water forms from the dissolution of minerals and rocks [Eq. (3.4)]. The foremost dissolution reaction is the Si−O−Si bond hydrolysis, leading to the formation of silicic acid (H_4SiO_4) (Park et al., 2020).

$$x. \, SiO_2(Solids) + 2H_2O \leftrightarrow (x-1)SiO_2 + H_4SiO_4 \tag{3.4}$$

Reverse osmosis is considered to be the most effective method for removing the dissolved and colloidal silica, which could be present in increased concentrations in brackish water. The existence of silica along with its capability to foul the membranes restricts the application of silica-containing waters for desalination use and when this water is used, it accompanies with several economic penalties.

The chemistry of SiO_2 is complicated and to some degree it is unpredictable. In some cases, silica is an anion. In a similar manner as total-organic carbon details the total organics (as carbon) concentration without specifying what the organic compounds are, silica details the total silicon (as silica) concentration without specifying the silicon compounds. The "total silica" content of water is made out of "unreactive silica" and "reactive silica." Reactive silica (e.g., silicates SiO_4) is dissolved silica which is marginally ionized and is not been polymerized into a lengthy chain. Reactive silica is the silica form that can be used in reverse osmosis projection programs. Although this reactive silica has anionic nature, this is not included as an anion with regard to balancing a water analysis; however it is considered as a part of the entire total dissolved solids. Unreactive silica is colloidal or polymerized silica, behaving more like a solid instead of a dissolved ion. Unreactive silica will be available in extremely less concentrations in seawater and might cause fouling as soon as its concentration in the reject stream goes beyond 100.0 ppm. SiO_2, in the colloidal form, could be separated by reverse osmosis unit; on the other hand, it can result in colloidal fouling of the reverse osmosis front-end. Colloidal silica, having very small sizes (0.0080 μm), could be estimated empirically by the SDI testing, however only that portion that is larger than 0.45 μm. Particulate silica compounds (e.g., sand, silts, and clays) are typically 1 micron or bigger and could be estimated using the SDI test. Polymerized SiO_2, which employs SiO_2 as the building block, exists in nature as agates and quartzes. This polymerized SiO_2 form also results from exceeding the saturation level of reactive silica. The reactive silica solubility is regularly restricted to 200%−300% with the utilization of a silica dispersant. The solubility of reactive silica increases with temperature, enhances at a pH under 7.0 or greater than 7.8, and diminishes in the presence of iron that functions as a catalyst during silica polymerization.

In a reverse osmosis operation, the level of silica in the reject stream increases and reaches saturation, which could lead to silica depositions or metal silicate precipitation on the surface of the membrane (scaling). Silica fouling is extremely hard to remove from reverse osmosis membrane, and ultimately results in performance degradation like a loss in permeability and premature shutdown of the system (Lisitsin et al., 2005). Even though strong cleaning chemicals such as hydrofluoric acid (HF) and ammonium bifluoride (NH_4HF_2) could be employed for removing the silica fouling, the application of harsh cleaning chemicals might lead to the damage of equipment along with environmental issues (Sheikholeslami et al., 2001). Thus reverse osmosis plant is frequently forced to run at low efficiency. In other words, the recovery rate of a reverse osmosis system is reduced for maintaining the silica level in the reject stream under its saturation point, on the basis of operational temperature and pH (Park et al., 2020).

For brackish water desalination, the silica scaling of reverse osmosis membranes is least understood as compared to hardness scaling because of the complicated silica behavior at the water/membrane interface. In a research work carried out by Lu and Huang (2019), the team introduced −OH, −NH_2, −SO_3H, and −COOH functional groups onto polyamide (PA) membrane for developing distinctive surface physicochemical characteristics. The resultant membranes were examined further under comparable scaling conditions for yielding temporal flux loss information that was empirically elucidated by a logistic growth model. The team observed that the flux loss of permeate was intensely correlated to the initial SiO_2 layer developed by straight interaction between reactive silanol and reciprocal groups on the surface of the membrane, instead of the complete scaling layer. Prominently, surface properties of the membrane dictated the preliminary silica layer development by means of three possible mechanisms, that is, interfacial energy change, competitive adsorption, and electrostatic repulsion. Out of these mechanisms, the electrostatic repulsion was recognized to be the main mechanism. Consequently, by the modification of the surface properties of the membrane, the three aforesaid mechanisms might be enhanced for favoring the formation of a loose, disordered initial silica scaling layer. As a result, the flux loss of the membrane might be minimized. The aforementioned outcome contributed significant insights into the design heuristics of antiscaling reverse osmosis membrane for brackish water desalination (Lu & Huang, 2019). In a work performed by Cob et al. (2014), aluminum hydroxide ($Al(OH)_3$) was observed to be the most efficient silica precipitant, eliminating almost the entire molecularly dissolved silica. Conversely, it was noted that a residual amount of aluminum was retained in solution, and the alumino-silicate colloids were not eliminated. The usage of the strong base anion exchange resin also exhibited a decent performance, eradicating up to 94.0% of SiO_2.

3.2.4 Microbes

Similar to fresh waters, the seawater and brackish water consist of pathogenic microorganisms inclusive of viruses, protozoa, and bacteria (Affam & Ezechi, 2019). Disinfection is applied at various locations for removing the microbes at the time of the treatment process. During the pretreatment process, a disinfectant (often chlorine) can be introduced to reduce the membrane biofouling and to protect the membrane from degradation. Also, membranes have the capability to remove microbes by stopping their passage to the permeate water. Ultrafiltration (UF) membranes having pores (~ 0.0010 to 0.10 microns) have been proved to accomplish substantial reductions of protozoa and viruses. Also, improved performance can be expected from the reverse osmosis membranes, as compared to UF membranes.

In the reverse osmosis membranes, microbial fouling is a critical issue (Nguyen et al., 2012). The colonies of bacteria will develop virtually at any place in the membrane module wherever favorable conditions exist. Concentration polarization also contributes to biofouling by providing an environment on the membrane surface that is nutrient-enriched for microorganisms. Satellite colonies could separate and start to develop somewhere else inside the membrane module, expanding the membrane surface area which is covered with microorganisms and their related biofilm. The possibility of membrane biofouling could be figured out by reviewing the assimilable organic carbon (AOC). This analysis is a bioassay that estimates the development capability of microbes in a specimen. The method of the test is detailed in section 9217 of the standard methods (Standard Methods for the Examination of Water and Wastewater). An AOC value of 10 ppm is a recommended standard for minimizing the membrane biofouling; however in certain instances, fouling might still happen even at this reduced value.

The extent of membrane fouling with microorganisms that has already happened is analyzed by monitoring the count of colonies that slough off the membrane into the reverse osmosis concentrate stream. The aforementioned phenomenon is commonly estimated using one of the two strategies stated below:

Strategy 1: Culture: This method can be performed easily and does not need any costly equipment. It is used for determining the number of colony-forming units (CFU) present in a sample of water, utilizing section 9000 of the standard methods. The CFU number in a specimen is the number of culturable microbes available. This process is comparatively economical; however the counted colonies may depict just around 1.0%–10.0% of the total bacterial count (TBC). On the other hand, the abovementioned strategy could be beneficial in tracing microbial fouling. In a reverse osmosis concentrate stream, concentrations of about 1000 CFU/mL or higher are regarded as a fouling issue that can considerably and adversely influence the reverse osmosis system performance.

Strategy 2: Total bacterial count (TBC): The TBC is estimated by literally checking the real quantity of microorganisms compiled on a filter after it is used for filtering a water sample at issue (Vrouwenvelder & van der Kooij, 2001). The specimen is stained using acridine orange and then observed with a fluorescent microscope of epi-illumination. This method is quicker and accurate relative to the culture process; however it is not useful for fieldwork.

Fig. 3.4 presents the possible mechanisms of the biofouling formation on the membrane surface. In the first stage, the biofoulants, like algae and bacteria, stick to the surface by means of weak van der Waals forces. In the second stage, the primary colonists simplify the arrival of additional cells by keeping more assorted hosting centers/adhesion sites and building the matrix that grips the biofilm together. Subsequent to the colonization in the third stage, the biofilm might develop by a combination of cell division and recruitment. The fourth stage is the development of biofilm formation and here the biofilm will be established and the cells turn out to be more antibiotic-resistant (Saleh & Gupta, 2016). The microbial fouling has to be properly managed before the biofilm turns out to be mature. The biofilm shields the microbes from the activity of shear forces and biocidal chemicals utilized for attacking them. Microorganisms could be destroyed by employing some non-oxidizing biocides, ultraviolet (UV) radiation, ozone, or chlorine. An efficient

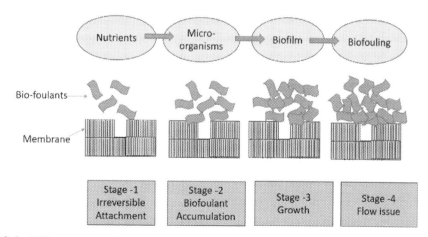

FIGURE 3.4   Different stages of biofouling formation on the surface of the reverse osmosis membrane. *Source: Adapted from Saleh, T.A., & Gupta, V.K. (2016). Membrane fouling and strategies for cleaning and fouling control. In: Nanomaterial and polymer membranes. Amsterdam: Elsevier, pp. 25–53.*

strategy for controlling bacteria and biofilm development typically involves a consolidation of these methods. In a study by Abushaban et al. (2017), microbial adenosine triphosphate (ATP) has been utilized for measuring the growth potential of bacteria using an indigenous bacterial consortium in an Australian-based seawater reverse osmosis desalination plant. A substantial decrease (55%) in the growth potential of bacteria was observed by using a dual media filtration, with 4.50 mg-Fe(III)/L coagulant included before the media filtration. A review paper by Matin et al. (2011) discussed the causes, consequences, and control of biofouling in reverse osmosis membranes used for seawater desalination. In another study by Matin et al. (2014), the surface modification of membranes with an initiated chemical vapor deposition technique was presented as a potentially effective technique for controlling the membrane biofouling.

3.2.5 Organics

Organics are compounds that consist of carbon (except carbonate, bicarbonate, and carbon dioxide). In the case of water treatment, organics are categorized as man-made or naturally occurring. The naturally occurring organic matter is usually negatively charged suspended solids or colloids, consisting of lignins, tannins, fulvic acid compounds caused by certain vegetative matter decay, or water-soluble humic acid compounds caused by certain vegetative matter decay. Naturally occurring organic matter can act as a foulant to reverse osmosis membranes, especially the negatively charged composite polyamides. The organics will be adsorbed on the surface of the membrane leading to a flux decrease that could be irreversible in certain instances. Adsorption is usually preferable at pH under 9.0 and where the organic compounds will be positively charged. The most problematic among the organics are the emulsified organics, which have the ability to form an organic film on the surface of the membrane. Due to the fact that several organics are nutrients for microbes, the organic fouling intensifies the microbial fouling.

Natural Organic Matter (NOM) is normally present in surface saline waters (open ocean seawater or brackish water) and comprises compounds that are formed by naturally decaying algae and other marine flora and fauna [i.e., pigments, oils, carbohydrates, proteins, humic substances (humic and fulvic acids)]. An increased concentration of NOM in the source water utilized for potable water production is unsuitable as it leads to water discoloration, develops hazardous disinfection byproducts while disinfected using chlorine, and leads to complexation with heavy metal, which consecutively aids in membrane fouling. In typical non-algal bloom conditions, brackish water or standard seawater brought together by open intakes does not have enough NOM concentration to contribute substantial challenge to desalination plant processes. An increased concentration of NOM is typically seen in the course of algal blooms or/and when the desalination facility intake is situated nearby wastewater treatment plant (WWTP) discharge, or nearby the convergence with river or other freshwater sources (Voutchkov, 2017).

Humic acids are polymeric substances and have the ability to form chelates with metal ions (e.g, iron) present in the saline water source. The previously stated characteristic of humic acid is extremely significant for surface brackish water or seawater pretreatment system using iron coagulants, as they could build a layer of chelates on the membrane surface, and lead to fouling. Usually, these types of fouling layers could be dissolved at pH 9.0 or higher, at conditions where both the humic substances and the membranes carry a negative charge. The aforementioned characteristic is employed for the cleaning of the membrane. Humic substances are considered to be hydrophobic in nature and consequently the hydrophilic membranes will be less exposed to fouling by humic acid. In their natural state, humic acids are not considered to be a source of food for the majority marine organisms. On the other hand, while oxidized with chlorine or any other oxidant, the humic acid can turn out to be definitely biodegradable and function as a source of food for marine bacteria developing on the surface of the reverse osmosis membrane. Consequently, constant chlorination of source water comprising a huge quantity of humic acids frequently leads to additional membrane-biofouling issues than it solves.

Total organic carbon (TOC) is the amount of carbon present in an organic compound and is regularly used as a non-specific indicator of water quality, and is represented in units of "ppm as carbon." Due to the fact that the TOC just estimates the quantity of carbon present in organic matter, the organic mass actual weight could be more up to three times greater in natural surface waters. The organic concentration, as determined by TOC, should be under 3.0 ppm for minimizing the fouling probability. The membrane organic fouling can reduce the overall membrane productivity. The reverse osmosis membranes with a neutral charge are considered to be more organic fouling resistant. The reverse osmosis system has a good capability of removing the organic compounds. In general, organic compounds with higher than 200 molecular weight are removed at levels higher than 99%. The rejection of compounds having lower than 200 molecular weight will vary on the basis of ionic charge, shape, and molecular weight. As a general rule, the alert levels for possible organic fouling in natural water sources are COD at 8 ppm and biochemical oxygen demand (BOD) at 5 ppm.

The concentration of oils, as well as greases, must be under 0.1 ppm in reverse osmosis feed water. The abovementioned materials would freely adsorb onto PA membranes and lead to a reduction in membrane efficiency. On the other hand, these materials could be removed from the membrane by employing alkaline cleaners in the case that the flux has not dropped by greater than 15.0% from startup. Organics present in reverse osmosis feed water could be reduced employing UV radiation, coagulation, or activated carbon (AC) filtration (Gerba & Pepper, 2019). Low molecular weight organics like acetone, isopropyl alcohol, and urea cannot be simply separated by means of the aforesaid methods. On the other hand, oxidation of these organics utilizing persulfate activated by UV light is demonstrated to be effective.

Lee et al. (2020) studied the impacts of marine organic matter characteristics on the fouling layer composition of seven parallelly organized reverse osmosis membrane modules. The study provided significant insights into the fouling behavior of marine organic matter in a full-scale seawater desalination facility in South Korea. Even though both hydrophilic and hydrophobic marine organic matter fractions controlled the fouling development of the reverse osmosis membranes, hydrophobic marine organic matter fractions complexed with multivalent metal ions such as magnesium, iron, copper, calcium, and aluminum were considered to be main contributors to the irreversible fouling of the reverse osmosis membranes. The hydrophilic marine organic matter fractions were preferentially deposited onto the surface of the membranes in the reverse osmosis first module, whereas residual hydrophobic marine organic matter fractions strongly contributed to the fouling formation of the reverse osmosis membranes in the seventh module.

3.2.6 Color

Color is normally composed of naturally found humic substances which are formed when the organic substances, for example, leaves, decay. This color gets adsorbed onto the exterior of the reverse osmosis membrane. The humic matters are themselves made out of three distinct kinds of organic compounds. The color of humic acid is that color which is precipitated at the time of acidification; and these organics will be dark-brownish to blackish in color. The fulvic acid will not precipitate at the time of acidification; and these substances will be yellowish to yellow-brownish in color. Finally, humin will be insoluble at all pH and it will be blackish in color. Turbidity is an additional factor that contributes to colored water. When there are suspended solids in a high level, then the turbidity increases and the water clarity gets affected and the color changes. Color might also be an indicator of the presence of other pollutants. As an illustration, when iron is present in water, the color of the water turns out to be reddish or brownish. If algae are present in water, then the water color will be green.

Color could be apparent or true. The apparent color is basically total color, consisting of dissolved and suspended organics and other suspended solids. True color is estimated by removing the suspended solids with the goal that the only color existing is because of dissolved organics. Color is assessed using the dimensionless units of APHA. Color adsorption onto reverse osmosis membranes is recommended when the compounds are positively charged or hydrophobic. Similarly, with other organics, an increased pH (greater than 9) limits fouling with color, but leads to other issues, including calcium carbonate scaling. The true color must be under 3.0 APHA to limit fouling because of color adsorption. The color adsorption onto the membrane would diminish the membrane efficiency.

Color could be reduced in reverse osmosis feed water by employing coagulation/clarification using hydroxide flocculants, UV radiation, adsorption of AC, NF, and UF. AC filter is considered to be the most frequently used type of treatment for removing color from water (Barnhart & Amamath, 1993). In addition to the removal of color from water, the above-stated conventional treatment technique can also absorb organic contaminants and particles present in water that might lead to unpleasant tastes and odors. Iron and turbidity could also be removed by means of this treatment technique. On the other hand, it is crucial to understand that AC filters will not eliminate bacteria. Consequently, this is characterized as filtration instead of purification. The AC filter is considered to be a less expensive treatment technique and is extremely efficient in eliminating the dissolved solids present in water.

3.2.7 Iron

Iron is present in groundwater as well as surface water, at variable concentration levels, typically up to 3.0–4.0 ppm and in certain instances up to 15.0 ppm. Iron is considered to be a water contaminant, and it takes two substantial forms, i.e., water-soluble form and water-insoluble form. The water-soluble form is referred to as the ferrous state and possesses a valence state of +2. When iron is available in water even at low concentrations, it can be associated with esthetic and operational issues like unpleasant taste, color, staining, and deposition in the water supply system resulting in high turbidity. The presence of iron in reverse osmosis foulants, frequently in substantial amounts, is normally realized in elemental composition analysis carried out together with membrane autopsy for the determination of fouling issues. The iron-containing colloidal oxides, phosphates, sulfides, and silicates in reverse osmosis feed water regularly contribute to the elemental composition of foulant acquired from the membrane surface. In the case of a membrane-based desalination process, the precipitation of iron could considerably decrease the efficiency of the process (Melliti et al. 2019). Correspondingly, iron also stimulates the formation of some kind of chlorine-tolerant microbes in water supply systems, which results in high costs for system cleaning and sterilizing along with taste and odor issues. The maximum allowable limit of iron concentration for potable water is 0.20 ppm. In the case of non-aerated well waters, the ferrous iron acts much like magnesium or calcium hardness in that it could be removed using softeners or its precipitation in the reverse osmosis system back-end could be regulated by the utilization of a dispersant chemical in the feed water of RO. The water-insoluble form is termed as the ferric state; and possesses a valence state of $+3$. Generally, the manufacturers of the reverse osmosis system will suggest that the united iron levels in the reverse osmosis feed be under 0.05 ppm (Hydranautics). If the entire iron is in the form of soluble ferrous, iron levels up to 0.5 ppm in the feed could be permitted if the pH is under 7.0 (however an iron dispersant is suggested). The air introduction into water-soluble ferrous iron will bring about the insoluble ferric iron oxidation. The soluble iron could be seen in deep wells; however, it can be changed into the more problematic insoluble iron by the air introduction by leaky pump seals or by being placed in tanks. Several scientific researches have been conducted to find advanced methods for iron removal like limestone filter treatment, oxidation using potassium permanganate, and adsorption technique.

Multimedia filters are considered to be one of the most recognized techniques to remove iron from water. The above-stated system is known for its ability to remove suspended particles, color, odors, turbidity, and iron to produce superior quality water. These filters consist of layers of media arranged from largest to smallest in grain size to enable efficient filtration of larger particles followed by smaller particles. When a coagulant is used simultaneously, small particles are joined together to create larger particles that can be more easily filtered out. This process assists to remove iron from water, including particles of up to 10 microns in size. A green sand filter system could also be employed as direct water filtration for dealing with iron, hydrogen sulfide, and manganese for the distribution of drinking water applications or it could be used as a pretreatment for reverse osmosis system, that needs lower levels of the above-stated ions.

The removal of iron by the physical–chemical method consists of iron oxidation by air succeeded by sand filtration. Soluble iron can be undergone treatment with dispersants or it can be removed by softeners, iron filters, or lime softening. The insoluble ferric hydroxides or ferric iron oxides, being colloidal, will cause fouling on the reverse osmosis system front end. The insoluble iron sources include surface sources, aerated well waters, and iron scale from unlined pipe and tanks. The insoluble iron could be removed by lime softening, iron filters, UF (with limits), multimedia filtration having polyelectrolyte feed (with limits), and softeners (with limits). Precautionary measures are needed with the usage of potassium permanganate ($KMnO_4$) in manganese greensand iron filters in that $KMnO_4$ is an oxidant that can cause harm to any PA-based membrane. Preventive measures are likewise needed with a cationic polyelectrolyte in which they could cause irreversible fouling to a negatively charged PA membrane. Corrosion-resistant vessels and piping (e.g., polyvinyl chloride, fiber-reinforced plastic, or stainless steels) are suggested for the entire reverse osmosis pretreatment, reverse osmosis systems, and distribution piping entering the reverse osmosis system. Iron as foulant will rapidly enhance reverse osmosis feed pressure necessities and increase permeate total dissolved solids. In certain cases, the existence of iron can lead to a biofouling issue by being the energy source for iron-reducing bacteria (FeRB). The FeRB can lead to the development of a slimy biofilm that could plug the reverse osmosis feed pathway.

Besides naturally derived colloidal matters, the iron colloidal fouling on the reverse osmosis membrane surface might be due to the corrosion of upstream piping and equipment, or by poor mixing or overdosing of iron-based coagulants employed for saline water conditioning. In the event that the saline water contains chlorine, then this colloidal iron shows a tendency for catalyzing the oxidation process due to chlorine, which in exchange increases the reverse osmosis membrane damage even though the residual chlorine present in the saline water is in extremely low concentrations. The iron fouling happens more often relative to manganese fouling because the iron oxidation takes place at a much lower pH. Consequently, a fouling issue could be generated even if the SDI is less than 5.0 and the iron concentration in the reverse osmosis feed water is less than 0.10 ppm. Water with low alkalinity typically has greater iron concentrations relative to water with high alkalinity, due to the fact that the Fe^{2+} concentration is typically limited by the ferrous carbonate ($FeCO_3$) solubility.

3.2.8 Manganese

Manganese is a water impurity seen both in surface water and in well water, with levels up to 3.0 ppm. Same as iron, manganese is present in organic complexes seen in surface waters. It is noted to be soluble in oxygen-free water. An alarming level for potential manganese fouling in an aerated reverse osmosis feed water is 0.05 ppm. The potable water guidelines limit manganese to 0.05 ppm because of its capability for causing dark stains. The dispersants utilized for controlling iron fouling could be employed to help control the fouling caused by manganese.

One of the methods to prevent fouling of the membrane is to avoid oxidation and precipitation of manganese by retaining the water in its reduced form. Water should not be exposed to any oxidizing agents (e.g., Cl_2) or to air over the entire reverse osmosis process. Reverse osmosis plants utilizing the customary oxidative filtration manganese removal pretreatment technique regularly suffer from serious fouling issues because of the inadequate removal of colloidal manganese particles formed by such pretreatment methods. The utilization of MnO_2-coated greensand as an iron removal filter deserves a distinctive mention as a source of reverse osmosis membrane fouling. Colloidal particles of MnO_2 either shed from the manganese greensand or transferred from the reduction of $KMnO_4$ used in the regeneration of manganese greensand are found to be present at variable degrees as a brown-black coating on reverse osmosis membrane. The impacts of membrane fouling due to colloidal manganese species start with flux drop from a thin coating, to pressure drop increase from denser depositions. Any carryover of traces of oxidants such as chloramine and hypochlorite in the feed water catalyzed by manganese as foulants on the membrane would lead to a serious drop in salt rejection because of free-radical reactions deteriorating the PA-selective layer of the thin-film composite (TFC) membranes.

For controlling such adverse impacts of manganese and iron species in water-influencing the reverse osmosis process, a recent approach found to be effective is to use antiscalants for manganese and iron sequestration, retaining manganese and iron in solution without aggravating colloidal fouling propensity that is already present in certain waters. In the case of manganese removal, the manganese dioxide (MnO_2) could be employed as an adsorbent as per the below reaction:

$$Mn + MnO_2(s) \rightarrow 2MnO(s) \tag{3.5}$$

The manganese oxides will be subsequently adsorbed on manganese dioxide grains. When the entire manganese dioxide has been used up, it could be regenerated using sodium hypochlorite (NaOCl). The removal of manganese by physical–chemical method (aeration along with sand filtration) could also be used; however it should be noted that the manganese oxidation kinetics will be extremely slow at pH < 9. Instead of media filtration, microfiltration or UF can also be used for removing small manganese hydroxide ($Mn(OH)_2$) particles developed from oxidation process. The above-stated technology is considered to be latest technology for the removal of manganese.

3.2.9 Aluminum

Based on its low solubility, aluminum (Al) is generally not seen in any remarkable concentrations in surface waters or wells. When present in a reverse osmosis feed water, aluminum is usually colloidal in nature (not ionic) and results from alum carryover by a municipal or on-site clarifier or lime softener. Aluminum sulfate, also termed as an alum, is a well-known coagulant which is very efficient in the absorption as well as precipitation of negatively charged, naturally occurring colloidal material (e.g., silt and clay) from the surface waters. Alum, when added into water, undergo dissociation into trivalent aluminum and sulfate. The hydrated Al ion reacts with water to form various complex hydrated Al hydroxides, which subsequently polymerize and begin absorbing the negatively charged colloids present in the water. The chemistry of Al is complex due to the fact that it is amphoteric. Al at low pH's could remain as an aluminum hydroxide compound or as a positively charged trivalent cation. Al at higher pH's could remain as a negatively charged anion compound. In general, the range of lowest solubility for Al compounds is in the pH range from 5.5 to 7.5.

Currently, there is increasing evidence that the simultaneous presence of Al cations at low concentration and silicate species has an unfavorable and expensive impact on the reverse osmosis membrane (Lagref & Haci, 2014). Aluminum silicates are observed to be deposited onto piping materials in supply systems (Kreiwall et al., 1996) in low-pressure (microfiltration/UF) and high-pressure water treatment processes set up in California (Norman et al., 1999), and in a hollow-fiber unit for the treatment of brackish water in Saudi Arabia, etc. Over the years, a few facilities on Djerba Island (Tunisia) were affected by alumino-ferro silicate scaling, and these are hard to clean as compared to calcium carbonate scaling (Farhat et al., 2012). The examinations using X-ray diffraction (XRD) technique confirmed a close relationship between silicate fouling and the presence of aluminum cations (Al^{3+}) traces. Technically, this type of membrane fouling was theorized to happen through soluble Al^{3+} reacting with ambient silica (H_4SiO_2) to develop kaolinite ($Al_2Si_2O_5(OH)_4$) inside the reverse osmosis unit. Alumina fouling can be prevented by (1) pH control and coagulation before multimedia filtration, or (2) using alumina-compatible antiscalants.

In addition to well-waters consisting of naturally derived Al cations, the increased occurrence of alumino-silicate fouling problems could be described by the wide utilization of aluminum salts in membrane filtration pretreatment steps. Alum (aluminum sulfate) could be employed as a coagulant for removing the NOM like humic acid. Non-optimal concentrations are probably the cause of increase in residual aluminum in feed water to the membrane. The scaling intensity could be explained by the low aluminum dosage ($<100.0\ \mu g/L$) required to decrease severely the silicate solubility, principally at pH 7. Simultaneously, the aluminum concentration released during a conventional alum pretreatment could reach $200.0\ \mu g/L$. With regard to process impacts, the reverse osmosis salt rejection and the specific flux could begin to drop subsequent to 300 h of functioning. Certain specific flux drop has been estimated up to 60.0% above 100 h of operation.

3.2.10 Calcium carbonate

The scaling issue caused by calcium carbonate ($CaCO_3$) is one of the most well-known problems in any reverse osmosis system. However, it is quite easy to identify and deal with. Fundamentally, if the ion product (IPc) of $CaCO_3$ in the reverse osmosis concentrate is more prominent relative to the solubility constant (K_{sp}) under the reject conditions, at that point, the $CaCO_3$ scaling will be developed. If IPc $<$ K_{sp}, scaling in improbable.

$$\text{The IPc at any saturation level is referred to as: IPc} = [cation]^a[anion]^b \qquad (3.6)$$

where [cation] = concentration of cation; [anion] = concentration of anion; IPc = ion product; superscripts a = cation amount in the salt; b = anion amount in the salt.

$$\text{At saturation,} K_{sp} = [cation]^a[anion]^b \qquad (3.7)$$

where K_{sp} is the solubility product.

Langelier saturation index (LSI) is a technique to describe the corrosive probability or the scaling of low TDS brackish water on the basis of the saturation level of $CaCO_3$. For the boiler water plants and municipal water plants, LSI is critical in deciding if the water is corrosive (possesses a negative LSI) or will have a tendency for $CaCO_3$ scaling (possesses a positive LSI). The LSI is significant to reverse osmosis chemists as an estimation of the $CaCO_3$ scaling potential. The value of LSI is determined by subtracting the estimated pH of $CaCO_3$ saturation from the real feed pH.

$$LSI = pH - pH_a \qquad (3.8)$$

where

$$pH_a = (9.30 + A + B) - (C + D) \qquad (3.9)$$

$$A = \frac{(\log_{10}[TDS] - 1)}{10}$$

$$B = -13.120 \times \log_{10}(°C + 273.0) + 34.55$$

$$C = \log_{10}\left[Ca^{2+}\right] - 0.40,$$

where $[Ca^{2+}]$ is in ppm as calcium carbonate.

$$D = \log_{10}\left[alkalinity\right],$$

where [alkalinity] is in ppm as calcium carbonate.

A positive LSI implies that scaling is favored; while a negative LSI implies that corrosion is favored. It is recommended to maintain the LSI almost zero (or less) in the reverse osmosis reject stream for minimizing $CaCO_3$ scaling. The solubility of $CaCO_3$ diminishes with an increase in temperature, higher pH and alkalinity levels, and a higher concentration of calcium. The value of LSI could be reduced by lowering pH by the injection of an acid (commonly hydrochloric or sulfuric) into the reverse osmosis feed water. A recommended LSI target in the reverse osmosis concentrate is a negative value of about 0.2 (that demonstrates that the concentrate is 0.2 pH units lesser than the point of $CaCO_3$ saturation). A " − 0.2" LSI permits for pH excursions in real plant activity. A polymer-based antiscalant could likewise be utilized to prevent $CaCO_3$ precipitation. Several antiscalant providers have stated the effectiveness of their product up to a positive LSI value of 2.5 in the reverse osmosis concentrate (however an increasingly stable LSI level is positive 1.8).

Calcium also forms scaling with phosphate, sulfate, as well as fluoride. The LSI would not help in predicting the aforesaid types of scales; and hence water quality analysis, by means of ion product and solubility constants, is needed for determining the possibility for scaling with calcium phosphate or calcium fluoride. The presently available antiscalants could resolve calcium sulfate and calcium fluoride scaling, and some also could resolve calcium phosphate scaling.

3.2.11 Trace metals—strontium and barium

Strontium and barium generate sulfate scaling which is not easily soluble. Barium sulfate ($BaSO_4$) scale in reverse osmosis causes flux decline and possibly serious membrane damage (Boerlage et al., 2002). As a matter of fact, barium is considered to be the least soluble among the entire alkaline-earth sulfates. Its solubility is 10^{-5} mol/L (equivalent to 2.33 ppm) in pure water. Besides, if not detected in time, barium sulfate (barite) scaling might age to form a hard adherent layer. At this point, cleaning might not be efficient and the membrane elements should be replaced. Barium can function as a catalyst for calcium and strontium sulfates scale (Dupont Water Solution, 2020). $BaSO_4$ can be formed from different sources. For example, in the offshore oil industries, seawater (consisting of high sulfate concentration) is injected into reservoirs, and the resultant mixture mostly features stable supersaturated barium sulfate, ready to precipitate. While alkaline scales can be controlled by altering the feed pH prior to filtration, sulfate scales are insensitive to pH and therefore will not be controlled by the addition of acid. $BaSO_4$ scaling is generally needle-shaped crystals, which could simply deteriorate the selective layer of the membrane, leading to loss of integrity.

Investigation of the ion product with the solubility constants for strontium sulfate ($SrSO_4$) and $BaSO_4$ is important to decide the scaling potential with the aforesaid species. In the event that the IPc for $BaSO_4$ is above the solubility constant, the scale would develop. If the IPc of strontium sulfate is >0.8 K_{sp}, then scaling is likely to occur. On the other hand, the induction period (the time taken for scale formation) is lengthier for these types of sulfate-based scaling relative to that for the $CaCO_3$-based scaling. The level of strontium and barium could be reduced in reverse osmosis feed water by sodium softening. Antiscalant can be utilized for controlling or inhibiting the scaling with no reduction in the concentration of both species.

The method adopted by DuPont is extensively used as a prediction technique for the $BaSO_4$ scale in reverse osmosis systems using the salt solubility at temperature 25°C and the solubility product, K_{sp}. This solubility product expresses the dynamic equilibrium between the crystal solid phase and the scalant crystal ions in solution. As an example, the K_{sp} for barium sulfate is expressed as

$$K_{sp} = \left[Ba^{2+}\right]\left[SO_4^{2-}\right] (\text{at equilibrium}) \tag{3.10}$$

K_{sp} can be evaluated graphically as a function of ionic strength for the required recovery (Fig. 3.5). The correlation between solubility product and ionic strength is derived from information on the influence of individual monovalent and divalent cations on barium solubility. When the concentration product of scalant goes beyond the solubility product, the solution will get supersaturated and scaling might happen. DuPont recommends a recovery in a manner that the scalant concentration product is 20% below the solubility product for preventing the scale.

3.2.12 Chlorine

Polyamide (PA)-based composite membranes are extremely sensitive to free chlorine. The PA composite membrane degradation happens very quickly upon exposure and can bring about a significant decrease in rejection when exposed to 200-1,000 hours of exposure to 1 ppm of free chlorine. The degradation rate depends on two significant factors: 1) the presence of transition metals, for example, iron, would catalyze the membrane oxidation; 2) degradation will be faster at higher pH, relative to neutral or lower pH value. The degradation mechanism is basically the loss of polymer cross-linking. The above-stated loss leads to the dissolving of membrane polymer, the same as a nylon stocking while exposed to chlorine bleach. This permanent destruction would proceed as long as the membrane is exposed to the oxidizer.

Additionally, chloramines will be dangerous to PA-based composite membranes. The chloramines are practically at all times in equilibrium with free chlorine. Despite the fact that the tolerance to chloramines of the FilmTec FT30 TFC membrane is 300,000 ppm-h,

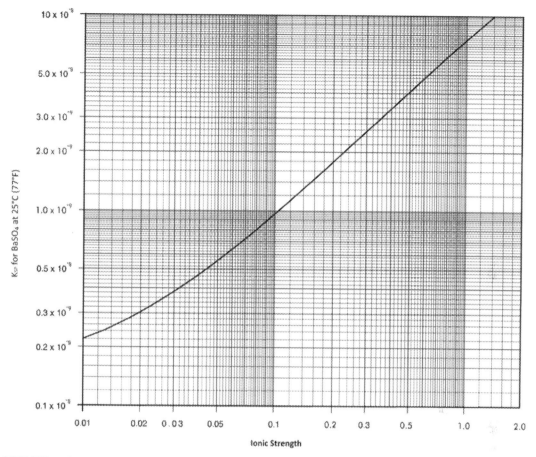

FIGURE 3.5 Solubility product for barium sulfate versus ionic strength. Source: *Courtesy: DuPont Water Solution. Reproduced from DuPont Water Solution's FilmTec™ reverse osmosis technical manual; Version 3; Form No. 45-D01504-en, Rev. 3, April 2020.*

still it is recommended by FilmTec that influent water having chloramines should be undergone dechlorination before the membrane (Dupont Water Solution, 2020). In majority of cases, ammonia (NH_3) will be added to chlorine for generating the chloramines. The best chloramine utilization is found in wastewater systems with an inhabitant convergence of smelling salts, to which chlorine will be introduced to form the chloramines.

Further cautionary note with chloramine is the requirement for proper pH control. In the event that the pH increases to 9.0, dissolved NH_3 gas, represented as $NH_3(g)$, swells at least some of the PA-based composite membranes. The aforesaid swelling could be sufficient for dropping the rejection of salt from 98.0% to almost 85.0%. Further reduction of the pH to around 7.0 changes the NH_3 gas to ammonium ion (NH_4^+), which does not cause membrane swelling, and subsequently rejection comes back to nominal. The usage of chlorine dioxide (ClO_2) is not suggested for application with PA-based composite membranes

Dupont Water Solution, 2020). This is due to the fact that free chlorine is constantly available with ClO_2 that is produced onsite from chlorine and sodium chlorate. Primarily, the PA-based composite membrane which has been degraded because of chlorine attack would display a flux reduction. This flux reduction is succeeded by an increase in the salt passage. Chlorine could be eliminated from reverse osmosis feed water using sodium bisulfite or carbon filtration. The carbon present in carbon filters can assist the microorganism growth; and hence the carbon filtration is usually not suggested for reverse osmosis feed water dechlorination except if the organic concentrations are sufficiently higher to warrant its utilization, or if the sodium bisulfite dosage is unreasonably less for precise control.

3.2.13 Other calcium-based compounds

In addition to $CaCO_3$, there are three other calcium-based compounds that would cause reverse osmosis membrane scaling. These calcium-based compounds are calcium fluoride, calcium sulfate, and calcium phosphate. Despite the fact that there is no predetermined feed water guiding principle for the abovementioned calcium compounds, they are worth examining.

- The scaling by calcium fluoride can develop when the fluoride concentration is as low as 0.10 ppm and the calcium concentration is high. Scaling would happen as the IPc surpasses the solubility constant. Also, sodium softening or antiscalants could be employed for controlling the scaling caused by calcium fluoride.
- The calcium phosphate has turned into a standard issue with the rise in municipal wastewater treatment for reuse. The surface waters might likewise consist of phosphate. The calcium phosphate compounds consist of fluoride, chloride, hydroxyl, iron, or/and aluminum. The scaling potential of reverse osmosis membranes with the calcium phosphate compounds is higher and will happen as the IPc surpasses the solubility constant. The aforesaid process can happen at concentrations of ortho-phosphate as low as 0.5 ppm. In order to control phosphate-based scaling, sodium softening or antiscalants along with low pH can be employed.
- Calcium sulfate is considered to be a sparingly soluble salt. Fig. 3.6 presents the scanning electron micrograph ($200 \times$) presenting three-dimensional rosette-like gypsum (calcium sulfate) crystals on a FILMTEC BW30 reverse osmosis membrane (Benecke et al., 2018). Same as strontium and barium sulfate, the possibility to scale with calcium sulfate is higher, as the IPc surpasses 80.0% of the solubility constant. Sodium softening or antiscalants for removing calcium could be utilized for controlling the scale caused by calcium sulfate.

Calcium has additionally been demonstrated to influence the NOM deposition. In a research carried out by Schafer et al. (1998), the team illustrated that NOM as humic substances get deposited preferably on hydrophilic membranes; for example, PA-based membranes. The presence of calcium brought about high flux decrease because of humic acid precipitation. The higher the concentration of calcium, the quicker the decline in flux (Seidel & Elimelech, 2002). Calcium bonds with the acidic functional groups present in NOM bringing about a dense fouling layer on the surface of the membrane. The bridging between NOM molecules deposited is upgraded when there is calcium, prompting extra compactness of the

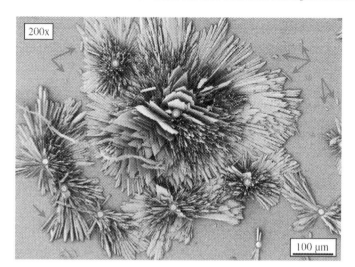

FIGURE 3.6 Scanning electron micrograph (200×) showing three-dimensional rosette-like gypsum (calcium sulfate) crystals on a FILMTEC BW30 reverse osmosis membrane. Source: *Reproduced from Benecke, J., Haas, M., Baur, F., & Ernst, M. (2018). Investigating the development and reproducibility of heterogeneous gypsum scaling on reverse osmosis membranes using real-time membrane surface imaging. Desalination, 428, 161—171.*

fouling layer. The operational process at lower flux, higher shear, and lower transmembrane pressure reduced the NOM deposition on the surface of the membrane and, in this manner the membrane fouling (Schafer et al., 1998; Seidel & Elimelech, 2002). As stated before, calcium could be reduced or removed in reverse osmosis feed water utilizing lime softening or sodium softening.

3.2.14 Magnesium

Magnesium is available in seawater at a concentration of about 1300 ppm. After sodium, the magnesium is considered to be the most commonly seen cation in seawater. Water hardness is due to the presence of magnesium and other alkali earth metals. Water with large amounts of alkali earth ions is known as hard water, and water with low amounts of alkali earth ions is termed as soft water. Magnesium will be mostly present as Mg^{2+}(aq) in aqueous solutions, however also as magnesium hydroxide ($Mg(OH)_2$ (aq)). In seawater, magnesium will be present as magnesium sulfate ($MgSO_4$). The water solubility of $Mg(OH)_2$ is about 12.0 ppm. Other magnesium compounds such as magnesium carbonate (600 ppm) are evidently more water-soluble. $MgSO_4$ can add a bitter taste to water, and it has a water solubility of 309.0 g/L at 10°C (Lenntech, 2005). Different magnesium compounds must be typically removed from water, due to their role in water hardness. This can be accomplished by means of water softening. Magnesium hydroxide is commonly used as a flocculant in water purification. Besides calcium scale, other significant scaling compounds that impact reverse osmosis system operations include magnesium hydroxide and magnesium carbonate.

3.2.15 Hydrocarbons

The most commonly seen organic colloidal foulants are oil-product-based hydrocarbons. These compounds are not present naturally in seawater or in majority brackish waters, and their presence signifies that the saline water intake zone or aquifer is under the impact of artificial sources of pollution, usually discharges or oil leaks from boats, ships or near-shore oil storage tanks in port zones, or discharges from a WWTP, or discharges from storm drains collecting surface runoff from urban regions. Even in extremely small quantities (almost 0.020 ppm), hydrocarbons can lead to increased fouling of reverse osmosis membranes. Consequently, it would be advisable that the total hydrocarbon content in the source feed water to the reverse osmosis unit should be maintained at less than 0.020 ppm always.

It must be noted that apart from dissolved air flotation clarification, which could decrease the total hydrocarbon content to lower than 95.0% of its source water concentration, all of the remaining pretreatment methods offer extremely limited (almost 5.0%–15.0%) hydrocarbon removal. Hydrocarbons lead to permanent fouling of the reverse osmosis membrane, and if saline feed water with hydrocarbon concentrations above 0.1 ppm is introduced to the reverse osmosis membrane, then it would be permanently damaged within 1–2 h after the hydrocarbons approach the reverse osmosis membrane.

3.2.16 Other chemical exposure

The exposure of TFC membranes to various organic compounds can lead to swelling or dissolution of the PSF support layer (RO/NF Polyamide Membrane Feed Water Requirements, 2013).

Suspected chemicals are the following:

- Solvents: Dimethylformamide, N-methyl-2-pyrrolidone (NMP), dimethyl sulfoxide, and so forth.
- Aromatic compounds: Diesel fuel, phenol, toluene, benzene, gasoline
- Others: Strong ethers, esters, aldehydes, ketones

3.3 Ionic strength of feed stream and concentrate stream

The calculations of scaling should be performed for determining whether a sparingly soluble salt causes possible scaling problem in a reverse osmosis system. For analyzing the potential of scaling, a comparison between the ion product of the reviewed salt in the reject stream and the solubility product (K_{sp}) of that particular salt under conditions in the reject stream should be made. In general, scaling control processes will not be required if IPc < Ksp.

The ion product of a particular salt A_cB_d is defined as

$$IP = [A]^c[B]^d \tag{3.11}$$

where [A] and [B] are the molal concentrations of the correspondent ions.

The ion species concentration in the reject stream is typically unknown; however, this could be calculated easily by the multiplication of feed stream concentration with the concentration factor (CF). The CF can be obtained from the recovery r (denoted as a decimal):

$$CF = 1/(1 - r) \tag{3.12}$$

where 100% salt rejection is presumed.

The Ksp is typically expressed in molal concentration and depends on temperature and ionic strength. The reject stream temperature is almost similar to the feed stream temperature. The feed water ionic strength can be calculated as follows:

$$I_f = (1/2)\sum m_i z_i^2 \tag{3.13}$$

where m_i = ion (i) molal concentration (mol/kg) and z_i = ion (i) ionic charge.

When the water analysis is not provided in molar (or molal) concentration, then it can be converted using the below expression:

$$m_i = c_i/(1000MW_i) \tag{3.14}$$

where c_i = ion (i) concentration in ppm and MW_i = ion (i) molecular weight.

After the calculation of feed water ionic strength (I_f) using Eq. (3.13), it is possible to obtain the concentrate stream ionic strength (I_c) by the following expression:

$$I_c = I_f\left(\frac{1}{1-r}\right) \tag{3.15}$$

3.4 Calculations

3.4.1 Example 3.1

Find the concentrate stream ionic strength (I_c), with an 80% recovery. Analysis of feed water

Sl. no.	Ion	Concentration (mg/L)	mol/L	mol/kg
1	Mg^{2+}	61	2.51	$\times 10^{-3}$
2	Ca^{2+}	200	5.0	$\times 10^{-3}$
3	Na^{2+}	388	16.9	$\times 10^{-3}$
4	SO_4^{2-}	480	5.0	$\times 10^{-3}$
5	HCO_3^-	244	4.0	$\times 10^{-3}$
6	Cl^-	635	17.9	$\times 10^{-3}$

Solution:

The feed water ionic strength (I_c) can be calculated as follows:

$$I_f = \frac{1}{2}\left\{4\left([Mg^{2+}] + [Ca^{2+}] + [SO_4^{2-}]\right) + \left([HCO_3^-] + [Na^+] + [Cl^-]\right)\right\}$$

$$I_f = \frac{1}{2}\left\{4\left[(2.51 + 5.0 + 5.0)\times 10^{-3}\right] + \left[(4.0 + 16.9 + 17.9)\times 10^{-3}\right]\right\}$$

$$I_f = 0.0444$$

After the calculation of feed water ionic strength (I_f), we can obtain the concentrate stream ionic strength (I_c).

For a recovery of 80%, $I_c = 0.0444 \left(\frac{1}{1-0.80} \right) = 0.222$.

3.4.2 Example 3.2

In an SDI test, if $T_i = 1$ min and $T_{15} = 4$ min, then calculate the SDI value.
Solution:

$$\text{SDI} = \frac{P_{30}}{T_t} = 100 \times \frac{\left(1 - \frac{T_i}{T_f} \right)}{T_t}$$

T_i = time required to collect the first 500 mL filtrate = 1 min, T_t = total elapsed test time = 15 min, T_f = time needed for collecting final 500 mL of filtrate = 4 min

$$\text{SDI} = \frac{P_{30}}{T_t} = 100 \times \frac{\left(1 - \frac{1}{4} \right)}{15} = \frac{75}{15} = 5$$

3.4.3 Example 3.3

Calculate the rejection percentage of reverse osmosis units after the consumer had operated the system for 30 min and observed that the hardness of the reverse osmosis water is 6 mg/L. Assume that the tap water hardness is 82 mg/L.

Solution:

Calculate the rejection percentage from

$$Rej = 1 - \left(\frac{C_p}{C_f} \right) = 92.7\%.$$

Therefore membrane hardness rejection is 92.7%. Rejection rates less than 95.0% might signify that the membrane must be replaced.

3.5 Case studies

3.5.1 Case study-1: Al-Khafji City Saudi Arabia SWRO plant

This study defines the evaluation of seawater quality at a location on the Gulf coast of Saudi Arabia for assessing its appropriateness for a seawater reverse osmosis plant. The site is situated on the Al-Khafji coast and this place is near to the site of oil and gas production platforms, marine oily water discharges, decommissioned pipelines, submerged pipelines, and associated equipment and facilities. The Arabian Gulf is an exceptional as well as profoundly used water body and its physical configuration poses distinct challenges (Saeed et al., 2002). The Gulf is a shallow and semienclosed water body with slight input from precipitation or inflow from rivers. Gulf water has a total dissolved solid concentration of almost

43,800 ppm or greater, normally greater than total dissolved solid concentration of normal seawater (almost 35,000 ppm). The average ionic ratio of Gulf seawater to that of normal seawater is around 1.25:1.0, and accordingly, the ionic concentration is correspondingly greater in Gulf seawater compared with normal seawater. The water temperature during summer could go beyond 36°C, whereas during winter, it can be less than 15°C (Saeed et al., 2002). Table 3.5 lists the methodology used for the quality analysis of seawater in the work with significant notes on the analytical approaches employed (Saeed et al., 2019).

TABLE 3.5 Different techniques and methods used in seawater analyses.

Sl. no.	Parameters	Techniques	Methods
1	Alkalinity (ppm)	Titration	Standard method 2320 B
2	Hardness (ppm)	Titrimetric, EDTA	Standard method 2340 C
3	TDS (ppm)	Drying at 180°C to a constant weight	Standard method 2540 C
4	SDI_{15} units	0.45 micron pore size filter with silt density index manifold	ASTM-D4189−95
5	Total suspended solids (TSS)	Filtration and drying	Standard method 2540 D
6	Turbidity (NTU)	Nephelometric technique	Standard method 2130 B
7	Conductivity (μS/cm)	Conductivity probe	Standard method 2510 B
8	Salinity (parts per thousand)	Temperature-compensated refractometer	—
9	Temperature (°C)	Mercury thermometer	—
10	pH	Calibrated pH meter	—
11	Sodium (mg/L)	Inductively coupled plasma (ICP)-atomic absorption emission spectroscopy (AES)	Standard method 3120 A
12	Chloride (Cl^-; mg/L)	Potentiometric titration using silver nitrate	Standard method 4500-Cl D
13	Calcium (Ca; mg/L)	ICP-AES	Standard method 3120 A
14	Potassium (K; mg/L)	ICP-AES	Standard method 3120 A
15	Strontium (Sr; μg/L)	ICP-AES	Standard method 3120 A
16	Iron (Fe; μg/L)	ICP-AES	Standard method 3120 A

(*Continued*)

TABLE 3.5 (Continued)

Sl. no.	Parameters	Techniques	Methods
17	Cobalt (Co; μg/L)	ICP-AES	Standard method 3120 A
18	Zinc (Zn; μg/L)	ICP-AES	Standard method 3120 A
19	Vanadium (V; μg/L)	ICP-AES	Standard method 3120 A
20	Magnesium (Mg; mg/L)	ICP-AES	Standard method 3120 A
21	Barium (Ba; μg/L)	ICP-AES	Standard method 3120 A
22	Boron (B; μg/L)	ICP-AES	Standard method 3120 A
23	Copper (Cu; μg/L)	ICP-AES	Standard method 3120 A
24	Manganese (Mn; μg/L)	ICP-AES	Standard method 3120 A
25	Nickel (Ni; μg/L)	ICP-AES	Standard method 3120 A
26	Mercury (Hg; μg/L)	Atomic absorption-cold vapor generation analyzer technique	Standard method 3122 B
27	Cadmium (Cd; μg/L)	Atomic absorption (AA) -graphite tube atomizer (GTA)	Standard method 3111 B
28	Chromium (Cr; μg/L)	AA-GTA	Standard method 3111 B
29	Lead (Pb; μg/L)	AA-GTA	Standard method 3111 B
30	Arsenic (As; μg/L)	AA-vapor generation analyzer (VGA)	Standard method 3114 B
31	Selenium (Se; μg/L)	AA-VGA	Standard method 3114 B
32	Fluoride (F$^-$; μg/L)	Ion chromatography	Standard method 4110 A
33	TOC (mg/L)	Total organic carbon in water	Standard method SM 5310 A
34	Phosphate (PO$_4^{3-}$; μ/L)	—	Standard method 4110 A
35	Nitrate (NO^{3-}; μ/L)	—	Standard method 4110 A

(Continued)

TABLE 3.5   (Continued)

Sl. no.	Parameters	Techniques	Methods
36	Sulfate (SO_4^{2-}; mg/L)	HACH	Standard method 4110 A
37	Biochemical oxygen demand (BOD_5, mg/L)	Initial and final dissolved oxygen	Winkler
38	Total viable bacterial count (colony-forming units/mL)	Heterotrophic plate count	Standard method 9215 B
39	Fecal coliform	Fecal coliform technique	Standard method 9221 E

Adapted from Saeed, M.O., Al-Nomazi, M.A., & Al-Amoudi, A.S. (2019). Evaluating suitability of source water for a proposed SWRO plant location. Heliyon, *5(1), e01119.*

Various biological and physicochemical parameters such as BOD, dissolved oxygen (DO), dissolved carbohydrates and proteins, TOC, TSS, turbidity, SDI, TDS, salinity, pH, temperature, conductivity, total hardness, total alkalinity, bacterial count, chlorophyll-a, major ions, in addition to trace metals were evaluated. Apart from TSS, physicochemical as well as chemical variables analyzed in the work showed concentrations typical of the Arabian Gulf water and seawater in a larger sense. Average TSSs values were marginally greater relative to those reported for Gulf coastal waters, and more prominently they were exceedingly variable. This variability might lead to occurrences of filtration issues for the seawater RO plant. From the study, it was recommended that plant must search for in-depth seawater distance from shore as the site of its intake. The work also offered recommendations for treatment possibilities for assuring a more effective operation of the reverse osmosis plant.

3.5.2 Case study-2: Data analysis from a reverse osmosis plant in the Gulf

In this study, the turbidity analysis of the Persian Gulf seawater (reverse osmosis plant intake) was carried out. Water samples were taken from the inlet of the reverse osmosis plant during three consecutive years and seven different days during the same year. The operation system consisted of a seawater intake unit, a pretreatment unit, a reverse osmosis unit, and a mineralization unit, as shown in Fig. 3.7.

The water samples were collected from the reverse osmosis plant on a specific day (temperature almost 25°C and humidity 70%), and the results for the conductivity, DO, TDS, pH, turbidity, TSS, chlorophyll estimations, cation concentrations, and anion concentrations of the seawater intake are given in Table 3.6.

Fig. 3.8 presents the variation of turbidity of seawater during each step in reverse osmosis process. The multiprobe results showed that the turbidity of the seawater intake was 22.95 NTU. At the DAF unit, a remarkable decrease in the turbidity was observed from 22.95 NTU to 6.7 NTU. Here, an almost 71% reduction in turbidity was observed. These results are in agreement with the results obtained by Khiadani et al. (2014). The reverse osmosis membrane also performs as an ultrafine filter eliminating 99% of colloidal and suspended solids, bacteria,

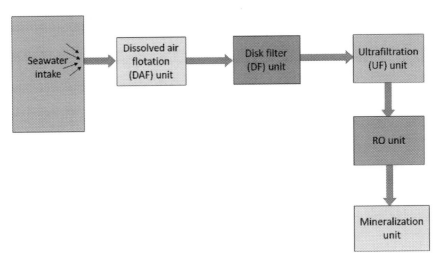

FIGURE 3.7   Schematic representation of the desalination process in this study.

TABLE 3.6   Analysis of different properties of seawater intake.

Sl. no.	Property	Value	Unit
1	Conductivity	64,889	μS
2	Dissolved oxygen	96.6	%
3	Total dissolved solids (TDS)	42,177	ppm
4	pH	8	
5	Turbidity	22.95	NTU
6	Total suspended solids (TSS)	77.8	ppm
7	Chlorophyll	1.2	ppm
8	TOC	30.03 ± 0.97	ppm
9	TIC	26.89 ± 0.85	ppm
10	Chloride	55,800	ppm
11	Sodium	30,800	ppm
12	Sulfate	8000	ppm
13	Magnesium	3701	ppm
14	Calcium	1265	ppm
15	Potassium	1104	ppm
16	Bromide	447	ppm

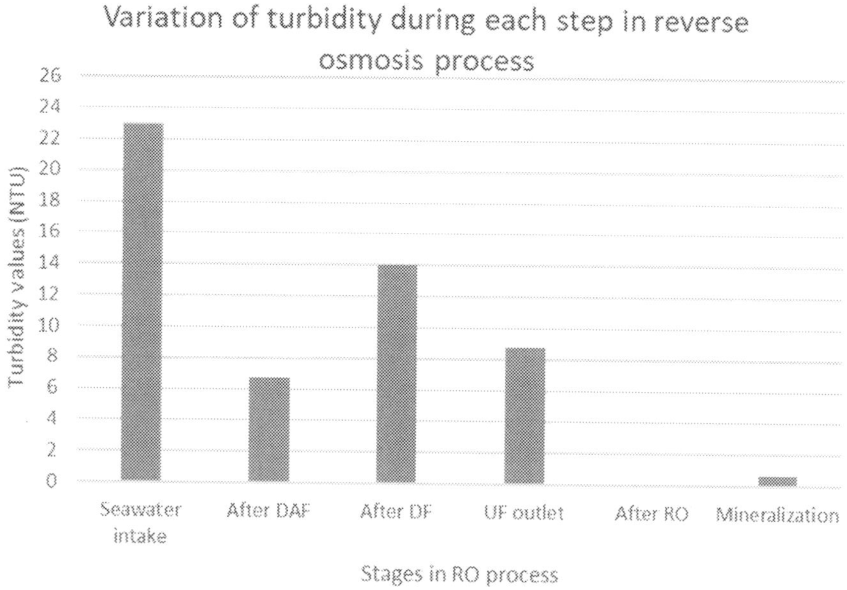

FIGURE 3.8   Variation of turbidity of seawater during each step in reverse osmosis process.

as well as organic molecules having molecular weights more than 200 Da. It was observed that the turbidity reached an extremely low value of about 0.06 NTU after the reverse osmosis process. Thus it was confirmed that the turbidity of seawater can be remarkably decreased by the above-stated processes in the reverse osmosis plant, approximately 99.7% reduction. After the mineralization process, the turbidity increased to 0.67 NTU. The remineralization process using the limestone chips will lead to an increase in turbidity (e.g., to 0.67 NTU), however must be lesser than 1 NTU in any case. The results in this study are intended to contribute to the growing body of ecological information on the unique conditions of the Gulf compared with the regions elsewhere in the world where identical researches have been conducted.

References

Abushaban, A., Mangal, M. N., Salinas-Rodriguez, S. G., Nnebuo, C., Mondal, S., Goueli, S. A., & Kennedy, M. D. (2017). Direct measurement of ATP in seawater and application of ATP to monitor bacterial growth potential in SWRO pre-treatment systems. *Desalination and Water Treatment*, *99*, 91−101.

Affam, A. C., & Ezechi, E. H. (Eds.), (2019). *Handbook of research on resource management for pollution and waste treatment*. IGI Global.

Barnhart, E.L., & Amamath, K.R. (1993). The efficacy of color removal techniques in textile wastewater. Research Project, 332941.

Benecke, J., Haas, M., Baur, F., & Ernst, M. (2018). Investigating the development and reproducibility of heterogeneous gypsum scaling on reverse osmosis membranes using real-time membrane surface imaging. *Desalination*, *428*, 161−171.

Boerlage, S. F., Kennedy, M. D., Bremere, I., Witkamp, G. J., Van der Hoek, J. P., & Schippers, J. C. (2002). The scaling potential of barium sulphate in reverse osmosis systems. *Journal of Membrane Science*, *197*(1−2), 251−268.

Bundschuh, J., & Hoinkis, J. (2012). Addressing freshwater shortage with renewable energies. In J. Bundschuh, & J. Hoinkis (Eds.), *Renewable energy applications for freshwater production (sustainable energy developments)* (pp. 1–23). CRC INC.

Cob, S. S., Hofs, B., Maffezzoni, C., Adamus, J., Siegers, W. G., Cornelissen, E. R., & Witkamp, G. J. (2014). Silica removal to prevent silica scaling in reverse osmosis membranes. *Desalination, 344*, 137–143.

DuPont Water Solution's FilmTec™ RO technical manual; Version 3; Form No. 45-D01504-en, Rev. 3; April 2020.

El-Dessouky, H. T., & Ettouney, H. M. (2002). *Fundamentals of salt water desalination.* Elsevier.

Farhat, S., Kamel, F., Jedoui, Y., & Kallel, M. (2012). The relation between the RO fouling membrane and the feed water quality and the pretreatment in Djerba Island plant. *Desalination, 286*, 412–416.

FilmTec™ Seawater RO Elements for Marine Systems, 2020. https://www.dupont.com/content/dam/dupont/amer/us/en/water-solutions/public/documents/en/45-D01519-en.pdf [Accessed on 7/7/2020].

FLUID SYSTEMS® TFC® - SS 8, 2004. ELEMENTS High Rejection, Seawater, RO Elements https://www.lenntech.com/Data-sheets/Kock-2822-TFC-SS-8-L.pdf [Accessed on 7/7/2020].

Gerba, C. P., & Pepper, I. L. (2019). *Municipal wastewater treatment. In Environmental and pollution science* (pp. 393–418). Academic Press.

HR5255PIElement Configuration: Hollow Fiber, 2006. https://www.toyobo-global.com/seihin/ro/spec-HR5255PI.htm [Accessed on 7/7/2020].

Hydranautics; Technical Application Bulletin No. 111 https://www.lenntech.com/Data-sheets/Hydranautics-TAB111-L.pdf [Accessed on 7/7/2020].

Hydranautics SWC4+ Membrane, 2005. https://www.pureaqua.com/hydranautics-swc4-plus-membrane/ [Accessed on 7/7/2020].

Khiadani, M., Kolivand, R., Ahooghalandari, M., & Mohajer, M. (2014). Removal of turbidity from water by dissolved air flotation and conventional sedimentation systems using poly aluminum chloride as coagulant. *Desalination and Water Treatment, 52*(4–6), 985–989.

Kreiwall, D., Harding, R., Maisch, E., & Schantz, L. (1996). The impact of aluminum residual on transmission main capacity. *Public Works, 127*(13), 28–30.

Kucera, J. (2015). *Water Quality Guidelines. Reverse osmosis: Industrial processes and applications* (2nd Edition, pp. 133–156). Wiley.

Lagref, J. J., & Haci, O. (2014). Membrane filtration: Managing aluminum in membrane filtration. *Filtration & Separation, 51*(4), 26–28.

Lee, Y. G., Kim, S., Shin, J., Rho, H., Lee, Y., Kim, Y. M., & Chon, K. (2020). Fouling behavior of marine organic matter in reverse osmosis membranes of a real-scale seawater desalination plant in South Korea. *Desalination, 485*, 114305.

Lisitsin, D., Hasson, D., & Semiat, R. (2005). Critical flux detection in a silica scaling RO system. *Desalination, 186* (1–3), 311–318.

Lu, K. G., & Huang, H. (2019). Dependence of initial silica scaling on the surface physicochemical properties of reverse osmosis membranes during bench-scale brackish water desalination. *Water Research, 150*, 358–367.

Magnesium (Mg) and water, 2005. https://www.lenntech.com/periodic/water/magnesium/magnesium-and-water.htm [Accessed on 7/7/2020].

Matin, A., Khan, Z., Gleason, K. K., Khaled, M., Zaidi, S. M. J., Khalil, A., & Yang, R. (2014). Surface-modified reverse osmosis membranes applying a copolymer film to reduce adhesion of bacteria as a strategy for biofouling control. *Separation and Purification Technology, 124*, 117–123.

Matin, A., Khan, Z., Zaidi, S. M. J., & Boyce, M. C. (2011). Biofouling in reverse osmosis membranes for seawater desalination: phenomena and prevention. *Desalination, 281*, 1–16.

Melliti, E., Touati, K., Abidi, H., & Elfil, H. (2019). Iron fouling prevention and membrane cleaning during reverse osmosis process. *International Journal of Environmental Science and Technology, 16*(7), 3809–3818.

Micale, G., Cipollina, A., & Rizzuti, L. (2009). *Seawater desalination for freshwater production. Seawater desalination* (pp. 1–15). Berlin, Heidelberg: Springer.

Nguyen, T., Roddick, F. A., & Fan, L. (2012). Biofouling of water treatment membranes: A review of the underlying causes, monitoring techniques and control measures. *Membranes, 2*(4), 804–840.

Ning, R. Y. (2015). *Reverse osmosis chemistry—Basics, barriers and breakthroughs. Desalination updates.* IntechOpen.

Norman, J.E., Hoang, T., & Leslie, G.L. (1999). Diagnosis and remediation of silicate scale fouling in microfiltration membranes: A case study. In Proceedings of AWWA Membrane Technology Conf., Long Beach, CA, USA.

Park, Y. M., Yeon, K. M., & Park, C. H. (2020). Silica treatment technologies in reverse osmosis for industrial desalination: A review. *Environmental Engineering Research, 25*(6), 819–829.

RO/NF Polyamide Membrane Feed Water Requirements, Hydranautics Technical Application Bulletin, TAB 116, October, 2013.

Saeed, M. O., Al-Nomazi, M. A., & Al-Amoudi, A. S. (2019). Evaluating suitability of source water for a proposed SWRO plant location. *Heliyon, 5*(1), e01119.

Saeed, M.O., Al-Thobaiti, E.S., Al-Daili, M.A., & Al-Hamza, A.A. (2002, December). Effect of the intake bay design on feed water quality for the Jubail desalination and power plants. In Proceedings of the 6th Saudi engineering conference, King Fahd University of Petroleum and Minerals, Dhahran, Saudi Arabia (pp. 3–16).

Saleem, H., Trabzon, L., Kilic, A., & Zaidi, S. J. (2020). Recent advances in nanofibrous membranes: Production and applications in water treatment and desalination. *Desalination, 476*, 114178.

Saleem, H., & Zaidi, S. J. (2020a). Developments in the application of nanomaterials for water treatment and their impact on the environment. *Nanomaterials, 10*(9), 1764.

Saleem, H., & Zaidi, S.J. (2020b). Innovative nanostructured membranes for reverse osmosis water desalination, https://doi.org/10.29117/quarfe.2020.0023.

Saleem, H., & Zaidi, S. J. (2020c). Nanoparticles in reverse osmosis membranes for desalination: A state of the art review. *Desalination, 475*, 114171.

Saleh, T. A., & Gupta, V. K. (2016). *Membrane fouling and strategies for cleaning and fouling control. Nanomaterial and polymer membranes* (pp. 25–53). Amsterdam: Elsevier.

Schafer, A. I., Fane, A. G., & Waite, T. D. (1998). Nanofiltration of natural organic matter: Removal, fouling, and the influence of multivalent ions. *Desalination, 118*.

Seidel, A., & Elimelech, M. (2002). Coupling between chemical and physical interactions in natural organic matter (NOM) fouling of nanofiltration membranes: Implications for fouling control. *Journal of Membrane Science, 203*.

Sheikholeslami, R., Al-Mutaz, I. S., Koo, T., & Young, A. (2001). Pretreatment and the effect of cations and anions on prevention of silica fouling. *Desalination, 139*(1–3), 83–95.

Standard Methods for the Examination of Water and Wastewater, 1998. 20th edition, published jointly by the American Public Health Association, the American Water Works Association, and the Water Pollution Control Federation.

Toray TM820C-370 Membrane, 2012. https://www.pureaqua.com/toray-tm820c-370-membrane/ [Accessed on 7/7/2020].

Voutchkov, N. (2017). Membrane foulants and saline water pretreatment. *Pretreatment for Reverse Osmosis Desalination,* 11–41. Available from https://doi.org/10.1016/b978-0-12-809953-7.00002-4.

Vrouwenvelder, J. S., & van der Kooij, D. (2001). Prediction and prevention of biofouling of NF and RO membranes. *Desalination, 139,* 65.

World Health Organization. (2011). Safe drinking-water from desalination (No. WHO/HSE/WSH/11.03). World Health Organization.

Yadav, S., Saleem, H., Ibrar, I., Naji, O., Hawari, A. A., Alanezi, A. A., . . . Zhou, J. (2020). Recent developments in forward osmosis membranes using carbon-based nanomaterials. *Desalination, 482,* 114375.

Zaidi, S. J., Fadhillah, F., Saleem, H., Hawari, A., & Benamor, A. (2019). Organically modified nanoclay filled thin-film nanocomposite membranes for reverse osmosis application. *Materials, 12*(22), 3803.

CHAPTER

4

Transport Models, Membrane Materials, and Basic Flow Patterns

4.1 Operating membrane systems

Membrane systems could be managed either through cross-flow filtration or dead-end filtration. The diagrammatic representation of cross-flow filtration and dead-end filtration are shown in Fig. 4.1.

4.1.1 Dead-end filtration

During the dead-end filtration, the total water that reaches the membrane surface will be pushed through the membrane. Subsequently, certain components and solids will be retained on the membrane, whereas water flows through the membrane. This process will be dependent on the membrane pore size. Eventually, the flow of water will experience an increased resistance to flow through the membrane (Cadotte et al., 1980). As the pressure of feed water is constant, the increased resistance will lead to a decrease in flux. After a certain time, the flux will decline and the membrane will require proper cleaning. The benefit of dead-end filtration is that no waste stream will be generated. Its drawback is that the impurities will accumulate on the membrane surface on the feed side, thereby forming a cake layer that will reduce the performance and will need regular filter changes. This type of filtration is used in ultrafiltration and microfiltration.

In the dead-end-type filtration, the loss of energy is lower relative to the cross-flow filtration. The pressure that is required for pushing water through a membrane is termed as transmembrane pressure (TMP). This TMP is the difference between the average feed pressure and the permeate pressure or the pressure gradient of the membrane. The feed pressure is regularly measured at the first point of a membrane module. However, this pressure will not be equal to the average feed pressure, due to the fact the flow through the membrane will lead to hydraulic pressure losses. During the membrane cleaning process, the impurities will be removed physically or chemically. During this cleaning process, the module will be out of order temporarily. So the dead-end filtration is considered to be a discontinuous process. The length of period that a module operates filtration is

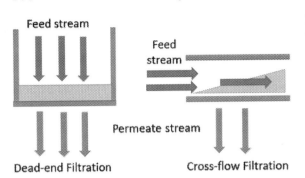

FIGURE 4.1 Representation of dead-end filtration and cross-flow filtration.

termed as filtration time, and the length of period for cleaning a module is termed as cleaning time. In practice, one always prefers to make filtration time last longer with the shortest possible cleaning time.

4.1.2 Cross-flow filtration

In a cross-flow filtration, a continuous turbulent flow along the surface of the membrane will prevent the buildup of matter on the membrane surface. The feed water that flows across the membrane will be at a higher pressure providing the driving force for the filtration process and has a higher flow speed for developing turbulent conditions. This process is termed as cross-flow due to the fact that the filtration flow will be in a perpendicular direction to the feed flow direction. So the permeate passes across the membrane in cross-flow. This type of filtration is an exceptional method for filtering liquids having high concentrations of filterable matter.

The benefits of cross-flow filtration are the reasonable cost of filtration modules, particularly for spiral wound membrane (SWM) modules, and the choice of membranes and module configurations. The disadvantages of this type of filtration are that its permeate flux will be limited while treating highly charged and viscous fluids and for attaining increased solid concentrations, of greater than 50% of dry solids. Also, cross-flow filtration needs high-pressure pumps for generating increased fluid velocities in membrane modules.

4.2 Transport models

The objective of a transport model is to mathematically relate the performance (usually flux of both solute and solvent) to operating conditions (normally pressure and concentration driving forces) (Luis, 2018). The main purpose of the transport model is the prediction of membrane performance under specific conditions (Sundaramoorthy et al., 2011). Various models have been developed for describing the mass transport across the reverse osmosis (RO) membranes, and they are classified into the following three types (Mujtaba et al., 2017):

1. Homogenous or nonporous membrane models (like solution-diffusion, extended solution diffusion models, as well as solution-diffusion imperfection models).
2. Porous models (like surface force-pore flow preferential sorption-capillary flow, and finely porous models).
3. Irreversible thermodynamic models (like Spiegler—Kedem and Kedem—Katchalsky models).

The main features of these models are briefly described below.

4.2.1 Nonporous models

4.2.1.1 Solution-diffusion nonporous model

Lonsdale et al. (1965) originally developed the solution—diffusion transport model in the year 1965. In this model, the membrane is assumed to be nonporous with no imperfections. Here, the transport across the membrane takes place as the molecule of interest gets dissolved in the membrane and subsequently gets diffused across the membrane as shown in Fig. 4.2. The main advantage of this model is its simplicity. In this model, the solvent transport and solute transport are not dependent on each other, as demonstrated in the below two equations. The solvent flux across the membrane is linearly proportional to the actual pressure difference across the membrane.

$$J_w = K_w(\Delta P - \Delta \pi) \tag{4.1}$$

in which J_w = solvent flux, K_w = solvent permeability coefficient, $\Delta \pi$ = osmotic pressure difference of the solution, and ΔP = applied pressure driving force (hydrostatic pressure difference).

The solute flux across the membrane is proportional to the effective concentration difference of the solute across the membrane:

$$J_s = K_s(C_{A2} - C_{A3}) \tag{4.2}$$

in which J_s = solute flux, K_s = solute permeability coefficient, C_{A2} = molar concentration of solute at the boundary layer, and C_{A3} = molar concentration of solute in the permeate.

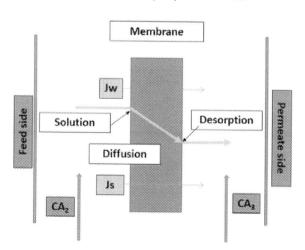

FIGURE 4.2   Solution diffusion model.

4.2.1.2 Solution-diffusion imperfection model

This model was developed by Sherwood et al. (1967). The solution-diffusion imperfection model is considered to be a modification of the solution–diffusion model, which includes pore flow as well as solute diffusion and solvent diffusion across the membrane. The aforesaid model recognizes that imperfections (pores) exist on the membrane surface through which solute and solvent can flow.

The solvent flux is given by the following equation:

$$N_w = J_w + K_3 \Delta P C_w = K_w (\Delta P - \Delta \pi) + K_3 \Delta P C_w \tag{4.3}$$

in which N_w = total solvent flux, C_w = solvent concentration on the membrane feed side, and K_3 = coupling coefficient.

The solute flux is given by the following equation:

$$N_s = J_s + K_3 \Delta P C_R = K_s (C_{A2} - C_{A3}) + K_3 \Delta P C_R \tag{4.4}$$

in which N_s = total solute flux, C_R = solute concentration on the membrane feed side, and K_3 = coupling coefficient.

4.2.2 Porous models

4.2.2.1 Preferential sorption–capillary flow porous model

In the year 1970, Sourirajan developed the preferential sorption–capillary flow (PSCF) model, and it considers that the separation is because of surface phenomena and fluid transport across the pores. Here, the membrane is regarded as microporous, and the barrier layer possesses chemical properties (preferential repulsion for solutes and preferential sorption for the solvent). Accordingly, a layer of a virtually pure solvent is preferentially sorbed in the pores and on the surface of the membrane. The solvent is forced across the membrane capillary pores under the pressure from that layer.

Solvent flux will be the same as given in Eq. (4.1).

The solute flux is given by the following equation:

$$N_s = (D_{AM} K_s / T)(C_{A2} - C_{A3}) \tag{4.5}$$

in which N_s = total solute flux, D_{AM} = solute diffusivity in membrane, K_s = solute permeability coefficient, T = effective membrane thickness, C_{A2} = molar concentration of solute at the boundary layer, and C_{A3} = molar concentration of solute in the permeate.

4.2.2.2 Surface force-pore flow model

This model was developed by Sourirajan and Matsuura in the year 1981, and this model is considered to be a modification of the PSCF model that permits the membrane characterization and membrane specification as a function of pore size distribution, together with a quantitative measure of the surface forces which develop between solvent–solute and the membrane wall within the transportation corridor. As per the surface force-pore flow (SFPF) model, the pore diameter has a strong bearing upon the separation factor and the mean separation factor could be determined from the pore size distribution.

4.2.3 Irreversible thermodynamic models

The solute and solvent transport equations in a reverse osmosis membrane are described by the irreversible thermodynamics. A general description of the reverse osmosis process is shown in Fig. 4.3. There is a recent work by Mondal and De (2020), which is dedicated toward a proper theoretical background to the physics behind the reverse osmosis membrane separation process, illustrating the modeling challenges and mathematical theories that can lead to an accurate prediction of the filtration rate and the product quality.

4.2.3.1 Kedem and Katchalsky model

The Kedem and Katcalsky model was developed by Kedem and Katchalsky in the year 1958 on the basis of irreversible thermodynamics. In the model, the membrane is considered as a black box. This model adopts that the membrane is close to equilibrium, and flow across the membrane is very slow. Here, the main principle is that there is a coupling between the fluxes of the components which transfer across the membrane. A new parameter named Staverman reflection coefficient was introduced, which links both the solvent flux and solute flux. The aforesaid model describes the solute flux and solvent flux in terms of osmotic and pressure differences for solvent flux, and average concentration and osmotic variation for the solute.

The volume flow (J_v) is given by the following equation:

$$J_v = L_p \Delta P - L_p \sigma \Delta \pi \tag{4.6}$$

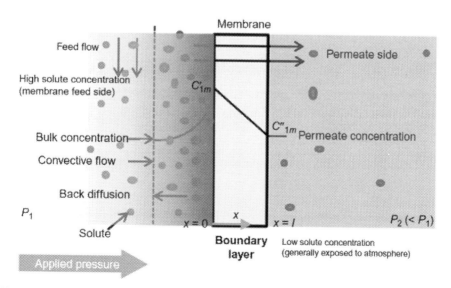

FIGURE 4.3 Thermodynamic principle of reverse osmosis. *Source: Reproduced from Mondal, S., & De, S. (2020). Reverse osmosis modeling, simulation, and optimization. In Current trends and future developments on (bio-)membranes (pp. 187–206). Elsevier.*

The solute flow (J_s) is given by the following equation:

$$J_s = \omega\Delta\pi - \mathfrak{c}(1 - \sigma)J_v \qquad (4.7)$$

in which $\mathfrak{c}$ = mean concentration, $\mathfrak{c} = 0.5(c1 + c2)$, L_p = coefficient of filtration, σ = coefficient of reflection, and ω = coefficient of permeation.

4.2.3.2 Spiegler and Kedem model

This model was developed by Spiegler and Kedem in the year 1966, by modifying the Kedem and Katcalsky model. It was assumed that the solute flux is a consolidation of convection and diffusion. Further, this model considered the convective coupling aspects of the transport of the solute. According to this model, the average concentration does not characterize the solvent flux and the solute flux, and hence, the equations for flux are developed with respect to local coefficients and subsequently integrated for the complete membrane layer. This Spiegler and Kedem model has been used to predict salt and organic compound transport across the membrane in single and binary solute systems. In addition to the above, it has also been used in the area of multicomponent separation description, and hence, it is very beneficial in reverse osmosis industrial processes.

The solute flux is given by the following equation:

$$J_s = P_s\Delta x \frac{dC}{dx} + (1 - \sigma)J_v C \qquad (4.8)$$

in which J_s = solute flux, C = solute concentration, J_v = permeate flux, P_s = solute permeability, σ = reflection coefficient, Δx = membrane thickness, and $\frac{dC}{dx}$ = concentration gradient.

Permeate flux will be given by the following equation:

$$J_v = L_p(\Delta P - \sigma\Delta\pi) \qquad (4.9)$$

in which L_p = membrane permeability, C = solute concentration, and $\Delta\pi$ = osmotic pressure difference.

Out of the above-discussed models, the most commonly used model to describe the transport in reverse osmosis membranes is the solution–diffusion mechanism model developed by Lonsdale.

4.3 Membrane materials

The reverse osmosis membrane flux, and rejection, mainly depends on the chemical composition and structure of the membrane. Therefore, the selection of membrane and membrane material plays a significant role in the efficiency improvement of reverse osmosis membranes. Generally, the flux will be inversely proportional to the membrane thickness. Thus, the flux could be increased by reducing the membrane thickness. The majority of the commercially available reverse osmosis membranes have an asymmetric structure with a top-thin selective layer and porous sublayer. This top thin layer will regulate the resistance towards water transport and performs the separation.

An ideal reverse osmosis membrane must have the following features:

- high permeate flux and salt rejection
- low fouling and economical
- high mechanical stability
- resistant to oxidation and biological degradation
- good thermal stability

The efficiency of a reverse osmosis process is directly dependent on the membrane material properties (Li, 2007). More precisely, the flux and rejection properties of a reverse osmosis system are determined by the chemical nature of the membrane polymer as well as membrane microstructure. Preferably, reverse osmosis membranes must contribute to superior rejection and increased flux, along with durability and increased strength. The resistance to flow through a membrane will be inversely proportional to the membrane thickness. For achieving significant water flux, the active membrane layer should be very thin, which in reverse osmosis membranes range from about 0.1 to 2 μm.

Cellulose acetate (CA) and aromatic polyamide (PA) are the two most commonly used materials for commercial reverse osmosis membranes (Byrne, 2002). The aromatic PA membrane is also commercially known as thin-film composite (TFC) membrane. The membranes developed from these polymers vary in several aspects, including their structure, physical properties, and performance. Cellulose triacetate (CTA) membrane is a paper by-product membrane attached to a synthetic layer and needs a limited quantity of chlorine in the source of water to avoid the formation of bacteria on it. The TFC membrane is made of synthetic material, and require chlorine to be separated before the water reaches the membrane. This is due to the fact that chlorine leads to permanent damage to the thin-film membrane. These membranes are very strong and can be used at an increased temperature of 45°C, relative to CA-based membranes (35°C). Figs. 4.4 and 4.5 are the chemical structure of CA and PA, respectively (Singh et al., 2020).

In the following section, the different membrane materials are briefly discussed.

4.3.1 Cellulose acetate membrane

The unique CA membrane, reported in the late 1950s by Loeb and Sourirajan, was prepared from cellulose diacetate polymer (Strathmann et al., 1971). The present CA membrane is typically prepared from a combination of CTA and diacetate. Fig. 4.6

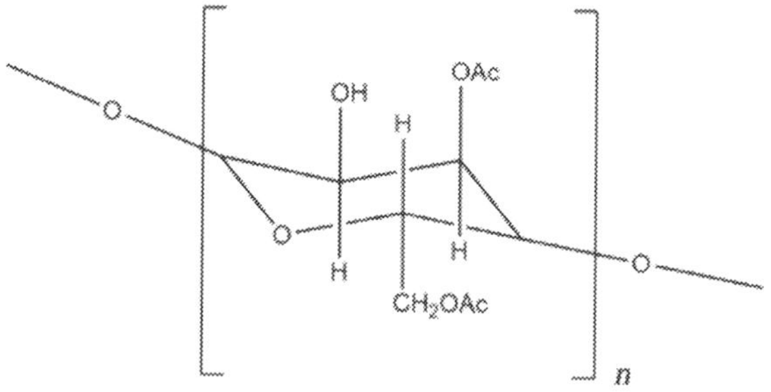

FIGURE 4.4 Cellulose acetate chemical structure. *Source: Reproduced from reference Singh, P.S., Ray, P., & Ismail, A.F. (2020). Synthetic polymer-based membranes for desalination. In Synthetic polymeric membranes for advanced water treatment, gas separation, and energy sustainability (pp. 23–38). Elsevier.*

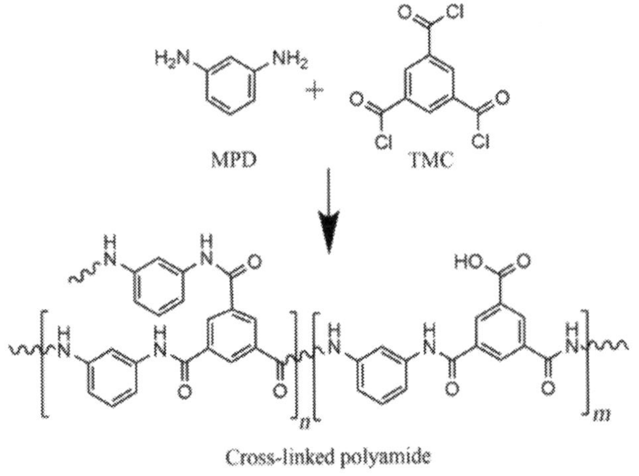

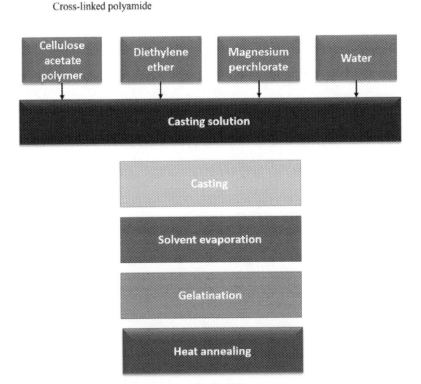

FIGURE 4.5 Chemical structure of polyamide formed by interfacial polymerization between metaphenylenediamine (MPD) and trimesoyl chloride (TMC). *Source: Reproduced from reference Singh, P.S., Ray, P., & Ismail, A.F. (2020). Synthetic polymer-based membranes for desalination. In Synthetic polymeric membranes for advanced water treatment, gas separation, and energy sustainability (pp. 23–38). Elsevier.*

FIGURE 4.6 Fabrication sequence of a cellulose acetate membrane. *Source: Courtesy: Mark Wilf. Adapted from Wilf, M. & Awerbuch, L. (2007). The guidebook to membrane desalination technology. Italy: Balaban Desalination Publications.*

presents the fabrication sequence of a CA membrane (Wilf & Awerbuch, 2007). This membrane is prepared by casting a thin-film of an acetone-based solution of CA polymer with swelling additives from a trough on top of nonwoven polyester fabric (Fig. 4.7). Two added processes, a cold bath succeeded by high-temperature annealing, finish the casting

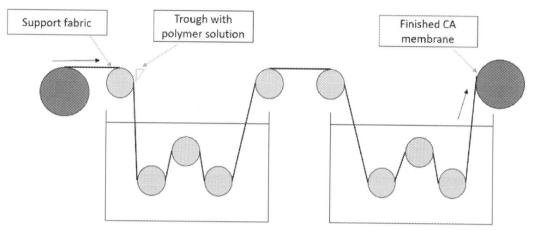

FIGURE 4.7 Fabrication process of a cellulose acetate (CA) membrane. *Source: Courtesy: Mark Wilf. Adapted from Wilf, M. & Awerbuch, L. (2007). The guidebook to membrane desalination technology. Italy: Balaban Desalination Publications.*

process. In the course of casting, the solvent will be moderately detached by evaporation. After the casting process, the membrane will be submerged into a cool water bath that separates the residual acetone and additional leachable compounds. Succeeding the cold bath phase, the membrane will be subjected to annealing in a warm water bath at 60°C−90°C temperature. The annealing process improves the semipermeable property of the CA membrane with a flux decline and a substantial reduction in the salt passage. Subsequent to the processing, this CA membrane forms asymmetric structure having a dense surface layer of around 0.10−0.20 microns, which will be responsible for the salt rejection ability. The water flux and salt rejection of a CA membrane could be controlled by changing the temperature and the duration of the annealing process.

The CA membrane polymers hydrolyze quickly at increased pH, and hence, the operational feed water pH range of this membrane will be almost 6−8. Consequently, the CA membrane element could be cleaned only in the narrow pH range nearly neutral (pH: 6−8). On the other hand, the CA membrane polymers have adequate tolerance to free chlorine which allows a process with chlorinated feed water along with online disinfection for controlling the growth of bacteria. Because of this reason, the CA is still the preferred membrane material for usage in which regular disinfection of reverse osmosis system with free chlorine is practiced, like in certain food applications and in pharmaceutical industries. Moreover, one of the membrane producers currently manufactures capillary reverse osmosis membranes for the desalination of seawater using the CA polymer. Excluding the formerly stated applications, the remaining desalination market is controlled by the composite PA membranes with SWM configuration.

4.3.2 Composite polyamide membrane

The fabrication process for a composite PA membrane comprises of two distinctive stages (Fig. 4.8). Initially, a support layer of polysulfone (PSF) is cast on top of a nonwoven polyester fabric. The polymer solution of PSF is applied from a trough onto a mobile polyester-backing

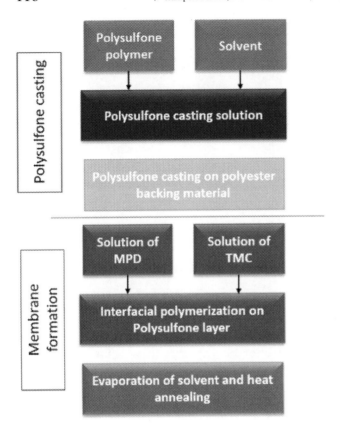

FIGURE 4.8 Fabrication sequence of a composite polyamide membrane. *Source: Courtesy: Mark Wilf. Adapted from Wilf, M. & Awerbuch, L. (2007). The guidebook to membrane desalination technology. Italy: Balaban Desalination Publications.*

fabric. After the application of PSF and the development of the ultrafiltration membrane layer, the fabric passes through a water bath for removing the solvent and is brought together on a drum. The PSF layer will be extremely porous and is considered to be not semipermeable; i.e., it will not have the capability for separating water from the solution of dissolved ions (Trimble, 2007). On the other hand, the PSF layer has increased water permeability and it provides the water pathways. During the subsequent process stage, the drum with the PSF membrane is transferred to the second machine in which interfacial polymerization (IP) takes place (Fig. 4.9). A semipermeable membrane layer will be formed on the PSF support by IP process of two monomers, namely metaphenylenediamine consisting of amine groups and the second one, trimesoyl chloride, providing carboxylic acid chloride functional groups (El-Aassar, 2012). The IP process will be extremely fast and occurs on the exterior of the PSF support forming a barrier of thickness 1000−2000 Å. This barrier layer will be responsible for the semipermeable property: the rejection of dissolved species and the flow of water. After the polymerization process, the membrane web reaches a rinsing bath. Then the membrane will be rinsed for removing additional reagents and subsequently transferred to the oven for drying.

The above-stated fabrication process allows autonomous optimization of the distinctive properties of the salt rejecting layer and the membrane support. The resultant composite membrane is characterized by low salt passage and greater specific water flux relative to

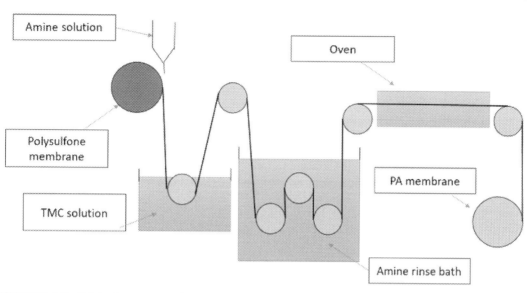

FIGURE 4.9 Fabrication process of a composite polyamide (PA) membrane barrier on polysulfone support. *Source: Courtesy: Mark Wilf. Adapted from Wilf, M. & Awerbuch, L. (2007). The guidebook to membrane desalination technology. Italy: Balaban Desalination Publications.*

the CA membranes. The difference in properties of CA membranes and PA membranes is shown in Table 4.1. The initial composite membranes prepared from aliphatic polymers were noted to be extremely sensitive to the presence of oxidants and experienced inadequate performance stability in field conditions (Asano, 1998). The subsequent development of composite membranes, made from aromatic PA fabricated by Cadotte et al. (1980), has better stability in a broad range of feed pH (2–10), tolerance to free chlorine, and displays outstanding longer-term performance stability with most of feed water types. This type of membrane material has been employed nowadays exclusively for manufacturing commercial reverse osmosis membranes. A variety of types of membranes made of composite aromatic PA are used in brackish water, seawater, and nanofiltration applications. These composite membranes are used in applications such as food processing, wastewater reclamation, brackish water and seawater desalting, potable water softening, and several other industrial applications.

In the following section, the different composite PA membranes are briefly discussed.

4.3.2.1 Thin-film composite membranes

A significant accomplishment in the reverse osmosis membrane field is the development of PA-based TFC membrane, prepared by Cadotte and his team members during the 1970s. This TFC membrane was manufactured by using the IP process, and this has contributed to significant progress in the development of the composite membrane. Correspondingly, this has turned out to be a standard among the most advanced innovations in water and wastewater purification processes. In spite of the fact that the advancement of reverse osmosis membrane technology brought about a reduction in the energy consumption for the desalination process to 3.4 kWh/m^3 from 8.0 kWh/m^3, still it is greater than the theoretical limit

TABLE 4.1   Comparison of cellulose acetate membranes and polyamide membranes.

Property	Cellulose Acetate membrane	Polyamide membrane
Membrane type	Homogenous asymmetric	Homogenous asymmetric, thin-film composite
Surface roughness	Smooth	Rough
Fouling tolerance	Good	Fair
Biological growth	Metabolizes membrane	Causes membrane fouling
Chlorine tolerance	Up to 1 ppm continuously	<0.02 ppm
Surface charge	Neutral	Negative (anionic)
Temperature tolerance	Up to 35°C	Up to 45°C
Feed pressure (brackish membrane)	200–400 psi	145–400 psi
pH range	4–6	2–12
Silica rejection (%)	~85	~96+
Salt rejection (%)	~95	~98+
Pretreatment requirements	Low	High
Organics removal	Low	High
Cleaning frequency	Lower (months to years)	High (weeks to months)

of 1.06 kWh/m^3, considering seawater salinity of 35,000 mg/L with 50% conventional recovery (Elimelech & Phillip, 2011). Present developments in the usage of energy recovery devices brought about a reduction in energy consumption of the reverse osmosis membrane technology to lower than 2.5 kWh/m^3 (Peñate & García-Rodríguez, 2011).

In the reverse osmosis industry, TFC membranes are predominant and industry-standard, as of now. Also the TFC membrane shows high rejection for silica, compared with the cellulosic membranes. Generally, the TFC membranes are made out of three different layers. Fig. 4.10 presents the schematic illustration of a TFC membrane showing the different layers of the composite film (Maruf et al., 2011). The topmost layer is a very thin PA film, formed by the IP process, and is responsible for the rejection of salt. The layer of PA is prepared onto a porous ultrafiltration membrane, that is located on top of nonwoven fabric support (Safarpour et al., 2015). In spite of the fact that TFC membranes have experienced tremendous growth in the past several years, there are additional difficulties for the extra advancement of the TFC membranes. Membrane fouling, scaling, and membrane degradation due to chlorination are the predominant restraints of these types of membranes. Chlorine is generally employed for the membrane cleaning application; however, the PA layer is very sensitive to chlorine, because it easily reacts with aromatic rings and amide nitrogen of PA. Hence, antibiofouling properties and chlorine resistance are required for resolving the biofouling problem in the PA TFC reverse osmosis membranes.

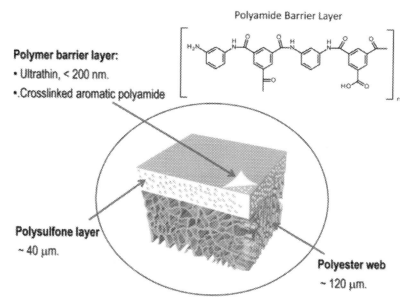

FIGURE 4.10 Schematic illustration of a thin-film composite membrane highlighting the different layers of the composite film; (Top) typical chemical structure of the cross-linked aromatic polyamide barrier layer. *Source: Reproduced from Maruf, S.H., Ahn, D.U., Greenberg, A.R., & Ding, Y. (2011). Glass transition behaviors of interfacially polymerized polyamide barrier layers on thin film composite membranes via nano-thermal analysis. Polymer, 52(12), 2643–2649.*

Boron compounds are quite common under natural conditions, in lower or higher concentrations. Adequate removal of boron is essential in all seawater desalination applications for drinking water production using reverse osmosis technology. Because of the current developments in the membrane technology, seawater reverse osmosis membranes which accomplish up to 95% boron rejection in the testing condition of manufacturers have been commercialized. On the other hand, this nominal level of boron rejection corresponds to 88%–90% boron rejection under the higher recovery conditions of seawater reverse osmosis units. Scientists are continuously carrying out studies to improve the boron rejection ability of the reverse osmosis membranes.

4.3.2.2 Thin-film nanocomposite membranes

Numerous studies have recommended various methods to overcome the problems associated with TFC membranes. The introduction of nanoparticles into the PA matrix is a comparatively new approach that was first suggested by Hoek and his coworkers in 2005 (Duan et al., 2015). Polymer nanocomposite membranes are the modified form of conventional polymeric membranes with the nanomaterials dispersed within the polymer lattices. The nanomaterials used in the nanocomposite membranes are categorized into organic material, inorganic material, hybrid material, and biomaterial.

During the past two decades, various hydrophilic nanomaterials (such as carbon nanotubes, graphene oxide, silica, titanium dioxide, zeolites, and alumina) have been employed for improving the efficiency of TFC membranes (Saleem & Zaidi, 2020a, 2020b, 2020c;

Saleem et al., 2020; Vatanpour et al., 2012; Wu et al., 2013; Yadav et al., 2020; Zaidi et al., 2019). Recent researches on nanoparticle incorporated thin-film nanocomposite membranes for application in water purification have been critically analyzed by Saleem and Zaidi (2020c). Surface chemical modification by the incorporation of hydrophilic carbonaceous nanomaterials like carbon nanotubes and graphene oxide, metals, as well as metallic oxide nanomaterials like silver, zeolite, silica, zinc oxide, aluminum oxide, and titanium dioxide generates the hybrid materials, which cause an increase in water flux, membrane hydrophilicity, selectivity, permeability, good mechanical stability, thermal stability, increase in rejection of salt, and chemical stability. Table 4.2 shows the comparison of the performances of different nanoparticle-incorporated thin-film

TABLE 4.2 Comparison of the performances of different nanoparticle-incorporated thin-film nanocomposite polyamide-based reverse osmosis membrane, relative to the thin-film composite membranes reported in the corresponding kinds of literature.

S. no.	Nanoparticle used	Flux improvement compared to pure membrane	Salt rejection improvement	Antifouling performance	Ref.
1	Multiwalled carbon nanotubes (MWCNT), 15.5 wt.%	More than double	90%	Increased antifouling performance and almost no iron was noticed on the exterior subsequent to the test	Inukai et al. (2015)
2	Carboxy-functionalized MWCNTs, 0.1 wt.%	Increased the water flux to 28.05 L/m^2/h from 14.86 L/m^2/h	>90%	Increased antifouling property against bovine serum albumin and inorganic foulant Ca (HCO$_3$)$_2$	Zhao et al. (2014)
3	MWCNT, 15.5 wt.%	Permeate flux of 0.16 m^3/m^2/day for a 10 mmol/L-NaCl feed concentration at 0.7 MPa	99.7%	Superior antifouling behavior for the bovine serum albumin model organic-foulant	Takizawa et al. (2017)
4	Carbon nanotube, 15.5 wt.%	9 L/m^2/h, for 0.2% NaCl aqueous solution at 0.7 MPa	96.0% for 0.2% salt aqueous solution at 0.7 MPa	—	Takeuchi et al. (2018)
5	MWCNT, 0.001 wt.%	30% increase in water flux relative to the other reverse osmosis membranes under all condition of the salt concentration	Salt rejection of CNT-polyamide thin-film nanocomposite reverse osmosis membrane was a little lower	The irreversible fouling resistance of the CNT polyamide thin-film nanocomposite membrane was of approximately 50% less	Baek et al. (2017)
6	Zwitterion functionalized single-walled CNT, 20 wt.%	The flux of water improved by greater than a factor of 4	98.6 (+1%)		Chan et al. (2013)

(Continued)

TABLE 4.2   (Continued)

S. no.	Nanoparticle used	Flux improvement compared to pure membrane	Salt rejection improvement	Antifouling performance	Ref.
7	Graphene oxide, 38 ppm	38 ppm GO-based thin-film composite membrane was almost 80% greater	Maintained similar level of NaCl salt rejection	98% enhancement in the antibiofouling property	Chae et al. (2015)
8	Graphene oxide, 100 ppm	39% increment in water flux	Small decrease of 1% in salt rejection	Better fouling resistance	Ali et al. (2016)
9	Reduced-graphene oxide/TiO_2, 0.02 wt.%	21% greater than that of the bare reverse osmosis membrane	99.45%	Better fouling resistance	Safarpour et al. (2015)
10	Graphene oxide quantum dots, 5 mg	51.8% enhancement in permeate flux	98.8%	Improved antifouling properties	Song et al. (2016)
11	Nitrogen-doped graphene oxide quantum dots, 0.02 wt./vol.%	Increased the permeability of water by almost three times	Same salt rejection (~93%)	—	Fathizadeh et al. (2019)

Reproduced from Saleem, H., & Zaidi, S.J. (2020c). Nanoparticles in reverse osmosis membranes for desalination: A state of the art review. Desalination, 475, 114171.

nanocomposite PA-based reverse osmosis membrane, relative to the TFC membranes reported in the corresponding kinds of literature.

The polymer nanocomposite membranes are of two types; the first one is the thin-film nanocomposite membranes and the other one is the blended nanocomposite membranes. For blended nanocomposite membranes, the polymer together with the nanoparticles is dispersed in casting solution during the membrane casting process. The prepared nanocomposite membranes are termed as nanoparticle-entrapped membranes or nanoparticle-blend membranes. However in the case of thin-film nanocomposite membrane, the nanoparticles generate a thin-film on the exterior of the membrane by means of self-assembly by dip-coating technique or deposition on the surface of membrane using the pressure (Vatanpour et al., 2012). Recently, the preparation of thin-film nanocomposite membranes with various nanoparticles and their preparation techniques, and examining the separation performance of membrane have been the focal point of several investigations. Till now, a few types of graphene oxides, carbon nanotubes, zeolites, metal, and metal oxide nanoparticles, etc. with different functionalities and properties have been used in thin-film nanocomposite membranes. It is suggested that the incorporation of nanoparticle contributes to the enhancement of thin-film nanocomposite membrane properties along with superior separation characteristics, for example, increased hydrophilicity and water permeability without reducing the salt selectivity, fouling resistance, and chlorination resistance.

Additional efforts are required for monitoring the long-standing stability of these thin-film nanocomposite membranes within actual processes, and to evaluate the environmental problems of application and possible discharge of nanoparticles. There are numerous research activities in lab as well as pilot-scale on the use of thin-film nanocomposite membranes; however very limited reports exist on commercial-scale manufacturing and real-world application. More research is needed to advance commercialization in order to confirm that the advantages thin-film nanocomposite membranes compensate their manufacturing and environmental expenses.

4.4 Membrane module configurations

The device in which membranes are installed is called a membrane module and it can be put in different configurations. Practically, individual membrane modules are called elements. The various commercial modules available today are designed for a specific application/process in which they offer the technically and commercially the best solution. Multiple membrane modules (or elements) joined each other will form a train, and numerous trains together form the reverse osmosis system. The reverse osmosis technology was launched with tubular and plate and frame configurations. Because of low packing density, these preliminary module configurations have been slowly removed from potable applications and at the present time are being very occasionally employed in standard reverse osmosis applications. Conversely, novel configurations of plate and frame modules are still being employed in food processing industries and for waste stream treatment. Earlier, the two main membrane module configurations used for reverse osmosis applications were spiral wound and hollow fiber. Presently, most of the reverse osmosis membrane manufacturers provide elements in spiral wound configuration.

4.4.1 Plate and frame configurations

The plate and frame configuration type was presented at the initial stages of development of reverse osmosis technology and afterward practically abandoned in support of higher packing density spiral wound as well as hollow fiber configurations (HFMs). The selective membrane layer is inserted between two support plates, which offer flow channels to the fluid on both membrane sides; they are collected leaving the plane sheet membrane layers. The inserted plates with the membrane can be built up vertically for increasing the membrane surface. The flow of fluid occurs parallel across the membrane module. Currently, the plate and frame modules are still used in applications in which the hollow fiber and spiral wound modules cannot offer adequate performance or reliability. In the current plate and frame configuration, the flow regime offers turbulent flow along with a shorter feed flow path. Consequently, the propensity for membrane fouling or scaling is considerably decreased. Because of the increased cost of membrane modules, this plate and frame configuration is not employed in commercial-scale drinking water production applications.

4.4.2 Hollow fiber membrane module

The conception of the hollow fine fiber membrane configuration module was presented by Mahon (1966). The HFM uses a semipermeable membrane made of hollow fibers extruded from non-cellulosic or cellulosic materials. These fibers are asymmetrically structured, about $85-150\mu$ ($0.0033-0.060$ in.) outside diameter and $40-80\mu$ ($0.0016-0.0030$ inch) inside diameter. A large number of such fibers will be developed to form a bundle and subsequently folded in half to a length of around 1.2 m. Fig. 4.11 depicts the general design and the geometry of the modules of interest (Wan et al., 2017).

A plastic perforated tube, functioning as the feed water distributor, will be introduced in the middle and extends the complete bundle length. This bundle will be enveloped and the two ends will be epoxy-sealed for developing a sheet-like product water tube end as well as a terminal end that prevents the feed stream from bypassing to the reject passage. The HFM bundle, about $4''-8''$ ($10-20$ cm) in diameter, is enclosed in a cylindrical shell or housing around $54''$ (137 cm) long and $6''-12''$ ($15-30$ cm) diameter. This assembly can be termed as a permeator. The pressurized feed water arrives at the permeator feed end by means of a center distributor tube, flows through the tube wall, and subsequently flows radially around the fiber bundle to the external permeator pressure shell. Water is permeated through the exterior wall of the fibers into the hollow core or fiber bore, through the fiber bore to the tube sheet or product end of the fiber bundle, and leaves through the product connection on the permeator feed end.

In a hollow fine fiber membrane module, the concentration polarization (CP) is not great at the surface of the membrane due to the fact that the product water flow per membrane unit area will be low. The ultimate outcome is that hollow fiber units operate in a laminar or nonturbulent flow regime. The hollow fine fiber membranes should function beyond a minimum reject flow for minimizing the CP and sustain uniform flow distribution across the fiber bundle. Normally, a single hollow fiber permeator could be functioned at until 50% recovery and meet the minimal concentrate flow needed. Also this type of module configuration permits a large membrane area per unit volume of permeator that leads to compact systems (El-Dessouky & Ettouney, 2002).

The commonly used membrane materials are aramids (exclusive PA-type material in an anisotropic form) and CA blends. In spite of its significant benefits, Hollosep manufactured by

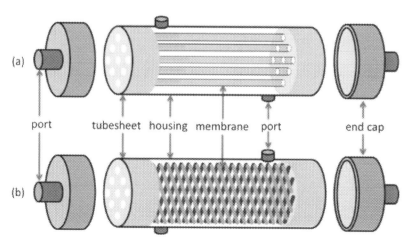

FIGURE 4.11 Hollow fiber membrane configuration modules with (a) a parallel membrane arrangement and (b) a criss-cross membrane arrangement. *Source: Reproduced from Wan, C.F., Yang, T., Lipscomb, G.G., Stookey, D. J., & Chung, T.S. (2017). Design and fabrication of hollow fiber membrane modules.* Journal of Membrane Science, 538, 96–107.

Toyobo is presently the only reverse osmosis hollow fiber module, that is based on CTA, specifically subsequent to the withdrawal of Permasep module of DuPont from the Middle East market. Due to the extremely close-packed fibers and convoluted feed flow within the module, these types of membrane modules need superior quality feed water (the concentration of suspended solids should be low) as compared to the spiral wound module configuration. These hollow fiber modules are commonly employed for the purpose of seawater desalting and treatment of improved quality brackish water (well water). Because of the fouling vulnerability of the typical HFM, the hollow fiber membrane module types are not used for municipal wastewater desalting.

4.4.3 Spiral wound membrane module

Despite the fact that there is an extensive history for the SWM module from the early stage of development of RO, the main structure of this type of module has never transformed. Here, a permeate spacer is inserted between two membranes, the porous support side facing the spacer. The three edges of the membranes are closed with glue for forming a membrane envelope, the open end connected with a percolated center tube. The membrane leaf thus formed is spirally wound around the center tube along with a feed spacer. For shortening the leaf length, numerous membrane leaves are wound concurrently. Industrial-scale SWM modules consist of numerous membrane envelopes as presented in Fig. 4.12, each with about $1-2$ m^2 area, enveloped around the central collection pipe (McKeen, 2017).

Fig. 4.13 shows the configuration of a SPW module for reverse osmosis (DuPont Water Solution, 2020). A polyethylene or a polypropylene net of about $0.2-2.0$ mm thickness is used for the feed spacer, while polyester cloth of almost $0.2-1.0$ mm thickness, hardened using epoxy or melamine resin will be employed for the permeate spacer. The feed stream passes along the feed spacer, by the side of the center tube, while the product water passes through the permeate spacer, in a spiral manner, at a right angle to the direction of feed passage and is brought together by the center tube.

The spiral-wound module is characterized by:

1. Low pressure drop in the permeate channel.
2. Lowest concentration polarization.
3. Least possible membrane contamination.
4. Compactness.
5. Higher-pressure durability.

The diameter of a module is about $4''-8''$; however a $16''$ module is also now available. The reverse osmosis module development is properly outlined in the paper by Johnson and Busch (2010). According to this paper, in the year 1971, Dow Chemical fabricated the CTA membrane in HFM configuration (DOWEX). Also, the manufacturer Toyobo fabricated the CTA hollow fiber modules for the application of seawater desalination (Hollosep). In 1969, DuPont manufactured a membrane module based on aromatic PA in the HFM (Permasep B-9 and B-10) for desalination. These modules had been ruling the Middle East region market till DuPont withdrew from the market in the year 2001. The main drawbacks of the hollow fiber module of DuPont were as follows: (1) the narrow range of pH in which the membrane operated securely, which brought about higher pretreatment expenses, (2) the foulant was hard to remove because

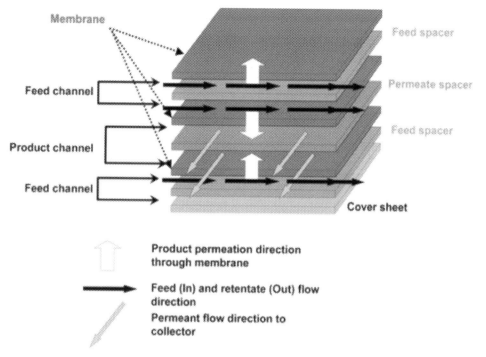

FIGURE 4.12 Flow details in the spiral-wound membrane configuration. *Source: Reproduced from McKeen, L.W. (2017). Markets and applications for films, containers, and membranes. In: Permeability properties of plastics and elastomers (pp. 61–82). https://doi.org/10.1016/b978–0–323–50859-9.00004-x.*

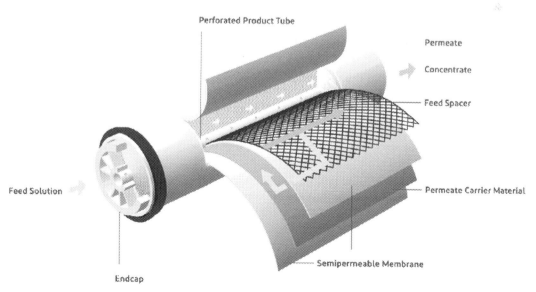

FIGURE 4.13 Configuration of spiral wound membrane module for reverse osmosis. *Source: Courtesy: DuPont Water Solution. Adapted from DuPont Water Solution's FilmTec reverse osmosis technical manual, 2020.*

of the lower cross-flow velocity (CFV), and (3) higher fouling and scaling propensity because of the thin spacing between hollow fibers and owing to the existence of the dead zone (Ismail et al., 2018).

In the year 1985, recognizing the limitations of the CTA membrane and the HFM, Dow acquired the TFC PA membrane of Film Tech cooperation. The TFC membrane was in the SWM configuration and, since then, the market moved slowly from the HFM module to the SWM module. Seawater facilities in Galilah (United Arab Emirates), Agragua Gran Canaria (Spain), and Agip Gela (Italy) (Gorenflo & Sehn, 2006; Gorenflo et al., 2005; Reverberi & Gorenflo, 2007) were the successful demonstrations of the spiral-wound module.

In the research work by Johnson and Busch (2010), it was stated that the capacity of the 8-inch module increased by two-times while the salt passage decreased by three-times during the last 20 years. The manufacture of a 16″-diameter module enabled the membrane surface 4.3 times as larger as that of the 8-inch module. Additionally, the increase in the operational pressure from the former 1000−1200 psig permitted 60% product recovery. The improved salt passage with an increase in the product recovery has been compensated by the increase in the rejection of salt.

4.5 Basic flow patterns

Information on the basic flow patterns for reverse osmosis systems is essential to understand properly the functioning of a reverse osmosis system. Arrays, recycle, double pass, and multiple trains are the different flow patterns employed in a reverse osmosis system.

4.5.1 Arrays

Focusing on the SWM modules as the frequent type of membrane modules operated in the desalination application nowadays, a reverse osmosis array or "train" or "skid" comprises several pressure vessels (PVs) organized in definite patterns. Fig. 4.14 illustrates an

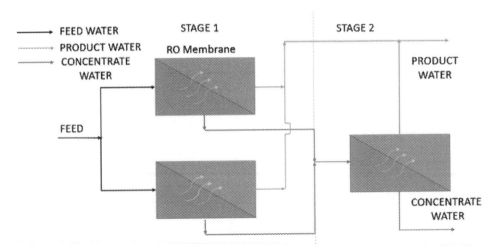

FIGURE 4.14   2:1, Two-stage array having two pressure vessels in the first stage and a single pressure vessel in the second stage.

array of three PVs. Here, the PVs are organized into two sets, with two PVs arranged parallel, succeeded by one PV. Both sets of PVs are arranged in series. Every set of PVs arranged parallel (although there is just a single vessel) is known as a STAGE.

Theoretically, incoming the feed water to the reverse osmosis system is divided equally amongst the PVs in the first stage. The product water from each PV in stage 1 is mixed and assembled in a common header. Stage 1 reject turns out to be the feed for the second stage. The permeate from the PVs in the second stage will be assembled and mixed with product water from the first stage to develop the overall product water from the system. The second stage concentrate turns out to be the concentrate for the whole system.

Reverse osmosis system presented in Fig. 4.14 is termed as a 2:1 array or a two-stage array, representing that there are two stages and the 1st stage with two PVs and the 2nd stage with single PV.

Example 4.1

4:3:1 means three stages with a total of eight vessels; stage 1 with four PVs, stage 2 with three PVs, and stage 3 with a single PV. The concentrate of each stage will be the feed stream for the subsequent stage.

Example 4.2

A 10:5 array will have two stages; the first stage with 10 PVs and the second stage with five PVs.

Example 4.3

An incoming flow of 80 GPM will need two PVs in the first stage (refer to Fig. 4.15). Initially, the first stage recovers around 50% of the incoming water (assuming six 8-inch

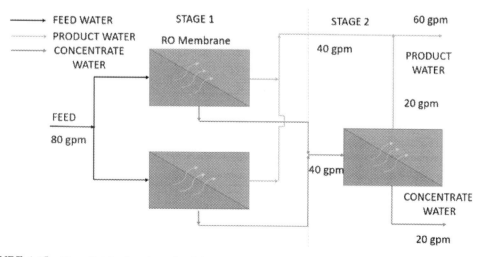

FIGURE 4.15   Flow distribution through a 2:1 array.

diameter membrane modules arranged sequentially), to facilitate 40 GPM product water from the first stage and 40 GPM concentrate/reject. This concentrate is subsequently passed to the second stage as the feed. As the flow here is just 40 GPM to the second stage, merely one PV is necessary. The reject in the second stage from the single PV will be 20 GPM, much higher than the 16-GPM minimal flow rate of concentrate per PV. The second-stage permeate will be almost 20 GPM, and totaling that to the 40-GPM first-stage permeate/product water makes the overall system recovery of 75% or 60 GPM.

In general, the overall recovery from a two-stage reverse osmosis system is around 75%. Greater recoveries (almost 80%) could be attained in the case that the incoming water is reasonably free of scale formers as well as suspended solids. Recoveries more than 80% usually need greater than two stages.

4.5.2 Recycle

Fig. 4.16 shows a reverse osmosis array with a reject/concentrate recycle. A reject/concentrate recycle is typically exploited in small reverse osmosis systems, in which the CFV is not sufficiently great enough to retain better scouring of the surface of the membrane. The returning of a portion of the reject to the feed enhances the CFV and decreases the individual module recovery, thus dropping the fouling possibility.

The reject recycle has certain disadvantages mentioned below:

- Greater consumption of energy, because the concentrate and influent streams flowing together and should be repressurized. This leads to increased operating expenses for the system.
- Greater feed-pump necessities, as the reverse osmosis feed pumps should pressurize both the recycled reject stream and the influent stream. Consequently, the reverse

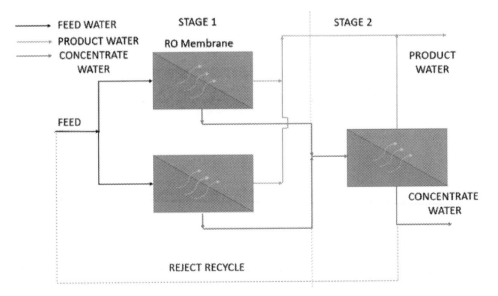

FIGURE 4.16   A 2:1 array with reject recycle.

osmosis feed pump should be larger, which might increase the pumping cost for the reverse osmosis system.
- Overall product quality will be lower. This is due to the fact that comparatively higher concentration reject is mixed with the low-concentration influent stream.

4.5.3 Double-pass/two-pass

Two-pass or double pass denotes the additional purification of permeate/product water from one reverse osmosis unit by passing it through an additional reverse osmosis unit. The initial reverse osmosis will be the first pass. The product water from the first pass is subsequently fed to a different reverse osmosis referred to as the second-pass reverse osmosis. The purpose of the second-pass reverse osmosis is to polish the product of the first-pass reverse osmosis to obtain superior-quality water. Fig. 4.17 represents a schematic diagram of a two-pass reverse osmosis system. The second pass design principles are mostly similar to the first pass principles. For the reason that the reject obtained from the second pass is comparatively cleaner (improved quality relative to the first-pass feed), it is essentially recycled to the front of the first pass. This reduces the waste from the reverse osmosis system and subsequently increases the quality of the feed water, as the first-pass feed is diluted with the moderately superior-quality concentrate from the second-pass (Kucera, 2015). The recovery of the second pass could be up to 90.0% with just two stages. The above-mentioned higher recovery could be attained due to the comparatively lower content of Total Dissolved Solids (TDS) in the feed of the second pass. The overall recovery of the system would be almost 73.0% with 75.0% first-pass recovery and 90% second pass recovery.

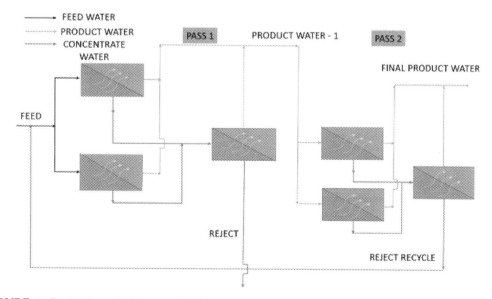

FIGURE 4.17   A schematic diagram of double-pass reverse osmosis.

In general, a tank is needed between the first-pass system and second-pass system, to equalize the flows between the passes. Nevertheless, if the count of first-pass skids is the same as the count of second-pass skids, a tank might be probably not needed. A few vendors locate the two passes on a single skid, thus eliminating the reverse osmosis feed pump to the second-pass reverse osmosis.

4.5.4 Multiple trains

When higher flow rates are required to be treated, multiple trains or skids positioned in parallel should be employed. As an example, an 800-GPM reverse osmosis may need just a single skid, if the vendor has the same size skid in their inventory. On the other hand, the vendor can use two 400-GPM reverse osmosis skids to make up the production rate of 800 GPM. There is a benefit of using multiple skids in the sense that multiple skids contribute to the system redundancy; one skid could be online whereas the other is offline for maintenance or cleaning purposes. Moreover, multiple skids are also used to juggle variable permeate requirements. The disadvantage of multiple skids is their operating expenses and capital expenses; the higher the number of skids, the greater the maintenance/operating and capital costs. The aforesaid expenses should be compared to the ability for providing water in the course of shutdown of any one skid for maintenance or cleaning.

4.6 Seawater reverse osmosis system configurations

The seawater reverse osmosis (SWRO) system configurations most extensively employed include single-pass reverse osmosis treatment, in which the source water will be processed by reverse osmosis just one time (Fig. 4.18), and two-pass reverse osmosis treatment, in which the seawater will be initially processed by a seawater reverse osmosis system, and subsequently, the product water produced by this system is reprocessed by brackish water reverse osmosis (BWRO) membranes (Fig. 4.19).

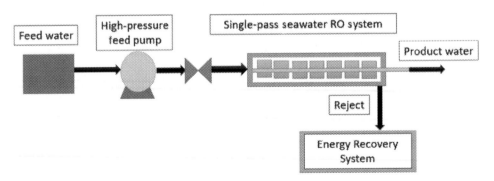

FIGURE 4.18   Single-pass seawater reverse osmosis (RO) system.

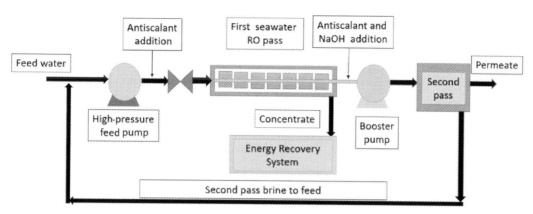

FIGURE 4.19 Conventional full two-pass seawater reverse osmosis (RO) system.

4.6.1 Single-pass seawater reverse osmosis systems

Single-stage seawater reverse osmosis systems are designed for producing desalinated seawater (product water) in a single step using just a single set of reverse osmosis trains working in parallel. Generally, between 800 and 900 seawater reverse osmosis membrane elements installed in 100–150 PVs are required for producing 10,000 m^3/day of product water appropriate for drinking purposes, while using a single-stage seawater reverse osmosis system. Single-stage seawater reverse osmosis systems are extensively used for producing potable water nowadays. On the other hand, the above-mentioned systems have only limited industrial applications primarily due to the water quality restrictions of the product water produced.

4.6.2 Two-pass seawater reverse osmosis systems

Two-pass seawater reverse osmosis systems are normally used when either the source seawater salinity is comparatively higher (that is, greater than 35,000 ppm) or/and the permeate water quality requirements are very strict. As an illustration, if high-temperature/high-salinity source water (like Persian Gulf seawater and Red seawater) is used in combination with standard-rejection (99.6%) seawater reverse osmosis membranes, then single-stage seawater reverse osmosis systems might not be able to generate product water appropriate for potable water use. In such a scenario, two-pass seawater reverse osmosis systems have been demonstrated to be an economical and efficient configuration for the production of potable water. Also, reverse osmosis systems with two or more passes are extensively employed for the production of extremely pure industrial water. The two-pass seawater reverse osmosis system usually consists of a combination of a single-pass seawater reverse osmosis system and a single- or multipass BWRO system coupled in series. Certain seawater reverse osmosis desalination plants are configured as two-stage SWRO with a high-recovery rate operation (Table 4.3; Kim et al., 2020).

TABLE 4.3   Two-stage seawater reverse osmosis desalination plants.

Plant	Country	Plant capacity (m3/d)	Overall seawater reverse osmosis recovery rate (%)	Hydraulic pressure (bar)	Energy recovery device type
Kindasa	Kingdom of Saudi Arabia	26,800	50	71	Pelton turbine
Valdelentisco	Spain	140,000	50	77	Pressure exchanger
Rambla Morales	Spain	60,000	58	83	Pelton turbine
Maspalomas II	Spain	26,200	60	90	Pelton turbine
Las Palmas III	Spain	86,000	50	N/A	Pressure exchanger
Fukuoka	Japan	50,000	60	N/A	Pelton turbine
Curacao	The Netherlands	10,200	58	N/A	Pelton turbine

Adapted from Kim, J., Park, K., & Hong, S. (2020). Optimization of two-stage seawater reverse osmosis membrane processes with practical design aspects for improving energy efficiency. Journal of Membrane Science, 601, 117889.

Calculation

Example 4.4

Reverse osmosis system will produce 410 m^3/h (1,805 gallons/min). The design average flux rate is 25 L/(m^2.h) (14.7 gfd). The membrane element type selected for this system has 37 m^2 of membrane area (400 ft^2) per element. Calculate the number of membrane elements required?

Solution

1 m^3/h = 1000 L/h.
Number of membrane elements required = 410,000 L/h/(25 L/(m^2.h) $\times$ 37 m^2) = 443 elements.
Number of membrane elements required = 1805 gal/min $\times$ 1440 min/day/(14.7 gallons/ft^2/day $\times$ 400 ft^2) = 442 elements.
The number of elements required will be rounded up according to the number of elements per vessel. Assuming seven elements per vessel:
442/7 = 63.14 pressure vessels.
Number of elements required: 63 $\times$ 7 = 441 elements.

4.7 Hybrid reverse osmosis desalination systems

Recently, there has been an upsurge in reverse osmosis desalination due to the fact that it is regarded as an economical as well as an energy-efficient alternate to thermal-based distillation for commercial desalination facilities. Hybrid reverse osmosis systems are recently being investigated for addressing certain challenges that are still related to reverse osmosis, such as feed pretreatment, brine treatment, and membrane fouling. The most frequently investigated hybrid reverse osmosis systems are reverse osmosis-multistage flash (RO-MSF), reverse osmosis-pressure retarded osmosis, reverse osmosis-reverse electrodialysis (RO-RED), reverse osmosis-electrodialysis (RO-ED), and reverse osmosis-capacitive deionization (RO-CDI). The application of RO-MSF and reverse osmosis-multieffect distillation in desalination plants and power plants like Ras Al-Khair (Saudi Arabia), Az-Zour (Kuwait), Fujairah I and II (United Arab Emirates), and Ras Al-Khair (Saudi Arabia) has revealed the applicability and the feasibility of reverse osmosis hybrid system on a large scale (Almulla et al., 2005). The implementation of the aforestated systems has led to an improvement in the efficiency of the desalination facilities by means of the advancement in permeate quality as well as concentrate quality. Furthermore, these hybrid desalination facilities are reported to contribute an increment in energy preservation along with a decrease in the operation expenses during the desalination operations.

Reverse osmosis-pressure retarded osmosis hybrid system in desalination has demonstrated virtuous possibilities for reverse osmosis concentrate management. In a pressure-retarded osmosis process, water is separated from a high-concentrated pressurized solution to a low-concentrated solution by means of an osmotic gradient, and electric power is generated from the hydraulic head attained, as pure water leaves the pressurized solution (Kim et al., 2015). As compared to stand-alone pressure-retarded osmosis, the energy produced from the hybrid reverse osmosis-pressure-retarded osmosis will be increased due to the fact that the denser reject generated from the reverse osmosis process could be recycled back to the concentrated solution (i.e., seawater) for improving the salinity gradient across the pressure-retarded osmosis membrane (Kim et al., 2013). Also, this hybrid system avails the prospect of eradicating the negative impact of concentrate disposal to the natural habitat and the sea; and it eliminates the necessity of extra expenditure on additional brine managing. In a model proposed by Achilli et al., (2009) for the hybrid RO-pressure-retarded osmosis desalination, it was noted that the hybrid method could have 50.0% recovery rate of water and will reduce energy usage by 65.0% relative to the reverse osmosis stand-alone system.

RO-MSF hybrid has also been used for managing the brine generated from reverse osmosis. Helal et al. (2003) performed an optimization study for nine scenarios, for the purpose of assessing the minimum expense of water that could be achieved from the RO-MSF hybrid system, based on theoretical modeling as well as experimental confirmation. It has been noted that reverse osmosis is generally desired for the desalination of feed consisting of a low concentration of salt, particularly for the brackish water desalination for minimizing the fouling limitations and improve energy efficiency. The RO-MSF hybrid system has been noted to be economically sustainable for the treatment of reject obtained from the less-saline feed reverse osmosis desalination (Helal, 2004).

Recently, the researches on electrically induced desalination by ED have been developing progressively. RO-ED hybrid system offers numerous advantages; and the topmost out of those advantages is the capability to accomplish increased water recovery from ED through the reverse osmosis treatment of ED reject. Electrodialysis is typically used for desalinating dilute streams or feed with low conductivity for minimizing the large electric power necessities of this process (Pellegrino et al., 2007). McGovern et al. (2013) used an ED system that was counter-flowed with reverse osmosis for concentrating 120.0 ppt saline feed. Operational conditions like current density have been optimized such that the optimal values of output variables could be attained. The current density of the hybrid system that is needed for minimizing the cost of water is higher relative to that needed for maximizing efficiency. The aforementioned points out that the effect of ED properties on the output variables will be greater, as compared to the properties of reverse osmosis unit per unit area of the membrane (McGovern et al., 2013).

The hybrid RO-CDI desalination system is also investigated as a potential substitute for an individual reverse osmosis desalination system. Recently, a unified method that merges reverse osmosis with constant-voltage-operated CDI or constant-current-operated CDI has been employed for producing superior quality ultrapure water from seawater. At the same feed concentration and flow rate, the RO-constant-current-operated-CDI hybrid system has more benefits relative to the reverse osmosis constant-voltage-operated CDI hybrid, like extended absorption times for the CDI cells when the dead volume or spacer volume and the capacitance is similar to that of constant-voltage-operated CDI (Minhas et al., 2014). In the meantime, the specific energy consumption (SEC) will be similar for both reverse osmosis constant-voltage-operated CDI hybrid and reverse osmosis constant-current-operated CDI hybrid, when desalinated water is produced at a similar feed concentration and flow rate. Consequently, energy usage will be similar regardless of whether a constant voltage or current is maintained. The consolidation of reverse osmosis and CDI in reverse osmosis constant-current-operated CDI hybrid has also been demonstrated to increase fresh water productivity.

RO-reverse electrodialysis hybrid is considered to be a novel concept in hybrid desalination technology. The reverse electrodialysis unit in the hybrid system yields energy from the salinity gradient existing between an extremely concentrated solution (e.g., seawater or concentrated brine) and a low-salinity solution. The reverse electrodialysis-treated solution will be the feed solution of the reverse osmosis unit due to the fact that the solution turns out to be diluted subsequent to reverse electrodialysis treatment. With the application of the diluted solution as the reverse osmosis feed, the pump work needed for transporting the solution to the reverse osmosis system will be reduced. The reject from the reverse osmosis unit could then function as an energy recovery source as it could be recirculated back to the reverse electrodialysis unit for maintaining the higher salinity gradient. Moreover, the reverse electrodialysis process offers a platform to control the concentration of brine in a manner that water recovery could be optimized (Li et al., 2013).

4.8 Case studies

4.8.1 Case Study 1: Reverse osmosis brackish water desalination facility, Arab Potash Company (APC), Jordan

Alsarayreh et al. (2020) explored the possibility of establishing the concentrate recycle design on the original design of an industrial medium-sized multistage and multipass

SWM brackish water reverse osmosis desalination facility (1200 m^3/day) of Arab Potash Company situated in Jordan. Particularly, the above-stated work analyzed the effect of recycling the higher salinity stream of the first-pass (at various recycled percentages) to the feed stream on the process performance indicators such as the overall recovery rate, fresh water salinity, and SEC.

The simulation was performed at a fixed raw water flow rate, temperature, pressure, and concentration. The aforestated study confirmed that an implementation of a 100% recycle could lead to an increase in the production capacity (even though with improved salinity of the permeate water) relative to 0% recycle mode that is presently used in the reverse osmosis process of Arab Potash Company facility. Remarkably, enhancing the first-pass recycle percent from 0% to 100% has increased the first-pass mass transfer coefficient of 13.0%, which has a beneficial impact on the amount of product water. Furthermore, an enhancement in the product water flow rate of the second pass by 3.0% was deduced to enhance the residence time of water inside the second-pass reverse osmosis modules. This is also related to a minor increase of the permeate salinity of 2.975 mg/L, which is less than the recommended drinking water salinity limit of ~200 mg/L set by different nations of the world.

4.8.2 Case Study 2: Al Taweelah A2 plant, Abu Dhabi, United Arab Emirates

The objective of the work by Al Bloushi et al. (2018) was to assess the efficiency of Al Taweelah A2 MSF distillation plant, and also to analyze the possibility of substituting the MSF distillation technology by reverse osmosis or integrating MSF distillation with reverse osmosis as a hybrid system, after the end of the commercial life of the plant. This plant is situated near Abu Dhabi city, and this facility began to function in the year 2001 with a concession period of 20 years. It consists of a 50-million imperial gallons per day MSF desalination plant and 720-MW combined cycle power plant.

From the above-mentioned study, it was noted that the effect of MSF distillation hybridization with reverse osmosis in Al Taweelah A2 facility has led to a remarkable decrease in the annual energy usage that is the same as 71.90%, retaining the same water production capacity of 56,750 m^3/day or 20,713,750 m^3/year. Correspondingly, the effect of MSF distillation hybridization with reverse osmosis in Al Taweelah A2 facility has led to a substantial decrease in the unit water expense from 0.90 to 0.42 $/m^3, which is, the same as 53.3% decrease. Further, the effect of MSF distillation hybridization with reverse osmosis in Al Taweelah A2 facility has led to a substantial decrease in greenhouse gas emissions, almost the same as 71.8%. Consequently, the overall results confirmed a remarkable reduction in water production expense (53.3%), energy use (71.90%), and annual gas emission (71.8%).

4.8.3 Case Study 3: Howtat Bani Tamim desalination plant, Saudi Arabia

The operating performance of a reverse osmosis-based desalination facility in Howtat Bani Tamim city 190 km south of Riyadh, Saudi Arabia was assessed by Khalil et al. (2015). The desalination units comprised two sets of reverse osmosis membranes both operating as a two-stage single-pass system. Additional stages increase the recovery from the system. In each set, the first stage consists of 36 vessels with 7 membranes per vessel giving a total of 252 membranes, whereas the second stage consists of 18 vessels with 7 membranes per vessel giving a total of 126 membranes. This gives an overall membrane capacity of 378 membranes for each set and 756 membranes for the whole plant.

The facility was examined for 40 months beginning from January 2011 and completing in April 2014, and the source feed water was attained from six wells. The entire operating parameters have been measured and analyzed. The source feed water was primarily treated for TDS and iron removal for reducing membrane fouling and the feed water was injected into two sets of reverse osmosis membranes both operating as a two-stage single-pass system. The overall product output of the facility was 26,282 m^3/day with an average TDS concentration of 500 ppm after blending the product water with feed water. The total dissolved solid content of the feed water was 2000 ppm and the product water was 28.79 ppm yielding 98.56% removal. The overall efficiency of the facility in terms of flow capacity was almost 87.60%. The reject water with 3,529.4 m^3/day average flow and total dissolved solid values of 13,170.40 ppm were disposed of in evaporation tanks.

Reference

Achilli, A., Cath, T. Y., & Childress, A. E. (2009). Power generation with pressure retarded osmosis: An experimental and theoretical investigation. *Journal of Membrane Science, 343*, 42–52. Available from https://doi.org/10.1016/j.memsci.2009.07.006.

Al Bloushi, A., Giwa, A., Mezher, T., & Hasan, S. W. (2018). Environmental impact and technoeconomic analysis of hybrid MSF/RO desalination: The case study of Al Taweelah A2 plant. *Sustainable desalination handbook*, 55–97. Available from https://doi.org/10.1016/b978-0-12-809240-8.00003-4.

Ali, M. E., Wang, L., Wang, X., & Feng, X. (2016). Thin film composite membranes embedded with graphene oxide for water desalination. *Desalination, 386*, 67–76.

Almulla, A., Hamad, A., & Gadalla, M. (2005). Integrating hybrid systems with existing thermal desalination plants. *Desalination, 174*, 171–192. Available from https://doi.org/10.1016/j.desal.2004.08.041.

Alsarayreh, A. A., Al-Obaidi, M. A., Al-Hroub, A. M., Patel, R., & Mujtaba, I. M. (2020). Performance evaluation of reverse osmosis brackish water desalination plant with different recycled ratios of retentate. *Computers & Chemical Engineering, 135*, 106729.

Asano, T. (Ed.), (1998). *Wastewater reclamation and reuse: Water quality management library* (Vol. 10). CRC Press.

Baek, Y., Kim, H. J., Kim, S.-H., Lee, J.-C., & Yoon, J. (2017). Evaluation of carbon nanotube-polyamide thin-film nanocomposite reverse osmosis membrane: Surface properties, performance characteristics and fouling behavior. *Journal of Industrial and Engineering Chemistry, 56*, 327–334.

Byrne, W. (2002). *Reverse osmosis: A practical guide for industrial users* (2nd (ed.)). Littleton, Colorado: Tall Oaks Publishing, Inc.

Cadotte, J. E., Petersen, R. J., Larson, R. E., & Erickson, E. E. (1980). A new thin-film composite seawater reverse osmosis membrane. *Desalination, 32*, 25–31.

Chae, H.-R., Lee, J., Lee, C.-H., Kim, I.-C., & Park, P.-K. (2015). Graphene oxide-embedded thinfilm composite reverse osmosis membrane with high flux, anti-biofouling, and chlorine resistance. *Journal of Membrane Science, 483*, 128–135.

Chan, W.-F., Chen, H.-Y., Surapathi, A., Taylor, M. G., Shao, X., Marand, E., & Johnson, J. K. (2013). Zwitterion functionalized carbon nanotube/polyamide nanocomposite membranes for water desalination. *ACS Nano, 7*(6), 5308–5319.

Duan, J., Pan, Y., Pacheco, F., Litwiller, E., Lai, Z., & Pinnau, I. (2015). High-performance polyamide thin-film-nanocomposite reverse osmosis membranes containing hydrophobic zeolitic imidazolate framework-8. *Journal of Membrane Science, 476*, 303–310.

DuPont Water Solution's FilmTec™ RO technical manual; Version 3; Form No. 45-D01504-en, Rev. 3; April 2020.

El-Aassar, A.-hM. A. (2012). Polyamide thin film composite membranes using interfacial polymerization: synthesis, characterization and reverse osmosis performance for water desalination. *Australian Journal of Basic and Applied Sciences, 6*(6), 382–391.

El-Dessouky, H. T., & Ettouney, H. M. (2002). *Fundamentals of salt water desalination.* Elsevier.

Elimelech, M., & Phillip, W. A. (2011). The future of seawater desalination: Energy, technology, and the environment. *Science (New York, N.Y.), 333*(6043), 712–717.

Fathizadeh, M., Tien, H. N., Khivantsev, K., Song, Z., Zhou, F., & Yu, M. (2019). Polyamide/nitrogen-doped graphene oxide quantum dots (N-GOQD) thin film nanocomposite reverse osmosis membranes for high flux desalination. *Desalination, 451,* 125–132.

Gorenflo, A., Redondo, J. A., & Reverberi, F. (2005). Basic options and two case studies for retrofitting hollow fiber elements by spiral wound. RO technology. *Desalination, 178,* 247–260.

Gorenflo, A., & Sehn, P. (2006). The 13,500 m^3/d plant in Galilah (UAE): Experiences and performance within the first 12 months of operation, Deutsche Meerwasserentsalzung(DME) Conference Seawater Desalination in United Arabic Emirates, Berlin.

Helal, A. M. (2004). Optimal design of hybrid RO/MSF desalination plants part II: Results and discussion. *Desalination, 160,* 13–27. Available from https://doi.org/10.1016/S0011-9164(04)90014-8.

Helal, A. M., El-Nashar, A. M., Al-Katheeri, E., & Al-Malek, S. (2003). Optimal design of hybrid RO/ MSF desalination plants part I: Modelling and algorithms. *Desalination, 154,* 43–66. Available from https://doi.org/10.1016/S0011-9164(03)00207-8.

Inukai, S., Cruz-Silva, R., Ortiz-Medina, J., Morelos-Gomez, A., Takeuchi, K., Hayashi, T., Tanioka, A., et al. (2015). High-performance multifunctional reverse osmosis membranes obtained by carbon nanotube· polyamide nanocomposite. *Scientific Reports, 5,* 13562.

Ismail, F., Khulbe, K. C., & Matsuura, T. (2018). *Reverse osmosis.* Elsevier.

Johnson, J., & Busch, M. (2010). Engineering aspects of reverse osmosis module design. *Desalination and Water Treatment, 15,* 236–248.

Khalil, K. H., Wali, F. M., & El-Dosarit, A. M. (2015). Operational performance and monitoring of a reverse osmosis desalination plant: A case study. *International Journal of Current Engineering and Technology, 5*(6), 3760.

Kim, D. I., Kim, J., Shon, H. K., & Hong, S. (2015). Pressure retarded osmosis (PRO) for integrating seawater desalination and wastewater reclamation: Energy consumption and fouling. *Journal of Membrane Science, 483,* 34–41. Available from https://doi.org/10.1016/j.memsci.2015.02.025.

Kim, J., Park, K., & Hong, S. (2020). Optimization of two-stage seawater reverse osmosis membrane processes with practical design aspects for improving energy efficiency. *Journal of Membrane Science, 601,* 117889.

Kim, J., Park, M., Snyder, S. A., & Kim, J. H. (2013). Reverse osmosis (RO) and pressure retarded osmosis (PRO) hybrid processes: Model-based scenario study. *Desalination, 322,* 121–130. Available from https://doi.org/10.1016/j.desal.2013.05.010.

Kucera, J. (2015). *Reverse osmosis: Industrial processes and applications.* John Wiley & Sons.

Li, K. (2007). *Ceramic membranes for separation and reaction.* John Wiley & Sons.

Li, W., Krantz, W. B., Cornelissen, E. R., Post, J. W., Verliefde, A. R. D., & Tang, C. Y. (2013). A novel hybrid process of reverse electrodialysis and reverse osmosis for low energy seawater desalination and brine management. *Applied Energy, 104,* 592–602. Available from https://doi.org/10.1016/j.apenergy.2012.11.064.

Lonsdale, H. K., Merten, U., & Riley, R. L. (1965). Transport properties of cellulose acetate osmotic membranes. *Journal of Applied Polymer Science, 9.*

Luis, P. (Ed.), (2018). *Fundamental modeling of membrane systems: Membrane and process performance.* Elsevier.

Mahon, H.I. (1966). Permeability separatory apparatus and process utilizing hollow fibers. United States Patent 3,228,877, issued January 11, 1966.

Maruf, S. H., Ahn, D. U., Greenberg, A. R., & Ding, Y. (2011). Glass transition behaviors of interfacially polymerized polyamide barrier layers on thin film composite membranes via nano-thermal analysis. *Polymer, 52*(12), 2643–2649.

McGovern R.K., Zubair S.M., & Lienhard V.J.H. (2013). Design and optimization of hybrid ED-RO systems for the treatment of highly saline brines. In: *International Desalination Association World Conference 2013,* Tianjin, China.

McKeen, L. W. (2017). Markets and applications for films, containers, and membranes. Permeability properties of plastics and elastomers, William Andrew, Elsevier Inc. 61–82, ISBN: 978-0-323-50859-9. https://doi.org/10.1016/b978-0-323-50859-9.00004-x.

Minhas, M. B., Jande, Y. A. C., & Kim, W. S. (2014). Combined reverse osmosis and constant-current operated capacitive deionization system for seawater desalination. *Desalination, 344*, 299–305. Available from https://doi.org/10.1016/j.desal.2014.03.043.

Mondal, S., & De, S. (2020). Reverse osmosis modeling, simulation, and optimization. In *Current trends and future developments on (bio-)membranes* (pp. 187–206). Elsevier.

Mujtaba, I. M., Srinivasan, R., & Elbashir, N. O. (Eds.), (2017). *The water-food-energy nexus: Processes, technologies, and challenges.* CRC Press.

Pellegrino, J., Gorman, C., & Richards, L. (2007). A speculative hybrid reverse osmosis/electrodialysis unit operation. *Desalination, 214*, 11–30. Available from https://doi.org/10.1016/j.desal.2006.09.024.

Peñate, B., & García-Rodríguez, L. (2011). Energy optimisation of existing SWRO (seawater reverse osmosis) plants with ERT (energy recovery turbines): Technical and thermoeconomic assessment. *Energy, 36*(1), 613–626.

Reverberi, F., & Gorenflo, A. (2007). Three year operational experience of a spiral-wound SWRO system with a high fouling potential feed water. *Desalination, 203*, 100–106.

Safarpour, M., Khataee, A., & Vatanpour, V. (2015). Thin-film nanocomposite reverse osmosis membrane modified by reduced graphene oxide/TiO$_2$ with improved desalination performance. *Journal of Membrane Science, 489*, 43–54.

Saleem, H., Trabzon, L., Kilic, A., & Zaidi, S. J. (2020). Recent advances in nanofibrous membranes: Production and applications in water treatment and desalination. *Desalination, 476*, 114178.

Saleem, H., & Zaidi, S. J. (2020a). Developments in the application of nanomaterials for water treatment and their impact on the environment. *Nanomaterials, 10*(9), 1764.

Saleem, H., & Zaidi, S.J. (2020b). Innovative nanostructured membranes for reverse osmosis water desalination. Available from: https://doi.org/10.29117/quarfe.2020.0023.

Saleem, H., & Zaidi, S. J. (2020c). Nanoparticles in reverse osmosis membranes for desalination: A state of the art review. *Desalination, 475*, 114171.

Sherwood, T. K., Brian, P. L. T., & Fisher, R. E. (1967). Desalination by reverse osmosis. *Industrial & Engineering Chemistry Fundamentals, 6*, 2–12.

Singh, P. S., Ray, P., & Ismail, A. F. (2020). Synthetic polymer-based membranes for desalination. In *Synthetic polymeric membranes for advanced water treatment, gas separation, and energy sustainability* (pp. 23–38). Elsevier.

Song, X., Zhou, Q., Zhang, T., Xu, H., & Wang, Z. (2016). Pressure assisted preparation of graphene oxide quantum dot-incorporated reverse osmosis membranes: Antifouling and chlorine resistance potentials. *Journal of Materials Chemistry A, 4*(43), 16896–16905.

Strathmann, H., Scheible, P., & Baker, R. W. (1971). A rationale for the preparation of Loeb-Sourirajan-type cellulose acetate membranes. *Journal of Applied Polymer Science, 15*(4), 811–828.

Sundaramoorthy, S., Srinivasan, G., & Murthy, D. V. R. (2011). An analytical model for spiral wound reverse osmosis membrane modules: Part I—Model development and parameter estimation. *Desalination, 280*, 403–411.

Takeuchi, K., Takizawa, Y., Kitazawa, H., Fujii, M., Hosaka, K., Ortiz-Medina, J., . . . Endo, M. (2018). Salt rejection behavior of carbon nanotube-polyamide nanocomposite reverse osmosis membranes in several salt solutions. *Desalination, 443*, 165–171.

Takizawa, Y., Inukai, S., Araki, T., Cruz-Silva, R., Uemura, N., Morelos-Gomez, A., . . . Endo, M. (2017). Antiorganic fouling and low-protein adhesion on reverse-osmosis membranes made of carbon nanotubes and polyamide nanocomposite. *ACS Applied Materials and Interfaces, 9*(37), 32192–32201.

Trimble, S. W. (2007). *Encyclopedia of water science.* CRC press.

Vatanpour, V., Madaeni, S. S., Khataee, A. R., Salehi, E., Zinadini, S., & Monfared, H. A. (2012). TiO$_2$ embedded mixed matrix PES nanocomposite membranes: influence of different sizes and types of nanoparticles on antifouling and performance. *Desalination, 292*, 19–29.

Wan, C. F., Yang, T., Lipscomb, G. G., Stookey, D. J., & Chung, T. S. (2017). Design and fabrication of hollow fiber membrane modules. *Journal of Membrane Science, 538*, 96–107.

Wilf, M., & Awerbuch, L. (2007). *The guidebook to membrane desalination technology.* Italy: Balaban Desalination Publications.

Wu, H., Tang, B., & Wu, P. (2013). Optimizing polyamide thin film composite membrane covalently bonded with modified mesoporous silica nanoparticles. *Journal of Membrane Science, 428*, 341–348.

Yadav, S., Saleem, H., Ibrar, I., Naji, O., Hawari, A. A., Alanezi, A. A., . . . Zhou, J. (2020). Recent developments in forward osmosis membranes using carbon-based nanomaterials. *Desalination*, *482*, 114375.

Zaidi, S. J., Fadhillah, F., Saleem, H., Hawari, A., & Benamor, A. (2019). Organically modified nanoclay filled thin-film nanocomposite membranes for reverse osmosis application. *Materials*, *12*(22), 3803.

Zhao, H., Qiu, S., Wu, L., Zhang, L., Chen, H., & Gao, C. (2014). Improving the performance of polyamide reverse osmosis membrane by incorporation of modified multiwalled carbon nanotubes. *Journal of Membrane Science*, *450*, 249–256.

CHAPTER

5

Pretreatment: Fouling and Scaling Control

5.1 Pretreatment techniques—an overview

Presently, the RO desalination plants include two important treatment processes designed for successively removing the suspended and dissolved solids present in the source water. The objective of the first process, that is, pretreatment, is to eliminate the suspended solids from the saline source water and to prevent certain naturally occurring soluble solids from turning into solid form, and precipitate on the reverse osmosis (RO) membranes at the time of the salt separation process. The second process, that is, the RO, separates the dissolved solids present in the pretreated feed water, thereby producing low-salinity freshwater appropriate for human consumption, agricultural uses, and industrial applications (Curcio & Drioli, 2009).

Usually, the pretreatment will be applied to the saline water source before reaching the RO membrane system for minimizing the membrane fouling. Pretreatment is extensively used for removing the foulants, optimize recovery and system productivity, and increase the membrane life. The principal objective of RO feed water pretreatment is to guarantee that the RO membrane is not damaged by scaling, as well as fouling, and degraded. Scaling is the formation of a mineral salt layer on the surface of the membrane as a result of direct surface crystallization and the precipitated salt crystal deposition onto the surface of the membrane. Fouling refers to the deposition of particulate matters; for example, iron flocs, silica, algae, biological slime, suspended solids, clay, silt, and other suspended matters on the surface of the membrane or even within the membrane matrix. Fouling usually happens in the lead membrane elements and advances progressively towards the tail elements. Fouling control involves the pretreatment of the feed water for minimizing fouling and regular cleaning for handling any fouling that still happens.

Because of the fouling sensitivity of the RO membrane units, superior-quality feed water is needed to guarantee the longer-term, stable performance of the RO system. A proper pretreatment unit providing high-quality feed water, irrespective of source water quality variation, is very important for efficient RO plant operation. The pretreatment processes help to reduce the fouling potential, limit scaling on the surface of the membrane, retain a satisfactory performance level, and increase the life of the RO membrane (Al-Malek et al., 2005). The silt density index (SDI) is commonly used to characterize the fouling capability of a feed stream. The membrane manufacturers recommend the SDI value of the RO feed water to be less than 3.0 for avoiding high pressure loss in the membrane module and to limit the membrane fouling. An ineffective pretreatment system results in frequent cleaning of the membrane, increased rates of membrane fouling, diminished membrane life, lower recovery rate, poor product quality, and increased operational pressure. All these factors will have a direct effect on the operational cost. Hence, careful consideration must be given to the pretreatment system while designing a desalination plant. In a recent study by Anis et al. (2019), the team provided an overview of the advancement and present trends in conventional as well as nonconventional RO pretreatment.

Pretreatment can be categorized into two classes, namely chemical pretreatment and physical pretreatment. The chemical pretreatment involves the incorporation of disinfectants, polyelectrolytes, coagulants, and scale inhibitors (Migliorini & Luzzo, 2004). Physical pretreatment involves mechanical filtration using cartridge filters, screening, membrane filtration, or sand filters. In the past, the majority of RO plants used conventional pretreatment, which involves physical and chemical pretreatments without the use of membrane technologies. In general, conventional pretreatment technology involves processes such as flocculation, sand filtration, settling, and cartridge filtration as physical pretreatment. With reducing quality of feed water and decreasing cost of membranes, in numerous projects, the membrane pretreatment process preceding the RO stage provides an alternative to conventional pretreatment (Vial & Doussau, 2002; Wolf & Siverns, 2004). The ultrafiltration (UF) and microfiltration (MF) membrane processes are practical choices and it is projected that membrane pretreatment will quickly develop in the years to come (Vial et al., 2003). The scope of pretreatment relies upon the quality of feed water that changes with the plant location and the intake system. In the case of feed water from well sources, the cartridge filtration is generally adequate, whereas the feed water from open seawater intakes requires extensive pretreatment. Badruzzaman et al. (2019) provided a critical review of both conventional and membrane-based pretreatment technologies by presenting water quality problems that influence their performances, critical design characteristics and their effects on pretreatment selection, and a theoretical decision matrix for selecting pretreatment technologies for site-specific conditions.

In RO wastewater treatment, the membranes are the cutting-edge technology solution currently in pretreatment because of the numerous benefits they offer. The scope of membranes in the pretreatment of seawater is anticipated to increase (Vial et al., 2003) in RO desalination due to the fact that it contributes to more effective pretreatment at reduced expenses and at a smaller footprint (Wolf & Siverns, 2004). Generally, the physical pretreatment comprises flocculation and multimedia filtration succeeded by cartridge filtration; however various setups, for example, membranes, dissolved air flotation (DAF), or laminar settlers are seen in plants globally. A simple pretreatment process plant is given in Fig. 5.1. The chemical pretreatment process relies upon the physical pretreatment process being used (Fritzmann et al., 2007).

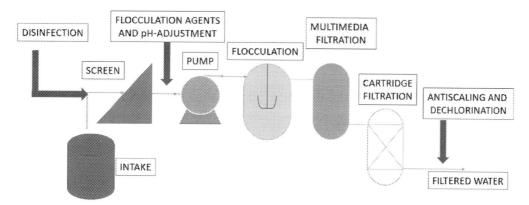

FIGURE 5.1 Simple process scheme of a conventional pretreatment process. Source: *Adapted from Fritzmann, C., Lowenberg, J., Wintgens, T., & Melin, T. (2007). State-of-the-art of reverse osmosis desalination. Desalination, 216(1), 1−76.*

5.1.1 Conventional pretreatment

In the past, RO pretreatment was carried out by conventional systems mostly by chemical addition and sand filtration followed by cartridge filtration (Wolf & Siverns, 2004). The conventional pretreatment involves chemical pretreatment and physical pretreatment which usually involves flocculation, sedimentation, as well as filtration to remove mechanically the algae and colloidal particles. Chemical pretreatment is responsible for disinfection, increase in the salt solubility, and pH adjustment, whereas the physical pretreatment removes the dispersed particles present in the feed water for preventing fouling, blocking, and membrane flux decrease. With the deterioration in the feed water quality and decreased cost of the membrane pretreatment, plant designers are considering the use of membrane-based pretreatment for replacing the less effective conventional pretreatment systems.

The conventional pretreatment in a seawater reverse osmosis (SWRO) plant involves the following steps (Wolf & Siverns, 2004):

- Separation of large particles using a coarse strainer
- Disinfection of water using chlorine
- Clarification without or with flocculation
- Clarification and reduction of hardness employing lime treatment.
- Media filtration
- Alkalinity reduction by an adjustment in pH
- Scale inhibitor addition
- Free chlorine reduction using sodium bisulfite or activated carbon filters
- Sterilization of water employing ultraviolet (UV) radiation
- Final separation of suspended particles using cartridge filters

The conventional filter systems will be backwashed using filtered water as well as air more than one time a day. The rate of filter replacement depends on the quality of source water and ranges between every 2−8 weeks (Wolf et al., 2005). Just a basic cartridge filtration or a one-stage sand filtration can accomplish SDI values less than 3.0, provided the system is supplied with water from beach well sources that have raw water of good quality (Wolf et al., 2005). In any case, when open seawater intakes are employed, poor quality feed water in the course of algae bloom or storms can lead to issues even in an appropriately tuned conventional pretreatment network. There are many drawbacks for a conventional pretreatment system that contribute to increased rates of RO membrane fouling and short-life of the RO membrane. The conventional pretreatment systems for the RO membrane operation are noted to have the following drawbacks (Wolf & Siverns, 2004):

- Troubles to supply a steady SDI < 3.0, particularly in the course of high-turbidity feed water conditions
- Large footprint because of slower filtration velocities
- Variations in feed water quality to the RO membrane
- Coagulant affects the performance of the membrane
- Problems in removing particles smaller than $10-15$ μm
- The possibility of breakthrough during filter backwash

The conventional pretreatment footprint is about $35-40$ m^2/1000 cubic meter per day product water (Wilf, 2004). This type of pretreatment needs more space and is labor-intensive and complicated (van Hoof et al., 2001). In addition to the frequently used mechanical separation stages of cartridge filtration as well as multimedia filtration, different other mechanical separation techniques have been recognized and employed globally (Van de Venter et al., 2005). In an investigation on the pretreatment of the RO desalination part of the hybrid plant at Fujairah (United Arab Emirates), the conventional pretreatment demonstrated a decrease in SDI$_5$ values from 12.0 to 19.0 in the feed water to SDI$_{15}$ 2.0$-$4.5 after pretreatment, and a turbidity decrease of up to 75%. The above was accomplished by employing chemical disinfection, flocculation, cartridge filtration, and multimedia sand filtration (Al-Malek et al., 2005).

5.1.2 Physical pretreatment

The different physical pretreatment techniques currently used in the RO-based desalination industries are discussed in detail in the following section. The physical pretreatment involves mechanical filtration by source water screening (bar racks, rotating screens, band screens, drum screens, microscreens, and cartridge filters), sedimentation, DAF, granular media filtration (GMF), and membrane-based filtration (Voutchkov, 2017).

5.1.2.1 Source water screening

The screening process is considered to be the first treatment step in each desalination plant. Based on the intake type and the saline water quality it brings together, screening equipment can range from simple cartridge filters to complex mechanical screens (configured for sequentially removing huge debris and marine organisms) and microscreens (designed to hold finer solids like plankton, silt, sand, and other solid debris present in the saline source water). The main objective of the source water screening is to protect the downstream pretreatment or RO facilities from structure damage, equipment damage, increased filter media clogging and fouling (Lema & Martinez, 2017).

Open ocean intakes will be normally equipped with bar racks or coarse bar screens succeeded by fine screens (small-sized screens) with openings of 1−10 mm that stop the majority of marine organisms (such as crabs and fish) from reaching the RO desalination facility. The coarse bar screens are stationary always, whereas the fine screens can be of two categories namely (1) stationary (passive) and (2) rotating (periodically movable) screens. A characteristic surface water intake unit for a large-scale or medium-scale membrane-based desalination facility with open intake will have a group of mechanically or manually cleaned coarse bar screens succeeded by traveling band screens or drum screens.

5.1.2.1.1 Bar racks (coarse bar screens)

Bar racks typically have a distance of 50−300 mm between the bars, and their objective is to restrict marine organisms and large debris from reaching the plant intake. In the case of offshore intakes, the screens will be set up on the vertical inlet tower of the intake. The design flow-through velocity for clean screens will be normally 3−4 cm/s. This design velocity is chosen for minimizing the intrusion of marine organisms on the screens and also to elucidate the loss of flow-through surface due to shellfish growth and debris buildup on the coarse bar surface.

5.1.2.1.2 Fine screens

The fine screens include rotating screens, drum screens, band screens, and wedge-wire screens.

Rotating screens Fine self-cleaning rotating screen normally has 3−10 mm openings. This screen is fitted vertically in water intake channels downstream of the bar racks and is well-equipped with a rotating cleaning apparatus, mostly connected to a water spray nozzle for removing the remains from the surface of the screen. This nozzle is provided with cleaning water by pumps, sized for a flow of 45−68 m^3/h and a pressure of 4−7 bar. As the foremost function of the fine screen is to shield the intake pumps against destruction, the actual screen openings must be smaller as compared to the spacing between the impellers of the intake pump.

Band screens The band screens are vertical traveling screens consisting of distinct screening panels having fine mesh openings that are connected to support roller chains, mounted on metal-framed guide tracks. Normally, the screens are designed to enter into a cleaning cycle at a water elevation differential range of 0.1−0.2 m, which typically corresponds to an almost 30% decrease in the screening area. The majority of the commercial band screens travel at a velocity range of 2.0−10.0 m/min.

Drum screens These screens have extensive application in intakes of large seawater desalination facilities in the Middle East. The drum screen consists of a rotating cylindrical frame covered using wire-mesh fabric. The frame will be positioned in a screen construction and supported with a horizontal center shaft that rotates gently on roller bearing. The most frequently used water pattern for the drum screen is in-to-out, where the source water arrives at the cylinder interior side and transfers radially outward, creating a converging flow pattern. As an illustration, in the Hamriyah seawater RO desalination plant, Sharjah (United Arab Emirates), the raw feed water for the desalination process is collected from an open sea intake arrangement and subsequently screened for coarse as well as fine impurities with rotating drum screens of 1.0 mm mesh size.

Wedge-wire screens These are passive screening apparatus situated offshore, which will be straightly joined to the suction end of the intake pump station, thus eliminating the requirement for extra fine-screening or coarse-screening processes. The wedge-wire screens have no mechanical movable parts and, hence, these screens are also known as passive screening techniques. These screens are made up of copper–nickel alloy and have 3.0-mm openings. These screens will have a flow-through velocity of around 0.150 m/s. Currently, the wedge-wire screen size most frequently employed for desalination processes is 3.0 mm.

5.1.2.1.3 Microscreens

In the event that the membrane filtration pretreatment system is chosen for the RO desalination facility, then the fine screens will not be sufficient to remove source water fine-particles in order to safeguard the reliability of the membrane filtration pretreatment facility. Normally, microstrainers, or microscreens, or disk filters could be used for this purpose. The majority microstrainers have screens with small openings (usually 80–400 μm) situated within filtration chambers. The source water reaches the interior side of the strainer, transfers radially outwards through the screen, and leaves through the outlet. As the microstrainers operate at lower differential pressure, they are expected to reduce the intrusion of aquatic organisms from the source water. The disk filtration system can be equipped with an organisms return pipe, through which the marine organisms can be sent back to the source waterbody, thus decreasing their entrainment.

5.1.2.1.4 Cartridge filters

The cartridge filters are fine micro-filters of nominal size between 1 and 25 μm, made up of thin plastic fibers or other fine-filtration media that is fitted around a central tube to develop normal-sized cartridges (Sanza et al., 2007). Mostly, the cartridge filters (Fig. 5.2) are the only screening device between the intake wells and the RO unit in seawater desalination facilities and brackish water desalination facilities, with well intakes producing superior-quality source water. These filters have been considered to be RO membrane safety facilities instead of just screening devices; and their foremost function is capturing particulates present in the pretreated source water that might have transferred through the upstream pretreatment units. This will help to prevent damage or premature fouling of the RO membranes. Typical cartridge filters for RO desalination facilities are usually 1.016–15.24 m long and are fixed in vertical or horizontal pressure vessels (filter housings). Cartridges are rated for the separation of particles of 1.0, 2.0, 5.0, 10.0, or 25.0 μm, with the most commonly used size being 5.0 μm. These filters are normally fixed

FIGURE 5.2 Cartridge filters. Source: *Reproduced from Sanza, M.A., Bonnélyea, V., & Cremerb, G. (2007). Fujairah reverse osmosis plant: 2 years of operation. Desalination, 203(1–3), 91–99.*

downstream of the GMF system (in the event that such a system is used for pretreatment) for capturing fine sand, silt, and particles that may be present in the pretreated water. When the source water is of superior quality, with an SDI value below 2.0, and does not require the removal of particulates by filtration before desalination, then the cartridge filters could be employed as the single pretreatment unit. In this case, cartridge filters serve as a barrier for capturing the fine silt as well as particulates that can intermittently reach the source water at the time of the startup of intake well pump or because of the failure of intake equipment or piping. A characteristic indication of the proper functioning of the pretreatment system of a particular desalination facility is the SDI decrease through the cartridge filters.

5.1.2.2 *Sand removal, sedimentation, and dissolved air flotation*

The objective of the pretreatment systems such as sand removal, sedimentation, and DAF is to reduce the amount of debris, grit, and suspended solids brought together by the plant intake and safeguard the downstream filtration systems from the overloading of solids (Voutchkov, 2017). A properly designed desalination plant intake typically produces source water of less sand as well as lower silt contents. Hence, such desalination facilities normally are not designed with separate sand removal systems. Minor amounts of sand and coarse silt present in the source water are restrained by the sedimentation or filtration systems. On the other hand, if the open intake of the desalination facility is situated nearby an area of extended seasonal wind or wave-driven turbulence, turbulent underwater currents, substantial ship traffic, or regular dredging activities, then a large quantity of sand as well as silt might enter the desalination facility constantly and should be removed using separate systems. Dredging is the process of removing silts and other materials from the bottom of water bodies.

5.1.2.2.1 Settling canals and retention basins

Certain large onshore intakes will be designed with elongated canals that transport the saline source water into retention basins, where it will be presettled and the silt, sand, and huge debris are collected. The saline source water from the reservoir will overflow into the forebay of the screening station or intake pump station, from where it will be transferred into the pretreatment unit of the desalination plant. These canals as well as retention basins will be dredged intermittently or well-equipped with sediment separation or flushing system for minimizing the accumulation of solids over time.

5.1.2.2.2 Strainers

Based on the desalination plant size, the grit removal systems most extensively used are 200–500 μm strainers. The above-mentioned sized strainers could separate sand and silt particles of 0.10 mm or bigger. Strainers have been normally used in medium-scale and small-scale desalination plants, that is, plants with a capacity of 20,000 m³/day or low.

5.1.2.2.3 Sedimentation tanks

Sedimentation is normally used upstream of GMF as well as membrane-based filtration when the source water has a daily average turbidity greater than 30 NTU or experiences turbidity spikes of about 50 NTU or higher that continue for a period of quite a few hours. In the event that sedimentation basins are not provided, then large turbidity spikes might stimulate the pretreatment filters to go beyond their solid holding capacity, which in return might influence filter pretreatment capability and lessen the duration of filter runs. In the design of an ideal sedimentation tank, one of the controlling parameters is the settling velocity (v_s) of the particle to be removed. When particles settle discretely, the particle settling velocity can be calculated, and the basin can be designed to remove a specific size particle. Stokes' law is a formula to determine the rate of sedimentation. The Stokes' law is as follows:

$$v_s = \frac{g(\rho_s - \rho)d^2}{18\mu} \tag{5.1}$$

where v_s = terminal settling velocity, g = acceleration due to gravity (m/s²), ρ_s = density of particle (kg/m³), ρ = density of fluid (kg/m³), d = diameter of sphere (m), μ = dynamic viscosity (Pa · s).

5.1.2.2.4 Dissolved air flotation clarifiers

DAF process is considered to be very appropriate for the separation of floating particulate foulants like grease, oil, algal cells, or other light solid contaminants that are not efficiently separated by sedimentation or filtration process. The DAF systems could normally produce effluent turbidity of less than 0.5 NTU, and could be united in a single structure with dual-media gravity filters for sequential pretreatment of the source water. The DAF process employs very small air bubbles to float organic substances (grease, oil) and light particles present in the source water. The floated solids will be accumulated at the topmost of the

DAF tank and subsequently skimmed off for disposal, whereas the lower turbidity source water will exit near the tank bottom (Rathoure & Dhatwalia, 2016). The different factors to be considered in the DAF system design are solids-loading rate, polymer addition, hydraulic loading, and air-to-solids ratio. The surface loading rate for removal of floatable substances and light particulates by DAF is almost one-tenth of that required for conventional sedimentation. One more benefit of DAF relative to conventional sedimentation is the increased density of the developed residuals. DAF can accomplish effluent turbidity of less than 0.5 NTU, and successfully removes higher concentrations of algae, and displays benefits in the treatment of extremely cold feed water (Van de Venter et al., 2005).

5.1.2.3 Granular media filtration

Currently, the GMF is considered to be the most frequently used saline water pretreatment process for RO desalination plants, apart from cartridge filtration (Voutchkov, 2010). Fig. 5.3 is a schematic representation of a typical seawater desalination plant with GMF pretreatment. The GMF process includes source water filtration by means of one or more layers of granular media (e.g., garnet, silica sand, and anthracite coal). The conventional filter employed for saline water pretreatment is normally a rapid single-stage dual-media (sand and anthracite) unit. On the other hand, in certain circumstances where the source water consists of increased concentrations of suspended solids and organics (total organic carbon (TOC) concentration greater than 6.0 ppm), then the two-stage filtration system will be used. In the above-stated configuration, the first filtration stage is primarily configured for removing coarse solids as well as organics in suspended form. The second stage filters are mainly designed for retaining the fine solids and silt and to separate a fraction of the soluble organics present in the saline feed water by biofiltration.

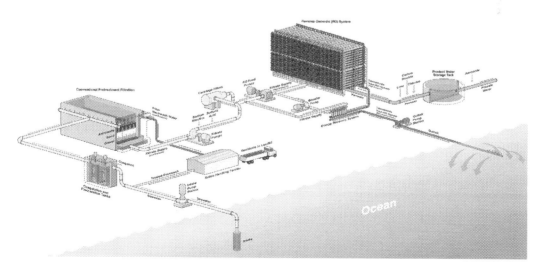

FIGURE 5.3 Typical seawater reverse osmosis desalination plant with granular media pretreatment. *Source: Reproduced from Voutchkov, N. (2010). Considerations for selection of seawater filtration pretreatment system. Desalination, 261(3), 354–364.*

The important filtration system components, filter cells, and filter media are discussed in the following section.

5.1.2.3.1 Filter cells

A typical GMF system comprises several distinct units (vessels or cells) that work in parallel. The filter cell number primarily depends on the total flow which the filters are configured to handle. The manufacturing cost of the filtration system is generally decreased when a lower number of distinct cells are used.

5.1.2.3.2 Filter media

The depth, size, uniformity, and type of filter media are extremely important for the proper functioning of pretreatment filters. Dual-media filter has two filtration media layers; a standard design includes 0.40–0.80 m of anthracite or pumice over 0.40–20 m of sand. Deep dual-media filters will be commonly used in the desalination plant if the filtration system is configured to accomplish an increased separation of soluble organics from source water by biofiltration. For the aforestated case, the anthracite level depth will be increased to 1.5–1.8 m range.

5.1.2.3.3 Models of filtration

The filtration mechanism has been modeled by two different methods: microscopic and macroscopic. Even though the two models are not effective to predict the long-standing filter performance, these models are beneficial to describe the relative importance of the filtration mechanisms and the significance of several design and operating parameters.

1. Microscopic models: Yao and his team, in 1971, developed a transport and attachment model that defines the buildup of particles on a single-media grain "collector" by sedimentation, interception, as well as diffusion process. A modification of this model, termed as trajectory analysis, was developed by Rajagopalan and Tein, in 1976, and it includes additional attractive forces because of van der Waals forces and reduced collisions because of viscous resistance. This model can be used for demonstrating the impact of the uniformity coefficient and dual-media filters on the efficiency of filtration.
2. Macroscopic models: These models do not describe the mechanisms of particle transport or attachment. They are based on a mass balance expression as well as an empirical rate expression for relating the duration of ripening, quality of water, time to terminal head-loss, and time to breakthrough.

A simplified steady-state model allows the calculation of the time to breakthrough (t_B) and the time to the limiting head (t_{HL}):

$$t_B = \frac{\sigma_B D}{v_a(C_0 - C_E)} \tag{5.2}$$

$$t_{HL} = \frac{(H_T - h_L)D}{(k_{HL})(v_a)(C_0 - C_E)} \tag{5.3}$$

where $D =$ depth of filter bed (m), $\sigma_B =$ specific deposit at breakthrough (mg/L), $v_a =$ filtration rate (m^3/s · m^2 of filter surface area; also m/s), $C_E =$ effluent concentration of particles (mg/L), C_O = influent concentration of particles (mg/L), $h_L =$ limiting head (m), $H_T =$ influent concentration of particles (mg/L), and $k_{HL} =$ headloss rate constant (L · m/mg).

A regression analysis of data collected from actual or pilot filters can be used to determine σ_B and K_{HL}.

5.1.2.4 Membrane filtration

Ultrafiltration and microfiltration membranes have been effectively used in the pretreatment of significantly more complex source water relative to seawater, for example, in municipal and industrial waste water for several years. These membranes form a barrier against the passage of bacteria, colloidal materials, and suspended particles. As a result a lower SDI value of the RO feed water is obtained; even with consistent variation in the quality of raw water, enabling operation with a stable as well as a high permeate flux even in a prolonged operation (Vial & Doussau, 2002; Wolf & Siverns, 2004). In a membrane-based pretreatment process, the source water is generally prefiltered using a mechanical screen before the water is fed to the membrane. Relative to the conventional pretreatment, chemical dosing will be remarkably reduced in membrane-based pretreatment (Wolf & Siverns, 2004). Currently, the commercial modules available are pressure-driven spiral wound, immersed hollow-fiber, pressure-driven capillary, and immersed plate modules. The hollow-fiber membrane (HFM) modules are the most commonly used for the pretreatment process (Wolf et al., 2005). Fouling which happens on the membrane surface must be removed. Backflushing using product water or chlorinated product water in combination with air sparging will be extremely effective in removing particles that deposit on the surface of the membrane (Wolf & Siverns, 2004).

5.1.2.4.1 Filtration process

The membrane-based pretreatment processes have four operating modes namely (1) filtration (source water processing), (2) backwash, (3) cleaning, and (4) integrity testing (Voutchkov, 2017). The above-mentioned operating modes are usually monitored and controlled using a programmable logic controller.

Filtration (processing of source water) The membrane filtration of saline source water occurs in the source water processing phase. Based on the specific membrane product and configuration, the filtration could take place either in cross-flow filtration (CFF) mode or direct-flow/dead-end filtration mode. In the case of direct-flow filtration mode, the entire source water will pass across the membrane. However, in the case of CFF mode, just a portion of the source water (normally 90%−95%) will pass across the membrane, and the remaining water (reject) will travel along the membrane feed side; and its movement along the surface of the membrane will generate shearing velocity that removes the solids separated from the saline water out of the membrane. Generally, in the CFF mode, some of the reject streams will be recycled back to the feed stream. The CFF design in these membranes is the same as in the RO membrane element. The main advantage of CFF mode is that membranes could function uninterruptedly.

Backwash   During the filtration process, the solids filtered out of the source water will get accumulated on the feed side of the surface of the membrane. The filtered solids will be intermittently separated from the filtration system by membrane backwashing using concentrate or filtered water. The backwash is commonly activated using a timer and takes place every 30–120 min for almost 30–60 s. Backwash can also be commenced when the transmembrane pressure reaches a specific maximum threshold, after which the membrane system will not be able to achieve the target flux and quality of filtered water.

Cleaning   Periodic membrane backwash and chemically-enhanced backwash will not completely remove the fouling of the membrane, and hence the transmembrane pressure required to generate filtered saline water of the targeted flux and quality increases with the passage of time. As soon as the transmembrane pressure reaches a preset level (usually 1.5–2.5 bar for a pressurized system and 0.7–0.8 bar for a submerged system), the membrane modules must be taken offline and subsequently cleaned using chemicals that will reduce the transmembrane pressure to a practical level. Cleaning of the membrane is normally required in every 1–3 months, and is carried out using a combination of a solution of low pH [sulfuric acid (H_2SO_4) or citric acid ($C_6H_8O_7$)] succeeded by a solution of high pH [sodium hypochlorite (NaClO) and sodium hydroxide (NaOH)].

Integrity testing   The membrane-based pretreatment system will be equipped with proper integrity testing features that permit the recognition of random breaks or punctures in the membrane leafs or membrane fibers; cracks happening in membrane modules, connectors, and piping; and other issues that might happen in the course of membrane manufacturing, operation, or installation. The most commonly used integrity testing for the membrane system is a pressure hold-visual assessment carried out when the system is offline.

5.1.2.4.2 Important components of filtration system

MF and UF membranes are arranged in distinct functional filtration units termed as modules. The most frequently employed membrane module configurations include hollow-fiber, spiral-wound, flat plate, as well as tubular configurations. The modules will be placed in shells, housings, or cassettes that are brought together into bigger membrane filtration system components—vessels and racks. Membranes employed for the pretreatment of saline water are normally made of polysulfone, polyvinylidene diflouride, or polyethersulfone (PES). The entire membrane made up of the above-stated polymers will be hydrophilic in nature, with PES being the most hydrophilic material. As recommended by the membrane manufacturers, the HFM elements might be operated in an outside-in or inside-out flow pattern. The tubular membranes will have an interior tube diameter which will be an order of magnitude greater relative to HFM.

5.1.2.4.3 Service support facilities and equipment

The membrane-based pretreatment system has three categories of service support facilities and equipment: (1) backwash system, (2) clean-in-place system, and (3) cleaning-chemicals feed system. Normally, the backwash unit consists of a filtered water storage tank, backwash pump, as well as air compressor for air backwash. The clean-in-place unit for the membrane-based pretreatment system will be the same as the configuration in the RO system. At times, the same clean-in-place system could be used for both pretreatment and RO membrane cleaning, even though this is not appropriate, particularly for large desalination plants. The cleaning-chemicals feed system generally includes base, acid, sodium bisulfite, and sodium hypochlorite storage and feed systems to service the chemically-enhanced backwash and clean-in-place membrane maintenance works.

5.1.2.4.4 Filter performance

Both UF and MF systems could remove four (99.99%) or more logs of pathogens like *Cryptosporidium* and *Giardia*. The latest generation membranes could also efficiently remove viruses. Characteristic UF membrane element having a pore size of $0.01-0.02\,\mu m$ will separate over four-logs of viruses. MF elements having pore sizes of $0.03\,\mu m$ or less could usually accomplish three-log removal of viruses. Also, both the UF and MF membranes can remove most algae. The operation will not be influenced by moderate or mild algae blooms, when the algae content in the saline source water is below 20,000 cells per milliliter. Further, the UF and MF membrane systems have been demonstrated to be extremely efficient for the removal of turbidity and the reduction of colloidal and nonsoluble organics from saline source waters. Turbidity could be dropped steadily below 0.1 NTU and filter effluent SDI values are frequently under 3 majority of the time.

In a research performed by Kumar et al. (2006), the team compared the UF and MF membranes used in the pretreatment process for determining the differences in the quality of filtrate. In the trials, feed water to the membrane has been prefiltered using a 1-μm filter to create comparable feed water quality in the entire trials. 100 and 20 kDa UF and 0.1 μm MF membranes were examined. The test results demonstrated no remarkable difference between 100 kDa UF and MF membranes with respect to flux decline in the RO element, confirming similar fouling capability of the filtrate. The 20-kDa UF membrane brought about a decreased flux in the RO element, proposing lower membrane fouling. On the other hand, increased pressures must be used to accomplish comparable flux (Kumar et al., 2006). The membrane-based pretreatment has several benefits with respect to the quality of RO feed water and has been effectively used in different desalination plants. In a study performed by Vial et al. (2003), the team used 0.1-μm HFMs for Mediterranean seawater pretreatment. The storms resulted in substantial peaks of turbidity as well as SDI of the source water along with minor increments in turbidity of filtrate and SDI. The membrane pretreatment offered good quality feed water to the RO membrane with an SDI constantly under 1.8, permitting operation at high rates of recovery thereby reducing the running cost of the complete system. In a work by Pearce et al. (2004), the group examined the possibility of membrane-based pretreatment in a reverse osmosis desalination plant at Port Jeddah (Saudi Arabia) as an alternate to its conventional pretreatment unit, which was not able to meet the target quality of feed water in the course of storms and algal bloom. The UF pretreatment membrane feed was obtained from the stream optimized for conventional pretreatment, that is, the stream was subjected to acidification and then ferric chloride ($FeCl_3$) was added. Membrane-based pretreatment with air-enhanced backwash accomplished 2.2 average filtrate SDI and all the values under 3, in the course of algal bloom and storms. Relative to the conventional pretreatment framework, there was an improvement in the quality of RO feed water with respect to the SDI by two units. Also, the increased quality of RO feed water brought about a decrease in the fouling of the RO element by 75%.

The problems with the pretreatment framework at Tampa Bay, Florida, in combination with the operational problems, have considerably increased the expenses for the desalination operation (Wolf & Siverns, 2004). The capacity of the plant is 94,000 m^3/day and primarily featured a two-stage dual-media sand filtration pretreatment framework. On the other hand, the pretreatment has not been sufficient and was not able to meet the targeted level of feed water SDI values, which brought about extreme fouling of the RO membrane elements. The resultant greater usage of chemicals, higher energy consumption, and more frequent replacement of RO essentially increased the operational expenses. The application of immersed hollow fiber UF membranes substituting the conventional pretreatment unit guaranteed stable operation by generating a reverse osmosis feed water with an SDI value of under 1.0 irrespective of feed water turbidity and algae bloom (Wolf & Siverns, 2004). Furthermore, the use of an UF membrane pretreatment unit allowed the RO plant to be operated at recovery rates and flux. The membrane pretreatment improved the permeate flux, provides superior quality feed water to the RO stage, and frequently makes the utilization of cartridge filters needless.

5.1.3 Chemical pretreatment

For reducing the fouling tendency of the saline source water, the water must be conditioned prior to the RO process using different chemicals such as scale inhibitors, flocculants, coagulants, pH adjustment chemicals (e.g., bases and acids), oxidants (e.g., chlorine dioxide and chlorine), and oxidant reduction compounds (e.g., sodium bisulfite) (Voutchkov, 2017). Flocculants along with coagulants will be added for improving the exclusion of particulate as well as colloidal foulants in the downstream pretreatment systems. Following the pretreatment filtration, scale inhibitors can be added to the saline source water for suppressing the mineral scale foulant crystallization on the RO membrane surface.

5.1.3.1 Coagulation and flocculation

Usually, the coagulants are added before the pretreatment sedimentation tank, DAF unit, or filter. The most frequently used coagulants for saline water conditioning in RO desalination, before sedimentation or filtration, are ferric salts (ferric chloride and ferric sulfate). The proper use of coagulant is very important for the efficient as well as the reliable performance of GMF units The coagulation process will reduce the surface charge of the particles present in the source water, and thereby facilitate their agglomeration to form bigger sized particles, called flocs, which will be easy to settle or/and filter by GMF process. The process of formation of flocs is called flocculation. The coagulation process will be very important for saline source water of high turbidity and/or high natural organic matter (NOM) concentration. In such cases, a standard method is to introduce a coagulant of a dose almost two times greater than the level of the actual saline source water turbidity. In the course of coagulation, the alkalinity decreases, and carbon dioxide will be produced (Migliorini & Luzzo, 2004). The addition of coagulating agents to the raw water before a membrane pretreatment reduces the fouling possibility in the membrane pretreatment and produces high-quality feed water to the RO membrane.

To improve the pretreatment, flocculants (polymers) are sometimes applied in addition to the coagulants. On the other hand, polymer addition, even if marginally overdosed, might also lead to organic fouling of the RO membrane. Generally, the possibility for RO membrane fouling because of the polymer overdosing is a major problem, as compared to the benefits of polymer usage.

5.1.3.2 Antiscaling agents

Scaling is the process of precipitation of salts on the surface of the membrane brought about by supersaturation. The scaling process reduces the productivity of the membrane and recovery of water. Different salts can lead to scaling, and using the solubility product, the limiting salt can be estimated. The different types of scalings are discussed in Chapter 3, Guidelines for water quality. Based on the limiting salt, various scale inhibitors are used. In the case of $CaCO_3$ scaling, the inclusion of H_2SO_4 is typically adequate. The scale inhibitors might control the scaling brought about by calcium fluoride, carbonates, and sulfates. In general, the incorporation of antiscaling agents is recommended for seawater RO systems which operate with recoveries of more than 35% (Migliorini & Luzzo, 2004). The antiscaling agents will be needed irrespective of the physical treatment methods chosen. The sodium hexametaphosphate (SHMP) had been applied as an antiscalant; however it was broadly substituted by polymer compounds because of the eutrophicating properties of SHMP and the related disposal issues.

5.1.3.3 Chlorination

Chlorine is a commonly used disinfectant for the disinfection of water. Chlorine is used for the deactivation of majority microorganisms and it is comparatively cheap. Chlorine is effective because it is soluble in water, highly oxidizing, and its germicidal efficiency lasts for days. Chlorination is independent of the physical pretreatment applied for disinfecting the water, and it restricts the biological growth that leads to fouling of filters as well as membranes, and reduces the treatment efficiency. Chlorine is added to the raw water as chlorine gas (Cl_2) or sodium hypochlorite (NaOCl), which in water undergo hydrolyses to hypochlorous acid:

$$NaOCl + H_2O \rightarrow NaOH + HOCl$$

$$Cl_2 + H_2O \rightarrow HCl + HOCl$$

In water, the hypochlorous acid dissociates to form hydrogen as well as hypochlorite ions.

$$HOCl \rightarrow OCl^- + H^+$$

The sum of OCl^-, HOCl, NaOCl, and Cl_2 is considered as free residual chlorine. In the case of continuous chlorination at the point of intake, a concentration of free residual chlorine of about 0.5−1.0 ppm must be retained along the pretreatment line for preventing biofouling (Water Chemistry and Pre-treatment: Biological Fouling Prevention). Chlorine reacts with organic matter and disintegrates it into small fractions that act as nutrients for increased biological growth at the surface of the RO membrane where no chlorine is available to restrict this formation. In order to overcome this problem, periodic chlorine shock injection with an offline RO phase is performed. Subsequent to this, entire chlorinated feed water must be washed away before the startup of operation, thus making sure that no chlorinated water arrives at the RO membranes.

The germicidal efficiency of chlorine depends on the concentration of undissociated HOCl, which is almost a hundred times more efficient than the dissociation product OCl^-. The germicidal effectiveness in this manner increases from 7.5 to 6.5 pH value, and temperature from 25°C to 5°C, however, diminishes with an increase in salinity. In general, the bromide dissolved in seawater will vary the reaction behavior of chlorine:

$$HOCl + Br^- \rightarrow Cl^- + HOBr$$

$$HOBr \rightarrow OBr^- + H^+$$

The primarily formed biocide in seawater is hence hypobromous acid (HOBr), relative to HOCl (hypochlorous acid) in brackish water. Due to the fact that hypobromous acid is a weak acid as compared to hydrochlorous acid, pretreatment could be carried out at increased values of pH than in brackish water pretreatment since more hypobromous acid is available at higher values of pH. During the shock dosing operation, the chlorine concentration and the dosing frequency appear to depend firmly on the site and the process. It was reported that the frequency of dosing will change from one to four times each day with a 5 ppm concentration of chlorine (Migliorini & Luzzo, 2004). Another study discusses about 2−3 ppm ideal shock dosing rates for 2 h a week (Al-Malek et al., 2005). Instead of chlorination, UV radiation can disinfect the raw water; however, till now it is seldom used and should be seen as challenging because of the absence of depot effect.

5.1.3.4 Sodium hydroxide (NaOH) addition

NaOH is normally used for adjusting the pH of the feed seawater to the first or second pass of RO systems designed for improved boron rejection. The addition of NaOH allows the conversion of boron (present as boric acid in seawater) into borate, which will have a bigger molecule of stronger charge, and hence the borate will be easier to be rejected by the RO membranes, as compared to boric acid. In seawater RO plants having two-pass membrane systems, NaOH is generally added into the first-pass permeate due to the fact that this water does not have a substantial quantity of scaling compounds, and hence antiscalants are not required to be added for its treatment in the second RO pass. NaOH dosage typically ranges between 2 and 20 ppm, and based on real-world experience, the majority seawater RO facilities use 12.0−15.0 ppm of NaOH for increased boron rejection.

5.1.3.5 Dechlorination

The dechlorination step must be carried out before the RO stage in view of the fact that the residual chlorine present in the feed water taken to the RO element might cause membrane damage by oxidation. The chlorine resistance changes on the basis of the membrane material. For composite PA membranes, the major membrane manufacturers anticipate membrane degradation after an exposure of 200–1000 h at 1 ppm of free chlorine. The membrane degradation is quicker in alkaline water relative to acidic or neutral water. Further, increased temperatures accelerate the degradation of the membrane by an oxidation process.

Generally, sodium metabisulfite (SMBS; chemical formula $Na_2S_2O_5$) is used for dechlorination because of its increased cost-efficiency (Saeed, 2002). The SMBS reacts with water to form sodium bisulfite ($NaHSO_3$):

$$Na_2S_2O_5 + H_2O \rightarrow 2NaHSO_3$$

The $NaHSO_3$ later reduces the hypochlorous acid (HOCl)

$$2NaHSO_3 + 2HOCl \rightarrow H_2SO_4 + 2HCl + Na_2SO_4$$

Other than SMBS, the activated carbon is also considered to be very efficient to reduce residual free chlorine. A residual chlorine concentration greater than 0.5 ppm in the cartridge filtration stage is needed for preventing the buildup of biofilm (Saeed, 2002).

5.2 Scale control

The scaling of RO membranes might happen once the sparingly soluble salts are concentrated inside the element above their solubility limit. For instance, if a reverse osmosis unit is operated at 50% recovery, then the concentrate stream concentration will be practically double the feed stream concentration. As the plant recovery increases, the scaling risk also increases. Scaling is the particle deposition on a membrane, thereby leading to plugging. Without any means of scale inhibition, RO membranes as well as flow passages inside the membrane elements might scale because of the precipitation of sparingly soluble salts, like strontium sulfate, barium sulfate, calcium sulfate, and calcium carbonate. The majority of the natural waters consist of comparatively higher concentrations of sulfate, calcium, and bicarbonate ions. In a recent study carried out by Matin et al. (2019), some approaches presently considered for the control and prevention of scale formation have been discussed. Fig. 5.4 is the schematic illustration of the key steps in scale formation onto the RO membrane surface over time (Piyadasa et al., 2017). In membrane-based desalination processes at a higher recovery ratios, the solubility limits of calcite and gypsum exceed saturation levels causing crystallization on the surface of the membrane. The surface blockage of the scale leads to a reduction in permeate flux, thereby dropping the process efficiency and raising the costs of operations (Kislik, 2010).

It is economically desirable to avoid scaling formation, although there are efficient cleaners for scale. The scale frequently plugs RO element feed passages, thereby making the cleaning process extremely time-consuming as well as difficult. There also exists the possibility that scaling can damage the surface of the membrane. The most commonly used scale control techniques are acid addition, antiscalant addition, and ion-exchange softening.

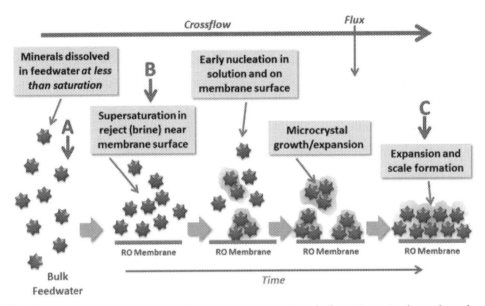

FIGURE 5.4 Schematic representation of the important steps in scale formation onto the surface of reverse osmosis membrane over time. Source: *Reproduced from Piyadasa, C., Ridgway, H. F., Yeager, T. R., Stewart, M. B., Pelekani, C., Gray, S. R., & Orbell, J. D. (2017). The application of electromagnetic fields to the control of the scaling and biofouling of reverse osmosis membranes-A review.* Desalination, 418, 19–34.

5.2.1 Scale control methods

The different scale control techniques are discussed in the below section:

5.2.1.1 Addition of scale inhibitor

The antiscalants or scale inhibitors can be used for controlling the calcium fluoride scaling, sulfate scaling, and carbonate scaling (Olabarria, 2015). In general, there are three types of antiscalants: organophosphonates, polyacrylates, and SHMP. Organophosphonates are considered to be more stable and efficient relative to SHMP. Organophosphonates act as antifoulants for insoluble iron and aluminum, maintaining them in solution. Polyacrylates have been normally recognized for decreasing the formation of silica scale by means of a dispersion mechanism. SHMP is generally of low cost; however, this will not be stable relative to polymer organic antiscalants. Negligible amounts adsorb to the microcrystal surface, stopping additional growth as well as precipitation of the crystals. Proper care should be taken for avoiding the hydrolysis of SHMP in the dosing feed tank. The hydrolysis process would reduce the scale prevention efficiency, and also develop a risk of calcium phosphate scaling. As a result, SHMP antiscalants are normally not suggested.

Polymer organic antiscalants are also more effective relative to SHMP. However, precipitation reactions might happen with antiscalants having a negative charge and polyelectrolytes having a positive charge or multivalent cations (e.g., iron or aluminum). The resultant products will be extremely hard to separate from the membrane elements. The antiscalant manufacturers should provide details regarding the dosage rates, and overdosing must be avoided. It should be made sure that no substantial quantities of cationic polymers are available during the addition of an anionic antiscalant. In the case of RO plants operating on seawater having total dissolved solids of 35,000 ppm range, scaling is not a major issue as in brackish water plants. This is due to the fact that the recovery of seawater plants is limited by the osmotic pressure of the concentrate stream to 30%—45%. On the other hand, for safety purposes, an antiscalant is recommended while operating beyond recovery of 35%.

5.2.1.2 Addition of acid

The addition of acid destroys carbonate ions, removing one of the reactants essential for the precipitation of calcium carbonate (Vedavyasan, 2015). The above-mentioned method is extremely efficient in inhibiting the calcium carbonate precipitation, however ineffective in avoiding other forms of scale. Further drawbacks include the acid corrosivity, the tank cost, monitoring equipment cost and the lowering of pH of the RO permeate.

The majority of natural surface waters and ground waters are nearly saturated with calcium carbonate. The $CaCO_3$ solubility is dependent on the pH, as shown in the below equation:

$$Ca^{2+} + HCO_3^- \leftrightarrow H^+ + CaCO_3 \tag{5.4}$$

By the addition of H^+ as acid, the equilibrium will be moved to the left-hand side for keeping the $CaCO_3$ dissolved. Sulfuric acid is simpler for handling and in several countries easily obtainable as compared to hydrochloric acid; on the other hand, extra sulfate introduced to the feed stream may possibly cause sulfate scaling.

Calcium carbonate has a tendency to dissolve in the concentrate stream instead of precipitating. This tendency could be expressed by the Stiff & Davis Stability Index (S&DSI) for seawaters and the Langelier Saturation Index (LSI) for brackish waters. At the pH of saturation (pH_s), the water will be in equilibrium with calcium carbonate.

The definitions of the Stiff & Davis Stability Index and Langelier Saturation Index are as follows:

$$S\&DSI = pH - pH_s$$

where total dissolved solids will be greater than 10,000 mg/L.

$$LSI = pH - pH_s$$

where total dissolved solids will be lesser than 10,000 mg/L.

Here the methods for predicting the pH_s will be different for S&DSI and LSI.

$$\text{For LSI, } pH_s = (9.30 + A + B) - (C + D).$$

Constant A is a correction for the ionic strength of the sample. This constant is based on TDS.

Constant B takes into account the temperature effect.

Value C is obtained from hardness corresponding to calcium hardness.

Value D is obtained from hardness corresponding to total alkalinity.

$$A = \frac{(\log_{10}[\text{TDS}] - 1)}{10}$$

$$B = -13.120 \times \log_{10}(^{\circ}\text{C} + 273.0) + 34.55$$

$$C = \log_{10}\left[Ca^{2+}\right] - 0.40,$$

where $[Ca^{2+}]$ is in ppm as calcium carbonate.

$$D = \log_{10}\left[\text{alkalinity}\right],$$

where [alkalinity] is in ppm as calcium carbonate.

If water has a negative Langelier Saturation Index value (pH < pH$_S$), it will be undersaturated with regard to CaCO$_3$, and is possibly corrosive. On the other hand, for waters with a positive Langelier Saturation Index value (pH > pH$_S$), the water will be supersaturated with calcium carbonate and the water has the possibility to develop scale. Saturated water will have a Langelier Saturation Index value of zero (pH = pH$_S$) (Al-Rawajfeh et al., 2005).

For controlling the CaCO$_3$ scaling by the addition of acid alone, the S&DSI or LSI in the concentrate stream should be negative. The introduction of acid will be beneficial for controlling just the carbonate scale.

Calculation:

Example 1

A water sample from a tap has the following parameters: TDS = 500 ppm; pH = 8.0, temperature = 25°C; alkalinity as CaCO$_3$ = 100 ppm; calcium as CaCO$_3$ = 250 ppm; calculate the water LSI.

Determine the values of A, B, C, and D first.

$$A = \frac{(\log_{10}[\text{TDS}] - 1)}{10} = A = \frac{(\log_{10}[500] - 1)}{10} = 0.17$$

$$B = -13.120 \times \log_{10}(^{\circ}\text{C} + 273.0) + 34.55 = -13.120 \times \log_{10}(25^{\circ}\text{C} + 273.0) + 34.55$$
$$= 2.08$$

$$C = \log_{10}\left[Ca^{2+}\right] - 0.40 = \log_{10}[250] - 0.40 = 2.0$$

$$D = \log_{10}\left[\text{alkalinity}\right] = \log_{10}[100] = 2.0$$

Now that we know the values of all the constants, we can determine pH$_S$:

$$\text{pH}_S = (9.30 + A + B) - (C + D)$$

$$\text{pH}_S = (9.3 + 0.17 + 2.08) - (2.0 + 2.0)$$

$$\text{pH}_S = 7.6$$

Now, we can determine the water LSI,

$$\text{LSI} = \text{pH} - \text{pH}_S = 8.0 - 7.6 = 0.4$$

5.2.1.3 Ion-exchange softening

Ion exchange is a process in which ions of like charge are exchanged between the solid resin phase and the water phase. The ion-exchange softening method uses sodium which is exchanged for calcium and magnesium ions that are concentrated in the RO feed water (Cheremisinoff, 2001), the following are ion exchange chemical equations:

$$Ca^{2+} + 2NaZ \rightarrow Na^+ + CaZ_2 \tag{5.5}$$

$$Mg^{2+} + 2NaZ \rightarrow Na^+ + MgZ_2 \tag{5.6}$$

where NaZ denotes the sodium exchange resin.

When the entire Na^+ ions have been replaced by magnesium and calcium, then the resin should be regenerated using a brine solution. The ion-exchange softening process removes the requirement for continuous feed of either antiscalants or acid. When properly operated, the ion exchange is totally effective in eliminating all hardness, whether carbonates or noncarbonates.

5.2.1.4 Preventive cleaning

Scaling can also be controlled by preventive membrane cleaning in certain cases. This cleaning will permit the system to operate without any chemical dosage or softening. Normally, these types of systems operate at a lower recovery of almost 25%, and the membrane elements should be replaced every one to two years. Therefore, these systems are mostly for small single-element facilities for producing drinking water from seawater or tap water. The easiest method of cleaning is by a forward flush by concentrate valve opening at lower pressure. Shorter cleaning intervals are noted to be more efficient as compared to the long cleaning times. Also, cleaning could be conducted by employing cleaning chemicals. In the case of batch processes such as wastewater treatment, the membrane cleaning after each batch is a regular exercise. The frequency of cleaning, the chemicals for cleaning, and cleaning procedures must be determined as well as optimized on an individual basis. Proper care must be taken not to allow the development of a scaling layer over time.

5.2.1.5 Operating variable adjustment

In the case where all the other scale-control approaches are ineffective, the plant operating variables must be manipulated in a manner that the scaling does not happen. The dissolved salt precipitation could be avoided by maintaining its concentration lower than the solubility limit. This can be achieved by decreasing the recovery of the system until the concentration of the concentrate is low enough. Also, the solubility will be dependent on pH and temperature. For silica, increasing the pH and temperature increases its solubility.

5.2.2 Economical analysis

The addition of acid is not very economical due to the cost of monitoring equipment, tanks, and acid. If not removed by a degasification process, additional carbon dioxide existing in the permeate of acid-fed systems increases the expense of ion-exchange regenera-tion. The scale inhibitors are comparatively inexpensive products and have no extra expenses. As compared to either antiscalant or acid addition, the main drawback of the soft-ening process is cost factoring in equipment expenses.

5.3 Fouling and its prevention

In general, fouling is the buildup of undesired deposits on the surface of the membrane or inside the membrane pores, leading to a reduction in water flux and salt rejection. As water is the operating environment for the majority RO applications, it is important to understand the behaviors of water and ion transport across the RO membrane, which could indicate how RO fouling happens. The intermolecular interactions of water and ions with the membrane are largely influenced by the membrane structure like the free volume size in the membrane. Otherwise stated, if the structure of the membrane is more compact, then additional energy would be needed for the permeation of water, and consequently, it would be easier to foul, since particles are more susceptible to mount up on the surface of the membrane, termed as surface fouling.

Fouling can be categorized into internal fouling and surface fouling, with respect to the fouling places. The fouling mechanisms of low-pressure membranes (i.e., UF and MF) are somewhat different from those of high-pressure membranes (i.e., RO and nanofiltration). In the case of UF and MF processes, pore adsorption as well as clogging are more frequent, whereas, for RO and nanofiltration processes, surface fouling will be relatively more recur-rent because of the relative compact and nonporous nature of the RO membrane. Relative to internal fouling, surface fouling could be controlled more easily by improving feed water hydrodynamic conditions or chemical cleaning. Scientists are continually carrying out stud-ies for improving the membrane performance and for reducing the RO membrane fouling (Matin et al., 2011; Matin et al., 2011; Misdan et al., 2012; Ozaydin-Ince et al., 2013; Saleem et al., 2020; Saleem & Zaidi, 2020a, 2020b; Shenvi et al., 2015; Zaidi et al., 2019). Based on the feed water compositions and their interactions with the membrane, both internal fouling and surface fouling could be irreversible. With respect to the foulant types, fouling can also be categorized into inorganic fouling, organic scaling, biofouling, and colloidal fouling (Jiang et al., 2017).

5.3.1 Fouling prevention techniques

In the following sections, different fouling prevention techniques are discussed (DuPont Water Solutions, 2020).

5.3.1.1 Colloidal and particulate fouling prevention

Colloids are fine suspended particles and its size range from some nanometers to a few microns. Colloidal fouling is the fouling of the membrane resulting from the colloids or particles deposited on the host materials. The commonly observed colloidal foulants could be categorized into two, namely (1) inorganic foulants and (2) organic macromolecules. The main organic macromolecules present in the water are proteins, polysaccharides, and some NOMs, whereas the major inorganic foulants include silica, iron oxides/hydroxides, and aluminum silicate minerals. Fig. 5.5 shows the diagrammatic illustration of colloidal fouling and scanning electron microscopy image of a fouled membrane (Saleh & Gupta, 2016).

Colloidal fouling of RO elements could significantly deteriorate the performance by losing productivity and also the rejection of salt (Ismail et al., 2019). The source of colloids or silt in RO feed waters is diverse and generally includes iron corrosion products, colloidal silica, clay, as well as bacteria. Pretreatment chemicals used in a clarifier (e.g., ferric chloride, aluminum sulfate) could be used for combining the fine particle sized colloids leading to its agglomeration to form larger particles that could be eliminated very easily by either cartridge or media filtration. Different methods or indices are proposed for predicting the colloidal fouling potential of feed waters, including SDI, turbidity, and modified fouling index (MFI). The SDI is considered to be the most extensively used fouling index.

Turbidity is a measure of the degree to which the water loses its transparency because of the presence of suspended particulates. Turbidity is due to particulate matters like silt, clay, finely divided organic and inorganic matter, plankton and other microbes. The turbidity of RO feed water must be less than 1 NTU. The SDI will be an indicator of the quantity of particulate matter present in water and it is related to the fouling tendency of RO systems. SDI is carried out on the basis of determining the rate of plugging a

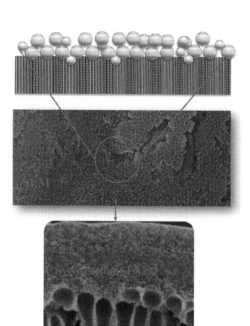

FIGURE 5.5 Diagrammatic illustration of colloidal fouling and scanning electron microscopy (SEM) image of a fouled membrane. Source: *Adapted from Saleh, T., A., & Gupta, V. K. (2016). Membrane fouling and strategies for cleaning and fouling control. In: Nanomaterial and polymer membranes (pp. 25–53). Amsterdam, The Netherlands: Elsevier.*

45-μm filter employing a constant 30 psig feed pressure for a definite time period. The MFI will be proportional to the suspended matter concentration and is considered to be a more accurate index relative to the SDI in order to predict the water tendency to foul RO membranes.

The different techniques for preventing colloidal fouling are discussed in the next section.

5.3.1.1.1 Media filtration

In media filtration, the removal of colloidal and suspended particles is in accordance with their deposition on the filter grain surface, when the water passes through a bed of these filter media. With a properly designed and functioned filter, an SDI value of less than 5 could generally be accomplished. The most commonly used filter media in the treatment of water are anthracite as well as sand. As compared to the single sand filter media, the dual-filter media with anthracite over sand will allow more penetration of the suspended particles into the filter bed, thereby leading to more effective filtration and extended runs between cleaning cycles. The design depth of the filter media is a minimum of 0.8 m. In a dual filter media, the filters will be generally filled with 0.5 m of sand covered with 0.3 m of anthracite.

There are two types of filters used, pressure and gravity filters (Ratnayaka et al., 2009). The filtration flow rate is usually 10–20 m/h, and the backwash rate will be in the range of 40–50 m/h. The available pressure is typically around 5 m of head for gravity filters, and 2 bar to greater than 4 bar for pressure filters. In the case of feed waters with an increased fouling potential, the flow rate of less than 10 m/h or/and a second pass media filtration are preferred. At the time of operation, the incoming source water to be filtered arrives at the filter top, subsequently gets percolated through the filter bed, and is carried away by a collector system at the filter bottom. At times, when the differential pressure increase between the outlet and inlet of the pressure filter is 0.3–0.6 bar, and almost 1.4 m for the gravity filter, then the filter should be backwashed and rinsed for carrying away the deposited matter. Normally the backwash time will be almost 10 min. Before placing a backwashed filter back into operation, it should be rinsed for proper draining, until the filtrate meets the required specification.

5.3.1.1.2 Oxidation–filtration

Some well waters, usually brackish waters, will be in a reduced state, and these types of waters might consist of divalent manganese and iron, and in certain cases ammonium and hydrogen sulfide. The above-mentioned water will have no oxygen and hence termed as anoxic. Generally, the oxygen has been already consumed by microbial processes, as the water is already contaminated with biodegradable organic substances. A technique for handling anoxic waters is oxidizing manganese and iron by air, potassium permanganate, or sodium hypochlorite. The hydroxides developed can be subsequently removed by the media filtration process. Also, the hydrogen sulfide present can be oxidized to elemental sulfur, and then removed by media-filtration. Oxidation along with filtration could be performed in a single-step by employing a filter media having the capability for oxidizing divalent manganese as well as iron by electron transfer. Greensand (green mineral glauconite) is considered to be such a granular medium, and it could be regenerated with potassium permanganate when the oxidizing capability is drained. Subsequent to regeneration, the residual potassium permanganate should be properly rinsed out for avoiding the oxidation damage of the membranes. The aforementioned method can be used when Fe^{2+} present in the source water is less than 2 ppm. In the case of greater Fe^{2+} concentration, potassium permanganate could be constantly dosed into the input water stream of the filter.

5.3.1.1.3 In-line filtration

The media filtration efficiency to decrease the SDI value could be significantly increased if the colloids in the source water have undergone coagulation or/and flocculation before the filtration process. In-line filtration could be used when the SDI value of the source water is just slightly above 5.0. A coagulant can be added into the source water stream, efficiently mixed, and the micro-flocs developed can be instantly removed by the media filtration process. Ferric chloride and ferric sulfate can be used for destabilizing the colloid negative surface charge and to entrap them into the ferric hydroxide micro-flocs formed. Also, aluminum coagulants are considered to be efficient, however, not suggested due to the possibility of having fouling issues with residual aluminum. Fast dispersion and mixing of the coagulant is very significant. For strengthening the hydroxide micro-flocs and thus improving their filterability, or/and for bridging the colloidal particles together, flocculants could be applied alone or along with coagulants. Flocculants are considered to be soluble high molecular weight (MW) organic compounds (Niaounakis & Halvadakis, 2006), and through the various active groups present, they might be negatively charged (anionic), positively charged (cationic), or close to neutral (nonionic).

5.3.1.1.4 Coagulation–flocculation

In the case of source water consisting of high contents of suspended matter leading to a high SDI value, the typical coagulation–flocculation technique is recommended. The hydroxide flocs will be permitted to develop as well as settle in explicitly developed reaction chambers. The hydroxide sludge will be subsequently separated, and the supernatant water will additionally undergo the media filtration process. In this coagulation–flocculation process, either a compact coagulation–flocculation reactor or a solids-contact type clarifier might be used.

5.3.1.1.5 Ultrafiltration or microfiltration

UF or MF membranes can remove almost all suspended matters in the source water. For UF, dissolved organic compounds will also be removed, in accordance with their molecular mass as well as on the membrane molecular mass cut-off. Therefore, an SDI value less than one can be achieved by using a properly designed and well-maintained MF or UF system. There are both modes, that is, CFF and dead-end/direct-flow filtration, in UF and MF. In dead-end filtration, 100% of the source water passes across the MF or UF filter medium (i.e., 100% recovery). For CFF, there will be three streams (1) feed, (2) permeate, and (3) reject or concentrate. In MF and UF HFMs, there will be two categories of configurations: (1) outside-in flow and (2) inside-out flow. In the case of outside-in flow mode, there will be added flexibility in the quantity of feed water to flow around the hollow fibers, while the inside-out mode will have to consider the pressure drop across the internal volume of the hollow fibers. However, the inside-out mode provides extra homogenous flow distribution across the hollow fiber bore, relative to the outside-in flow. The CFF in MF and UF systems will operate at a high flux rate and increased recovery, and hence backwashing will be commonly employed for reducing the fouling. Membrane backwashing is the reversal of flow through a membrane system compared with the normal flow direction needed for permeate production.

5.3.1.1.6 Cartridge microfiltration

The cartridge filter is considered to be a protection device for shielding the RO membranes and the high-pressure pumps (HPPs) from suspended particles. This filter, having an absolute pore size under 10 μm, is the recommended minimal pretreatment needed for each RO system. In general, cartridge MF is the final step in the pretreatment process. It is recommended to have 5 μm absolute pore size. With better prefiltration techniques, only low RO membrane cleaning will be needed. In the case of a high possibility of fouling with metal silicates or with colloidal silica, it is recommended to have a cartridge filtration with one to 3 μm absolute pore size.

This cartridge filter must be made of a synthetic nondegradable material such as polypropylene or nylon (Scott, 1995), and well-equipped with a pressure gauge for indicating the differential pressure, thus signifying the fouling extent. Frequent examinations of used cartridges provide useful information as regards the fouling risks and cleaning requirements. In the event that the differential pressure across the cartridge filter increases quickly, it is considered to be a sign of potential problems in the pretreatment process or in the source water supply. Replacing these filters more frequently than every 1−3 months generally specifies a problem with the pretreatment process.

5.3.1.1.7 Operational and design considerations

The colloidal fouling prevention not only just depends on the proper pretreatment selection, but also on the system operation and design. For example, when surface water is pretreated by coagulation−flocculation and UF processes, then the RO unit can function with an increased product water flux, and practically no cleaning will be needed. On the other hand, if the water is only cartridge filtered, then the RO unit will require additional membrane area, and very frequent cleaning as well as maintenance will be needed. An ineffective pretreatment system could be somewhat compensated by increasing the membrane area, by system modification, and by very regular or/and harsh cleaning. On the other hand, improving the pretreatment system implies lower membrane costs. For minimizing the pretreatment work and to improve the quality of the feed water, it is recommended to use the source water of the best quality. Source water contamination by waste water effluent might lead to severe problems in the RO plant.

5.3.1.1.8 Other methods

There are also several other techniques, different from the previously mentioned techniques, for preventing the colloidal fouling. Lime softening can also be used for the removal of iron as well as colloidal matter from the source water. The strong acid cation (SAC) exchange resin softening will remove hardness and also it will remove low concentrations of aluminum and iron, which could cause membrane fouling. The softened water is considered to demonstrate a lower fouling tendency relative to unsoftened water due to the fact that the multivalent cations will stimulate the adhesion of naturally occurred colloids, which will be generally negative charged. The iron separation efficiency will be dependent on the species of iron present. Fe^{3+} and Fe^{2+} are removed properly by the SAC exchange resins, if more than 0.05 mg/L, have a propensity to cause membrane fouling, and catalyze its degradation. Some scaling inhibitors, also known as antifoulants, will be able to handle iron. The above-stated technique is used for a comparatively low concentration of iron.

5.3.1.2 Biological fouling prevention

Biofouling is the process in which microorganisms adhere as well as propagate on the surface of the membrane. Otherwise stated, biofouling is the development of biofilm to an undesirable degree, which can lead to high operational expenses. Fig. 5.6 shows different factors influencing the attachment of bacteria to the membrane surface. Biofouling is considered to be more complex relative to other fouling types. There are two main components of biofilms, namely (1) the bacteria and (2) the extracellular polymeric substances excreted by bacteria in the course of the metabolism process.

The different biofouling prevention methods are discussed in the below section.

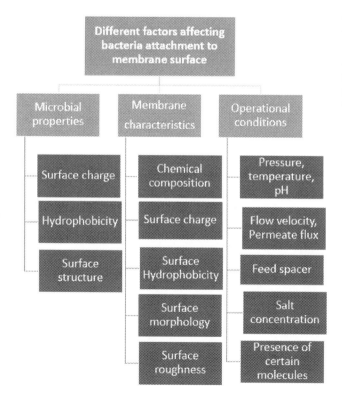

FIGURE 5.6   Different factors affecting bacteria attachment to the membrane surface. Source: *Adapted from Jiang, S., Li, Y., & Ladewig, B. P. (2017). A review of reverse osmosis membrane fouling and control strategies. Science of the Total Environment, 595, 567–583.*

5.3.1.2.1 Chlorination/dechlorination

For biological fouling prevention, chlorination can be applied as a reverse osmosis pre-treatment technique. At the intake, chlorine is continuously added, and about 20–30 min reaction time must be allowed. A free residual chlorine concentration of almost 0.5–1.0 ppm must be maintained over the entire pretreatment line. However, dechlorination upstream of the membranes is needed for protecting the membranes from oxidation. Continuous chlorination along with dechlorination of the feed water has been an ordinary practice for years. The chlorine will react with the organic substances present in the source water and disintegrates it to form biodegradable fractions. As chlorine is not available on the membranes, microbes could develop with an enriched nutrient supply, except the situation where the system will be sanitized regularly. Consequently, the continuous chlorination–dechlorination technique is turning out to be less popular.

Rather than continuous chlorination, chlorine can be rather applied offline periodically to the pretreatment section. The entire chlorine-containing feed water should be rinsed out cautiously before the system is operated again, and the chlorine absence should be verified (e.g., by monitoring the oxidation–redox potential).

5.3.1.2.2 Addition of sodium bisulfite (NaHSO₃)

During normal plant operation, sodium bisulfite ($NaHSO_3$) could be introduced into the feed stream, for a limited duration. The aforesaid intermittent application is generally termed as shock treatment. About 500–1000 ppm sodium bisulfite can be dosed for about 30 min, in a standard application. This treatment could be performed every 24 h or just when the biological growth is suspected. The water produced at the time of sodium bisulfite dosage might consists of certain bisulfite, depending on the operating conditions, the type of membrane and the concentration of the feed. Based on the requirements of product water quality, this water produced during shock treatment can be either used or discarded. The bisulfite will be very effective against aerobic bacteria; however it is not effective against anaerobic microbes.

5.3.1.2.3 Addition of DBNPA (2,2, dibromo-3-nitrilo-proprionamide)

In RO systems with bioactive source feed water, a biofilm could form within 3–5 days after inoculation with possible organisms. Therefore, the widely used sanitization frequency is every 7 days at the time of low biological activity (winter) and about 3–5 days at the peak time of biological activity (summer). The optimum sanitization frequency would be dependent on the location and should be identified by the operation features of the RO unit. The typical approach to use 2,2-dibromo-3-nitrilo-propionamide (DBNPA) is by intermittent dosing. The 2,2-dibromo-3-nitrilo-propionamide quantity used will depend on the biological fouling severity. Due to the fact that DBNPA can be deactivated by reducing agents, an increased concentration of DBNPA is needed if there is residual reducing agent present in the feed water. Alkaline cleaning is also recommended for removing the dead biofilm.

5.3.1.2.4 Biofiltration

Biofiltration is the process of biological water treatment for reducing the organic components that either causes organic fouling or provide carbon sources for the biofilm growth on the surface of the membranes. Biofiltration processes include bank filtration for river source, soil passage as well as slow sand filtration. Filter beds of bio-active granular activated carbon (GAC) have been extensively employed in public waterworks, where the carbon filter biological activity can be additionally increased by treating the feed with ozone. When such filters work at adequately low filter velocities (2.0–10.0 m/h) and with satisfactorily high beds (height 2.0–3.0 m), the majority of the biolife activity will take place in the topmost section of the filter bed, and the filtered water will be practically without any nutrients or bacteria.

5.3.1.2.5 Ultrafiltration/microfiltration

UF and MF processes can eliminate microbes and particularly algae that are very problematic in certain cases to remove using normal techniques (Gupta et al., 2019). The pretreatment processes using UF and MF membranes will help to prevent and control the start of biofouling. The UF and MF membranes must be prepared from a chlorine-resistant material for tolerating the intermittent treatment using biocides. On the other hand, these membranes will not remove low-MW fragments of organic substances and other compounds that serve as nutrients for microbes.

5.3.1.2.6 Ultraviolet irradiation

UV irradiation at a wavelength of 254 nm is considered to possess a germicidal effect (Kowalski, 2009). Currently, UV irradiation is used particularly in small-scale plants. In this process, no chemicals will be required, and the equipment requires very little attention except for intermittent cleanings or the mercury-vapor lamp replacement. However, the UV irradiation treatment is limited to moderately clean waters due to the fact that organic matter and colloids can reduce radiation penetration.

5.3.1.2.7 Other sanitization agents

Ozone is considered to be a stronger oxidizing agent as compared to chlorine. On the other hand, ozone will decompose readily. For killing the entire microbes, a specific level of ozone should be maintained. Also, the resistance of the construction materials against ozone must be taken into account. Typically, stainless steel can be used. Moreover, ozone removal should be carried out carefully for protecting the membranes, and for this purpose UV irradiation has been successfully used.

For controlling the algae growth, copper sulfate can also be employed. Usually, copper sulfate can be fed uninterruptedly at concentrations of 0.10–0.50 mg/L. The water pH should be less for preventing copper hydroxide precipitation.

5.3.1.3 Organic fouling prevention

Organic substances' adsorption on the surface of the membranes can cause water flux loss, which will be irreversible in certain instances (Pandey et al., 2012). The process of adsorption is favored with high MW compounds, when the compounds are positively charged or hydrophobic. Higher value of pH will help to inhibit fouling, due to the fact that both the membrane and several organic substances will be negatively charged at a pH greater than 9. Organics present as an emulsion will be forming an organic film on the surface of the membrane. Hence, the previously stated organics should be eliminated in the pretreatment process.

Organics occurring in natural water will be normally humic substances with total organic carbon concentration in the range of 0.5–20 ppm. When total organic carbon concentration goes beyond 3 ppm, the pretreatment process must be considered. Humic substances could be removed using a coagulation process with hydroxide flocs, adsorption on activated carbon, or by UF process. Activated carbon or coagulation process can be applied when greases and oils contaminate the RO feed water at a concentration above 0.1 ppm. These substances will be freely adsorbed onto the surface of the membrane. However, these substances could be cleaned off, using alkaline cleaning agents only when the water flux has not decreased by more than 15%.

5.3.1.4 Membrane degradation prevention

In addition to the fouling possibility of some matters present in the RO feed water, the membrane element chemical resistance against these substances should also be properly considered. In general, all oxidizing agents have the ability to damage the membrane and should be rejected. The membrane elements will be stable against a majority of other chemicals in the pH range of 2.0–11.0, provided that such chemicals are not in an organic phase and are dissolved.

5.3.1.5 Prevention of iron, manganese, and aluminum fouling

Iron fouling is considered to be extremely frequent, and the same as any other fouling, it will also cause a performance reduction in the membrane system, precisely the loss in flux. Furthermore, the presence of iron makes the membrane more prone to oxidation damage. Fortunately, this type of fouling is relatively easy to clean. This iron fouling occurs more commonly relative to manganese fouling due to the fact that the iron oxidation takes place at an extremely low pH. Therefore, fouling could be a problem even if the SDI is under 5, and the iron concentration in the RO feed water is less than 0.1 ppm. Water with low alkalinity typically has a higher iron concentration, as compared to water with high alkalinity, because the Fe^{2+} content is typically limited by the $FeCO_3$ solubility.

One method for preventing the membrane fouling is to inhibit the oxidation as well as precipitation of manganese and iron, by maintaining the water in its reduced state. The water exposure to any oxidizing agents or air during the entire RO should be avoided. A low pH value is advantageous to retard the Fe^{2+} oxidation. At pH value less than 6 and oxygen less than 0.5 ppm, the maximum permissible Fe^{2+} concentration will be 4.0 ppm.

Aluminum silicate fouling is observed in the primary stage and final stage of RO plants. Even small concentrations of aluminum (e.g., 50 ppb) may lead to a performance decline. For minimizing the aluminum fouling, the aluminum concentration in the feed water should be below 0.05 ppm.

5.4 Summary of pretreatment options

A summary of pretreatment options for scaling and fouling is provided in Table 5.1.

TABLE 5.1 Summary of pretreatment options for scaling and fouling.

Pretreatment	$CaCO_3$	$CaSO_4$	$SrSO_4$	$BaSO_4$	CaF_2	SiO_2	SDI	Fe	Al	Bacteria	Oxi. agent	Org. matter
Addition of acid	E							P				
Scale inhibitor antifoulant		E	E	E	E	P		P				
Softening with ion exchange	E	E	E	E	E							
Dealkalization with ion exchange	P	P	P	P	P							
Lime softening	P	P	P	P	P	P	P	P				P
Preventive cleaning	P					P	P	P	P	P		P
Operation parameter adjustment	P	P	P	P	P	E				P		
Media filtration						P	P	P	P			
Oxidation filtration							P	E				
In-line coagulation						P	P	P	P			P
Coagulation—flocculation						P	E	P	P			E
Microfiltration—ultrafiltration						E	E	P	P	P		E
Cartridge filtration						P	P	P	P	P		
Chlorination										E		
Dechlorination											E	
Shock treatment										P		
Preventive biological treatment										P		
GAC filtration										P	E	E

$CaCO_3$, calcium carbonate; $CaSO_4$, calcium sulfate; $SrSO_4$, strontium sulfate; $BaSO_4$, barium sulfate; CaF_2, calcium fluoride; SiO_2, silicon dioxide; SDI, silt density index; Fe, iron; Al, aluminum; Oxi agents, oxidizing agents; Org. matter, organic matter; E, very effective; P, possible.
Courtesy: DuPont Water Solutions (2020, April). FilmTec™ RO technical manual; Version 3; Form No. 45-D01504-en, Rev. 3. Reproduced from DuPont Water Solutions.

5.5 Pretreatment system in some of the commercial desalination plants

5.5.1 Hamriyah seawater reverse osmosis desalination plant, United Arab Emirates

The Hamriyah desalination plant is part of the first phase of the Hamriyah Station for Power Generation and Water Desalination project, which generates 2500 MW of electricity along with 140 million gallons of potable water per day. The main components of this facility include an open seawater intake system, pretreatment system, two treated-water storage tanks, and a product pumping station. Support infrastructure of this facility includes a DAF unit, UF unit, chemical dosing systems and low-voltage switchgear, RO unit, external storage tanks, chemical store, and main electrical power and control building (Hamriyah S.W.R.O. Desalination plant, Sharjah).

The desalination plant uses UF during pretreatment and RO for removing dissolved salts and impurities present in the saline source water. Pretreatment techniques used by the plant helps to prevent the clogging of the RO membranes. This desalination facility uses integrated membrane solutions (IMS) from hydranautics in the treatment process, for removing the impurities and dissolved salts, while decreasing the contaminants like microorganisms. The IMS technology comprises hydranautics energy-saving and high-rejection SWC5 SWRO membranes and hydranautics HydraCap60 UF modules.

Saline water for the process is collected from an open sea intake framework and then screened for coarse and fine impurities using rotating drum screens of mesh size 1 mm. Two intake pumps (one working and the other standby), each with a capacity of $11.0 \, \text{m}^3/\text{h}$ flow, are fitted downstream the drum screen. One pump transfers the seawater to the DAF system. Individual lines transfer the feed water to the intake pump station. The water will get mixed with ferric chloride in the flocculation basins of the DAF unit, which is followed by the DAF process for capturing the suspended particles present in the water. Intermediate tanks are covered with concrete slabs and/or glass reinforced plastic (GRP) covers for preventing any debris from passing to the UF feed pumps and to the recirculation pumps. UF feed pumps, four operating and one standby, will draw the water from the intermediate tank downstream of the DAF unit. The UF fibers are installed with an automatic strainer system upstream of the UF plant for preventing the clogging. The UF unit, which removes the contaminants, comprises two lines consisting of 12 UF blocks each and connected to auxiliary equipment for backflush, chemical-enhanced backflush as well as air integrity testing.

5.5.2 FUJAIRAH 2 RO plant, United Arab Emirates

The pretreatment in Fujairah 2 RO plant consists of the four following processes (Veolia):

Step 1: Open seawater intake: The Gulf of Oman seawater is collected using three parallel pipelines from a distance of 500 m offshore to a deep intake well. Before entering the intake well, the source water is screened employing static and traveling band screens for preventing any aquatic life or debris from reaching the desalination facility. The intake well is properly equipped using four vertical shaft turbine pumps (three operating and one on standby) for supplying the appropriate quantity of source water to the RO facility at all times and on the basis of the production demand. The source water will be chlorinated on an intermittent basis for preventing marine growth in the plant structures.

Step 2: DAF: The proper pretreatment of source water aims at removing the solids before the water reaches the RO system. The primary pretreatment step includes coagulation and flocculation, succeeded by the DAF process. The microscopic air bubbles produced will be attached to the small flocs as well as suspended solids, and float to the surface of the water, where the buildup happens as a brown layer. The layer must be intermittently separated as waste and is undergone treatment in the sludge treatment plant. Sixteen DAF process units have been applied with the specific objective of maximizing the removal of algal blooms and red tides. The exclusive DAF process will act as a barrier and ensures the RO plant production capacity at all times regardless of the quality of the source water. This distinctive process has turned out to be the standard in RO pretreatment in the Gulf Region; however it was initially introduced in the Fujairah 2 RO facility.

Step 3: Dual-media gravity filtration: A dual layer of pumice and quartzite sand functions as the filtration media inside the 12 filters. Subsequent to filtration times of 40−60 h, filters are subsequently backwashed individually for removing all the accumulated flocs as well as suspended materials. The unclean backwash water is then transported and treated in the sludge treatment plant. With the use of ten booster pumps, the filtered source water is pumped through the cartridge filters.

Step 4: Cartridge filtration: This seawater polishing step consists of 16 cartridge filter vessels, each with 224 cartridge filters. In this step, the entire suspended materials greater than 10 μm in size will be separated. The spent cartridge filters will be replaced and disposed of regularly in accordance with the strict disposal regulations.

5.6 Case studies

In a review paper by Kavitha et al. (2019), the conventional process, MF, UF, and integrated membrane systems for the seawater pretreatment were analyzed. Also, their benefits as well as limits, and the influence of the pretreated water on the efficiency of RO membranes, were examined by considering the studies carried out by researchers and by considering certain case studies.

5.6.1 Case Study 1: Pretreatment process in Jeddah desalination plant, Saudi Arabia

The pretreatment processes at the Jeddah desalination plant, Saudi Arabia was investigated by Abdul Hadi (1997), and at this desalination plant, the SWRO systems were introduced to the already present multistage flash units. The source seawater is treated with disinfectant NaClO and coagulant iron(III) chloride before pumping the water to the dual-media filter. This filter consists of sand and anthracite. For controlling the pH and preventing the scale formation, sulfuric acid is injected to the filtered source water. For removing particles bigger than 10 μm and for preventing the device blockage, the filtered water then passes through the micro-cartridge filter. For eliminating the residual chlorine, the stream is injected with SMBS downstream of cartridge filters. Subsequently, with the use of HPPs, the filtered water is pumped to the RO modules. In the course of plant operation, it was observed that the RO system experienced problems in the pretreatment systems. The problems were as follows: (1) oxidation caused by the chlorine reaction led to the membrane degradation, (2) uncontrolled SDI limit during seasonal variations of source seawater SDI.

Subsequent to 18 months of operation, a drop in the membrane efficiency was confirmed by the unusual increase in the permeate conductivity signifying a rise in the salt passage across the membranes. From the analysis of different test results, it was confirmed that the oxidation by chlorine was the main cause for the degradation of the membrane. It was recommended to have a dechlorination process prior to the HPP for preventing the system from oxidation. By performing different studies and finding the values of the permeate conductivity and the bacterial colony count, it was demonstrated that the intermittent chlorine injection mode prevented the membrane degradation and the biofouling. An appropriate solution was established in the form of a coagulant aid. Tests with an anionic, a cationic, and a nonionic polyelectrolyte were conducted in a smaller-scale setup of a dual-media filter unit. From the test results obtained, it was confirmed that dosing cationic polyelectrolyte at a rate of 0.1 mg/L with 0.3 mg/L of ferric chloride helped to increase the feed SDI values.

5.6.2 Case Study 2: Pretreatment process in Addur desalination plant, Bahrain

The pretreatment process in the Addur desalination plant in Bahrain was investigated by Khalid and Hussain (2004). The facility suffered from several problems and working at its lowermost production capacities. This facility experienced problems like RO membrane fouling, heavy chemical cleaning programs, severe production loss, and the replacement of the RO membranes within a short time of commissioning. This desalination facility is based on RO process and is intended to produce 10 million imperial gallons per day of permeate. Water undergoes pretreatment processes such as chlorination, pH adjustment, MF, multimedia filtration, coagulation, and dechlorination.

The inefficiencies found in this plant were illustrated by increased salt passage, quick increase in the differential pressure across the membrane, decline in permeate flow, and fast water flux reduction. The different factors that caused the reduced performance of the plant were the insufficient pretreatment processes, inadequate operational experience, improper structural design, inefficient membrane cleaning systems, RO unit location, plant location, and deficiency in the process plant design. For plant rehabilitation, it was recommended to change the pretreatment processes. For this saline water feed with wide-ranging SDI values from season to season, an extensive case-to-case basis pretreatment process was recommended which included comprehensive chlorination as well as backwashing of dual-media filters, pipeline sterilization by applying a strong oxidizing agent, moving the dosing point of antiscalants just before the HPP prior to the RO module, and changing the chemical dosage (coagulants, coagulant aids, chlorination chemicals, dechlorination chemicals, and pH adjustment chemicals). Furthermore, the number of chemical cleaning cycles were increased from one in the first year to 21 times in the second year, and in the following year the cleaning cycles were 46 times. The product water flushing of RO membranes was carried out to improve the pressure difference across the RO membranes. After the application of the aforestated modifications, the plant was anticipated to operate at a capacity of 5 million gallons per day of superior water quality.

Reference

Abdul Hadi, Hasan Al. Sheikh (1997). Seawater reverse osmosis pretreatment with an emphasis on the Jeddah plant operation experience. *Desalination, 110*, 183−192.

Al-Malek, S., Agashichev, S. P., & Abdulkarim, M. (2005). *Techno-economic aspects of conventional pretreatment before reverse osmosis (Al-Fujairah hybrid desalination plant)*. Singapore: IDA World Congress.

Al-Rawajfeh, A. E., Glade, H., & Ulrich, J. (2005). Scaling in multiple-effect distillers: The role of CO2 release. *Desalination, 182*(1−3), 209−219.

Anis, S. F., Hashaikeh, R., & Hilal, N. (2019). Reverse osmosis pretreatment technologies and future trends: A comprehensive review. *Desalination, 452*, 159−195.

Badruzzaman, Mohammad, Nikolay, Voutchkov, Lauren, Weinrich, & Joseph, G. Jacangelo (2019). Selection of pretreatment technologies for seawater reverse osmosis plants: A review. *Desalination, 449*, 78−91.

Cheremisinoff, N. P. (2001). *Handbook of water and wastewater treatment technologies*. Butterworth-Heinemann.

Curcio, E., & Drioli, E. (2009). Membranes for desalination. In G. Micale, L. Rizzuti, & A. Cipollina (Eds.), *Seawater desalination. Green energy and technology*. Berlin, Heidelberg: Springer, https://doi.org/10.1007/978−3−642−01150-4_3.

DuPont Water Solutions (2020, April). FilmTec™ RO technical manual; Version 3; Form No. 45-D01504-en, Rev. 3.

Fritzmann, C., Lowenberg, J., Wintgens, T., & Melin, T. (2007). State-of-the-art of reverse osmosis desalination. *Desalination, 216*(1), 1−76.

Fujairah 2 Reverse Osmosis Desalination Plant <https://www.veolia.com/middleeast/sites/g/files/dvc2506/files/document/2017/03/Fujairah_II_Brochure.pdf> Accessed 2.08.2020.

Gupta, S. K., Dhandayuthapani, K., & Ansari, F. A. (2019). *Techno-economic perspectives of bioremediation of wastewater, dewatering, and biofuel production from microalgae: An overview. In Phytomanagement of polluted sites* (pp. 471−499). Elsevier.

Hamriyah S.W.R.O. *Desalination plant, Sharjah* <https://www.water-technology.net/projects/hamriyah-swro-desalination-plant-sharjah/> Accessed 2.08.2020.

Ismail, A. F., Khulbe, K. C., & Matsuura, T. (2019). RO membrane fouling. *Reverse Osmosis, 189*−220. Available from https://doi.org/10.1016/b978-0-12-811468-1.00008-6.

Jiang, S., Li, Y., & Ladewig, B. P. (2017). A review of reverse osmosis membrane fouling and control strategies. *Science of the Total Environment, 595*, 567−583.

Kavitha, J., Rajalakshmi, M., Phani, A. R., & Padaki, M. (2019). Pretreatment processes for seawater reverse osmosis desalination systems—A review. *Journal of Water Process Engineering, 32*, 100926.

Khalid, Burashid, & Hussain, Ali Redha (2004). Seawater RO plant operation and maintenance experience: Addur desalination plant operation assessment. *Desalination, 165*, 11−22.

Kislik, Vladimir S. (2010). *Progress in liquid membrane science and engineering. In Liquid membranes* (pp. 401−437). Elsevier.

Kowalski, W. (2009). *UVGI disinfection theory. Ultraviolet germicidal irradiation handbook* (pp. 17−50). , 10.1007/978-3-642-01999-9_2.

Kumar, M., Adham, S., & Pearce, W. (2006). Investigation of seawater reverse osmosis fouling and its relationship to pre-treatment type. *Environmental Science and Technology, 40*, 2037−2044.

Innovative wastewater treatment & resource recovery technologies: Impacts on energy, economy and environmentIn J. M. Lema, & S. S. Martinez (Eds.), IWA Publishing.

Matin, A., Khan, Z., Zaidi, S. M. J., & Boyce, M. C. (2011). Biofouling in reverse osmosis membranes for seawater desalination: phenomena and prevention. *Desalination, 281*, 1−16.

Matin, A., Ozaydin-Ince, G., Khan, Z., Zaidi, S. M. J., Gleason, K., & Eggenspiler, D. (2011). Random copolymer films as potential antifouling coatings for reverse osmosis membranes. *Desalination and Water Treatment, 34*(1−3), 100−105.

Matin, A., Rahman, F., Shafi, H. Z., & Zubair, S. M. (2019). Scaling of reverse osmosis membranes used in water desalination: Phenomena, impact, and control; future directions. *Desalination, 455,* 135–157.

Migliorini, G., & Luzzo, E. (2004). Seawater reverse osmosis plant using the pressure exchanger for energy recovery: A calculation model. *Desalination, 165,* 289–298.

Misdan, N., Lau, W. J., & Ismail, A. F. (2012). Seawater reverse osmosis (SWRO) desalination by thin-film composite membrane—Current development, challenges and future prospects. *Desalination, 287,* 228–237.

Niaounakis, M., & Halvadakis, C. P. (2006). *Physico-chemical processes, olive processing waste management.* Amsterdam, The Netherlands: Elsevier Science.

Olabarria, P. M. G. (2015). *Constructive engineering of large reverse osmosis desalination plants.* Chemical Publishing Company.

Ozaydin-Ince, G., Matin, A., Khan, Z., Zaidi, S. J., & Gleason, K. K. (2013). Surface modification of reverse osmosis desalination membranes by thin-film coatings deposited by initiated chemical vapor deposition. *Thin Solid Films, 539,* 181–187.

Pandey, S. R., Jegatheesan, V., Baskaran, K., et al. (2012). Fouling in reverse osmosis (RO) membrane in water recovery from secondary effluent: A review. *Reviews in Environmental Science and Biotechnology, 11,* 125–145, https://doi.org/10.1007/s11157–012–9272–0.

Pearce, G. K., Talo, S., Chida, K., Basha, A., & Gulamhusein, A. (2004). Pre-treatment options for large scale SWRO plants: Case study of UF trials at Kindasa, Saudi Arabia, and conventional pre-treatment in Spain. *Desalination, 167,* 175–189.

Piyadasa, C., Ridgway, H. F., Yeager, T. R., Stewart, M. B., Pelekani, C., Gray, S. R., & Orbell, J. D. (2017). The application of electromagnetic fields to the control of the scaling and biofouling of reverse osmosis membranes-A review. *Desalination, 418,* 19–34.

Toxicity and waste management using bioremediationIn A. K. Rathoure, & V. K. Dhatwalia (Eds.), Engineering Science Reference.

Ratnayaka, D. D., Brandt, M. J., & Johnson, K. M. (2009). Water filtration granular media filtration. *Water Supply,* 315–350. Available from https://doi.org/10.1016/b978-0-7506-6843-9.00016-0.

Saeed, M. O. (2002). Effect of dechlorination point location and residual chlorine on the biofouling in a seawater reverse osmosis plant. *Desalination, 143,* 229–235.

Saleem, H., & Zaidi, S. J. (2020a). Developments in the application of nanomaterials for water treatment and their impact on the environment. *Nanomaterials, 10*(9), 1764.

Saleem, H., & Zaidi, S. J. (2020b). Nanoparticles in reverse osmosis membranes for desalination: A state of the art review. *Desalination, 475,* 114171.

Saleem, H., Trabzon, L., Kilic, A., & Zaidi, S. J. (2020). Recent advances in nanofibrous membranes: Production and applications in water treatment and desalination. *Desalination, 478,* 114178.

Saleh, T. A., & Gupta, V. K. (2016). *Membrane fouling and strategies for cleaning and fouling control. Nanomaterial and polymer membranes* (pp. 25–53). Amsterdam, The Netherlands: Elsevier.

Sanza, M. A., Bonnélyea, V., & Cremerb, G. (2007). Fujairah reverse osmosis plant: 2 years of operation. *Desalination, 203*(1–3), 91–99.

Scott, K. (1995). Air and gas filtration and cleaning. *Handbook of Industrial Membranes,* 309–327.

Shenvi, S. S., Isloor, A. M., & Ismail, A. F. (2015). A review on RO membrane technology: Developments and challenges. *Desalination, 368,* 10–26.

Van de Venter L., S. Williams, E. Garaña and W. Clunie, Large scale desalination demonstration project feasibility study Corpus Christi, Texas, IDA World Congress, Singapore, 2005.

van Hoof, S. C. J. M., Minnery, J. G., & Mack, B. (2001). Dead-end ultrafiltration as alternative pre-treatment to reverse osmosis in seawater desalination: A case study. *Desalination, 139,* 161–168.

Vedavyasan, C. V. (2015). Scale inhibitor. In E. Drioli, & L. Giorno (Eds.), *Encyclopedia of membranes.* Berlin, Heidelberg: Springer, https://doi.org/10.1007/978–3–642–40872–4_527–4.

Vial, D., & Doussau, G. (2002). The use of microfiltration membranes for seawater pre-treatment prior to reverse osmosis membranes. *Desalination, 153,* 141–147.

Vial, D., Doussau, G., & Galindo, R. (2003). Comparison of three pilot studies using Microza® membranes for Mediterranean seawater pre-treatment. *Desalination, 156,* 43–50.

Voutchkov, N. (2010). Considerations for selection of seawater filtration pretreatment system. *Desalination, 261*(3), 354–364.

Voutchkov, N. (2017). *Membrane filtration. Pretreatment for reverse osmosis desalination* (pp. 187–219). , 10.1016/b978-0-12-809953-7.00009-7.

Voutchkov, N. (2017). *Pretreatment by screening. Pretreatment for reverse osmosis desalination* (pp. 95—111). , 10.1016/b978-0-12-809953-7.00005-x.

Voutchkov, N. (2017). *Sand removal, sedimentation, and dissolved air flotation. Pretreatment for reverse osmosis desalination* (pp. 137—152). , 10.1016/b978-0-12-809953-7.00007-3.

Voutchkov, N. (2017). *Conditioning of saline water. Pretreatment for reverse osmosis desalination* (pp. 113—135). , 10.1016/b978-0-12-809953-7.00006-1.

Water chemistry and pre-treatment: Biological fouling prevention, Dow FILMTEC Membranes, Tech Manual Exerpt, Form No. 609—02034-1004.

Wilf, M. (2004). Fundamentals of RO—NF technology. *International conference on desalination costing.* Limassol.

Wolf, P.H. and Siverns, S. (2004). The new generation for reliable RO pre-treatment. *International conference on desalination costing.* Limassol.

Wolf, P. H., Siverns, S., & Monti, S. (2005). UF membranes for RO desalination pretreatment. *Desalination, 182*, 293—300.

Zaidi, S. J., Fadhillah, F., Saleem, H., Hawari, A., & Benamor, A. (2019). Organically modified nanoclay filled thin-film nanocomposite membranes for reverse osmosis application. *Materials, 12*(22), 3803.

Important Design and Operation Parameters

6.1 Introduction

There is one simple but very important fact in maintaining the membranes at their ultimate performance:

"Keep the Membrane Surface Clean"

All the impurities present in the water are removed at the surface of the membrane. The dynamics of this separation phase should ensure that concentrated materials are not accumulated at the membrane surface. In the event that concentrated materials are permitted to accumulate near the membrane surface (boundary layer), then fouling of the membrane surface and precipitation of low-soluble substances will occur, leading to a reduction in the membrane performance (Drioli et al., 2017).

RO systems can be periodically monitored using different parameters which are considered at the time of RO design. A membrane system must be designed in a way that all elements in the system operate within the frame of recommended operating conditions for minimizing the rate of fouling and for reducing mechanical damage (DuPont Water Solution, 2020). To precisely measure the RO system efficiency, we need to have information on water quality (feed water source, feed water salinity, feed water constituents, feed water temperature, and feed water flow), membrane characteristics [design configuration, pH tolerance, membrane flux, membrane compaction, membrane fouling, membrane scaling, and concentration polarization (CP)] system setup [operating pressure, recovery, concentrate flow, average permeate flux, salt rejection, salt passage, pretreatment, and normalized permeate flow (NPF) rate], operation and monitoring techniques, cleaning and maintenance practices.

6.1.1 Water

A good and reliable raw water analysis is of utmost importance for the design of RO systems. Understanding the water analysis and the potential problems that can arise from the sparingly soluble salts are crucial for the design and operational success of the RO system (Bucher, 2015). RO systems that are designed and constructed with unreliable or incomplete water analysis data are bound to fail. These types of mistakes are difficult to correct once the plant is put into operation.

It is also essential that the temperature should be provided as a range (temperature for winter and summer) instead of an absolute value. The variation in temperature can influence the scaling potential of a reverse osmosis (RO) system, particularly when silica and bicarbonate concentrations in the feed water are high. Measurements of pH, alkalinity, and gases must be taken at the point of sampling or pressurized samples need to be taken if these analyses are to be done in the laboratory. Analysis of suspended solids in the presence of soluble iron or hydrogen sulfide should be performed at the point of sampling. On the other hand, if the analysis is to be done in the laboratory then pressurized samples should be taken. The RO system design should consider water parameters such as feed water source, feed water salinity, feed water constituents, pH, feed water temperature, and feed water flow, as discussed in the following section (McKetta, 1998).

6.1.1.1 Source of feed water

The source of feed water is noted to have a substantial effect on the possibility of water causing the fouling of the RO membrane. Good-quality source water, like well water with a silt density index (SDI) value less than three, has a lesser chance of fouling a reverse osmosis membrane as compared to low-quality source water, like surface water with a silt density index value of five (Saji et al. 2020). A reverse osmosis unit designed to work on high-quality water sources can have a high flux relative to the RO system working on a low-quality water source. This is due to the fact that as the flux rate increases, it would bring the contaminants (hardness, suspended solids) to the membrane surface much faster as compared to a lower flux rate process. The aforesaid contaminants will be collected in the CP layer at the surface of the membrane, which results in faster scaling or fouling of the membrane. Therefore, when the hardness and suspended solid concentration in the feed water are higher, then the flux should be lower for reducing the possibility of scaling and fouling the membranes.

Further, the source water also impacts the design array of the RO system. This is for the reason that the feed water flow rate, and the reject flow rate, are also established on the basis of the quality of feed water. In general, high-quality feed water permits high feed water flow and low reject flow. Therefore, the high feed water flow and low reject flow reduce the number of RO membrane modules needed in the system.

6.1.1.2 Salinity of feed water

The salinity or total dissolved solids (TDS) is a measure of the total ionic concentration of dissolved minerals in the water. The TDS concentration of the feed water influences the salt rejection and the permeate flux of a reverse osmosis unit (Holloway et al., 2012). The osmotic pressure will be a function of the concentration and type of salts or organics present in the feed water. As the TDS content in the feed water increases, the osmotic pressure will also increase. Therefore, the amount of feed water driving pressure needed for reversing the natural osmotic flow direction is mainly determined by the salinity of the feed water. Normally, the feed water total dissolved solid concentration can be measured in parts per million (ppm) or micro Siemens/cm (μS/cm) using a TDS meter or a conductivity meter. Fig. 6.1 presents the effect of feed concentration on water flux and salt rejection, at constant feed pressure. It can be noted that higher salt concentration results in lower membrane water flux, when the feed pressure remains constant. When the feed water concentration increases, the salt rejection and permeate flow of the RO membrane drops, provided that other parameters like operating pressure, feed water temperature, and percentage recovery remain the same. Also, it can be observed that the salt passage across the membrane will increase (the salt passage is the inverse of the salt rejection) as the water flux decreases. Sometimes the RO system may experience variation in the composition of feed water at the time of the operation. The above-stated variation might be because of seasonal changes in source water or because of the alternating operation of several water sources of different salinity. The variations in the composition of feed water will influence the required feed pressure and also the salinity of the product water.

6.1.1.3 Temperature of feed water

Temperature is considered to be a critical design parameter as it will affect the system flux as well as the rejection performance. The flow rate of the product water and the salt passage amount will be proportional to temperature. The performance of the membrane will be very sensitive to variations in the feed water temperature (Wang et al., 2008). Fig. 6.2 presents the effect of feed water temperature on water flux and salt rejection. As the figure demonstrates, when the temperature of water increases, the permeate flow will also increase almost linearly, due to the increased diffusion rate of water across the membrane. On the other hand, the salt rejection reduces slightly with an increase in temperature. Higher feed water temperature also leads to higher salt passage or lower salt rejection (Raghavan & Reddy, 2014). This is because the salt diffusion across the membrane will be greater at elevated water temperature. When the temperature rises, the permeate flow rate will increase and also the salt passage. Hence, during winter-time, as the feed water is colder, the product water flow rate and the salt passage through the RO membrane will be lower (leading to improved water quality) provided other parameters like operating

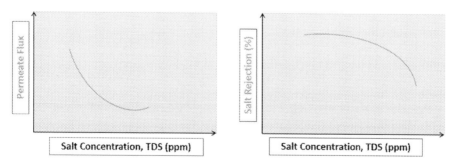

FIGURE 6.1 Effect of salt concentration on permeate flux and salt rejection. *Source: Courtesy: MICRODYN-NADIR GmbH; Adapted from MICRODYN-NADIR GmbH (2020). "Effects of varying operating parameters – RO & NF systems," MICRODYN-NADIR, Goleta, CA, USA, 2020. Accessed on: Aug. 12, 2020 [Online]. Available: https://www. microdyn-nadir.com/wp-content/uploads/TSG-O-017-Effects-of-Varying-Operating-Parameters-RO-NF-Systems.pdf.*

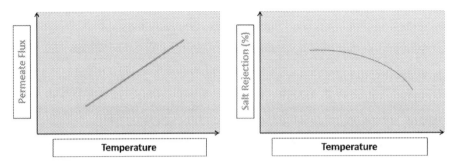

FIGURE 6.2 Effect of feed water temperature on permeate flux and salt rejection. Source: *Courtesy: MICRODYN-NADIR GmbH; Adapted from MICRODYN-NADIR GmbH (2020). "Effects of varying operating parameters – RO & NF systems," MICRODYN-NADIR, Goleta, CA, USA, 2020. Accessed on: Aug. 12, 2020. [Online]. Available: https://www.microdyn-nadir.com/wp-content/uploads/TSG-O-017-Effects-of-Varying-Operating-Parameters-RO-NF-Systems.pdf.*

pressure, feed water TDS, and percent recovery remain the same. However, when the feed water is hotter, during summer-time, the product water flow rate and the percent salt passage will be increased provided other parameters like operating pressure, feed water TDS, and percent recovery remain the same. This also signifies that it will take a longer time for filling the product water tank during winter-time relative to summer-time. In the field, the operating pressure is normally varied during the various seasons to account for these temperature changes: lower pressures in the hot summer months and higher pressures in the cold winter months (MICRODYN-NADIR GmbH, 2020).

Variation in the feed water temperature leads to a variation in the diffusion rate through the membrane. The variation in the permeate flux rate with temperature is described by the below equation:

$$\text{Temperature Correct Flow} = \frac{\text{Rated Permeate Flow}}{\text{Temperature correction factor}} \tag{6.1}$$

$$\text{Temperature correction factor} \, (T_{CF}) = \exp\left(K * \left(\frac{1}{(273 + t)} - \frac{1}{298} \right) \right) \tag{6.2}$$

where t is the feed water temperature in degrees Celsius, and K is a constant characteristic for a particular membrane material. In the aforestated equation, 25°C temperature is employed as a reference point, with $T_{CF} = 1$ (Hydranautics, 2019).

As the RO systems are designed to function at a constant output, the feed pressure must be adjusted for compensating the water flux variations because of variation in temperature. The salt passage (salt diffusion) through the membrane also varies almost at the same rate as the permeate flux. Since the product water flux is kept constant, the product water salinity varies as per the temperature variations.

6.1.1.4 Presence of suspended solids

Suspended solids and colloidal materials present in feed water are one of the major problems in RO desalination facilities (Ho & Sirkar, 2012). Although the majority facilities have some pretreatment systems such as 5-μm prefilters, the above-mentioned fine particles will be responsible for RO membrane fouling.

For measuring the degree of this fouling problem, a test method known as silt density index (SDI) is used. SDI is an empirical test used to characterize the fouling potential of a feed water stream. It is one of the most important parameters for the design and the operation of RO systems and should be carried out correctly. The test is on the basis of assessing the plugging rate of a 0.45-μm filter using a constant feed pressure of 30 psi for a particular time period. SDI_{15} is the SDI test which is run for 15 min. An SDI value below 5 is acceptable for the RO systems; however RO unit performance will be better in a lower SDI value. It signifies that at SDI values below 5, the RO membranes will foul at a very low rate only. There are some exemptions when a lower SDI value (below 3) is desired because of the nature of the suspended solids present in the source water.

6.1.1.5 Presence of natural organic matter and calcium

Divalent cations, mostly calcium, together with natural organic matter (NOM) inclusive of tannic acid, humic acid, and fulvic acid, can increase the membrane fouling (Boerlage, 2001). Due to the acidic nature, NOM can develop complexes with dissolved metal ions.

The formation of complex is dependent on the electronic charge of the metal ion, the metal ion size, and the energy required for breaking the water molecule shell that normally hydrates the metal ions present in water. As this shell breaks off from calcium, many negatively charged NOM molecules will get simultaneously attached to this calcium ion, thus forming a bigger particle (Kucera, 2015). These particles will be capable of developing the biofouling layers on the surface of the membrane by supplying nutrients for microorganisms. Fig. 6.3 is the picture of the feed spacer with biofilm. The factors that influence fouling with calcium–NOM complexes are cross-flow rate and permeate flux. At greater flux across the membrane, the calcium content increases in the CP boundary layer at the surface of the membrane. Also low cross-flow rates increase the calcium level in the CP layer. Therefore the increased calcium level at the surface of the membrane will increase the membrane fouling by the calcium–NOM aggregates.

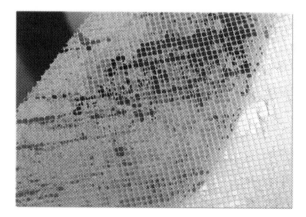

FIGURE 6.3 Picture of feed spacer with biofilm. Source: *Courtesy: DuPont Water Solutions; Reproduced from DuPont Water Solution's (2020). FilmTec™ RO technical manual; Version 3; Form No. 45-D01504-en, Rev. 3; April 2020.*

6.1.1.6 Feed water flow

In general, a minimum flow of feed water should be maintained throughout the membrane. Feed velocity will help to decrease the accumulation of concentrated materials at the surface of the membrane (LaGrega et al., 2010). When numerous membranes are being employed, the arrangement of the membranes is very important in maintaining suitable flow velocities (see the example, Fig. 6.4). The membrane arrangements should be properly checked against other associated factors like recycle flow, high pumping costs, etc.

6.1.2 Membrane

A complete understanding of membrane properties is essential for a discussion about the effect of operating conditions as well as module design on the membrane separation efficiency. In this section, we discuss the important RO design parameters such as membrane pH tolerance, chemical tolerance, design configuration, membrane fouling/scaling tendency, and CP.

6.1.2.1 Membrane design configuration

Membrane assembly units consist of pressure vessels (PVs) placed on RO blocks or racks, which support the PVs together with the interconnecting piping, and feed-permeate manifolds. The membranes are fixed inside the PVs.

Depending on whether maximum water recovery or maximum water quality is the objective, different membrane staging configurations are used. If maximum water recovery is the aim, "concentrate staging" is used, whereas, if maximum water quality is the aim, "permeate staging" is used (McMordie Stoughton et al., 2013).

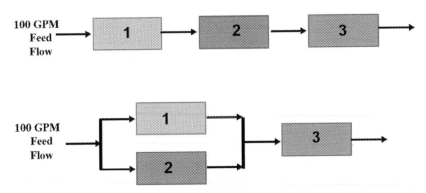

FIGURE 6.4 Examples of different membrane arrangements. High feed flow helps to reduce membrane fouling.

6.1.2.1.1 Concentrate staging

Fig. 6.5 shows a typical two-stage RO concentrate staging membrane configuration. The system is divided into two groups of PVs known as concentrate stages. In each stage, the PVs are connected in parallel with respect to the direction of the feed concentrate flow.

In this particular case, the first stage consists of six PVs connected in parallel with each PV containing six membrane elements inside (6 × 6 membrane array). The second stage consists of three vessels connected in parallel, with each PV containing six membrane elements inside (3 × 6 membrane array). The concentrate/reject from the first stage is used as the feed to the second stage. The product water of all vessels from each stage is combined into a common product water manifold. The concentrate from the second stage is the final concentrate, which is sent to waste for disposal.

A two-stage concentrate staging configuration is suitable for brackish RO systems where the water is of relatively low salinity and the objective is to obtain high recovery (usually 75% recovery). If seawater is to be desalinated, then a single stage is normally adopted.

6.1.2.1.2 Permeate staging

For certain applications, the single-pass RO system might not be capable of generating product water of the required quality (McMordie Stoughton et al., 2013). Such a situation may be encountered in a brackish RO application when the quality of permeate required is to supply make-up water for boilers. In this case, the permeate staging is adopted.

Fig. 6.6 shows a typical permeate staging. In order to accomplish an additional decrease in permeate salinity, the product water generated in the first pass can be further desalinated in the second pass. The concentrate from the second pass is of very good quality

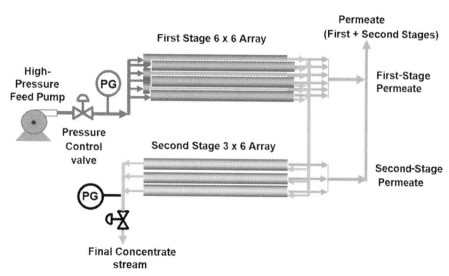

FIGURE 6.5 A typical two-stage reverse osmosis concentrate staging membrane configuration.

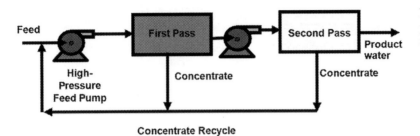

FIGURE 6.6 A typical permeate staging in a reverse osmosis system.

(better quality than the feed to first pass) and hence it can be recycled to the suction of the feed pump.

6.1.2.2 pH tolerance

Usually, pH influences the stability of both cellulose acetate (CA) and polyamide (PA) composite membranes. Because of hydrolysis at lower and higher pH, the CA membranes will have stability over a pH range of 4.0–6.0. Also, PA composite membranes undergo reaction with water; however the pH range of practical application is considered to be extremely wider, ranging from 2.0 to 12.0, based on the particular membrane and manufacturer. The standard operating pH is dependent on temperature, with elevated temperatures demanding a more limited pH range of operation. Moreover, pH influences the salt rejection abilities of PA composite membranes. At about pH 7.0–7.5, the salt rejection of the majority species will be highest. Salt rejection decreases at higher pH and at lower pH value. However the decrease will be very slow at a low pH value. This might be due to the ionic state of the ions being rejected, and certain variations on the membrane molecular level (Kucera, 2015). The water flux through a PA composite RO membrane will be comparatively constant over the pH range.

Due to the presence of carbon dioxide (CO_2) in the majority waters, the pH of RO permeate is usually very low as compared to the feed water pH, unless the CO_2 is totally separated from the feed water. In the case that CO_2 is available in feed water, then it would also be available in product water. This is due to the fact that gases will not be rejected by RO membranes. On the other hand, the membranes can reject carbonates as well as bicarbonates. The CO_2 passage will disturb the equilibrium among these compounds in the product water. CO_2 will readily pass through the membrane whereas bicarbonates do not. Therefore, the product water would be comparatively low in bicarbonates and high in CO_2. Therefore, a different equilibrium takes place in the product water, thereby dropping its pH:

$$CO_2 + H_2O \leftrightarrow HCO_3^- + H^+ \tag{6.3}$$

6.1.2.3 Chemical tolerance

It is essential to consider the chemical tolerance of membranes while designing a reverse osmosis unit (Scott & Hughes, 1996). Not all membranes are chemically resistant to the entire chemicals available in the feed water. Chemical damage happens when an impurity present in the feed water is incompatible with the polymer present in the RO membrane. In addition to oxidizers that deteriorate the crosslinking of a thin-film composite (TFC) membrane, there exist other numerous chemicals that can swell or dissolve the polysulfone substrate.

- Compounds such as strong ethers, esters, aldehydes, ketones, and aromatic hydrocarbons
- Basic or acidic pH may deteriorate some membranes, and if such membranes are employed, then it is important to make sure that that the feed water and the clean-in-place procedure will not harm the product
- Organic chemicals, like solvents in feed water, may lead to membrane damage.
- Chlorine from chlorinated feed water may cause membrane damage.

Low molecular-weight (MW) solvents like isopropanol, propanol, and methanol are considered to be acceptable.

6.1.2.4 Concentration polarization

CP is a phenomenon that adversely influences the overall efficiency of a membrane-based process. When water flows through the membrane, salts will get rejected and a boundary layer will be formed adjacent to the surface of the membrane where the salt concentration will be more compared with the salt concentration in the bulk solution. This increase in the concentration of salts is termed as CP. The CP factor or beta factor is defined as the ratio of the concentration of salt at the membrane surface (C_s) to the bulk concentration (C_b).

$$\text{CP factor or beta factor} = \frac{C_s}{C_b} \qquad (6.4)$$

Correct prediction of CP phenomena is very important for accurately designing RO processes, as the CP increases the transmembrane osmotic pressure and salt passage, and also it increases the membrane surface fouling and scaling. In the majority membrane-based water desalination systems, CP is considered to be a loss term and several resources and efforts are used for improved understanding and preventing the polarization effects. The value of the CP factor 1.2, recommended by Hydranautics for a 40″ long membrane element, corresponds to 18% permeate recovery (Wilf & Awerbuch, 2007). The CP is discussed in detail in Chapter 2, Reverse Osmosis Principles and System Components.

The CP phenomenon will decrease the NPF rate as well as the normalized salt rejection. The impacts of CP are as follows:

- Higher osmotic pressure at the surface of the membrane compared with the bulk feed solution.
- Decreased net driving force across the membrane.
- Decreased water flow across the membrane.
- Higher salt passage across the membrane.
- Higher probability of exceeding the solubility of sparingly soluble salts at the surface of the membrane, causing the precipitation and thereby membrane scaling.

When permeate flux is increased, it will also increase the transport rate of ions to the surface of the membrane and thereby increases the value of C_s. As the feed flow increases, the turbulence also increases and thereby decreases the thickness of the high concentration layer adjacent to the surface of the membrane. Hence, the CP factor is directly proportional to product water flow and inversely proportional to the average feed flow.

CP factor or beta factor is not a membrane property, nor the RO system design engineer directly selects it. CP factor is considered to be a function to estimate how rapidly the influent stream is dewatered through the RO unit. Therefore, the CP factor is an outcome of the selected system design. CP factor influences both the flux through a reverse osmosis membrane and the rejection of salt (Gallab et al., 2017). In practice, the CP factor for the RO unit will always be higher than 1.0, and therefore, the CP phenomenon always exists. The CP phenomenon is not possible to be excluded; however it could be reduced by the proper designing of the RO system. In a reverse osmosis system design, beta values should be less than 1.2 for minimizing membrane fouling and scaling.

6.1.2.5 Membrane compaction

During the RO system operation, the membrane material will be exposed to high pressure of the feed water. This will cause an increase in the membrane material density (known as compaction), which would reduce the diffusion rate of water and dissolved constituents through the membrane. Because of this compaction, more pressure must be applied for maintaining the design permeate flow. Simultaneously, a lower salt diffusion rate will give rise to a low product water salinity. The impact of membrane compaction is very critical in asymmetric CA membranes relative to composite PA membranes. In the seawater RO process, the feed pressure will be much greater as compared to the pressure in brackish water application, and hence, the compaction process will be very important. Also, the elevated feed water temperature can cause an increased rate of compaction. Generally, the compaction of the membrane brings about a certain percentage flux decline and has a very strong impact during the initial operating period of time.

6.1.2.6 Membrane fouling/scaling

In RO systems, membrane fouling is considered to be the foremost cause of permeate flux reduction as well as a decline in product quality. Hence, fouling control dominates the RO system design and operation. Generally, the performance of all membranes will lose with time. One of the main reasons for the decline in membrane efficiency is the deposition of some materials on the surface of the membrane. Even though the term "fouling" is used for material deposition on the membrane surface, this coating can be due to the fouling and scaling.

In a reverse osmosis system, membrane fouling is due to the presence of the suspended or emulsified materials present in the feed water (Ismail et al., 2018). These types of foulants include sulfur, iron, clay, oil, silica, and humic acids. These materials are present in an extremely fine or colloidal form. Even the standard 5-μm cartridge filters employed upstream of a reverse osmosis system will not be able to fully remove these foulants. The concentration of dissolved and suspended materials in the feed water is highest near the surface of the membrane. Membrane fouling negatively influences the performance of the membrane, and in extraordinary cases, it can lead to permanent damage of the membrane. Membrane fouling is normally the deposition of organic or inorganic substances on the surface of the membrane thereby causing feed channel blockage. In the primary phases of membrane fouling, the variations in performance are the same as those brought about by the membrane compaction process. The membrane fouling is generally related to a pressure drop increase. An uncontrollable fouling process might result in severe degradation of performance and even full damage of membrane elements. In a study by Saleem and Zaidi (2020), it was confirmed that the incorporation of nanomaterials into the TFC membrane can reduce the fouling possibility of the membranes. The best method for controlling the membrane fouling is to recognize the source of the fouling process and prevent it by the modification of operating conditions or pretreatment processes. Foulant deposits could be eliminated from the surface of the membrane by chemical cleaning techniques. On the other hand, the efficiency of the cleaning process relies upon the age of the foulant deposit, and on the appropriate choice of the cleaning solution.

Also, the feed spacer influences the flux, pressure losses, and fouling in the membrane process and consequently the product water unit cost. In a study by Haidari et al. (2018), the team reviewed the role of the feed spacer in spiral wound membrane (SWM) modules and offers an overview of studies carried out in narrow spacer-filled channels for determining the influence of various geometric characteristics of the feed spacer on hydraulic conditions. It is essential to follow the recommendations of the membrane manufacturers on maximum element flux, minimum brine flow, maximum element recovery, and minimum feed flow. The recommendations will be based on membrane element size and feed water quality. The membrane performance is controlled by the concentrations of the suspended and dissolved solids in the boundary layer. Increased concentrations signify higher osmotic pressure, increased tendency of suspended solids to coagulate and coat the surface of the membrane, and a greater possibility of scaling to occur. Scaling is the precipitation of dissolved salts present in the feed water on the surface of the membrane. The main preventative step for minimizing membrane fouling is maintaining proper operating conditions for the membrane. The membrane fouling, scaling, and its prevention techniques are discussed in Chapter 5, Pretreatment: Fouling and Scaling Control.

6.1.3 System

6.1.3.1 Feed pressure

Feed water pressure influences both the water flux and salt rejection of RO membranes (Scott, 1995). Osmosis is the process of flow of water across a membrane from the dilute solution side towards the concentrated solution side. RO technology involves the application of pressure to the feed water stream for overcoming the natural osmotic pressure. Pressure greater than the osmotic pressure should be applied to the concentrated solution for reversing the flow of water. By the application of pressure, a portion of the feed water (concentrated solution) is forced to flow through the membrane to obtain purified product water (permeate) in the dilute solution side. The details of the osmosis process, RO process, and the associated equations are provided in Chapter 2, Reverse Osmosis Principles and System Components.

Fig. 6.7 shows the effect of operating pressure on water flux and salt rejection. It can be noted that the water flux across the membrane increases with an increase in the feed water pressure. As the operating pressure increases, more product water will be produced and salt rejection of the membrane will also increase if other parameters like feed water temperature, feed water TDS, and percent recovery remains the same. Salt rejection increases with increasing feed pressure because water is being pushed through the membrane more quickly than salt is transported across it, thus diluting the salt. The system must be operated as per the design guidelines of the manufacturer; or else, operating at increased pressure would lead to premature membrane fouling or scaling, and thereby decrease the membrane life. Higher feed water pressure also leads to higher salt rejection however, as Fig. 6.7 shows, the relationship is less direct as compared to that of water flux. As the RO membranes act as imperfect barriers to dissolved salts present in feed water, there is always some salt passage across the membrane. Also, there is a maximum limit for the amount of salt that could be rejected by increasing the feed water pressure. As the plateau in the salt rejection curve specifies, beyond a certain pressure level, salt rejection will no longer increase and some salt passage remains coupled with water flowing across the membrane.

The RO systems with SWM elements function at a constant flux rate, that is to generate a constant product water flow. During the operating time, the feed pressure might be adjusted to compensate for the variations in feed water salinity, temperature, and product water flux reduction because of membrane fouling or compaction. In order to specify the high-pressure pump (HPP), it is generally assumed that the membrane flux would reduce by almost 20% over three years, and therefore the pump should be designed accordingly to compensate for this flux decline. If the RO plants use centrifugal pumps, it is better to use an oversize pump and control the feed pressure by throttling (partially closing the feed valve). Alternatively, electric motors with variable speed drives can be used, which allow adjustment of flow and feed pressure of the pump over a broad range with an extremely small loss in productivity. The variable speed drive decreases unproductive pressure losses which were frequently observed in the past. Certain RO units use positive displacement pumps (plunger or piston pumps) as HPPs (Asano, 1998). These pumps

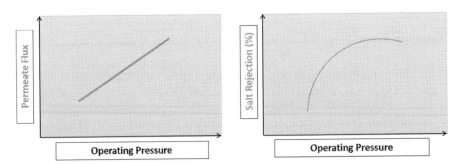

FIGURE 6.7   Effect of feed pressure on water flux and salt rejection. Source: *Courtesy: MICRODYN-NADIR GmbH; Adapted from MICRODYN-NADIR GmbH (2020). "Effects of varying operating parameters — RO & NF systems," MICRODYN-NADIR, Goleta, CA, USA, 2020. Accessed on: Aug. 12, 2020. [Online]. Available: https://www.microdyn-nadir.com/wp-content/uploads/TSG-O-017-Effects-of-Varying-Operating-Parameters-RO-NF-Systems.pdf.*

enable regulation of feed pressure at constant pump output, with a small difference in the productivity of the pump. These types of pumps are less common in RO units because of the capacity restriction of this type of pump, strong vibrations, noisy operation, and increased frequency of required maintenance. In a work by Peñate and García-Rodríguez (2012), the team presented a comprehensive review of the major innovations as well as future trends in the design of seawater RO desalination technology.

6.1.3.2 Flux

Flux shall determine the total size of the RO system with respect to the membrane area needed to accomplish the preferred separation. The flux of a membrane is defined as the quantity of permeate produced per unit area of membrane surface per unit time (Mohanty & Ghosh, 2011). Flux is used to express the rate at which water permeates a reverse osmosis membrane. Generally, flux is expressed as liters per square meter per hour or gallons per square foot per day (GFD).

$$\text{Permeate flux} = \frac{\text{Permeate flow rate}}{\text{Area of the membrane}} \qquad (6.5)$$

The water flux for a specified application is dependent on the source of the feed water. Cleaner source water offers increased flux, which signifies that a lesser membrane area will be needed for achieving the required separation.

In general, water flux is influenced by a number of operating variables, as mentioned below:

- It will be directly proportional to feed pressure and water temperature.
- It reduces with increasing feed TDS.
- It reduces to some extent as recovery increases till the feed water osmotic pressure equals the driving pressure.
- It will be relatively constant over a pH range; however for certain latest PA membranes, the flux will also be a function of pH.

The membranes have one common limitation, that is they can only produce a maximum flow of a certain maximum permeate flow for a given quantity of water. The aforesaid limit is controlled by the feed water quality and not by the brand of the membrane. As an example, a maximum permeate flow for brackish water treated by conventional filtration and of SDI 4.0 is 14 GFD. If the membranes are operated at fluxes higher than this value, fouling will take place.

Average permeate flux is another important parameter of the RO process design (Wilf, 2010). The average permeate flux is the combined permeate flow divided by the total membrane area installed in the RO unit. Recommended values of average product water flux are suggested by the membrane manufacturers. The upper limits of average permeate flux values are related to feed water source and projected feed water quality. The product water flux rate from membrane elements situated at the membrane feed end will be much greater than the average flux value. On the other hand, the practical experience from field operation of membrane systems strongly specifies that RO system design based on average flux rate, in the range suggested by membrane manufacturers, leads to stable membrane performance.

Calculation: Example 6.1

A reverse osmosis unit system produces 420 m^3/day (110,952 gallons/day). The membrane array has three PVs, each housing six membrane elements. Each element has a membrane area of 37 m^2 (400 ft^2). Calculate the average permeate flux.

Average permeate flux is determined as follows:

$$\text{Average permeate flux} = \frac{420,000L/\text{day}}{3 \times 6 \times 37 \times 24} = 26.27 \, L/m^2.h$$

$$\text{Average Average flux} = \frac{110,952 \text{ gallons}/\text{day}}{3 \times 6 \times 400} = 15.41 \text{ gfd}$$

On the other hand, the design average permeate flux is used for determining the required number of membrane elements in the RO system for necessary permeate capacity.

6.1.3.3 Recovery

Recovery is a critical parameter in the design and the operation of RO systems. The permeate recovery rate has a major effect on investment and operating costs in a reverse osmosis system. Percent recovery is the percentage of feed water that becomes the product water (American Society of Civil Engineers, 1996). A higher recovery rate means less feed water is sent to the drain as a concentrate. On the other hand, if the recovery rate is very high in the system design, then significant problems may arise due to membrane scaling and fouling. The percentage recovery should be established in the RO system design, considering different factors like feed water chemistry and pretreatment process. As a result, the proper recovery at which the system should operate depends on what the system was designed for. By determining the system recovery, it is possible to determine if the system is operating away from the intended design. Percent recovery is the ratio of product water flow to feed flow. The operation of membranes at greater than the design recovery will lead to membrane fouling and scaling.

$$\text{Percentage Recovery}(r) = \frac{\text{Permeate flow rate}}{\text{Feed flow rate}} \times 100 \tag{6.6}$$

$$\text{Percentage Recovery}(r) = \frac{Q_p}{Q_f} \times 100 = \frac{Q_P}{Q_{P}+Q_C} \times 100 = \frac{C_c - C_F}{C_c - C_p} \tag{6.7}$$

where r is the recovery rate (in %), Q_p is the product water flow rate, Q_f is the feed water flow rate, Q_c is the concentrate flow rate, C_f is the feed concentration, C_p is the permeate concentration, C_c is the concentrate concentration.

Calculation: Example 6.2

The designed recovery rate (r) of a reverse osmosis system is 75%. The feed concentration, reject concentration, and permeate concentration are noted to be 1000, 4000, and 200 mg/L of Cl$^-$, respectively. Calculate the actual recovery rate of the system

$$\text{Percentage Recovery}(r) = \frac{Q_p}{Q_f} \times 100 = \frac{Q_P}{Q_{P}+Q_C} \times 100 = \frac{C_c - C_F}{C_c - C_p}$$

Feed (C_f) = 1000
Concentrate (C_c) = 4000
Permeate C_p = 200
Actual recovery rate, R = $\frac{C_c - C_F}{C_c - C_p} = \frac{4000 - 1000}{4000 - 200} = \frac{3000}{3800} = 0.789 = 78.9\%$

Fig. 6.8 shows the effect of increased recovery on flux and salt rejection. It can be noted that the percentage recovery is inversely proportional to permeate flux and percent salt rejection, provided other parameters like operating pressure, feed water salinity, and feed water temperature remain the same. When the system recovery is increased, a higher percentage of the feed water passes across the membrane and into the product water stream. The residual feed will be more concentrated, resulting in higher osmotic pressure. This higher osmotic pressure will decrease the net driving pressure, and correspondingly the product water flux reduces. With increased recovery, salt rejection also reduces. As the higher recovery causes the residual feed osmotic pressure to increase, the driving force for water decreases while the driving force for salt transport does not vary. When percent recovery is low, more water will be drained and wasted; however the RO membrane efficiency will be better. Normally, the RO system manufacturers recommend a percent recovery rate for balancing performance and economics.

The maximum % recovery possible in a reverse osmosis system usually depends on the limiting osmotic pressure, feed water salinity, and the tendency of salts to precipitate on the surface of the membrane as a mineral scale. Apart from decreased salt rejection, the major issue that occurs when recovery is forced too high is scaling. As the recovery increases, the water at the tail end of the PV becomes more and more concentrated with salts. If the concentration of any salt increases past its solubility limit, it will precipitate out of solution, and can cause a number of problems for the membrane system. Recovery through specific membrane modules varies, depending on the module location in the PV. Most SWM modules function with single module recoveries varying from 10.0% to 15.0%, with an average of 11% to accomplish a total 50% recovery in a single, six-module PV. The module at the feed end of the PV usually demonstrates the lowermost recovery of all membrane modules in the PV. However the module at the reject end of the PV operates at the maximum recovery in the PV. In a study by Li et al. (2019), the team provided a critical review of the treatment techniques that enable intensification of the water recovery in

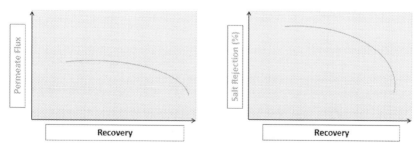

FIGURE 6.8 Effect of increased recovery on flux and salt rejection. Source: *Courtesy: MICRODYN-NADIR GmbH; Adapted from MICRODYN-NADIR GmbH (2020). "Effects of varying operating parameters − RO & NF systems," MICRODYN-NADIR, Goleta, CA, USA, 2020. Accessed on: Aug. 12, 2020. [Online]. Available: https://www.microdyn-nadir.com/wp-content/uploads/TSG-O-017-Effects-of-Varying-Operating-Parameters-RO-NF-Systems.pdf.*

brackish water RO desalination by the intermediate concentrate demineralization methods.

It is possible to reduce the recovery by increasing the feed flow. A different technique to decrease recovery is to reduce the operating pressure. Decreased operating pressure will produce a lesser quantity of product water. If the feed flow is maintained close to the initial value, then a reduced recovery is attained. The result of lower recovery is to decrease the overall concentration of all materials in the RO system, thus attaining more advantageous boundary layer conditions. The RO system designer has at their choice, the capability to manipulate the system recovery for minimizing the possibility of membrane scaling, both on the system level and the individual module level. Higher recovery can lead to the separation of more water in the lead modules in the PV, thereby increasing the possibility of membrane fouling and scaling (increasing beta factor). Reducing the recovery would increase both the salt rejection and permeate flux as shown in Fig. 6.8.

The recovery ratio influences the performance of the system, for example, feed pressure and permeate salinity, by assessing the average feed salinity. The average feed salinity could be expressed as a function of recovery rate, assuming that the entire dissolved species present in the fees water are separated by the RO membranes (Lior et al., 2012).

The average feed salinity is calculated either as arithmetic [Eq. (6.8)] or logarithmic mean [Eq. (6.9)]:

$$\text{Average feed salinity} = \text{Feed salinity} \times 0.5 \times \left(1 + \frac{1}{(1 - \text{Recovery rate})}\right) \tag{6.8}$$

$$\text{Average feed salinity} = \text{Feed salinity} \times \frac{\ln\frac{1}{(1 - \text{Recovery rate})}}{\text{Recovery rate}} \tag{6.9}$$

In general, arithmetic mean is usually used for calculations in cases of lower recovery rates (single-element calculations). For a high recovery rate, a logarithmic mean is applied for the system performance calculations (Lior et al., 2012).

This average feed salinity is also determined from feed salinity using the average concentration factor (ACF). In order to determine the ACF, a logarithmic dependence on recovery ratio (r) is presumed:

$$\text{ACF} = \ln\left(\frac{1}{1 - r}\right)r \tag{6.10}$$

where "ln" is the natural log.

Since the recovery rate strongly influences process economics, there is a trend to design the operation of RO units at the uppermost practical value.

6.1.3.4 Concentrate/Reject flow

There should be a minimum brine flow to flash the concentrated stream which contains foulants and scalants away from the membrane surface. This minimum brine flow depends on the feed water quality in relation to sparingly soluble salts and SDI. The typical minimum brine flow per vessel for RO permeate (second pass) is 8−10 gallons/min (GPM) whereas, for any other water sources, for example, brackish well softened is 12−14

GPM and brackish well not softened is 12–16 GPM, for seawater without microfiltration or ultrafiltration treatment is 12–16 GPM, for seawater with microfiltration or ultrafiltration treatment is 12–14 GPM, etc.

As the feed water source is cleaner, the reject flow might be lower, leading to smaller systems and lower overall operational expenses. At low reject flow rates, appropriate cross-flow velocity (CFV) will not be retained, and contaminants, like scale-formers and colloids, contribute an increased possibility of scaling or fouling a membrane. This is due to the fact that the CP boundary layer will be more thick at low CFV than it will be at greater reject flow rates. As the bulk concentration of contaminants towards the reject end of the PV could be three, four, or even five times the concentration as noted in the feed water, and as the contaminant concentration is even greater in the CP layer, the possibility of membrane scaling or fouling could be extremely high at low reject flow rates.

6.1.3.5 Salt rejection and salt passage

Percent salt rejection reveals how effective the membrane elements are at removing particular dissolved solids and other contaminants (MICRODYN-NADIR GmbH, 2020). It will not reflect the performance of each individual membrane element, but it will show the system performance as a whole. A well-designed system with properly functioning elements will reject the majority of feed water impurities depending on the membrane type. To determine the efficiency of the membrane elements at removing contaminants, we can use the following equation:

$$\text{Salt Rejection (\%)} = \frac{\text{Concentration of Feed water} - \text{Concentration of Permeate Water}}{\text{Concentration of Feed water}} \times 100$$

$$(6.11)$$

The high salt rejection signifies improved system efficiency. A low salt rejection for the particular element indicates that the membranes need cleaning or replacement.

The salt passage is the inverse of salt rejection; and it is the amount of salts (expressed in %) that are passing through the membrane elements. The salt passage is the ratio of salt concentration on the product water side of the membrane relative to the average feed concentration. The system performance will be good at lower salt passage. A high salt passage signifies that the membranes need cleaning or replacement.

Mathematically, the salt passage can be determined as the following:

$$\text{Salt Passage (\%)} = \frac{\text{Concentration of Permeate Water}}{\text{Mean salt concentration in Feed water}} \times 100 \qquad (6.12)$$

Salt rejected can also be calculated as follows:

$$\text{Salt Rejection (\%)} = 100\% - \text{Salt Passage} \qquad (6.13)$$

The salt rejection and salt passage calculations for a single RO element are demonstrated in Example 6.3, and for a membrane unit in Example 6.4. As shown in the below examples, typically, there will be a difference between apparent results of salt rejection for a single element and membrane unit, even while using the same membrane elements and treating feed water with similar salinity. The salinity of product water, generated by a reverse osmosis unit, will be a function of the number of process parameters (feed temperature, feed composition, permeate flux, and recovery rate), along with the salt rejection ability of the membrane elements.

Calculation: Example 6.3

In a reverse osmosis system, the RO element is tested at a 15.0% recovery rate. The feed water salinity and permeate salinity are 1600 and 4.5 ppm NaCl, respectively. Calculate the salt passage and salt rejection of the single RO element

Recovery, $r = 0.15$

Feed salinity: 1600 ppm NaCl

Permeate salinity: 4.5 ppm NaCl

First, we need to calculate the average feed salinity,

$$\text{Average feed salinity} = \text{Feed salinity} \times 0.5 \times \left(1 + \frac{1}{(1 - \text{Recovery rate})}\right)$$

$$\text{Average feed salinity} = 1600 \times 0.5 \times \left(1 + \frac{1}{(1 - 0.15)}\right) = 1741.17 \text{ ppm}$$

$$\text{Salt Passage (\%)} = \frac{\text{Concentration of Permeate Water}}{\text{Mean salt concentration in Feed water}} \times 100 = \frac{4.5}{1741.17} \times 100 = 0.25\%$$

$$\text{Salt Rejection (\%)} = 100\% - \text{Salt Passage} = 100\% - 0.25\% = 99.75\%$$

Calculation: Example 6.4

In a desalination plant, the RO unit operates at 80.0% recovery rate. The feed water salinity and permeate salinity are 1600 and 4.5 ppm NaCl, respectively. Calculate the salt passage and salt rejection of the membrane unit.

Recovery, $r = 0.80$; feed salinity: 1600 ppm NaCl; permeate salinity: 4.5 ppm NaCl

First, we need to calculate the average feed salinity,

$$\text{Average feed salinity} = \text{Feed salinity} \times \frac{\ln \frac{1}{(1 - \text{Recovery rate})}}{\text{Recovery rate}} = 1600 \times \frac{\ln \frac{1}{(1 - 0.80)}}{0.80} = 1600 \times 2.01 = 3216 \text{ ppm}$$

$$\text{Salt Passage (\%)} = \frac{\text{Concentration of Permeate Water}}{\text{Mean salt concentration in Feed water}} \times 100 = \frac{4.5}{3216} \times 100 = 1.3\%$$

$$\text{Salt Rejection (\%)} = 100\% - \text{Salt Passage} = 100\% - 1.3\% = 98.7\%$$

6.1.3.6 Pretreatment

In a reverse osmosis system, the proper pretreatment of the feed water can minimize the fouling possibility of the membrane and thus increase the overall system recovery rate. Depending on the source, the feed water may consist of different concentrations of suspended solids as well as dissolved matter, including both inorganic and organic substances. Dissolved solids have the ability to precipitate out of the solution and result in membrane scaling (Obotey Ezugbe & Rathilal, 2020). Suspended particles could settle on the surface of the membrane, thereby blocking the feed channels and increasing friction losses across the system. Also, the proper pretreatment of water, before reaching the RO unit, can decrease the amount of work performed by the RO pump, as a result decreasing the energy utilization.

The most frequent cause of a complete failure of a reverse osmosis system is insufficient pretreatment of the RO feed water (Ning, 2012). The RO units should be protected from incompatible contaminants, extreme fouling, and the possibility of scale formation. Adjustments made in the pretreatment techniques, equipment quality, or the monitoring instrumentation will typically lead to operational problems in the downstream RO unit. The pretreatment processes are detailed in Chapter 5, Pretreatment: Fouling and Scaling Control.

6.1.3.7 Normalized permeate flow rate

The relative ability of water to permeate the RO membrane could be monitored by employing a variable known as the normalized permeate flow (NPF) rate, which is the RO permeate flow rate standardized for the effects of water temperature, dissolved salt concentration, and operating pressures. The feed-to-reject pressure drop can track the resistance to the passage of water through the flow channels of different membrane elements. This value can be determined for the complete RO vessel array, and also for the individual vessel stages if interstage pressures are available.

Smaller particles that form a coating on the RO membrane surface can result in the dropping of RO NPF rate. Larger particles that are trapped inside the membrane flow channels later block the water flow and result in an increased normalized pressure drop. If any water constituent chemically reacts with the RO membrane, the impact will possibly be obvious in the NPF rate, and probably in the rejection of salt. As an example, if chlorine present in the water is allowed to react with the RO membrane, then the membrane oxidation happens, which is indicated by an increase in the NPF rate, shortly succeeded by a decrease in the rejection of salt. Comprehensive knowledge of the state of the RO system could be obtained by regularly computing and graphing the rejection of salt and the normalized performance variables.

6.1.4 Operation and monitoring

For understanding the proper performance of a reverse osmosis system, it is important to take the appropriate data, to employ the correct analytical techniques for interpreting these data, and to carry out regular maintenance for keeping the RO system and pretreatment system functioning as designed (Wes Byrne, 2020). Large-scale RO systems will have a number of membrane PVs, which are staged so that the reject stream from one set of parallel-connected vessels is connected into a smaller number of PVs, and then probably connected to a next stage with an even smaller number of membrane PVs. The aforestated staging will be based on retaining flow velocities adequate for keeping the suspended particles moving and to support the dissolved salts in diffusing back into the bulk stream from the surface of the membrane.

RO systems are typically operated by fine-tuning the membrane feed pressure as required for achieving the anticipated RO product water flow rate. This could be accomplished using a variable frequency drive (VFD) for controlling the rotational speed of the HPP motor, or by employing a throttle valve positioned downstream of the pump. With VFD control, the adjustment could be automatic. The RO system also has a reject stream throttle system for achieving the anticipated reject flow rate. In addition to the product water flow and reject flow meters, pressure sensors have also been fixed in the system piping for monitoring the pressure reaching the membrane elements, the reject pressure leaving the membrane, and probably the pressures inside the pipe manifolds that connect the membrane vessel stages. A product water pressure sensor might be required, particularly when there is substantial or variable product water backpressure on the RO system.

The electrical conductance of the water streams can also be used for monitoring the efficiency of the RO unit in removing the dissolved salts. A salt rejection percent of the RO system can be determined by subtracting the product water conductivity from the feed stream conductivity and then dividing it by the feed conductivity. Presently, the percent salt rejection is possibly the most frequently observed performance variable (M46 Reverse Osmosis and Nanofiltration, 2007).

Other instruments might also be required if there exists inconsistency in the feed water characteristics, or if any chemicals are injected. If the acidity of water varies through chemical injection or naturally, then the pH of the water must be constantly monitored. The water pH has a significant impact on salt rejection in a reverse osmosis system. If chlorine is being removed upstream, then an online chlorine-monitoring system or probably an oxidation reduction potential (ORP) monitor can be employed for detecting its availability.

6.1.5 Cleaning and maintenance

In the majority RO systems, the chemical cleaning is considered to be a repetitive requisite, and the cleaning frequency is dependent on the efficiency of the pretreatment process/equipment (Olabarria, 2015). When the scale particles or the fouling solids build up on the membrane surface, their features frequently vary and these materials turn out to be very resistant against cleaning. Materials such as clay and biological matter will get compressed against the surface of the membrane and turn out to be chemically resistant due to the fact that water is removed from their structure. The formation of scale might alter from being primarily calcium carbonate (can be cleaned easily) to calcium sulfate (cleaning will be difficult). The variation in normalized RO performance variables can be used for determining the cleaning requirements. The majority of membrane manufacturers recommend cleaning process before the performance variables vary by almost 15% (Singh et al., 2008).

Moreover, some scaling salts or fouling solids might have a significant influence on the quality of the product water. Aluminum salts might come from suspension as a fouling particle, only to redissolve in the event that the water acidity varies. The aforesaid can cause an increase in the aluminum passage from the surface of the membrane through the membrane and subsequently into the product water. Calcium carbonate ($CaCO_3$) scaling might leach a comparatively higher level of $CaCO_3$ through the membrane into the product water stream and influence the conductivity. The majority of other fouling solids do not have a substantial influence on the rejection of salt in a reverse osmosis system, unless the fouling is extreme.

In a membrane cleaning process, a cleaning solution will be passed through the membrane at conditions that help the delamination or dissolution of the fouling solids from the surface of the membrane, or from the spacing material along the flow channels in the membrane. The ideal solution will be dependent on the specific scale particles or fouling solids, and the cleaning ability will be often limited by the chemical tolerance of the membrane. The majority of strong oxidizing agents that are usually effective in cleaning biological solids will not be compatible with the RO membranes. There also exist some restrictions on the pH limits that must be used. As water temperature increases, water flux increases, and also it enhances the salt passage. As increased temperatures result in a higher cleaning rate, the temperature of the solution should be limited to less than 105F or as recommended by the membrane manufacturer.

Accomplishing a high cleaning flow rate that is balanced throughout the entire membrane vessels typically needs the cleaning of each vessel stage separately. It will also help to reduce the pressure needed for pushing the solution through the membrane elements. The cleaning solution must be pumped at higher flow rates, as suggested by the membrane manufacturer, and pumped at the maximal pressure needed to attain the targeted flow rate; however this might be limited to 60 psi for reducing the possibility of crushing or else destroying the elements. The solution will be directed in the standard feed-end flow direction and the exiting concentrate stream will be sent back to the cleaning tank at minimum backpressure. Also, there might be a small flow of product water that must also be sent back to the cleaning tank by employing a distinct line. Data have to be recorded in the course of cleaning. The success of cleaning is confirmed when the NPF rate and the normalized pressure drop return to their start-up values.

6.2 Pilot study for a reverse osmosis system

A reverse osmosis unit and its pretreatment unit designed exclusively on one water analysis might not be completely optimized for the fouling characteristic of the source. It may be oversized or it may not be perfect for water having an increased membrane-fouling potential. This could be best determined by carrying out a pilot study. A properly designed pilot study will use components that are scaled down; however still it offer the same media type, and use comparable flow velocities and exposure times. The RO pilot must duplicate the product water recovery, the product water flux rate, and reject stream vessel exit velocities, together with the scale inhibitor dose and shutdown flush techniques.

When the pretreatment techniques are piloted together with the RO, the system operation could be adjusted for minimizing the fouling rate of the RO membrane. Adjustments like modifications in the product water flux rate, or modifications in the rate at which water flows across the surface of the membrane and through the membrane elements can be performed. With the proper selection of equipment and sizing, it could be possible to prevent membrane fouling, and this could subsequently reduce the operating costs as well as increase the membrane life.

Moreover, the selection of the membrane should also be assessed. In large RO systems, establishing that a low-fouling membrane element accomplishes better as compared to a standard element will help to justify the increased cost. Low-energy elements may also be assessed for their possibility to decrease the pump sizing and the related power usage. A comprehensive review of the energy consumption of seawater RO desalination plants was carried out by Kim et al. (2019). The pilot study also allows understanding more details, especially about what can cause the fouling in a reverse osmosis system. A membrane element from the pilot study can be taken and subsequently perform autopsy and analysis of the solids. This would help to select the best-suited cleaning solution for taking out the fouling materials. The efficiency of the cleaning solutions as well as cleaning methodology can then be proved in the pilot unit. As the pilot unit is operated for a longer time, more data can be obtained. Hence, it is recommended to perform the pilot study for a minimum of several months.

6.2.1 Pilot-plant study-1: Chlorination-induced transport property variation in a seawater reverse osmosis membrane

A desalination pilot-scale study was carried out by Ettori et al. (2013) for assessing the variations in RO membrane water permeance and salt retention induced by chlorination. Also, a comparison was done between the pilot-scale results and the lab-scale results. The pilot plant was situated close to the bay area of Toulon city, in France. The pilot-scale unit consisted of a pretreatment stage and two RO modules operating in parallel and at an average flow rate of about 1 m^3/h. Fig. 6.9 is the diagrammatic process flow diagram of this pilot plant. A chlorination procedure was adapted for exposing just the surface-active

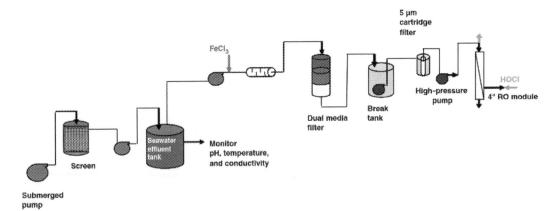

FIGURE 6.9 Diagrammatic process flow diagram of the pilot-scale plant. Hypochlorous acid (HOCl) solution was introduced through the retentate outlet and circulated from the bottom to the top of the module, placed vertically, using a peristaltic pump. *Source: Reproduced from Ettori, A., Gaudichet-Maurin, E., Aimar, P., & Causserand, C. (2013). Pilot scale study of chlorination-induced transport property changes of a seawater reverse osmosis membrane. Desalination, 311, 24—30.*

aromatic PA layer of composite RO membrane to successive free chlorine dosages ranging from 40 to 4000 ppm h, at pH 6.9. During the long-term seawater filtration, carried out using a 4-inch spiral-wound RO module, the team monitored the changes in water permeance, a reduction in salt retention, and the rate of increase of water permeance with time. These results were quantitatively the same as those reported in earlier studies carried out at the laboratory scale under the accelerated exposure conditions. The elemental analysis of the feed stream and product water stream confirmed that the divalent ion rejection continued to be constant (approximately 100%), irrespective of the free chlorine dosage reached, while the monovalent ion rejection of the seawater (mostly, chloride, sodium, and bromide ions) reduced with the increase in the exposure dosage. As a whole, altering the characterization technique to the pilot-scale additionally supported that PA chlorination, under pH conditions typically noted in desalination plants (6.9–8.0), is controlled by the concentration of hypochlorous acid, as noted from the surface elemental analysis by X-ray photoelectron spectroscopy.

6.2.2 Pilot-plant study-2: Ceramic membrane pretreatment for seawater reverse osmosis desalination in Tianjin Bohai Bay, China

An important factor for the effective seawater RO process is the proper source water pretreatment so that a steady, superior-quality feed water can be provided to the RO membranes. Recently, membrane-based pretreatment has been recognized as the ideal pretreatment technique in the RO process. Ceramic membranes are also developed for microfiltration and ultrafiltration applications. The ceramic membranes are mainly driven by the requirement to develop membranes with higher thermal and chemical tolerance. Cui et al. (2011) conducted a pilot-scale study of the pretreatment process by employing a ceramic membrane for the seawater desalination. The site was located in Tianjin Bohai Bay, in the Bohai Sea of China. Various coagulation techniques were compared, and the flocculation-natural sedimentation process was found to be the best technique for ceramic membrane filtration. The schematic process diagram of the ceramic membrane filtration system is shown in Fig. 6.10. It was found that the ceramic membrane can operate effectively in the

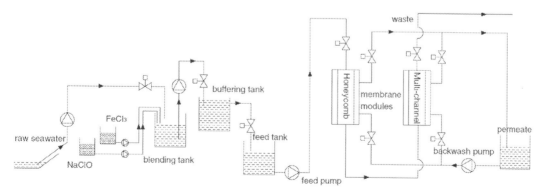

FIGURE 6.10   Flow diagram of the ceramic membrane filtration system. Source: *Reproduced from Cui, Z., Xing, W., Fan, Y., & Xu, N. (2011). Pilot study on the ceramic membrane pretreatment for seawater desalination with reverse osmosis in Tianjin Bohai Bay. Desalination, 279(1–3), 190–194.*

long run, even under reduced temperature conditions (3.0°C−6.0°C). The following results were obtained from this pilot study:

- The removal ratios and turbidity of the product water are reliable, and the quality of the product water produced by the ceramic membrane is appropriate for feeding into the seawater RO process.
- The best coagulation technique for ceramic membrane filtration is the flocculation-natural sedimentation method. Process without coagulation is not suitable for ceramic membrane pretreatment of the water in Tianjin Bohai Bay due to the resulting membrane fouling, which reduces the water permeability.
- The ceramic membrane pretreatment system maintained a steady permeability during the long-term experiment, even at extremely lower temperatures of 3.0°C−6.0°C.

6.3 Case studies

6.3.1 Case study-1: Determination of optimal design factors and operating conditions in a commercial seawater reverse osmosis desalination plant in Korea

A commercial seawater RO facility was established in Gijang-gun, Busan, Korea, and designed with a capacity of 10.0 million imperial gallons per day (MIGD). For optimizing the design factors as well as the operating conditions of the RO desalination facility, different assessments were carried out. A study was performed by Chu et al. (2020), which included split partial simulations of RO membranes, pretreatment studies, valuation of the product water quality with regard to production capacity, and a comparison of performance with other seawater RO facilities using the normalized specific energy consumption (SEC). This plant is composed of the source water intake, dissolved air bio-ball flotation, two different pretreatment processes (dual media filtration-based 8 MIGD, and ultrafiltration-based 2 MIGD trains), first-pass seawater RO, second-pass brackish water RO, and remineralization processes. Table 6.1 presents the source seawater quality of the Gijang seawater RO desalination facility for 1 year. Table 6.2 shows the different design factors for this seawater RO desalination facility.

TABLE 6.1 Quality of raw seawater of the Gijang seawater reverse osmosis desalination facility for 1 year.

Parameters	Ranges
Boron concentration, ppm	3.71−4.90
Dissolved organic matter, ppm	0.94−2.46
Turbidity, NTU	0.40−17.4
TDS, ppm	34,400−37,300
Temperature, °C	11.8−25.5
pH	7.3−8.8

Reproduced from Chu, K.H., Lim, J., Kim, S.J., Jeong, T.U., & Hwang, M.H. (2020). Determination of optimal design factors and operating conditions in a large-scale seawater reverse osmosis desalination plant. Journal of Cleaner Production, 244, 118918.

TABLE 6.2 Design factors for the Gijang seawater reverse osmosis desalination facility.

Parameters	Values
Capacity, MIGD	10
Train system	Two-pass systems
The number of trains	2 (8 MIGD + 2 MIGD)
Recovery rate (%)	48.3
Maximum boron concentration in final product water, ppm	1.0
Maximum total dissolved solid concentration in final product water, ppm	250.0

Reproduced from Chu, K.H., Lim, J., Kim, S.J., Jeong, T.U., & Hwang, M.H. (2020). Determination of optimal design factors and operating conditions in a large-scale seawater reverse osmosis desalination plant. Journal of Cleaner Production, 244, 118918.

The optimization of design factors and operating conditions of the seawater RO facility was attained by carrying out different assessments as follows:

1. Pretreatment studies: The turbidity values of the product water of the two pretreatment operations were slightly lower, whereas the SDI values varied significantly. Therefore, ancillary treatments (e.g., coagulation) for the dissolved air bio-ball flotation-dual media filtration process were required to maintain an SDI value less than 5 for the finest operation of the 8.0 MIGD train.
2. Split partial simulations of the RO membranes: These simulations were performed for confirming the effective performance of the RO membranes under various ratios and temperature conditions. The optimal energy-saving split partial ratio has been confirmed to be 4.5:5.5 which fulfilled the final product water quality standard of TDS 300 ppm or less.
3. Final product water qualities with numerous capacities: Based on the split partial simulation results, TDS and boron concentrations were 65.0–81.0 and 0.10–0.30 ppm, and 61.0–138.0 and 0.10–0.60 ppm in the 8.0 MIGD and 5.7 MIGD trains, respectively. Even though the concentrations at 10 MIGD also fulfilled the Korean water standards, it was essential for controlling the split partial ratio at increased temperatures because of the fixed capacity of the second pass.
4. Assessment and normalization of SEC: The total SECs were extremely suitable for total operation of the plant: 5.7 MIGD (4.13 kWh/m^3) > 8.0 MIGD (3.69 kWh/m^3) > 10.0 MIGD (3.66 kWh/m^3). Particularly, the normalized SEC value was less than 0.1 kWh/m^3/bar with an increased recovery rate in proportion to the operating pressure because of the size of the 8.0 MIGD train being designed by a combination of 16-inch RO membranes and 85% higher efficiency HPP.

Therefore, these specific performance assessments would be beneficial in establishing the optimal design factors and operating conditions for other commercial seawater RO desalination plants.

6.3.2 Case study-2: Design of reverse osmosis desalination plant in Egypt

In Egypt, because of the increasing fresh water needs in deserted and remote areas, the establishment of nonconventional water sources is very essential. A case study for a 3000 m^3

product water per day RO desalination plant in Ain El Sokhna-Suez, Egypt was studied and examined by Abdel-Fatah et al. (2016). Considering the plant location and site characteristics, several factors were evaluated in the RO plant design. The plant design was adopted using ROSA software and basic design equations for RO system design. A thorough economic study was also performed for assessing the feasibility of the plant.

The RO desalination facility includes a pretreatment system, a reverse osmosis unit, a post-treatment system, and other facilities like brine blow tank, product water storage tank, etc. The source water pumped up from wells has undergone disinfection by adding sodium hypochlorite (NaClO), and the disinfected water is stored in a storage tank. This water is subsequently pumped to multimedia filters using feed pumps; for reducing the SDI value of the water stream, typically coagulant is introduced to the raw water stream and mixed well. Using a properly designed multimedia filter, the SDI value below 5 could be obtained. A cartridge filter of pore size 5 μm is also available in this pretreatment system. TFC SWM configuration is employed in the RO unit. Dow Filmtec Type BW30−400 is chosen in the RO unit as this membrane has minimum replacement cost, easy maintenance, and simplicity in the plumbing system. The membrane elements are housed in fiber-reinforced plastic construction PVs rated at 450 psi design pressure. The RO membrane unit consists of a two-stage array, the first array PVs are 20 and the second array PVs are 10, where six elements are available in each PV. Almost 65% of the feed water will permeate through RO elements and turns out to permeate. TDS of permeate is less than 100 ppm. Reject water from the RO unit is transferred to a 50 m^3 brine tank. Sodium hydroxide is added for adjusting the product water pH to the set pH value.

By the use of ROSA software for designing the RO desalination system, the below design calculations were carried out.

Volume flow rate of element = 40 m^3/day

Recovery (considered in ROSA trial) = 60%

Volume flow rate of product water = 3000 m^3/day (needed to attain) = 3000 × 10^3 L/day

Volume flow rate of feed (calculated) = 5000 m^3/day

Therefore, number of elements = 5000/40 = 125 element

For safety reasons the team decided to have just six elements in each PV

∴ Number of vessels = 125/6 = 20.33333 ≈ 21 vessel

$$\text{Design flux permeate} = \frac{\text{Design permeate flow rate}}{\text{Total number of elements} \times \text{Active membrane surface area of element}}$$

$$\text{Design flux permeate} = \frac{3000}{125 \times 37} = 0.648 \text{ m}^3/\text{day}/\text{m}^2 = 27 \text{ L/m}^2\text{h}$$

The cost examination of the RO plant confirmed that the main factors influencing the permeate cost are the membrane cost and power consumption cost, while the chemical treatment signifies approximately 10% of the total cost.

6.3.3 Case study-3: Experimental and analytical study of a reverse osmosis desalination plant

In a work by Djebedjian et al. (2009), a 5000 m^3/day RO desalination plant in the city of Nuweiba, Egypt was considered for the case study. The measured data of the plant were recorded during 5 years of its normal operation. Moreover, experimental tests were performed on the site for investigating the influence of the main design as well as operating parameters on the performance of the RO plant.

The feed seawater salinity was in the order of 44,000 mg/L. The desalination plant includes five units, each with a capacity of 1000 m^3/day. This plant has a system such as the seawater intake system, the source water pretreatment unit, cartridge filters, the HPP, the RO membrane unit, the post-treatment system, and the turbocharger. The pretreatment system involves chemical injection and filtration for removing the particulate matter and to minimalize the biological fouling in addition to scaling. The seawater chlorination unit consists of chemical feed pumps and a polyethylene solution tank. In sand filters, the sand will trap the residual suspended materials and bacteria, and offers a physical matrix for the microbial decomposition of nitrogenous material into nitrogen gas. This plant in Nuweiba city is operating in the pressure range of 60−70 bar. The membrane stainless steel skid comprises 15 vessels, each one with five membrane elements of SWM type. The permeate post-treatment comprises chlorination and pH adjustment within the tolerable range of 7.5−8.5 mg/L.

In the course of the normal operation of the RO plant, the below data was recorded daily from years 2001 to 2006:

- Temperature after and before each RO element.
- Pressure after and before each filter, turbocharger, pump, and RO element.
- Flow rates of the concentrate and product water.
- Thermal conductivity after and before each RO element. Stream salinity was determined from the measured thermal conductivity.

The 5 RO units of the desalination plant were typically in continuous operation and showed similar performance and trends. From this study, it was confirmed that the RO system is found to be sensitive to the changes in the feed water temperature, salinity, and pressure;

1. Higher temperature of feed water improves the plant permeate flow rate.
2. As the feed water pressure increases the plant permeate flow rate also increases.
3. Increased feed water salinity leads to lower product flow rate and increased permeate salinity.

The cost analysis of the RO plant revealed that the main factors influencing the product water cost are the capital cost (33.6%) and power consumption cost (35.1%), whereas the chemical treatment represents 10.6% of the complete cost. The used maintenance schedule (for 5 years operation) was observed to be appropriate for the plant, as the variation in plant performance at the time of the operation period was not noticeable.

References

Abdel-Fatah, M. A., Elsayed, M., & Bazedi, G. A. (2016). Design of reverse osmosis desalination plant in Suez City (case study). *The Journal of Scientific and Engineering Research*, 3(4), 149−156, Corpus ID: 212584911.

American Society of Civil Engineers, American Water Works Association, and the USEPA. (1996). Technology transfer handbook: Management of water treatment plant residuals. American Society of Civil Engineers.

Asano, T. (1998). *Wastewater reclamation and reuse: Water quality management library* (Vol. 10). CRC Press.

Boerlage, S. F. (2001). *Scaling and particulate fouling in membrane filtration systems* (Vol. 26). CRC Press.

Bucher, W. (2015). Renewable energy systems and desalination—Volume III: Reverse osmosis by solar energy. *Encyclopedia of Desalination and Water Resources* (DEWARE).

Chu, K. H., Lim, J., Kim, S. J., Jeong, T. U., & Hwang, M. H. (2020). Determination of optimal design factors and operating conditions in a large-scale seawater reverse osmosis desalination plant. *Journal of Cleaner Production, 244*, 118918.

Cui, Z., Xing, W., Fan, Y., & Xu, N. (2011). Pilot study on the ceramic membrane pre-treatment for seawater desalination with reverse osmosis in Tianjin Bohai Bay. *Desalination, 279*(1–3), 190–194.

Design Parameters Affecting Performance, Hydranautics http://membranes.com/docs/trc/desparam.pdf [Accessed 27 July 2019].

Djebedjian, B., Gad, H., Khaled, I., & Rayan, A. M. (2009). Experimental and analytical study of a reverse osmosis desalination plant. *Mansoura Engineering Journal, 34*, 71–90.

Drioli, E., Giorno, L., & Fontananova, E. (2017). *Comprehensive membrane science and engineering.* Elsevier.

DuPont Water Solution's FilmTec™ RO technical manual; Version 3; Form No. 45-D01504-en, Rev. 3; April 2020.

Ettori, A., Gaudichet-Maurin, E., Aimar, P., & Causserand, C. (2013). Pilot scale study of chlorination-induced transport property changes of a seawater reverse osmosis membrane. *Desalination, 311*, 24–30.

Gallab, A. A. S., Ali, M. E. A., Shawky, H. A., & Abdel-Mottaleb, M. S. A. (2017). Effect of different salts on mass transfer coefficient and inorganic fouling of TFC membranes. *Journal of Membrane Science and Technology, 7*(175), 2.

Haidari, A. H., Heijman, S. G. J., & Van Der Meer, W. G. J. (2018). Optimal design of spacers in reverse osmosis. *Separation and Purification Technology, 192*, 441–456.

Ho, W., & Sirkar, K. (2012). *Membrane handbook.* Springer Science & Business Media.

Holloway, M. D., Nwaoha, C., & Onyewuenyi, O. A. (2012). *Process plant equipment operation, reliability, and control.* Wiley.

Ismail, F., Khulbe, K. C., & Matsuura, T. (2018). *Reverse osmosis.* Elsevier.

Kim, J., Park, K., Yang, D. R., & Hong, S. (2019). A comprehensive review of energy consumption of seawater reverse osmosis desalination plants. *Applied Energy, 254*, 113652.

Kucera, J. (2015). *Reverse osmosis: Industrial processes and applications.* John Wiley & Sons.

LaGrega, M. D., Buckingham, P. L., & Evans, J. C. (2010). *Hazardous waste management.* Waveland Press.

Li, X., Hasson, D., Semiat, R., & Shemer, H. (2019). Intermediate concentrate demineralization techniques for enhanced brackish water reverse osmosis water recovery—A review. *Desalination, 466*, 24–35.

Lior, N., El-Nashar, A., & Sommariva, C. (2012). *Advanced instrumentation, measurement, control, and automation (IMCA) in multistage flash (MSF) and reverse-osmosis (RO) water desalination. Advances in water desalination* (p. 494) John Wiley & Sons, Ltd., Chap, 6.

M46 Reverse Osmosis and Nanofiltration, 2nd edition; ISBN: 9781583214916; American Water Works Association, 2007.

McKetta, J. J., Jr. (1998). *Encyclopedia of chemical processing and design: Volume 64-Waste: Hazardous: Management guide to waste: Nuclear: Minimization during decommissioning.* CRC Press.

McMordie Stoughton, K., Duan, X., & Wendel, E. M. (2013). *Reverse osmosis optimization (No. PNNL-22682).* Richland, WA (United States): Pacific Northwest National Lab (PNNL).

MICRODYN-NADIR GmbH, "Effects of varying operating parameters – RO & NF systems," MICRODYN-NADIR, Goleta, CA, USA, 2020. Accessed on: Aug., 12, 2020. [Online]. Available: https://www.microdyn-nadir.com/wp-content/uploads/TSG-O-017-Effects-of-Varying-Operating-Parameters-RO-NF-Systems.pdf.

Mohanty, K., & Ghosh, R. (2011). *Gas-sparged ultrafiltration: recent trends, applications and future challenges. In Membrane and desalination technologies* (pp. 669–697). Totowa, NJ: Humana Press.

Ning, R. Y. (2012). *Chemistry in the operation and maintenance of reverse osmosis systems. Advancing Desalination,* Chap 4, (pp. 85-95), Robert Y. Ning, IntechOpen, Available from: http://doi.org/10.5772/39385. Available from: https://www.intechopen.com/books/advancing-desalination/chemistry-in-the-operation-and-maintenance-of-reverse-osmosis-system.

Obotey Ezugbe, E., & Rathilal, S. (2020). Membrane technologies in wastewater treatment: A review. *Membranes*, *10*(5), 89.

Olabarria, P. M. G. (2015). *Constructive engineering of large reverse osmosis desalination plants.* Chemical Publishing Company.

Peñate, B., & García-Rodríguez, L. (2012). Current trends and future prospects in the design of seawater reverse osmosis desalination technology. *Desalination, 284,* 1–8.

Raghavan, K. V., & Reddy, B. M. (2014). *Industrial catalysis and separations: Innovations for process intensification.* CRC Press.

Saji, V. S., Meroufel, A. A., & Sorour, A. A. (2020). *Corrosion and fouling control in desalination industry.* Springer International Publishing.

Saleem, H., & Zaidi, S. J. (2020). Nanoparticles in reverse osmosis membranes for desalination: A state of the art review. *Desalination, 475,* 114171.

Scott, K. (1995). *Handbook of industrial membranes.* Elsevier.

Scott, K., & Hughes, R. (1996). *Industrial membrane separation technology.* Springer Science & Business Media.

Singh, R., Hoffman, E. J., & Judd, S. (2008). *Membranes technology ebook collection: Ultimate CD.* Elsevier.

Wang, L. K., Chen, J. P., Hung, Y. T., & Shammas, N. K. (2008). *Membrane and desalination technologies* (Vol. 13). Springer Science + Business Media, LLC.

Wes Byrne, U.S. water, design and care of reverse osmosis systems, https://www.powermag.com/design-and-care-of-reverse-osmosis-systems/ Accessed on 20 August 2020.

Wilf, M. (2010). *The guidebook to membrane for wastewater reclamation.* Balaban Desalination Publications.

Wilf, M., & Awerbuch, L. (2007). *The guidebook to membrane desalination technology: Reverse osmosis, nanofiltration and hybrid systems: process, design, applications and economics.* Balaban Desalination Publications.

C H A P T E R

7

Designing of a Reverse Osmosis System

7.1 System design

Well-designed and accurately operated reverse osmosis systems offer a trouble-free performance over longer periods of time. Alternatively, errors made at the time of the design and operation of reverse osmosis systems can result in continuous problems and shorter membrane life. Taking water samples for laboratory analysis is a good starting point in the reverse osmosis design process. Moreover, a pilot study also contributes an opportunity to understand more about the reverse osmosis system along with its performance (MICRODYN-NADIR). In chapter 6, we already discussed the important parameters to be considered in reverse osmosis system design and operation. Also, we discussed the importance of conducting a pilot-scale study in the RO desalination plant in chapter 6.

7.1.1 Analysis of a water sample

The most important aspect in any reverse osmosis design process is the required water flux. This water flux must be chosen by the plant engineer or designer based on the feed water quality and source. A complete analysis of the water sample provides information on the dissolved salts (anions and cations); metals present in the water, like manganese, iron, and aluminum; probably the inorganic total suspended solids (TSS) in the water; and the water pH (acidity). The total organic carbon (TOC) measurement is the amount of carbon present in an organic compound and is generally used as a non-specific indicator of water quality (Hocking, 2005). The total suspended solids analysis confirms the concentration of filterable solids present in the water (Total Solids, Monitoring and Assessment). The dissolved metal concentration in the water varies in the sample as it reacts with oxygen brought by an air contact. This might trigger some of the metals to oxidize and subsequently turns out to be insoluble. The metals that remain suspended might increase the total suspended solids concentration significantly (Nie et al., 2008). Biological fouling solids are not properly represented in total suspended solid results. The mass of biological fouling solids usually becomes negligible if the total suspended solid filter is dried before weighing for the results. Moreover, the water can be tested for its silt density index (SDI) in the event that the metals are initially separated from the sample.

The SDI value relates to the fouling probability of a reverse osmosis membrane unit (Benjamin & Lawler, 2013).

The quality of water can vary over time, and hence no analysis is perfect. Sampling techniques can also influence the test results, and certain concentrations can also vary during the time between the sample collection and analysis. Also, the metals might have a tendency to attach to the inner surface of the container. Carbon dioxide (CO_2) and ammonia might degas or CO_2 might dissolve from air exposure (Moran, 2010). Any of the above-mentioned variations can lead to a change in the water pH. It is always recommended to measure the water pH on-site. A properly designed reverse osmosis system can effectively remove all the contaminants present in the water. Rejection rates of some contaminants by reverse osmosis single membrane element at the standard conditions are given in Fig. 7.1 (Frenkel, 2015). Thus proper water analysis must be conducted in the reverse osmosis system designing for projecting the product water quality and for evaluating the influence of the salts on the RO unit hydraulics.

Contaminant	% Nominal rejection	Contaminant	% Nominal rejection
Aluminum	96-98	Ammonium	80-90
Bacteria	99+	Borate	30-50
Boron	50-70	Bromide	90-95
Cadmium	93-97	Calcium	93-98
Chloride	92-95	Chromate	85-95
Copper	96-98	Cyanide	85-95
Fluoride	92-95	Hardness Ca & Mg	93-97
Iron	96-98	Lead	95-98
Manganese	96-98	Magnesium	93-98
Mercury	94-97	Nickel	96-98
Nitrate	90-95	Orthophosphate	96-98
Phosphate	95-98	Polyphosphate	96-98
Potassium	93-97	Radioactivity	93-97
Silica	80-90	Silicate	92-95
Silver	93-96	Sodium	92-98
Sulfate	96-98	Thoisulfate	96-98
Zinc	96-98		

FIGURE 7.1 Characteristic contaminant rejection rates from the water by reverse osmosis system. *Source: Reproduced from Frenkel, V.S. (2015). Planning and design of membrane systems for water treatment.* Advances in Membrane Technologies for Water Treatment, *329–347.*

7.1.2 Reverse osmosis system design guidelines

A complete reverse osmosis water treatment system comprises pretreatment section, reverse osmosis filtration units, and the post-treatment section. The different pretreatment practices are described in Chapter 5, Pretreatment: Fouling and Scaling Control. The post-treatment process is used for achieving the necessary product quality. In the case of seawater desalination, the post-treatment process will be usually an adjustment of pH, mineralization, and disinfection. The RO membrane system consists of a number of membrane elements, housed in pressure vessels (PVs) and arranged in a specific configuration. Feed water is supplied to the pressure vessels through high-pressure pumps. Maintenance equipment such as spare parts, tools, and instrumentation for system servicing are included as necessary. The membrane cleaning is conducted by a clean-in-place (CIP) system. Major advances as well as future anticipated developments in the design of seawater RO desalination technology have been reviewed by Peñate and García-Rodríguez (2012). In this study, the feasibility of utilization of renewable energy in moderate and large production capacities was discussed. Alghoul et al. (2009) demonstrated that the brackish water reverse osmosis single-stage system with a module connected to concentrate water is the best optimal design both ecologically and economically.

The membrane system has one inlet for feed water and two outlets for concentrate and permeate (Ramasamy, 2019). The performance of a reverse osmosis system is usually characterized by two parameters: product water (permeate) quality and product water flow. These parameters must always be related to a particular feed water analysis, temperature, recovery, and feed pressure. The main objective of a reverse osmosis system designer for a specifically required product water flow is to reduce feed pressure as well as the cost of the membrane whereas maximizing recovery and product water quality. The feed pressure required for producing the necessary product water flow for a specific membrane is dependent on the designed product water flux. High feed pressure will be required for the increased product water flow per unit of active membrane area. For seawater systems, the product water flux will be comparatively less even at maximal allowed pressure. Membrane elements for a specific system are chosen in accordance with (i) application, (ii) system capacity, (iii) fouling potential of feed water, (iv) total dissolved solids, (TDS) of feed water, and (v) required quality of permeate and energy requirements. The study by Kim et al. (2019) reviewed and analyzed seawater RO plants for a complete understanding of their specific energy consumption. This study confirmed that specific seawater RO designs could improve the RO system for efficiently achieving the established goals.

7.1.2.1 Guidelines for the design of membrane system

The factor that has the utmost effect on the design of the membrane system is the feed water fouling probability (Goh et al., 2018). The fouling of the membrane is due to the particles and colloidal materials, which are present in the feed water and subsequently get concentrated on the surface of the membrane (Fane et al., 2011). A system having higher permeate flux rates possibly has increased fouling probability, and hence additional repeated chemical cleaning will be required. The SDI of pretreated water is an indicator of the fouling possibility of colloidal materials or particles in a reverse osmosis unit (Williams, 2018; Xia et al., 2019). The fouling material concentration at the surface of the membrane increases with increasing element recovery and increasing permeate flux. The problems related to the reverse osmosis membrane processes and approaches have been discussed in detail by Hailemariam et al. (2020). Several studies have been conducted by researchers for improving membrane performance (Saleem & Zaidi, 2020a, 2020b, 2020c; Saleem et al., 2020; Zaidi et al., 2019).

A membrane system has to be designed in such a way that all the elements of the system function within a range of suggested operational conditions for minimizing the rate of fouling and eliminating mechanical deterioration (Dupont Water Solution). These operating conditions of the elements are limited by (a) the minimum concentrate flow rate, (b) the maximum permeate flow rate, (c) the maximum recovery, and (d) the maximum feed flow rate per element. The average flux of complete system is a representative number of the design. The system flux helps to estimate the necessary number of elements in a new venture (Hendricks, 2018). Systems working on high-quality feed water will be typically designed at higher flux value, while systems working on low-quality feed water will be designed at lower flux value. A continuous reverse osmosis process designed as per the system design guidelines and with a properly designed as well as operated pretreatment unit would be able to demonstrate steady performance with just four cleanings each year in regular applications. Going beyond the suggested limits might lead to decreased capacity, frequent cleanings, higher feed pressure, and short life of the membrane. A reasonable shift in the limits for a shorter period might be tolerable on the condition that the physical limits (maximal feed pressure and maximal pressure drop) are not exceeded.

The following guidelines must be considered while designing a membrane system (Singh, 2015):

- The product water flow rate will be in accordance with the net driving pressure (NDP) differential across the membrane.
- The flow rate of salt will be in accordance with the concentration difference across the membrane and is not dependent on the applied pressure.
- Product water total dissolved solids depend on the relative mass transfer rates of water as well as dissolved solute through the membrane.
- The physical and chemical nature of the membrane defines the preferred water transport over dissolved solutes.
- The greater the product water flux, the higher is the possibility of increased concentration polarization. As concentration polarization increases, the solution osmotic pressure in the feed-reject channel increases, which results in a higher salt passage and increases the possibility of fouling or/and scaling.

7.1.2.2 Recommended range of element operating conditions

For any RO design, recommended design limits should be adopted. The recommended design limits for the RO system are given in Table 7.1. The recommended design limit may vary based on the membrane manufacturer. Table 7.2 presents the range of different parameters as per the feed water type and the pretreatment type, recommended by

TABLE 7.1 Recommended design limits.

Feed water parameters		RO permeate	Brackish well softened	Brackish well not softened	Seawater surface conventional
SDI₁₅	Maximum	1	2	3	4
Turbidity (NTU)	Typical	0.1	0.1	0.1	0.1
Particle count 2 μm particles/mL	Typical	100	100	100	100
System average flux (gallons per square foot per day—GFD)	Conservative	18	14	14	7
	Typical	21	16	16	8
	Aggressive	24	20	18	10
Lead element flux (GFD)	Conservative	29	24	21	17
	Typical	30	27	24	20
	Aggressive	35	29	27	24
Percent flux drop (per year)	Conservative	7	10	10	10
	Typical	5	7	7	7
	Aggressive	3	5	7	7
Percent salt passage rise/per year	Conservative	7	15	15	15
	Typical	5	10	10	10
Feed GPM (maximum per vessel) 8 inch	Conservative	75	70	65	60
	Typical	75	75	75	75
	Maximum	75	75	75	75
Reject GPM (minimum per vessel) 8 inch	Conservative	10	12	12	15
	Typical	7.1	9.6	9.6	10.3
DeltaP (psi) 6 M vessel membrane element	Typical	25	25	25	25
	Maximum	40	40	40	40
	Minimum	10	10	10	10
Saturation limits with antiscalant (Langelier. saturation index & Stiff-Davis Index)	Typical	<1.8	<1.8	<1.8	<1.8
	Aggressive	<2.5	2.5	2.5	2.5

LANXESS (LANXESS). The pretreatment of membrane increases the quality of feed water in consideration of the fouling. The range of different parameters given in Table 7.2 is the suggested value for minimizing the possibility of fouling; on the other hand, it does not mean that the design of the system beyond the parameter range is not possible. It implies that the fouling probability will be more intense.

TABLE 7.2 Example of reverse osmosis system design guideline recommended by LANXESS.

Feed water type	Avg. product water flux (range) (L/m²h)	Lead element product water flux (L/m²h)	Concentrate flow rate per vessel (m³/h)	Feed flow rate per vessel (m³/h)	Pressure drop per vessel (bar)	Element recovery rate (%)	Salt passage increase (%)
Brackish wells	27 (23−29)	<34	8 inches: >3.0 4 inches: >0.6	8 inches: <16.0 4 inches: <3.2	<3.0	<20	>10
Municipal supply	23 (20−26)	<31	8 inches: >3.6 4 inches: >0.7	8 inches: <15.0 4 inches: <2.8	<2.0	<15	>10
Surface water microfiltration/ ultrafiltration pretreatment	27 (23−29)	<34	8 inches: >3.0 4 inches: >0.6	8 inches: <16.0 4 inches: <3.2	<3.0	<20	>10
Surface water media filtration	23 (20−26)	<31	8 inches: >3.6 4 inches: >0.7	8 inches: <15.0 4 inches: <2.8	<2.0	<15	>10
Secondary waste media filtration	17 (14−20)	<24	8 inches: >4.1 4 inches: >0.8	8 inches: <14.0 4 inches: <2.6	<2.0	<12	>15
Secondary waste microfiltration/ ultrafiltration pretreatment	20 (17−23)	<28	8 inches: >3.6 4 inches: >0.7	8 inches: <14.0 4 inches: <2.8	<2.0	<17	>10
Seawater intake media filtration	14 (11−17)	<30	8 inches: >3.6 4 inches: >0.7	8 inches: <14.0 4 inches: <2.8	<2.0	<13	>10
Seawater intake microfiltration/ ultrafiltration pretreatment	17 (14−20)	<35	8 inches: >3.4 4 inches: >0.7	8 inches: <16.0 4 inches: <3.0	<3.0	<15	>10
Seawater beach wells	17 (14−20)	<35	8 inches: >3.4 4 inches: >0.7	8 inches: <16 4 inches: <3.0	<3.0	<15	>10
Reverse osmosis product water	37 (32−42)	<48	8 inches: >2.4 4 inches: >0.5	8 inches: <17 4 inches: <3.6	<3.0	<30	>5

Courtesy: LANXESS Deutschland GmbH. Reproduced from LANXESS Deutschland GmbH SYSTEM DESIGN—Guidelines for the design of RO Membrane Systems Retrieved from http://lpt.lanxess.com/uploads/tx_lanxessmatrix/lew_lewabrane_manual_03_system_design_01.pdf [Accessed 25 February 2020].

7.1.3 Step-by-step procedure for the design of reverse osmosis membrane system

The following steps should be considered for designing a reverse osmosis membrane system (DuPont Water Solution, 2020):

7.1.3.1 Step 1: Considering the source of feed water, quality of feed water, feed/ permeate flow, and required quality of permeate

The membrane system design is greatly dependent on the accessible feed water quality and the application (FILMTEC Membranes). Hence, the information on the system design (customer/original equipment manufacturer, operating pressure limit, required product water quality, application, pretreatment, water source, annual water temperature, expected recovery rate, required product flow rate, etc.) and the analysis of feed water must be carefully analyzed and taken into consideration first.

1. Choose the source of feed water.
2. Choose the overall concentration of feed water as individual (specific) ions or as TDS (ppm).
3. Individual ion concentration from water analysis is favored always.

7.1.3.2 Step 2: Selecting the flow configuration and the number of passes

In the case of water desalination, the standard flow configuration will be plug flow, in which the feed solution will be passed through the system just once (Olabarria, 2015). Reject recirculation is observed in small units employed in industrial applications, and in large units when the number of elements is very limited for achieving a satisfactorily higher system recovery with plug flow. These units are also used in exceptional applications such as wastewaters and process liquids (Vu, 2017).

In general, a reverse osmosis system is intended for continuous operations, and the operational conditions of all membrane elements in the desalination facility are fairly consistent with time. However, in some applications, batch mode of operation is employed (Barello et al., 2015). For example, in the treatment of wastewater or industrial process solutions, where comparatively smaller feed water volumes are non-continuously released, batch operation mode can be used. The feed water will be assembled in a tank and later intermittently will undergo treatment. Semibatch mode is considered to be an adaptation of the batch mode, in which the feed tank is refilled using feed water in the course of operation.

If the product water quality required is very high that the quality could not be accomplished by a single-pass reverse osmosis system, then a two-pass reverse osmosis system must be considered (Kim et al., 2020b). A double-pass (permeate-staged) system is the consolidation of two standard reverse osmosis units, where the permeate/product water of the first pass (first system) turns out to be the feed for the second pass (second system) (Reverse Osmosis Optimization Prepared for the United States Department of Energy Federal Energy Management Program). The two reverse osmosis systems might be of the multistage or single-stage configuration, either with concentrate/reject recirculation or with plug-flow. The water production for medical and pharmaceutical applications is considered to be distinctive application of permeate-stage systems. As a substitute to the two-pass reverse osmosis system, an ion exchange resin system might also be a feasible design choice (Ward, 2007).

7.1.3.3 Step 3: Selecting the membrane element type

Elements are carefully chosen as per the fouling tendency of feed water, the salinity of feed water, energy requirements, necessary rejection, system capacity, and application. The typical size of element for systems producing product water greater than 10 gallons per minute (GPM) (2.3 m^3/h) will be 40 inches in length and 8 inches in diameter. In the case of superior-quality water utilization, where extremely less product salinity is essential, the electrodeionization (EDI) or the ion exchange resins are normally used for polishing the reverse osmosis permeate. The relation between the general selection of reverse osmosis element and feed salinity, as recommended by LANXESS, is presented in

Table 7.3. LANXESS also recommends that for feed water with low TDS (less than 500 ppm), a low-energy brackish water element can be used, and if the reverse osmosis feed is wastewater, then the low-fouling reverse osmosis element can be considered.

7.1.3.4 Step 4: Selecting the average membrane flux

As the water source, pretreatment type and reverse osmosis element type are fixed by the design engineer, the recommended value of the average product water flux (also termed as "design flux") is provided. The design flux (L/(m^2.h) or GFD (gallons/ft^2/day)) has to be selected based on customer experience, pilot data, or the characteristic design fluxes as per the feed source. In Table 7.4, the recommended average permeate flux versus different sources of water are given. For reverse osmosis/ ultrafiltration (UF) permeate with SDI less than one, high permeate water flux, that is, about 21–30 GFD is recommended, whereas for open intake seawater source with SDI less than five, low permeate water flux, that is, about 7–11 GFD is recommended. Table 7.5 shows the design recommendations for DuPont's FilmTec Elements in small-scale commercial application. The maximum feed flow rate, maximum pressure drop, and maximum feed pressure for different FilmTec Elements are shown in Table 7.6.

TABLE 7.3 Reverse osmosis element selection according to the feed salinity as recommended by LANXESS.

Feed salinity	Selection of reverse osmosis element
Brackish water—till 500 ppm	Brackish water RO (low energy)
Brackish water—till 5,000 ppm	Brackish water RO (standard)
Seawater and Brackish water—more than 5,000 ppm	Seawater RO (low energy)

Courtesy: LANXESS Deutschland GmbH. Reproduced from LANXESS Deutschland GmbH SYSTEM DESIGN—Guidelines for the design of RO Membrane Systems Retrieved from http://lpt.lanxess.com/uploads/tx_lanxessmatrix/lew_lewabrane_manual_03_system_design_01.pdf [Accessed 25 February 2020].

TABLE 7.4 Recommended average product water flux versus different sources of water.

Water source	Recommended product water flux (in GFD)
Reverse osmosis/ultrafiltration (UF) permeate (silt density index <1)	21–30
Well water (silt density index <3)	13–18
Surface water (silt density index <3)	13–17
Surface water (silt density index <5)	12–16
High salinity well water (silt density index <3)	8–12
Seawater, open intake (silt density index <5)	7–11
Wastewater pretreated by ultrafiltration (silt density index <3)	10–14
Wastewater (silt density index <5)	8–12

TABLE 7.5 FilmTec Elements design guidelines in small commercial applications.

	Reverse osmosis permeate	Softened municipal water	Well water	Surface or municipal water
Feed source				
Feed SDI	Silt density index <1	Silt density index <3	Silt density index <3	Silt density index <5
Typical target flux, L/m²h (GFD)	51 (30)	51 (30)	42 (25)	34 (20)
Maximum element recovery (%)	30	30	25	20
Maximum permeate Flow rate, m3/d(GPD)				
2.5-inch diameter	4.2 (1,100)	4.2 (1,100)	3.4 (900)	2.7 (700)
4.0-inch diameter	11.7 (3,100)	11.7 (3,100)	9.8 (2,600)	7.9 (2,100)
Minimum concentrate flow rate, m3/d(GPM)				
2.5-inch diameter	0.11 (0.5)	0.11 (0.5)	0.16 (0.7)	0.16 (0.7)
4.0-inch diameter	0.5 (2)	0.5 (2)	0.7 (3)	0.7 (3)

Courtesy: DuPont Water Solutions; reproduced from DuPont Water Solution's FilmTec™ RO Technical Manual; Version 3; Form No. 45-D01504-en, Rev. 3; April 2020.

TABLE 7.6 Maximum feed flow rate, maximum pressure drop, and maximum feed pressure for different FilmTec elements.

Element type	Maximum Feed flow rate in m³/h (GPM)	Maximum pressure drop per element in bar (psig)	Maximum feed pressure in bar (psig)
Tape-wrapped 2540	1.4 (6)	0.9 (13)	41 (600)
Fiber-glassed 2540	1.4 (6)	1.0 (15)	41 (600)
Seawater 2540	1.4 (6)	0.9 (13)	69 (1,000)
Tape-wrapped 4040	3.2 (14)	0.9 (13)	41 (600)
Fiber-glassed 4040	3.6 (16)	1.0 (15)	41 (600)
Seawater 4040	3.6 (16)	1.0 (15)	69 (1,000)

Courtesy: DuPont Water Solutions. Reproduced from DuPont Water Solution's FilmTec™ RO Technical Manual; Version 3; Form No. 45-D01504-en, Rev. 3; April 2020.

7.1.3.5 Step 5: Calculation of the number of membrane elements required

The design product water flow rate (Q_P) should be divided by the membrane surface area of the element chosen (S_E in m² or ft²) and by the design flux (f) for obtaining the number of elements N_E (Ezzeghni, 2018).

$$N_E = \frac{Q_P}{f \cdot S_E} \tag{7.1}$$

The number of reverse osmosis elements calculated might be marginally changed on the basis of the choice of element arrangement, specifically, the number of PVs and reverse osmosis elements per pressure vessel.

7.1.3.6 Step 6: Calculation of the number of pressure vessels required

The number of elements N_E must be divided by the number of elements per PV, N_{EpV}, for obtaining the N_V, the number of PVs, rounding up to the nearest integer. In the case of larger systems, the six-element vessels are considered to be standard; however, vessels with maximum of eight elements are obtainable. Further, for compact or/and smaller systems, short vessels might be carefully chosen.

$$N_V = \frac{N_E}{N_{EpV}}. \tag{7.2}$$

Smaller systems having just one or a limited number of elements will be typically designed with the elements arranged sequentially and a reject recirculation to retain the suitable flow rate through the brine/feed channels.

Practically all the RO design parameters can be changed by increasing the number of reverse osmosis elements per PV (LANXESS). Reverse osmosis design parameters include pressure drop per vessel, individual element recovery rate, feed flow rate per vessel, concentrate flow rate per vessel, lead element permeate flux, average permeate flux, etc. Certain factors turn out to be necessary while some other factors turn out to be unwanted. Table 7.7 shows design parameters of the reverse osmosis system and the relationship of an increase in the number of reverse osmosis element per PV, and the variation in design parameters of the reverse osmosis system. It is suggested that the aforestated parameters are in agreement with the design guidelines.

TABLE 7.7 Effects of changing the reverse osmosis design as recommended by LANXESS.

Sl. No.	Reverse osmosis system design parameters	Effect of the increasing number of elements/vessel	Judgment of the effect
1	Number of vessel	Smaller	Desirable
2	System recovery rate	Same	No variation
3	Recovery rate of element	Smaller	Desirable
4	Pressure drop per element	Larger	Undesired
5	Pressure drop per vessel	Larger	Undesired
6	Feed flow rate per vessel	Larger	Undesired
7	Concentrate flow rate per vessel	Larger	Desirable
8	Lead element flux	Larger	Undesired
9	Average permeate flux	Same	No variation

Courtesy: LANXESS Deutschland GmbH. Reproduced from LANXESS Deutschland GmbH SYSTEM DESIGN—Guidelines for the design of RO Membrane Systems Retrieved from http://lpt.lanxess.com/uploads/tx_lanxessmatrix/lew_lewabrane_manual_03_system_design_01.pdf [Accessed 25 February 2020].

7.1.3.7 Step 7: Selection of the number of stages

The number of stages outlines the number of PVs arranged sequentially; the feed will flow through till it departs the system and will be released as the reject. Each stage comprises a definite number of PVs arranged in parallel (Flow Configuration). The number of stages will be a function of the feed water quality, the number of elements per PV, and the planned system recovery.

Greater the recovery of system and lower the quality of feed water, the system will be longer with more elements connected sequentially. As an example, a system having four 6-element PVs in the first stage along with two 6-element PVs in the second stage will be having twelve elements arranged sequentially. A system having three stages and 4-element PVs in a 4:3:2 arrangements will also have 12 elements arranged sequentially. Normally, the number of elements in series is related to the number of stages and the system recovery, as given in Table 7.8. A one-stage system could also be designed for higher recovery in the event that concentrate recycling is used. The relationship between the concentration factor and the recovery rate is given in Table 7.9.

For an overall recovery of 45% and seven elements per vessel, one stage will be sufficient. For an overall recovery of 50% or higher and seven elements per vessel, two stages will be needed. With eight elements per vessel, it will be possible to get a 50% recovery with just one stage.

7.1.3.8 Step 8: Selection of the staging ratio

Staging ratio (R_s) is the relation of the number of PVs in successive stages (Sarbatly, 2020).

$$R_s = \frac{Nv(i)}{Nv(i+1)}. \tag{7.3}$$

TABLE 7.8 Number of stages of seawater and brackish water systems.

Sl. No.	System recovery (%)	No. of serial element positions	No. of stages (six-element PVs)	No. of stages (seven-element PVs)	No. of stages (eight-element PVs)
		Seawater RO system			
1	55–60	12–14	2	2	-
2	50	8–12	2	2	1
3	45	7–12	2	1	1
4	35–40	6	1	1	-

Sl. No.	System recovery (%)	No. of serial element positions	No. of stages (six-element PVs)
		Brackish water RO system	
5	85–90	18	3
6	70–80	12	2
7	40–60	6	1

TABLE 7.9 Relationship between the concentration factor and the rate recovery as recommended by LANXESS.

Sl. No.	Recovery rate (%)	Concentration factor
1	50%	2
2	75%	4
3	80%	5
4	90%	10

Courtesy: LANXESS Deutschland GmbH. Reproduced from LANXESS Deutschland GmbH SYSTEM DESIGN—Guidelines for the design of RO Membrane Systems Retrieved from http://lpt.lanxess.com/uploads/tx_lanxessmatrix/lew_lewabrane_manual_03_system_design_01.pdf [Accessed 25 February 2020].

In the case of a system that has four vessels in the first stage and two vessels in the second stage, then the R_s will be 2: 1. For a three-stage system with 4, 3, and 2 vessels in the first, second, and third stages, respectively, it will have a 4:3:2 staging ratio. For brackish water systems, the staging ratios between two succeeding stages will be typically around 2:1 in the case of six-element PVs and lesser than that for the smaller PVs. In two-stage seawater reverse osmosis systems having six-element PVs, the characteristic staging ratio will be 3:2.

The perfect staging of a system will be in a manner that every stage functions at similar fraction of the system recovery, on condition that the entire PVs consist of the identical number of elements. The system-staging ratio with a recovery "r" and "n" stages could be determined as follows:

$$R_s = \frac{1}{(1-r)}^{\frac{1}{n}}.$$ (7.4)

The pressure vessel number in the first stage $N_{v(1)}$ could be computed with the R_s value from the total number of vessels N_v.

In the case of a three-stage system ($n = 3$) and a two-stage system ($n = 2$), the number of PVs in the first stage can be determined as follows:

$$N_{V(1)} = \frac{N_V}{1 + r^{-1} + r^{-2}}, \text{ for } n = 3, \text{etc.}$$ (7.5)

$$N_{V(1)} = \frac{N_V}{1 + r^{-1}}, \text{ for } n = 2.$$ (7.6)

Further, the number of vessels in the second stage is calculated as follows:

$$N_{V(2)} = \frac{N_{V(1)}}{r}.$$ (7.7)

An alternative perspective for choosing a specific arrangement of PVs is the feed flow rate per PV of the first stage and the reject flow rate per PV of the final stage. The pressure vessel number in the first stage must be carefully chosen subsequently for providing a feed flow rate of 8–12 m³/h (35–55 GPM) per 8" vessel. Similarly, the vessel number in the final stage must be chosen in a way that the resulting reject flow rate is higher than the lowest of 3.6 m³/h (16 GPM).

7.1.3.9 Step 9: Balancing the product water flow rate

The product water flow rate of the tail elements (elements situated at the reject end) of a reverse osmosis system is usually lower as compared to the lead element flow rates. The ratio of the product water flow rate of the lead element and the tail element could turn out to be very high in some specific conditions such as:

- New membranes
- High water temperature
- Low-pressure membranes
- High feed salinity
- High system recovery

The objective of a perfect design is balancing the element flux in the various locations. This could be accomplished by the following ways:

- Improving the feed pressure between stages: desired for effective application of energy
- Application of a permeate back-pressure only to the first stage of a dual-stage system: low system cost substitute
- Hybrid system: Use membranes having high water permeability in the final points and membranes having low water permeability in the initial points, for example, in the first stage of a seawater reverse osmosis system, higher-rejection seawater membranes and in the second stage with higher-productivity membranes.

7.1.3.10 Step 10: Analyzing and optimizing the membrane system

The selected system must be subsequently analyzed and refined by applying the RO design software program. The different RO design software programs are discussed in Chapter 8.

7.2 Designing examples

7.2.1 Example 7.1:

Fig. 7.2 shows the details of the example used for step 10 in "how to design a reverse osmosis system." It assumes that the TDS of the well water is 2,800 mg/L. The flows of permeates and rejects of the first and second stages were obtained using the IMSDesign Hydranatics software as well as the final permeate of 73 mg/L in TDS. On the basis of Fig. 7.2, the following calculations can be carried out.

1. % Recovery of 1st stage $= \left[\frac{Q_p}{Q_p + Q_c}\right] \times 100 = \left[\frac{162.9 \text{ GPM}}{162.9 \text{ GPM} + 103.8 \text{ GPM}}\right] \times 100 = 61.1\%$

2. % Recovery of 2nd stage $= \left[\frac{Q_p}{Q_p + Q_c}\right] \times 100 = \left[\frac{37.1 \text{ GPM}}{37.1 \text{ GPM} + 66.7 \text{ GPM}}\right] \times 100 = 35.7\%$

3. % Overall system recovery $= \left[\frac{Q_p}{Q_p + Q_c}\right] \times 100 = \left[\frac{200 \text{ GPM}}{200 \text{ GPM} + 66.7 \text{ GPM}}\right] \times 100 = 75.0\%$

4. First stage flux $= \left[\frac{Q_p}{A}\right] = \left[\frac{162.9 \text{ GPM} \times 60 \times 24}{6 \times 6 \times 365 \text{ ft}^2}\right] = 17.85 \text{ GPD/ft}^2$

5. Second stage flux $= \left[\frac{Q_p}{A}\right] = \left[\frac{37.1 \text{ GPM} \times 60 \times 24}{3 \times 6 \times 365 \text{ ft}^2}\right] = 8.13 \text{ GPD/ft}^2$

6. Average RO flux $= \left[\frac{Q_p}{A}\right] = \left[\frac{200 \text{ GPM} \times 60 \times 24}{9 \times 6 \times 365 \text{ ft}^2}\right] = 14.6 \text{ GPD/ft}^2$

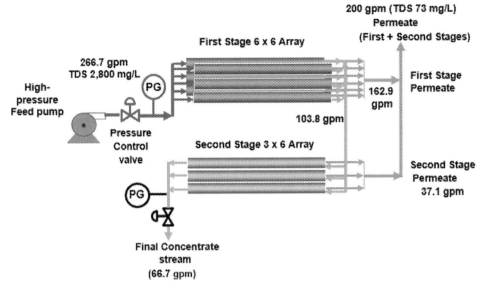

FIGURE 7.2   Schematic diagram of how to design a reverse osmosis system.

7. Percent overall system salt passage(% SP) = $\left[\frac{Permeate\ TDS}{Feed\ TDS}\right] \times 100 = \left[\frac{73\ mg/L}{2,800\ mg/L}\right] \times 100 = 2.6\%$

8. Percent Overall system salt rejection(% SR) = $100 - (\% SP) = 100 - 2.6\% = 97.4\%$

Where,

Q_P = Flow rate of permeate
Q_C = Flow rate of concentrate/reject
A = Membrane surface area (surface area of hydranautics membrane CPA2 = 365 ft²)

- The first-stage recovery is 61.1%, which is higher than the second-stage recovery. This is what a good design requires, because the TDS in the first stage is low and the system can operate at higher recovery without scaling.
- The second-stage recovery of 35.7% is lower than the first-stage recovery. This is what a good design requires since the TDS in the second stage is greater than in the first stage and the second stage needs to operate at lower recovery to avoid scaling.
- Despite the difference in recoveries between the first and second stages of the RO system, the overall system recovery is 75% as per design.
- The first-stage flux is 17.85 GPD/ft², which is high. However, this is acceptable because the TDS in the first stage is low and the membrane can still operate without scaling.
- The second-stage flux is only 8.13 GPD/ft² because the TDS in the second stage is high and operation at higher flux will cause scaling.

- Despite the difference in fluxes between first and second stages of the RO system, the average reverse osmosis system flux is 14.6 GPD/ft^2 as per design.

7.2.2 Example 7.2:

Conditions specified:

- Feed water source: Brackish surface supply water, with SDI value <5
- Six-element pressure vessels to be used
- Product water flow required = 720 m^3/day (132 GPM)

Steps:

1. Brackish surface supply water, with SDI value <5 and product water flow required = 720 m^3/day (132 gallons per minute)
2. Selection of plug flow
3. Brackish water element (Filmtec) with active membrane area 37.20 m^2 (400 ft^2)
4. The average flux recommended for surface supply water feed having silt density index <5 is 23.8 L/(m^{2}h) or 14.0 GFD (see Table 7.4)
5. Total number of elements $= \frac{\left(\frac{720 \text{ m}^3}{\text{day}}\right)\left(\frac{1,000 \text{ L}}{1 \text{ m}^3}\right)\left(\frac{1 \text{ day}}{24 \text{ h}}\right)}{\left(\frac{23.8 \text{ L}}{\text{m}^2 \cdot \text{h}}\right)(37.2 \text{ m}^2)} = 34$ or $\frac{\left(\frac{132 \text{ gal}}{\text{min}}\right)\left(\frac{1,440 \text{ min}}{1 \text{ day}}\right)}{\left(\frac{14 \text{ gal}}{\text{ft}^2 \cdot \text{day}}\right)(400 \text{ ft}^2)} = 34$
6. Complete number of pressure vessels $= \frac{34}{6} = 5.67$; we can round that to nearest integer, that is, 6
7. Number of stages for six-element PVs and the recovery of 75% = 2 (see Table 7.8)
8. The chosen staging ratio is 2:1. Suitable stage ratio = 4:2.
9. The selected system should subsequently be examined using the RO system design computer program. The program will calculate the permeate quality and feed pressure of the system and the operating data of all distinct elements. Subsequently, it will be easy for optimizing the system design by varying the type and number of elements and the arrangement.

7.2.3 Example 7.3:

Conditions specified:

- Feed water source: well water with silt density index value <2.0.
- Product water flow required = 200 gallons per minute (1,090 m^3/day).
- Six-element PVs to be employed.

Steps:

1. Feed water source: well water with silt density index value <2.0 and product water flow required = 200 gallons per minute (1,090 m^3/d).
2. Selection of plug flow.
3. Hydranautics CPA2 (membrane with an active membrane area of 365 ft^2 [33.90 m^2])

4. Average flux recommended for well water feed with silt density index $<2 = 14.6$ GFD $(24.3 \text{ L/m}^2\text{h})$
5. Total no. of elements $= \frac{200 \text{ GPM} \times 1,440 \text{ GPD}}{14.\text{GFD} \times 365 \text{ ft}^2} = 54$
6. Complete number of PVs $= 54/6 = 9$
7. Number of stages for six-element PVs and the recovery of 75% $= 2$
8. Staging ratio chosen: 2:1. Suitable stage ratio (array) $= 6:3$.
9. The selected system should subsequently be analyzed by applying the RO Integrated Membrane Solutions Design Software (IMSDesign) computer program. The program software calculates the feed pressure required, the product water quality of the RO system, and the operating information of all specific elements. Then it will be simple for optimizing the design of RO system by varying the type and number of elements and the arrangement.

7.3 System design considerations for controlling the microbiological activity

Biological fouling is considered to be severe and very common issue observed in the operation of reverse osmosis units (Al-Juboori & Yusaf, 2012; Pandey et al., 2012). A correctly designed RO system is a prerequisite for its proper operation and the following should be considered to control the microbial activity (DuPont Water Solution, 2020).

- In the case that intermediate open basins are employed, then provisions must be made for suitable disinfection at that open source and the system part downstream from it.
- If using intermediate sealed tanks, then proper ventilation arrangements must be provided with bacteria-retaining devices.
- Dead legs in the piping must not be allowed in design; however, in inevitable cases, it must be intermittently sanitized. Dead legs are sections of potable water piping systems that have been altered, abandoned, or capped such that water cannot flow through them. The pretreatment system components like retention tanks, filters, manifolds, and pipes must be opaque to sunlight for avoiding the increase in biological growth.
- Standby devices having larger surfaces, like cartridge or sand filters, must not be allowed. If such devices are inevitable, then drains must be set up for allowing discharge of the sanitization chemicals after the devices are sanitized, and before connecting them to the active unit.
- The RO section must be physically isolated from the pretreatment section employing a flange. The isolation will allow the use of chlorine to sanitize the pretreatment unit, whereas the RO membranes are secured from chlorine attack.
- Proper choice of elements: Elements with characteristics such as an exceptional surface combined with a feed spacer thickness or/and geometry that turns the elements extremely biofouling resistant are available. These antibiofouling membrane elements have been normally chosen for surface waters and tertiary effluent treatment.

7.4 Case studies

7.4.1 Case study 1: Design of 13.3 million gallons per day (MGD) seawater RO desalination facility

A 13.3 million gallons per day (MGD) seawater RO desalination facility in Yanbu industrial city, Saudi Arabia was designed and its main design features were discussed by Khawaji et al. (2007). The seawater reverse osmosis plant consists of six trains of about 2.2 million gallons per day capacity each. This plant has five main systems: seawater intake unit, pretreatment unit, high-pressure pumps, reverse osmosis modules, and the post-treatment unit. This study also discusses technical issues and different parameters related to the design of the RO plant and progresses made in the seawater RO desalination technology.

Main design concerns of seawater reverse osmosis facilities include feed water salinity, feed water temperature, water flux, power consumption, recovery ratio, membrane life, product water salinity, etc. The different flow rates for the reverse osmosis facility are presented in Fig. 7.3. Moreover, the main plant design features have been demonstrated in Table 7.10.

The seawater design flow rate for the reverse osmosis facility is 5,804 m^3/h. In the system design, the seasonal fluctuations in the seawater temperature from 22°C to 33°C happening at the desalination plant site were taken into consideration. The seawater salt concentration is 46,400 mg/L of dissolved salts with 23,500 mg/L chloride concentration. The seawater turbidity of 0.5−1.0 NTU and the pH varied from 8.1 to 8.3. A characteristic seawater analysis has been carried out, and the results are shown in Table 7.11. The feed water pressurized was fed to the six reverse osmosis trains arranged in parallel. The reverse osmosis trains will have a capacity of 350 m^3/h (2.22 million gallons per day) and the design pressure is 64.0−76.0 kg/cm^{2}g. For the above-stated system, 38.50% of the filtered seawater, about 350 m^3/h, is recovered as the product water. An operational pressure of 64.0 kg/cm^{2}g (gauge pressure) is appropriate on new membranes, and the pressure will be increased progressively. As the semipermeable membrane, the cellulose triacetate (CTA) in a hollow-fine fiber (HFF) configuration was chosen. The membranes restricted the salt passage whereas allowing the water to flow through. The reverse osmosis operation is of the single-stage configuration, as the product water is supplied as

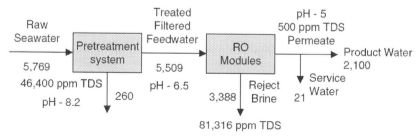

FIGURE 7.3 Flow rates for the seawater reverse osmosis facility in Yanbu Industrial City, Saudi Arabia. The flow rate in m^3/h. *Source: Reproduced from Khawaji, A.D., Kutubkhanah, I.K., & Wie, J.M. (2007). A 13.3 MGD seawater RO desalination plant for Yanbu Industrial City.* Desalination, *203(1−3), 176−188.*

TABLE 7.10 Plant design features of the seawater reverse osmosis facility in Yanbu Industrial City, Saudi Arabia.

Design features	Values
Plant capacity	50,400 m^3/d [13.3 MGD] at seawater temperature of 22°C
No. of trains	6
Permeate chloride as Cl^-	250 ppm
Permeate TDS	500 ppm
Seawater pH	8.1−8.3
Seawater temperature	22°C−33°C
Seawater conductivity at 25°C	57,000−64,000 microsiemens/centimeter (μS/cm) or micromhos/centimeter (umhos/cm)
Seawater TDS	46,400 ppm
Residual chlorine at reverse osmosis plant intake as Cl_2	0.1−0.25 ppm
Pretreatment recovery rate	95%
Permeate recovery ratio	38.5%
Pretreatment methods	Filtration and chemicals injection
SDI of dual media filter effluent	4
Reverse osmosis pump pressure	64−76 kg/cm^2g
Reverse osmosis membrane	Cellulose triacetate double element hollow-fine fiber
Total no. of membrane elements	1,824
No. of membrane modules per train	152
No. of membrane elements per module	2
Permeator arrangement	Horizontal position in parallel
Pretreated filtered seawater pH	6.5
Chlorination method	Intermittent
Power consumption	5.2 kWH/m^3
Life of membrane	Five years with 12.0% annual replacement
Permeate pH	5

Reproduced from Khawaji, A.D., Kutubkhanah, I.K., & Wie, J.M. (2007). A 13.3 MGD seawater RO desalination plant for Yanbu Industrial City. Desalination, 203(1−3), 176−188.

potable water after proper treatment. Each reverse osmosis train has 152 hollow-fine fiber membrane modules, and each reverse osmosis module consists of two reverse osmosis membrane elements. This RO facility is established with 1,824 cellulose triacetate hollow-fine fiber membrane elements. The rejection of salt by the cellulose triacetate membrane is

TABLE 7.11 Design conditions of raw seawater in reverse osmosis plant in Yanbu Industrial City (Saudi Arabia).

	Values
Conductivity at 25°C	57,000–64,000 (umhos/cm)
Total dissolved solids (TDS)	41,300–46,400 (mg/L)
Residual chlorine as Cl_2	0.5–0.7 (mg/L)
pH	8.1–8.3 (at 25°C)
Total suspended solids	1.0 (mg/L)
Turbidity	0.5–1.0 (NTU)
Sulfate as SO_4^{2-}	3,000–3,200 (mg/L)
Chloride as Cl^-	21,600–23,500 (mg/L)
Bicarbonate alkalinity as $CaCO_3$	85–95 (mg/L)
Total alkalinity as $CaCO_3$	120–130 (mg/L)
Nitrate as NO_3^-	<0.1 (mg/L)
Fluoride as F^-	1.5 (mg/L)
Potassium as K^+	425–650 (mg/L)
Sodium as Na^+	11,700–12,500 (mg/L)
Phosphate as HPO_4^{3-}	<0.1 (mg/L)
Magnesium as Mg^{2+}	1,500–1,600 (mg/L)
Calcium as Ca^{2+}	490–560 (mg/L)
Total iron as Fe^{2+} and Fe^{3+}	0.01 (mg/L)
Ammonia as NH_4	0.2 (mg/L)
Silica as SiO_2	0.5(mg/L)
Total dissolved oxygen as O_2	3.5 (mg/L)

Reproduced from Khawaji, A.D., Kutubkhanah, I.K., & Wie, J.M. (2007). A 13.3 MGD seawater RO desalination plant for Yanbu Industrial City. Desalination, 203(1–3), 176–188.

almost 99.40%. The product water generated by the reverse osmosis process flows through the permeate basin and subsequently to the product basin. The leftover 61.50% of the filtered seawater, concentrated brine, returns to the energy recovery turbine and subsequently discharged to the brine basin for disposal to the prevailing seawater return header.

The 13.30 million gallons per day new reverse osmosis plant design was based on the seawater reverse osmosis technology, along with the experience obtained from the former smaller plant. The reverse osmosis plant energy consumption was almost 6.0–8.0 kWh/m³ without energy recovery. Hence, it was suggested to install an energy recovery device for reducing the energy usage to 4.0–5.0 kWh/m³. In the Middle East, the main issue faced by reverse osmosis facilities is in the pretreatment process. Hence, the team suggested that an appropriate pretreatment is very critical to accomplish the effective operation of the reverse osmosis plant. The team also suggested that a method for reducing the desalinated water production expense is to use a hybrid system comprising two or more desalination technologies. In another study by Baig and Al Kutbi (1998), salient design features of a 20 million imperial gallons per day seawater RO desalination plant in Al Jubail, Saudi Arabia are discussed.

7.4.2 Case study 2: Effective design of a reverse osmosis desalination process for varying salt concentration and seawater temperature

The design of a reverse osmosis network for desalination for a wide range of salt concentrations and seawater temperatures was discussed in a study by Sassi and Mujtaba (2012). In the case of a fixed reverse osmosis design, the production of potable water from the reverse osmosis process can considerably vary with seasonal temperature variations in seawater. The reverse osmosis process has to be constantly adjusted with the seawater temperature variation for maintaining the permeate requirements. In this study, the main goal was to examine the influence of varying seawater salinity and temperature on the design and operation of a reverse osmosis process network for a specified water quality and quantity. A model optimization on the basis of a flexible superstructure that consists of all potential alternatives of a prospective reverse osmosis network has been established, and was employed in the RO network synthesis. The optimal design problem was formulated as a mixed-integer nonlinear programming (MINLP) problem formulation, which reduces the total annual expense. The MINLP problem formulation was also used in a different study by Lu et al., (2007), for formulating the optimum design problem.

The optimum structural and operating parameters of reverse osmosis process have been estimated for a specific water requirement of 520 m^3/d with a maximam salt concentration of 500 mg/L in the desalinated water by using the MINLP optimization problem formulation. The optimization method was implemented on the base superstructure presented in Fig. 7.4, refined on the basis of the task at hand (feed conditions, demand, and quality). The membrane modules used in this work were DuPont B-10 HFF reverse osmosis modules. It was presumed that the life of the membrane module is 5 years. The various parameters used in the optimization calculations are shown in Table 7.12.

The MINLP model presented the capability of handling the trade-offs between optimizing parameters (operating and design variables) and costs for various practical process alternatives. Numerous reverse osmosis designs and operation conditions were obtained for varying feed water concentrations and seawater temperature. The optimum reverse osmosis structures varied from single stage to dual stages without and with reject bypass and interstage pump. In majority of situations, the unit production cost of the optimal reverse osmosis design is quite reduced as feed temperature increased. For higher feed concentration and also higher temperatures, the unit cost will be increased as the feed temperature rises because of the trouble in maintaining the water quality constraint. The aforementioned work revealed that the salinity constraint showed a remarkable influence in the design of reverse osmosis plant, especially for higher salinity and higher temperature water resources. Moreover, the results

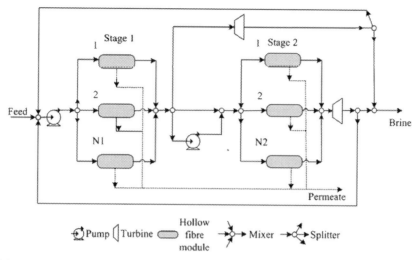

FIGURE 7.4 Two-stage reverse osmosis system superstructure in the study by Sassi et al. *Source: Reproduced from Sassi, K.M., & Mujtaba, I.M. (2012). Effective design of reverse osmosis based desalination process considering wide range of salinity and seawater temperature. Desalination, 306, 8–16.*

TABLE 7.12 Input data for mixed-integer nonlinear programming optimization in the study by Sassi et al.

	Values
Reference temperature T_0	25 (°C)
Turbine efficiency	80 (%)
Feed pump efficiency	75 (%)
Load factor	0.9
Module vessel cost	200 ($)
Membrane unit cost	800 ($)
Electricity unit cost	0.08 ($/kWh)

Reproduced from Sassi, K.M., & Mujtaba, I.M. (2012). Effective design of reverse osmosis based desalination process considering wide range of salinity and seawater temperature. Desalination, 306, 8–16.

demonstrated that the variation in the number of modules needed for the reverse osmosis process in low and high temperature seasons provides the chance of adjustable scheduling of cleaning and maintenance of membrane module.

7.4.3 Case study 3: Design of two-stage reverse osmosis unit for higher-temperature and higher-salinity seawater desalination having improved performance and stability

Even though seawater RO is an energy-efficient desalination technology, it may not be advantageous for the treatment of extreme seawater because of the operational restrictions related to the single-stage seawater RO configuration. For overcoming the above-stated limitation, the suitability of two-stage seawater RO configuration was studied. Kim et al. (2020a) assessed the suitability of two-stage seawater RO configuration under extreme feed conditions. The operational issues associated with single-stage seawater RO configuration were examined under the extreme feed conditions by concentrating on attainable recovery, energy consumption, water flux in the front elements, and product water quality (Fig. 7.5). In order to further increase the efficiency of two-stage seawater RO, modifications of the two-stage seawater RO configurations through split partial second-pass (SPSP) design and internally staged design (ISD) were implemented in the work. This study presented a clear picture on the selection of suitable reverse osmosis designs for extreme feed conditions.

This project involved the desalination of seawater with higher temperature and salinity same as the seawater from the Arabian Gulf. The study modeled a reverse osmosis train with a capacity of 10,000 m³/day, and this capacity is adequate for reverse osmosis train modeling to represent the system efficiency of reverse osmosis unit due to the fact that a seawater RO desalination plant consists of multiple trains. The number of trains could be adjusted in order to fulfill the required capacity. For both one-stage and two-stage seawater RO units, 120 pressure vessels were employed, with seven seawater RO elements in each pressure vessel. With this configuration, the average water flux was noted to be 13.35 L/(m².h) irrespective of reverse osmosis configuration, which is in appropriate water flux range for seawater RO (e.g., 12.0–14.0 L/(m².h)). In the meantime, 36 pressure vessels were installed in the brackish water RO system using split partial second-pass as the reverse osmosis design, such that the average water flux for the brackish water RO system was 18.22–21.94 L/(m².h), dependent on the feed flow rate supplied to the unit.

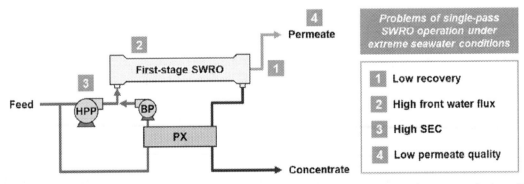

FIGURE 7.5   Problems of one-stage seawater reverse osmosis (SWRO) operation under extreme feed conditions. *Source: Reproduced from Kim, J., Park, K., & Hong, S. (2020a). Application of two-stage reverse osmosis system for desalination of high-salinity and high-temperature seawater with improved stability and performance. Desalination, 492, 114645.*

For designing a two-stage seawater RO, a harmony search (HS) was used for determining the optimum ratios for product water flow rate and number of PVs for each stage. Roughly, the design ratios (i.e., product water flow rate and number of pressure vessels) was 2:1 (i.e., first stage $\approx$ 80 pressure vessels; second stage $\approx$ 40 pressure vessels) for dual-stage seawater RO (Pelton turbine) and 1:2 (i.e., first stage $\approx$ 40 pressure vessels; second stage $\approx$ 80 pressure vessels) for two-stage seawater RO (pressure exchanger). When split partial second-pass was used along with the two-stage seawater RO system, dual-stage brackish water RO was employed for desalinating the seawater RO rear permeate. For the brackish water RO, the number of pressure vessels were 12 and 24 for the second and first stages, respectively.

From this study, it was noted that split partial second-pass design and internally staged design can be adopted for the two-stage seawater RO unit to additionally increase recovery and obtain high-water quality. Internally staged design lowered the water fluxes in the front elements by almost 3.0−4.0 L/(m^2.h); therefore the achievable recovery can be increased to 40.0%. SPSP design could competently decrease the product water total dissolved solids by selectively desalinating high salt concentration seawater RO rear permeate, and 78.0−87.0 ppm of final permeate was obtained with an extra 0.22−0.28 kWh/m^3. Moreover, this work contributed to a theoretical basis for the design and function of seawater desalination facilities, which involves the challenge of processing extreme seawater with increased temperatures and salt concentration regularly experienced in the Middle East, mainly in the Arabian Gulf area.

7.4.4 Case study 4: Novel design and control of integrated ultrafiltration (UF)-reverse osmosis (RO) system with RO reject backwash

An innovative design for a reverse osmosis-based desalination system directly integrated with a UF pretreatment system was demonstrated by Gao et al. (2016). The integration involved direct reverse osmosis feed from the ultrafiltration filtrate and ultrafiltration backwash employing the reverse osmosis reject. This arrangement reduces the overall RO facility footprint, whereas the application of the reverse osmosis reject for ultrafiltration backwash permits 100% ultrafiltration recovery and execution of flexible backwash approaches. The existing system design uses a control scheme, by which reverse osmosis productivity could be prescribed independently of the ultrafiltration unit, which self-adapts for providing the reverse osmosis system with its required feed flow rate at the specified reverse osmosis pump inlet pressure. Direct ultrafiltration-reverse osmosis system integration, which is different from conventional ultrafiltration-reverse osmosis system (Fig. 7.6), involves feeding ultrafiltration filtrate to the reverse osmosis high-pressure pump and reverse osmosis reject straight for ultrafiltration backwash (Fig. 7.7)

Pilot plant study: An integrated ultrafiltration-reverse osmosis facility was designed with a product water production capacity of 45.4 m^3/d. The ultrafiltration pretreatment unit comprises three hollow-fiber ultrafiltration modules each consisting of 50 m^2 ultrafiltration membrane elements. A self-cleaning 200 μm screen filter was fixed upstream of the ultrafiltration unit. A centrifugal low-pressure ultrafiltration pump with variable-frequency drive control served for both ultrafiltration feed and directing the ultrafiltration filtrate to the reverse osmosis feed pump. The ultrafiltration filtrate was fed to the

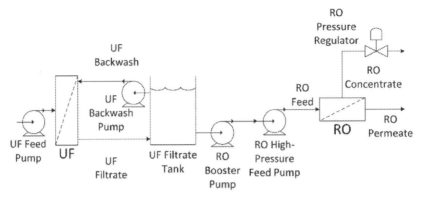

FIGURE 7.6 Process illustration of a conventional integrated ultrafiltration-reverse osmosis (UF-RO) system design that uses an intermediate ultrafiltration filtrate storage tank for ultrafiltration backwash water, ultrafiltration backwash pump, and reverse osmosis booster pump. *Source: Reproduced from Gao, L.X., Rahardianto, A., Gu, H., Christofides, P.D., & Cohen, Y. (2016). Novel design and operational control of integrated ultrafiltration-reverse osmosis system with RO concentrate backwash. Desalination, 382, 43–52.*

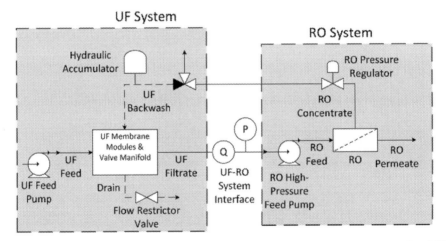

FIGURE 7.7 Process flowchart of a directly integrated ultrafiltration-reverse osmosis (UF-RO) system. Pressure (P) and flow rate (Q) at the ultrafiltration-reverse osmosis system interface are retained by the control system. *Source: Reproduced from Gao, L.X., Rahardianto, A., Gu, H., Christofides, P.D., & Cohen, Y. (2016). Novel design and operational control of integrated ultrafiltration-reverse osmosis system with RO concentrate backwash. Desalination, 382, 43–52.*

high-pressure pump, which then delivered the reverse osmosis feed to three spiral-wound membrane elements arranged sequentially. Each element had an 8-inch diameter and 40-inch length housed in a fiberglass PV. The reverse osmosis element salt rejection reported by the manufacturer was 99.65% (at 32,000 mg/L sodium chloride, 5.5 MPa or 800 psi) with a maximal water recovery per element of 15.0% enabling total recovery till 38.6% with the three elements arranged sequentially.

Field examination: The directly integrated ultrafiltration-reverse osmosis pilot plant was deployed at the NAVFAC Seawater Desalination Test Facility in the United States. The source seawater feed was collected from an open-sea intake by means of a strainer to the ultrafiltration-reverse osmosis pilot plant. The feed water salt concentration (33,440–36,800 ppm TDS) and pH (7.5–8.2) changed within a comparatively narrow range; on the other hand, variations in the feed TSS (0.1–5.2 ppm), temperature (11.2°C–19.7°C), and turbidity (0.4–14 NTU) were substantial. The feed pretreatment unit (200 μm screen filter and ultrafiltration) turned the water turbidity to 0.1 NTU, which was less than the maximum limit recommended. Field studies comprise the demonstration of the ultrafiltration-reverse osmosis control unit, especially the decoupled nature of the ultrafiltration and reverse osmosis control units and its capability for maintaining satisfactory reverse osmosis pump inlet pressure in the course of different ultrafiltration operations. Successively, the efficiency of ultrafiltration pulse backwash using the reverse osmosis reject delivered from the reverse osmosis system to the ultrafiltration unit was assessed. Self-adjustable backwash that includes the above-mentioned technique was also applied and its efficiency was compared with the usage of self-adjustable freshwater (i.e., reverse osmosis product water) backwash.

Seawater desalination field examination confirmed that the initiation of ultrafiltration backwash with reverse osmosis concentrate, based on the membrane resistance threshold, was better as compared to the fixed frequency backwash, prolonging the projected ultrafiltration operation from around 16 to 143 days before requiring chemical cleaning in place. The results also recommended that there is an advantage in exploring the increase in ultrafiltration and backwash efficiency by integrating coagulation with the self-adjustable ultrafiltration backwash.

7.4.5 Case study 5: Design and performance of solar photovoltaic brackish water reverse osmosis desalination unit

The integration of renewable power systems with small-scale units will theoretically help in the plant operation and following commercial success. In reality, the reverse osmosis systems are built using a modular methodology, and this will permit them to properly adapt to a renewable power unit. In remote areas, where brackish water is very common, small-scale photovoltaic-reverse osmosis unit would be a potential system for desalination.

The work by Alghoul et al. (2016) provided information on the effects of climate, design, and operation conditions on the brackish water RO and photovoltaic system efficiency and durability. Hence, a small-scale brackish water RO powered by a 2 kilowatts peak (kW$_p$) photovoltaic system has been designed, built, and tested. The reverse osmosis system has been designed for treating a feed salt concentration of a maximun 5,000 ppm for obtaining product water TDS of less than 50 ppm. Experimental work with respect to this project was performed at the solar field of the National University of Malaysia, Malaysia. Table 7.13 shows the major design assumptions used in designing and sizing a brackish water RO-photovoltaic desalination system. The Reverse Osmosis System Analysis (ROSA) simulation tool was employed for determining the design configuration, the optimum membrane type, and the number of membrane elements for producing potable water under 600 watts load and a product water total dissolved solids of below 50 ppm.

In the Malaysian climate, the design results confirmed that a 2 kilowatts peak photovoltaic system can only power a 600-watt reverse osmosis load. Subsequently, the brackish

TABLE 7.13　The design scope/ assumptions/limitations adopted in the work by Alghoul et al.

Parameters	Values
Feed TDS	2,000 ppm
Photovoltaic array power capacity (max)	2 kW$_p$
Unmet electrical load (%)	0.50–4.00
Annual solar radiation (kW/m^2)	4.794
Capacity shortage	Maximum 5%
Implemented membranes	Five types

Reproduced from Alghoul, M.A., Poovanaesvaran, P., Mohammed, M.H., Fadhil, A.M., Muftah, A.F., Alkilani, M.M., & Sopian, K. (2016). Design and experimental performance of brackish water reverse osmosis desalination unit powered by 2 kW photovoltaic system. Renewable Energy, 93, 101–114.

water RO desalination system was designed on the basis of a 600 W load, feed total dissolved solids of 2,000 ppm, and product water total dissolved solids of below 50 ppm. The results presented that 4″ × 40″ TW30–4040 membrane type and dual stage are the optimal brackish water RO design choices. The experimental results also confirmed that although the brackish water RO unit is designed under a feed total dissolved solids of 2,000 ppm, it is still able to generate product water flow with acceptable permeate salt concentration (below 50 ppm) even when the feed total dissolved solids is below 5,000 ppm. Under experimental assessment, the simulated and measured values showed good agreement under various feed salt concentration tests for product water salinity, product water flow, and reverse osmosis unit pressure. Therefore the results confirmed the validity of assumptions and procedures implemented in the design and sizing of a brackish water RO-photovoltaic desalination system.

7.4.6 Case study 6: Design of reverse osmosis systems with multiple product and multiple feed

In the work by Lu et al. (2012), a reverse osmosis-based desalination process was adopted for the fresh water production from three different water feeds (brackish water, seawater, and regenerated water). A systematic strategy was adopted for the optimum design of reverse osmosis desalination system that processes multiple feed streams concurrently and simultaneously delivers different permeate streams of dissimilar qualities. The implementation of this method can result in an affordable and desirable desalination system. This would result in substantial saving of energy and raw materials and help to make earning from the sale of multiple grades of product waters.

For the reverse osmosis system design and optimization, it is essential to adopt the proper modeling equations that could reasonably forecast the efficiency of the membrane with sensible computational complexity. In the aforementioned study, there are three water permeate outlets that fulfill various product water quality and quantity requirements. The minimal anticipated permeate flow rate for these outlets are 50, 100, and

200 m^3/h, while the corresponding maximum allowable permeate salt concentration are 500, 300, 100 mg/L, respectively. Four kinds of FilmTec RO membrane elements were included in the design studies of the above-mentioned study. The RO membrane elements used were high productivity, high rejection BW RO element BW30−400, the high rejection, fouling-resistant element SW30HR-320, the high rejection, high productivity element SW30HR-380, and the low-energy, high-productivity element SW30XLE-400. The results of the RO system optimization design have been shown in Table 7.14. The three-stage reverse osmosis configuration has been used in the design (presented as Fig. 7.8).

TABLE 7.14 Design as well as optimization results for the case study by Lu et al.

Parameters	Values
Process flow	Three-stage reverse osmosis system
Feed flow of seawater, $Q_{f,1}$	495 (m^3/h)
Feed flow of brackish water, $Q_{f,2}$	100 (m^3/h)
Feed flow of regenerated water, $Q_{f,3}$	50 (m^3/h)
Salinity of the first permeate, C^P_1	100 (mg/L)
Flow rate of the first permeate, Q^P_1	200 (m^3/h)
Salinity of the second permeate, C^P_2	300 (mg/L)
Flow rate of the second permeate, Q^P_2	100 (m^3/h)
Salinity of the third permeate, C^P_3	300 (mg/L)
Flow rate of the third permeate, Q^P_3	50 (m^3/h)
Membrane type in stage 1	SW30XLE-400
In stage 1, no. of elements per pressure vessel	5
In stage 1, no. of PVs	82
Operating pressure in stage 1, P_1	7.3 (Mpa)
Membrane type in stage 2	SW30HR-380
In stage 2, no. of elements per pressure vessel	3
In stage 2, no. of PVs	33
Stage 2 operating pressure, P_2	4.4 (Mpa)
Stage 3 membrane type	SW30XLE-400
In stage 3, no. of elements per pressure vessel	4
In stage 3, no. of pressure vessels	25
Stage 3 operating pressure, P_3	4.5 (Mpa)
Total annual cost, ($)	1,412,000

Reproduced from Lu, Y., Liao, A., & Hu, Y. (2012). The design of reverse osmosis systems with multiple-feed and multiple-product. Desalination, 307, 42−50.

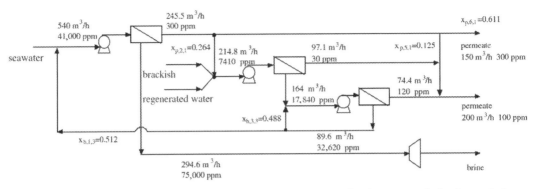

FIGURE 7.8 The optimum configuration of reverse osmosis system for the case study by Lu et al. *Source: Reproduced from Lu, Y., Liao, A., & Hu, Y. (2012). The design of reverse osmosis systems with multiple-feed and multiple-product. Desalination, 307, 42–50.*

The optimal design problem was formulated as an MINLP problem, which reduces the total annual cost of the reverse osmosis system. The cost equation associated with the capital cost and operating cost to the design variables, as well as the structural variables, was introduced in the objective function. The problem solution included the optimum system structure and operating conditions, as well as the optimum streams distribution. The design method can also be applied for the optimum selection of the type of membrane elements in each stage and the optimum number of membrane elements in each PV. The comparisons of numerous alternate schemes indicated that, for the design of reverse osmosis system, the feed position of streams as well as outlets of the system are the important variables that must be optimized.

7.4.7 Case study 7: Design and optimization of batch reverse osmosis desalination unit with a free piston for brackish water treatment

Batch reverse osmosis is a concept for accomplishing the minimal conceivable energy usage in desalination, even at higher recoveries. In a work by Park et al. (2020), the team described and explained the design as well as operation of a free-piston batch reverse osmosis desalination system that functions in two phases by employing only two pumps and three valves. The team also put forward a model along with a design procedure for the sizing of this reverse osmosis unit for meeting the targeted performance (i.e., specific energy consumption, rejection, recovery, and productivity) using explicit algebraic equations applied in a step-wise way. Furthermore, applied the model to a particular case of brackish groundwater desalination and assessed the efficiency.

A complete design procedure was developed using just the algebraic equations, making it easy for its application. A system was designed as well as optimized for BW feed at 3 kg/m^3 sodium chloride concentration. At the recovery of 80%, the optimized design having a high-flux membrane leads to specific energy consumption of 0.39 kWh/m^3, a rejection of 94.20%, and total production of $17.3 \text{ m}^3/\text{day}$, while with the high-rejection membrane, specific energy consumption increased to 0.525 kWh/m^3 and rejection improved to 98.50%. The batch reverse osmosis employing a free-piston design could produce potable water from brackish groundwater feed solution with lower energy consumption and second law efficiency of 33.20%—considerably greater than standard BW reverse osmosis, which is normally in the range of 10.0—15.0%. This study also proved that the free-piston batch-reverse osmosis is an effective solution for BW desalination at higher recovery (almost up to 90%) with productivity of up to 20 m^3/d by employing a single 8-inch reverse osmosis module.

References

Alghoul, M. A., Poovanaesvaran, P., Mohammed, M. H., Fadhil, A. M., Muftah, A. F., Alkilani, M. M., & Sopian, K. (2016). Design and experimental performance of brackish water reverse osmosis desalination unit powered by 2 kW photovoltaic system. *Renewable Energy, 93*, 101—114.

Alghoul, M. A., Poovanaesvaran, P., Sopian, K., & Sulaiman, M. Y. (2009). Review of brackish water reverse osmosis (BWRO) system designs. *Renewable and Sustainable Energy Reviews, 13*(9), 2661—2667.

Al-Juboori, R. A., & Yusaf, T. (2012). Biofouling in RO system: Mechanisms, monitoring and controlling. *Desalination, 302*, 1—23.

Baig, M. B., & Al Kutbi, A. A. (1998). Design features of a 20 migd SWRO desalination plant, Al Jubail, Saudi Arabia. *Desalination, 118*(1—3), 5—12.

Barello, M., Manca, D., Patel, R., & Mujtaba, I. M. (2015). Operation and modeling of RO desalination process in batch mode. *Computers & Chemical Engineering, 83*, 139—156.

Benjamin, M. M., & Lawler, D. F. (2013). *Water quality engineering: Physical/chemical treatment processes*. John Wiley & Sons.

DuPont Water Solution's FilmTec™ RO Technical Manual; Version 3; Form No. 45-D01504-en, Rev. 3; April 2020.

Ezzeghni, U.A. (2018). Designing and optimizing 10,000 m3/day conventional SWRO desalination plant. First Conference for Engineering Sciences and Technology (CEST-2018), 25-27 September 2018, Libya, https://dspace.elmergib.edu.ly/xmlui/handle/123456789/92.

Fane, A.G., Tang, C., & Wang, R. (2011). Membrane technology for water: Microfiltration, ultrafiltration, nanofiltration, and reverse osmosis. In Wilderer, PA (Ed.), Treatise on Water Science, v. 4, p. 301-335. Amsterdam; Hackensack, NJ: Elsevier Science, https://doi.org/10.1016/B978-0-444-53199-5.00091-9.

FILMTEC Membranes—Steps to Design a Reverse Osmosis System, DuPont Water Solution. Retrieved from https://water.custhelp.com/app/answers/detail/a_id/2209/related/1 [Accessed 2 February 2019].

Flow Configuration, Hydranautics. Retrieved from http://membranes.com/docs/trc/flowcon.pdf, [Accessed on 01/09/2020].

Frenkel, V. S. (2015). Planning and design of membrane systems for water treatment. *Advances in Membrane Technologies for Water Treatment*, 329—347.

Gao, L. X., Rahardianto, A., Gu, H., Christofides, P. D., & Cohen, Y. (2016). Novel design and operational control of integrated ultrafiltration-reverse osmosis system with RO concentrate backwash. *Desalination, 382*, 43—52.

Goh, P. S., Lau, W. J., Othman, M. H. D., & Ismail, A. F. (2018). Membrane fouling in desalination and its mitigation strategies. *Desalination, 425*, 130—155.

Hailemariam, R. H., Woo, Y. C., Damtie, M. M., Kim, B. C., Park, K. D., & Choi, J. S. (2020). Reverse osmosis membrane fabrication and modification technologies and future trends: A review. *Advances in Colloid and Interface Science, 276*, 102100.

Hendricks, D. W. (2018). *Water treatment unit processes: Physical and chemical*. CRC Press.

Hocking, M. B. (2005). 4-Water quality measurement. In M. B. Hocking (Ed.), *Handbook of chemical technology and pollution control* (3rd (ed.), pp. 105—135). San Diego, California, USD: Elsevier Inc.

Khawaji, A. D., Kutubkhanah, I. K., & Wie, J. M. (2007). A 13.3 MGD seawater RO desalination plant for Yanbu Industrial City. *Desalination, 203*(1—3), 176—188.

Kim, J., Park, K., & Hong, S. (2020a). Application of two-stage reverse osmosis system for desalination of high-salinity and high-temperature seawater with improved stability and performance. *Desalination, 492*, 114645.

Kim, J., Park, K., & Hong, S. (2020b). Optimization of two-stage seawater reverse osmosis membrane processes with practical design aspects for improving energy efficiency. *Journal of Membrane Science, 601,* 117889.

Kim, J., Park, K., Yang, D. R., & Hong, S. (2019). A comprehensive review of energy consumption of seawater reverse osmosis desalination plants. *Applied Energy, 254,* 113652.

LANXESS Deutschland GmbH SYSTEM DESIGN—Guidelines for the design of RO Membrane Systems. Retrieved from http://lpt.lanxess.com/uploads/tx_lanxessmatrix/lew_lewabrane_manual_03_system_design_01.pdf [Accessed 25 February 2020].

Lu, Y., Liao, A., & Hu, Y. (2012). The design of reverse osmosis systems with multiple-feed and multiple-product. *Desalination, 307,* 42–50.

Lu, Y. Y., Hu, Y. D., Zhang, X. L., Wu, L. Y., & Liu, Q. Z. (2007). Optimum design of reverse osmosis system under different feed concentration and product specification. *Journal of Membrane Science, 287*(2), 219–229.

Moran, D. (2010). Carbon dioxide degassing in fresh and saline water. I: Degassing performance of a cascade column. *Aquacultural Engineering, 43*(1), 29–36.

Nie, F. H., Li, T., Yao, H. F., Feng, M., & Zhang, G. K. (2008). Characterization of suspended solids and particle-bound heavy metals in a first flush of highway runoff. *Journal of Zhejiang University-Science A, 9*(11), 1567–1575.

Olabarria, P. M. G. (2015). *Constructive engineering of large reverse osmosis desalination plants.* Chemical Publishing Company.

Pandey, S. R., Jegatheesan, V., Baskaran, K., & Shu, L. (2012). Fouling in reverse osmosis (RO) membrane in water recovery from secondary effluent: A review. *Reviews in Environmental Science and Bio/Technology, 11*(2), 125–145.

Park, K., Burlace, L., Dhakal, N., Mudgal, A., Stewart, N.A., & Davies, P.A. (2020). Design, modelling and optimisation of a batch reverse osmosis (RO) desalination system using a free piston for brackish water treatment. *Desalination, 494,* 114625.

Peñate, B., & García-Rodríguez, L. (2012). Current trends and future prospects in the design of seawater reverse osmosis desalination technology. *Desalination, 284,* 1–8.

Ramasamy, B. (2019). Short review of salt recovery from reverse osmosis rejects. In *Salt in the earth.* IntechOpen.

Reverse Osmosis Optimization Prepared for the United States Department of Energy Federal Energy Management Program. Retrieved from https://www.energy.gov/sites/prod/files/2013/10/f3/ro_optimization.pdf [Accessed on 01/09/2020].

Saleem, H., Trabzon, L., Kilic, A., & Zaidi, S. J. (2020). Recent advances in nanofibrous membranes: Production and applications in water treatment and desalination. *Desalination, 476,* 114178.

Saleem, H., & Zaidi, S. J. (2020a). Developments in the application of nanomaterials for water treatment and their impact on the environment. *Nanomaterials, 10*(9), 1764.

Saleem, H., & Zaidi, S. J. (2020b). Innovative nanostructured membranes for reverse osmosis water desalination Qatar University Annual Research Forum and Exhibition (QUARFE 2020), Doha, 2020, https://doi.org/10.29117/quarfe.2020.0023.

Saleem, H., & Zaidi, S. J. (2020c). Nanoparticles in reverse osmosis membranes for desalination: A state of the art review. *Desalination, 475,* 114171.

Sarbatly, R. (2020). *Membrane technology for water and wastewater treatment in rural regions.* IGI Global. Available from http://doi:10.4018/978−1−7998−2645-3.

Sassi, K. M., & Mujtaba, I. M. (2012). Effective design of reverse osmosis based desalination process considering wide range of salinity and seawater temperature. *Desalination, 306,* 8–16.

Singh, R. (2015). Water and membrane treatment. *Membrane technology and engineering for water purification (pp. 81178).* Butterworth-Heinemann, Elsevier.

Spiral-Wound Elements Technical Manual, MICRODYN-NADIR | REVISION DATE: 04/07/2020, Retrieved from https://www.microdyn-nadir.com/wp-content/uploads/Microdyn-Nadir-Spiral-Wound-Element-Technical-Manual.pdf [Accessed on 01/09/2020].

System Design Membrane System Design Guidelines for 8″ FilmTec™ Elements, DuPont Water Solution. Retrieved from https://www.dupont.com/content/dam/dupont/amer/us/en/water-solutions/public/documents/en/45-D01695-en.pdf [Accessed 25 February 2020].

Total Solids, Monitoring and Assessment, United States Environment Protection Agency. Retrieved from https://archive.epa.gov/water/archive/web/html/vms58.html [Accessed on 01/09/2020].

Vu, T.T.N. (2017). Impacts of reverse osmosis concentrate recirculation on MBR performances in the field of wastewater reuse. Chemical and Process Engineering. INSA de Toulouse. English.NNT: 2017ISAT0011.

Ward, R.A. (2007). Water treatment equipment for in-center hemodialysis: Including verification of water quality and disinfection. *Handbook of dialysis therapy*. E-Book, Vol. 143.

Williams, D. E. (2018). Design and construction of subsurface intakes. In *Sustainable desalination handbook* (pp. 227–258). Butterworth-Heinemann.

Xia, L., Vemuri, B., Saptoka, S., Shrestha, N., Chilkoor, G., Kilduff, J., & Gadhamshetty, V. (2019). Antifouling membranes for bioelectrochemistry applications. In *Microbial electrochemical technology* (pp. 195–224). Elsevier.

Zaidi, S. J., Fadhillah, F., Saleem, H., Hawari, A., & Benamor, A. (2019). Organically modified nanoclay filled thin-film nanocomposite membranes for reverse osmosis application. *Materials, 12*(22), 3803.

CHAPTER

8

Reverse Osmosis Design Software Programs

8.1 Introduction

Reverse osmosis (RO) design software programs are developed by the membrane manufacturers to help the designer in building up an appropriate RO system design. Some RO membrane manufacturers have their design software programs accessible to the public, which is specific to their membranes. Every software package, however diverse in presentation, provides a similar outcome: RO unit design, inclusive of arrays, scaling indices, operating pressure, concentrate water quality and product water quality. The design software depends on stabilized, nominal membrane performance chosen under the design conditions. The actual performance might change within ± 15% of nominal, as indicated by the FilmTec Technical Manual of DuPont Water Solutions (DuPont Water Solution, 2020). The design software will provide alerts when the essential element operating parameters are surpassed, for example, low concentrate flow or high element recovery. The RO design software programs contribute to the potential for estimation of one and two-staged RO plants, subsequent to defining plant structure, flow rates, and feed composition. Furthermore, the fouling and scaling potential can likewise be assessed.

Uses of common sense and experience are very important in decision-making in order to make sure that the design chosen using the software is practical and meaningful, especially for feed streams apart from clean water sources like well water with silt density index less than value 3. Table 8.1 presents the list of manufacturers of membranes along with their RO design software programs. The design software programs listed in Table 8.1 are regularly updated. The updated version of the design software must be obtained before developing the design criteria for a new RO process. The majority of design programs are examined and discussed in this chapter in detail with specific examples. Most of the software programs can also be downloaded from the respective websites of the manufacturers. All the listed software programs use a graphical user interface that permits the designer for visualizing the physical configuration of the subject membrane array. The design software typically includes separate tabs or screens for various design components. The screens/tabs are organized in a logical sequence in such a manner that the designer can "step" through the design process one screen or tab at a time.

Reverse Osmosis Systems
DOI: https://doi.org/10.1016/B978-0-12-823965-0.00003-1

TABLE 8.1　Different RO system design software programs.

Sl. No.	Company	Design software package
1	DuPont Water Solution	WAVE https://www.dupont.com/water/resources/design-software.html
2	Hydranautics	Integrated Membrane Solutions (IMSDesign) https://membranes.com/solutions/software-imsdesign/
3	Toray Industries, Inc.	Toray Design System (TorayDS/DS2) https://ap3.toray.co.jp/toraywater/
4	LANXESS	Lewaplus https://lpt.lanxess.com/lewaplus-software/
5	SUEZ Water Technologies & Solutions	Winflows https://www.suezwatertechnologies.com/resources/winflows
6	Microdyn-Nadir	ROAM Ver. 2.0 https://www.microdyn-nadir.com/software-downloads/
7	SimTech Simulation Technology	IPSE software http://simtechnology.com/CMS/index.php/ipsepro
8	CSMPRO v6.0	CSM http://www.csmfilter.com/csm/03result/Software.asp

The software program selection depends totally on the membrane manufacturer, as indicated by the client. Every single RO system designer might have a preferred program, that they apply to get the projection data, if the membrane at the choice not be indicated. In majority events, it is reasonable to run a few programs and compare their differences, to discover which membrane performance fulfills the requirements of the particular application. Despite the fact that each program is specific to its manufacturer's membrane, there are harmonies amongst these design programs (RO Design and Design Software, 2015). Several research groups have carried out RO simulation studies using commercially available software programs (Agarwa et al., 2016; Altaee, 2013; Gaublomme et al., 2020; Haryati et al., 2017; Jiang et al., 2014; Talaeipour et al., 2017). In the following sections, details about different RO design software programs are described.

8.2 WAVE Design Software (Dupont Water Solutions)

Water treatment generally requires more than one technology for accomplishing the required water quality. The majority of software programs for designing of water-treatment plants do not permit the design engineer for optimizing multiple-technology systems, demanding separate software and more time for setup and management.

DuPont Water Solutions' Water Application Value Engine (WAVE) is considered to be the industry's first entirely integrated modeling software program for integrating the three leading technologies—RO, ultrafiltration (UF), and ion exchange (IX)—into one complete tool. It simplifies the designing process using a common interface and eventually assists in reducing the time required for managing the water-treatment system.

The latest WAVE software can be downloaded using the below link:
https://www.dupont.com/water/resources/design-software.html

By using the new version of WAVE 1.77a, the design engineer will have the opportunity of:

- Single intuitive interface for RO, UF, and ion exchange technologies, compatible with Microsoft Windows.
- Harmonized data for all products and processes.
- Expert modeling tool.
- Four language options: English, Portuguese, Spanish, and Chinese.

In a completely integrated software, modifications made to any operation will automatically propagate all through the system design.

WAVE is an integrated expert modeling software for the design of water-treatment plant offering the following characteristics:

- A powerful calculation engine with the ability to run complex designs at increased accuracy levels.
- Flexible design employing the three technologies, with multiple-unit operation combinations, and the opportunity to specify system-feed or net-product flow rate.
- True mass-balance volumes and flows that reflect variations in density because of water compressibility, water composition, and temperature.
- Upgraded water-equilibrium calculations and interface.
- Reliable hydraulic constraints and regeneration parameters, which reflect best practices and state-of-the-art product performance and application.
- The ability to introduce project-specific parameters for increasing the accuracy of operating-expense calculations.
- Default values for most parameters, permitting the designer to develop a design rapidly.

RO advanced features include:

- The latest RO products database with new calculation improvements and products, including FilmTec.
- The RO products database has been advanced to include the innovative seawater fouling resistant FilmTec SW30XFR-400/34.
- Users can now design systems with concentrate stream recycle simultaneously from Pass 2 to Pass 1 and 2.
- Improved internally staged design (ISD) capability; each element in a pressure vessel (PV) can be exclusively defined.
- Split permeate available with internally staged design.
- Pressure-drop reporting.
- Design warnings by element position.
- Batch modeling with the possibility of simultaneous batch-case creation.
- A new system-level hydraulic-flow calculator.
- Bypass, pass, and stage level concentrate recycle loops.

The WAVE software RO component has the succeeding main sections:

- Project data entry
- Units of measure specification
- Chemicals specification

- Specification of feed water
- Configuration of RO system
- Chemical adjustment and scaling risk calculation
- Generation of report and review
- Batch operation and case management

Here the step-by-step procedure for the RO system design using the WAVE software of DuPont water solutions is presented. This has been done as per our understanding, and not DuPont official explanation. As presented in Fig. 8.1, the Project data input screen of WAVE permits the design engineer to enter data about the project including customer name, project name, date, case name, and other applicable data. The yellow highlighted sections in the figure indicate the points to be accessed. Another important type of entries used in this software are the units of measure (as presented in Fig. 8.2) for the modeling. Further, Fig. 8.3 presents the specification of the chemical list available for use, whereas Fig. 8.4 shows the costs of raw water, wastewater disposal, electricity, and chemicals. The specifications of pump efficiencies, currencies, and user information in WAVE are shown in Figs. 8.5, 8.6, and 8.7, respectively.

In the WAVE design software, as displayed in Fig. 8.8, a reverse osmosis design begins with the selection of the RO process icon from the main menu. The icon should be dragged and then

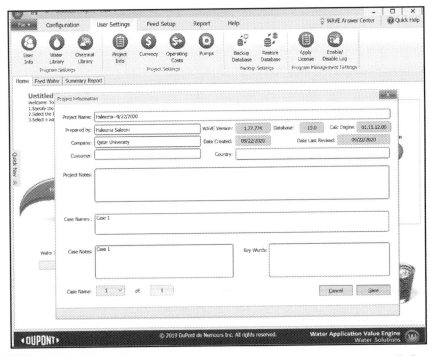

FIGURE 8.1 Project information input screen in WAVE design software. Source: *Courtesy: DuPont water solutions (Personal Communication with Dupont Water Solutions' team. Retrieved from https://www.dupont.com/water/resources/design-software.html).*

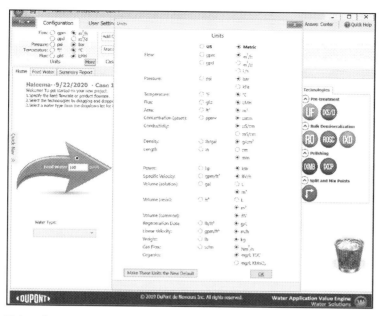

FIGURE 8.2   Units of measure specification in WAVE. Source: *Courtesy: DuPont water solutions (Personal Communication with Dupont Water Solutions' team. Retrieved from https://www.dupont.com/water/resources/design-software.html).*

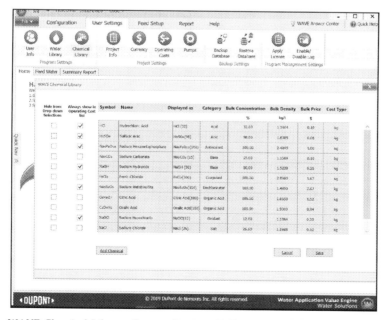

FIGURE 8.3   WAVE Chemical Library. Source: *Courtesy: DuPont water solutions (Personal Communication with Dupont Water Solutions' team. Retrieved from https://www.dupont.com/water/resources/design-software.html).*

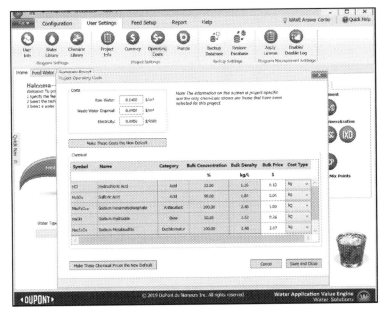

FIGURE 8.4 Costs of raw water, wastewater disposal, electricity, and chemicals. Source: *Courtesy: DuPont water solutions (Personal Communication with Dupont Water Solutions' team. Retrieved from https://www.dupont.com/ water/resources/design-software.html).*

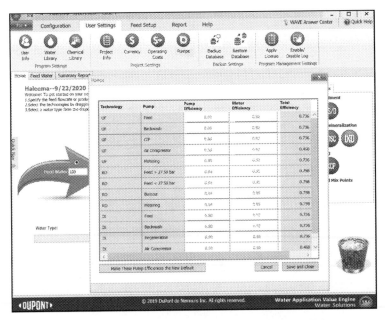

FIGURE 8.5 Pump efficiency specification in WAVE. Source: *Courtesy: DuPont water solutions (Personal Communication with Dupont Water Solutions' team. Retrieved from https://www.dupont.com/water/resources/design-software.html).*

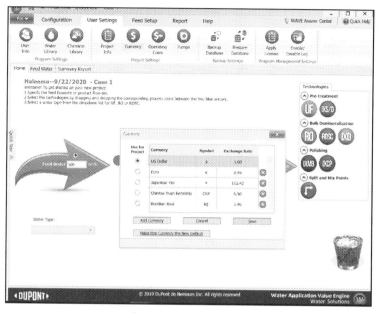

FIGURE 8.6 Currency specification in WAVE. Source: *Courtesy: DuPont water solutions (Personal Communication with Dupont Water Solutions' team. Retrieved from https://www.dupont.com/water/resources/design-software.html).*

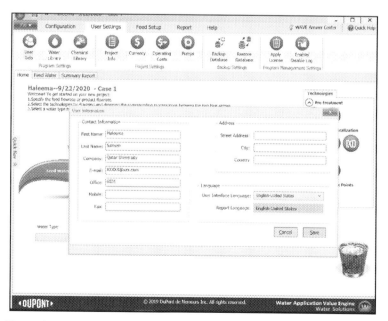

FIGURE 8.7 User information and language specifications in WAVE. Source: *Courtesy: DuPont water solutions (Personal Communication with Dupont Water Solutions' team. Retrieved from https://www.dupont.com/water/resources/design-software.html).*

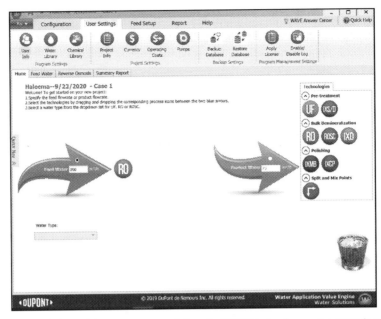

FIGURE 8.8   Drag-and-drop of the reverse osmosis icon in WAVE to begin reverse osmosis modeling. Source: *Courtesy: DuPont water solutions (Personal Communication with Dupont Water Solutions' team. Retrieved from https:// www.dupont.com/water/resources/design-software.html).*

dropped to the empty circle located between the two large blue arrows. With this option, 8-inch elements designs can be performed. ROSC stands for Reverse Osmosis Small Commercial. With this option, 4-inch and smaller elements designs can be performed. WAVE does not allow the inclusion of both RO and ROSC in the same system design.

8.2.1 Specification of feed water

As presented in Fig. 8.9, the feed water characteristics will be entered in the Feed Setup-Feed Water section. The anions, cations, and neutrals are listed in three different tables. Further, this design software possesses a Quick-entry option for entering the desired concentration of sodium chloride (NaCl). Besides the ionic composition, the designer can also include information on the solid contents of water such as turbidity in NTU (Nephelometric Turbidity Unit), total suspended solids (TSS), silt density index (SDI), and organic matter in total organic content (TOC). These will be used for identifying the suitable design guidelines for the designer.

In WAVE, the feed water must be charge-balanced before the user is allowed to either simulate or specify the system design. There are several paths for charge balancing the feed water. As presented in Fig. 8.10, charge balancing could be performed by the additions of anions or cations, or both, or by varying pH (which influences $HCO_3/CO_3/CO_2$ equilibrium). As displayed in Fig. 8.11, numerous water profiles are included by default in the WAVE Water Library. The designer could then save the water information he or she has just entered. Moreover, this design software permits the designer to blend numerous feed water streams.

FIGURE 8.9 Ionic composition specification in WAVE. Source: *Courtesy: DuPont water solutions (Personal Communication with Dupont Water Solutions' team. Retrieved from https://www.dupont.com/water/resources/design-software.html).*

FIGURE 8.10 Charge balance adjustment options in WAVE. Source: *Courtesy: DuPont water solutions (Personal Communication with Dupont Water Solutions' team. Retrieved from https://www.dupont.com/water/resources/design-software.html).*

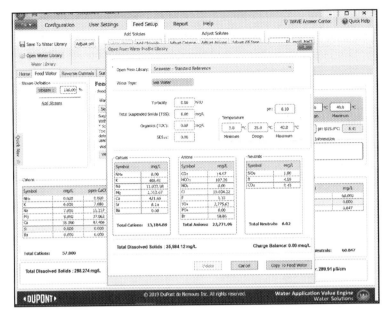

FIGURE 8.11   Water Library of WAVE. Source: *Courtesy: DuPont water solutions (Personal Communication with Dupont Water Solutions' team. Retrieved from https://www.dupont.com/water/resources/design-software.html).*

8.2.2 Configuration of a reverse osmosis system

As presented in Fig. 8.12, the design software WAVE of DuPont water solutions takes the system permeate flow rate or system feed flow rate in the RO configuration section. The below details are used for specifying a reverse osmosis system:

- No. of passes: Presently, this software permits for a maximum of two passes.
- No. of stages: Presently, this software permits for a maximum of five stages.
- Flow Factor: It is used for simulating worst scenarios with respect to energy demand and rejection. Flow Factor is used in this design software to account for flow loss because of fouling. The Flow Factor should be in the range of 0−2.
- The number of PVs for each stage.
- The no. of elements per PV: This number should be between one and eight.
- The elements of interest. Here, the standard FilmTec elements are chosen. When the "Specs" link is clicked, each element's specifications will be shown.
- ISD: WAVE permits the specification of different elements within a PV—essentially creating stages within a stage. This is called Internally Staged Design. As shown in Fig. 8.13, WAVE simulates the performance of various types of elements in single PV.
- Permeate backpressure for each stage and boost pressure.

This design software of DuPont assumes 75% system recovery by default for RO and that there will be no recycles, splits, or bypasses in the product water stream. As displayed in Fig. 8.14, the details will be specified in the Flow Calculator tab, and the designer could state concentrate recycle flow and bypass flow.

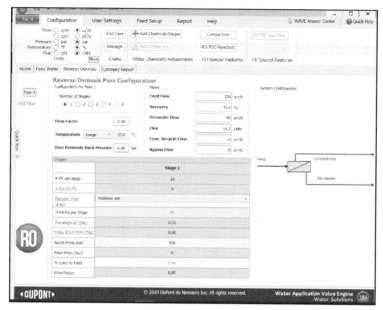

FIGURE 8.12 Reverse Osmosis system configuration of WAVE. Source: *Courtesy: DuPont water solutions (Personal Communication with Dupont Water Solutions' team. Retrieved from https://www.dupont.com/water/resources/design-software.html).*

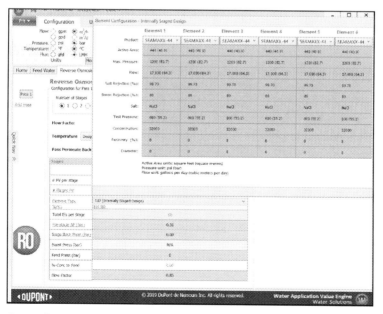

FIGURE 8.13 Internally staged design (ISD) specification in WAVE. Source: *Courtesy: DuPont water solutions (Personal Communication with Dupont Water Solutions' team. Retrieved from https://www.dupont.com/water/resources/design-software.html).*

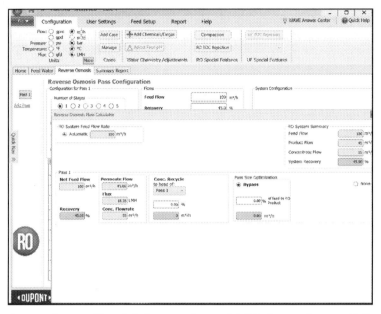

FIGURE 8.14   Reverse Osmosis Flow Calculator window in WAVE. Source: *Courtesy: DuPont water solutions (Personal Communication with Dupont Water Solutions' team. Retrieved from https://www.dupont.com/water/resources/design-software.html) (Gaublomme, D., Strubbe, L., Vanoppen, M., Torfs, E., Mortier, S., Cornelissen, E., ... Nopens, I. (2020). A generic reverse osmosis model for full-scale operation.* Desalination, 490, 114509).

8.2.3 Specifying pH adjustment/degasification and calculating the risk of scaling

This design software of DuPont makes possible adjustments in the water chemistry ahead of the 1st pass, between the two passes, and in the product water. This window includes options for adjusting pH, degasification, addition of antiscalants/SMBS (only the first pass), specifying the process temperature and pass recovery (Fig. 8.15). This Window also includes a Table with the Feed composition (before adjustment) and the Concentrate composition calculated by WAVE. Scaling risk in the Chemical Adjustment Popup Window is calculated in terms of the Langelier Saturation Index (LSI), Stiff & Davis Index (S&DI) and % saturation for some salts ($CaSO_4$, $BaSO_4$, $SrSO_4$, CaF_2), $Mg(OH)_2$, and SiO_2.

8.2.4 WAVE special features

Presently, this RO design software of DuPont offers two special features such as total organic carbon rejection specification and compaction specification, as shown in Figs. 8.16 and 8.17, respectively.

Total Organic Carbon (TOC) Reduction: Various organic compounds will be rejected by RO membranes, and in the case of WAVE, the rejection level is set for 80% by default, however, could be adjusted by the designer.

Compaction: The RO membranes has the ability to compact when subjected to a combination of moderate to higher temperatures and higher pressure. Membrane compaction could lead to a decrease in membrane transport properties, which might result in greater

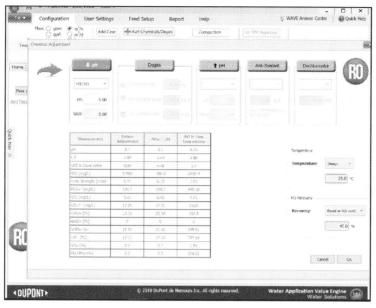

FIGURE 8.15 Chemical Adjustment in WAVE for reverse osmosis. Source: *Courtesy: DuPont water solutions (Personal Communication with Dupont Water Solutions' team. Retrieved from https://www.dupont.com/water/resources/design-software.html).*

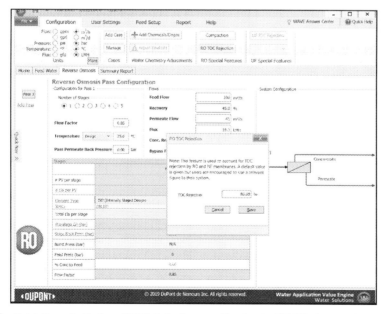

FIGURE 8.16 Total Organic Carbon (TOC) Rejection specification in WAVE. Source: *Courtesy: DuPont water solutions (Personal Communication with Dupont Water Solutions' team. Retrieved from https://www.dupont.com/water/resources/design-software.html).*

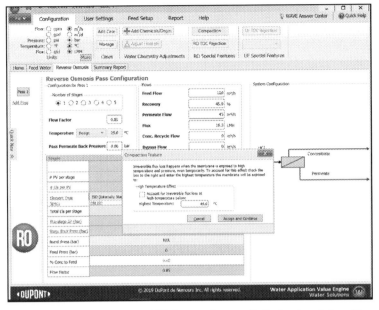

FIGURE 8.17 Compaction specification in WAVE. Source: *Courtesy: DuPont water solutions (Personal Communication with Dupont Water Solutions' team. Retrieved from https://www.dupont.com/water/resources/design-software.html).*

feed pressures to compensate, and the software permits designer to consider membrane compaction in the designing.

8.2.5 Generation of summary report, detailed report, and data exporting

After entering all the required information in WAVE, the designer should click on the "Summary Report" tab for generating a summary report (as displayed in Fig. 8.18), which includes the following:

- RO System Flow Diagram
- RO System Overview
- RO Flow Table (Stage Level)
- RO Solute Concentrations
- RO Design Warnings
- Special Comments
- RO Flow Table (Element Level)
- RO Solubility Warnings
- RO Chemical Adjustments

The summary report provides a satisfactory output for supporting the iterative design. As soon as the designer is satisfied with the design, the designer can click on the "Detailed Report" button for generating a detailed report. This report (Fig. 8.19) will have the below added details:

- Detailed flow at the pass level, including bypasses and recycles.

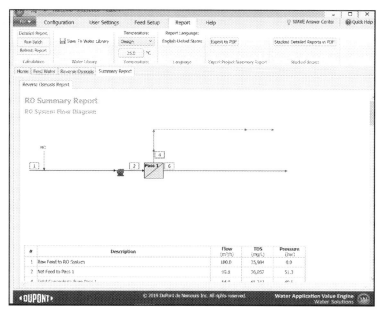

FIGURE 8.18 Summary report in WAVE. Source: *Courtesy: DuPont water solutions (Personal Communication with Dupont Water Solutions' team. Retrieved from https://www.dupont.com/water/resources/design-software.html).*

FIGURE 8.19 A detailed report in WAVE. Source: *Courtesy: DuPont water solutions (Personal Communication with Dupont Water Solutions' team. Retrieved from https://www.dupont.com/water/resources/design-software.html).*

- Flow table at the stage level inclusive of flows, total dissolved solids, and pressure drops in each stage for each pass.
- Flow table at the element level inclusive of flows, total dissolved solids, and pressure drops in each element in each stage for each pass.
- Solute concentrations in the permeate stream, feed stream and concentrate stream for each stage in each pass.
- RO utility cost and chemical cost inclusive of unrecovered water cost, wastewater disposal, electrical cost, and chemical cost.

As presented in Fig. 8.20, the designer will be able to export the detailed report as a word document, Excel, or PDF. As soon as the designer confirms the design, it could be saved in a database as a project or exported as DWPX file, which could be sent by e-mail to colleagues (Fig. 8.21). On the other hand, the designer can either import a DWPX file from a colleague for opening as a project or simply open a project from the database.

8.2.6 Batch feature and case management

At the time of designing a reverse osmosis system, it is beneficial to examine the influence of multiple combinations of some factors. As an example, it is advantageous to model the RO unit at higher temperature and a low flow factor for simulating complicated operational conditions. Rather than defining various individual cases, this design software of DuPont permits the designer for defining a possibly large number of flow factors and temperature combinations and generate important performance metrics in a single batch as presented in Fig. 8.22.

FIGURE 8.20 Detailed report exporting in WAVE. *Source: Courtesy: DuPont water solutions (Personal Communication with Dupont Water Solutions' team. Retrieved from https://www.dupont.com/water/resources/design-software.html).*

FIGURE 8.21 Detailed report exporting or saving in WAVE. Source: *Courtesy: DuPont water solutions (Personal Communication with Dupont Water Solutions' team. Retrieved from https://www.dupont.com/water/resources/design-software.html).*

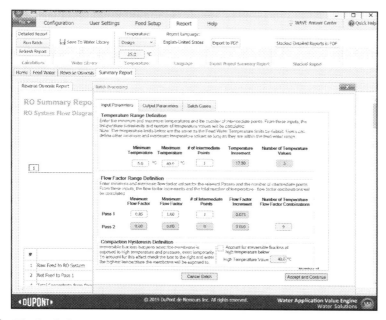

FIGURE 8.22 The batch feature in WAVE. Source: *Courtesy: DuPont water solutions (Personal Communication with Dupont Water Solutions' team. Retrieved from https://www.dupont.com/water/resources/design-software.html).*

8.3 IMSDesign (Hydranautics)

8.3.1 IMSDesign Cloud

IMSDesign Cloud is the Hydranautics's cloud-based membrane projection software (Hydranautics). The software is used to simulate RO, UF, and nanofiltration membrane projections in different combinations (Membrane projection software, 2020). This thoughtfully designed web simulator is supported with an array of advanced features that makes it a value-added tool. Users could use this auto-updated tool to run projections, anytime, from anywhere, and share or forward them to other people.

The latest IMSDesign (Integrated Membrane Solutions Design) software can be downloaded using the below link: https://membranes.com/solutions/software-imsdesign/

The following are the intelligent features of IMSDesign Cloud:

User-friendly Interface: A user-friendly interface offers seamless use with minimum learning time.
Backward Compatibility: It is compatible with the previous versions of the software. Successfully runs projection files created in the previous software versions from 2008 onwards. Once saved, they will remain in the Cloud forever, thus offering easy access.
Cross-Platform: Seamlessly functions across notebooks, tablets, and across Windows, OS, and Android operating systems.
Auto Update: The IMSDesign Cloud is always auto updated and thus enables to do more.
Integrated Simulator: Run with various combinations of RO, UF, and nanofiltration products in a seamless manner.
Feed Pressure Input: By entering the feed pressure as input the simulator will predict performance—useful in analyzing an operating system.
Simulate Custom Organics: Predict performance with organic chemicals and biochemical oxygen demand (BOD), chemical oxygen demand (COD), total organic carbon (TOC)—useful in wastewater applications.
Go Hybrid: Hybrid feature is now available for all six stages of both passes.
Sharing & Forwarding Projections: The projections can be shared with others very easily.

The RO design page of IMSDesign cloud is shown in Fig. 8.23. The step-by-step procedure for the RO designing is discussed in the following section for the IMSDesign Desktop.

8.3.2 IMSDesign Desktop

IMSDesign Version 2.226.84 (Integrated Membrane Solutions Design) is the most recent version of Hydranautics membrane projection software. This software estimates the degree of efficiency of a reverse osmosis plant and is simple and sophisticated. IMSDesign contributes upgraded program features, enhanced graphics and incorporates advanced features, which improve the user's capabilty to rapidly and precisely design and analyze membrane-based systems. It gives the user full control over the data used in the membrane selection process. The aforementioned control guarantees the user complete confidence in the anticipated performance of any Hydranautics membrane.

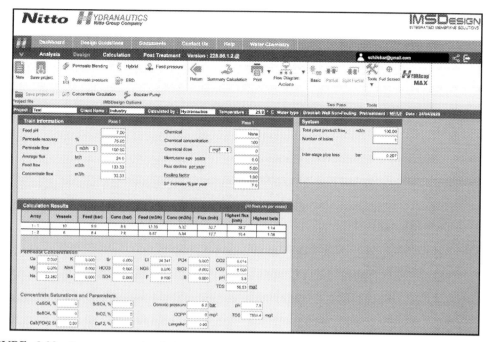

FIGURE 8.23 Reverse osmosis design page of IMSDesign cloud of Hydranautics. *Source: Courtesy: Hydranautics—A Nitto Group Company (Personal Communication with Hydranautics' team. Retrieved from https://membranes.com/solutions/software/).*

This program is intended to figure the throughput of a reverse osmosis plant and enables the user to move the water information through a series of stages. Every single stage performs as an input for the following stage. Contingent upon the results on the entire stages, the IMSDesign software makes an examination and design that is most reasonable for the user's RO plants.

In order to make the RO design, the user should perform the subsequent major steps.

8.3.2.1 Analysis

This is the primary stage where the user computes how the raw water sample is treated. The user will be able to enter the values of the water analysis report and assess the analysis by utilizing different methods, for example, pH equilibrium, measure saturation levels, and water balancing. During the performance of an analysis, the user should enter the raw water report values and then investigate its ion combinations. Moreover, IMSDesign allows the user to identify the ion and chemical composition of permeate, feed, and reject water.

Fig. 8.24 is the screen where the user enters the details of water analysis, either as mg/L ion or mg/L calcium carbonate. The screen also has inputs for pH, CO_3, CO_2, NH_3, and electrical conductivity. There are separate sections for cations, anions, and saturations.

The "online help" menu opens up a long list of information that the user can find help with, as presented in Fig. 8.25. When the "Analysis" page is finished by the user, the "Design" input screen will be appeared by clicking on the DESIGN button.

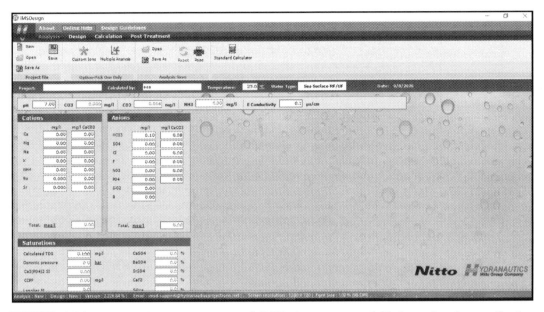

FIGURE 8.24 Water analysis input screen of IMSDesign program of Hydranautics. Source: *Courtesy: Hydranautics—A Nitto Group Company (Personal Communication with Hydranautics' team. Retrieved from https://membranes.com/solutions/software/).*

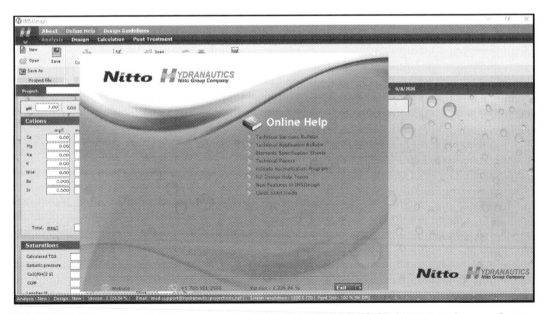

FIGURE 8.25 "Online Help" contents of IMSDesign program of Hydranautics. Source: *Courtesy: Hydranautics—A Nitto Group Company (Personal Communication with Hydranautics' team. Retrieved from https://membranes.com/solutions/software/).*

8.3.2.2 Design

"Design" is considered to be the second stage in the IMSDesign RO design software. In order to carry out this step, the user can indicate parameters, for example, permeate recovery, permeate flow, etc. First, the user needs to enter the project name. There are separate sections for trains, systems, and system specifications. The user will also be able to specify rates of flux decrease, increase in salt passage, fouling factor, membrane age, chemical dosing rate, etc. Feed pH and chemical feed are located at the upper left and center sections of the screen, respectively. The pH is entered and for the chemical feed, the user can select sulfuric acid (H_2SO_4) or HCl. The rate of chemical dosing required will also be shown in this section.

Membrane age: In the case of general projections, three years should be chosen, which assumes a membrane life of three years. This information works closely with the flux decrease and increase in salt passage inputs mentioned below.

Just below the membrane age, the flux decline percentage per year data is entered. It can be noted that there will be an increase in the flux decline as the quality of water deteriorates. Fig. 8.26 presents the primary design input screen of the IMSDesign program of Hydranautics. By clicking on the button "Recalc array" at the bottom, it will change the chosen array to the one that is more suitable for the conditions indicated in the program. After entering all the details in the "Design" screen, the user has to perform "RUN" to get the design calculations (Fig. 8.26). In system specification, the element type can be selected from the list provided (Fig. 8.27) and run the program as shown in Fig. 8.28.

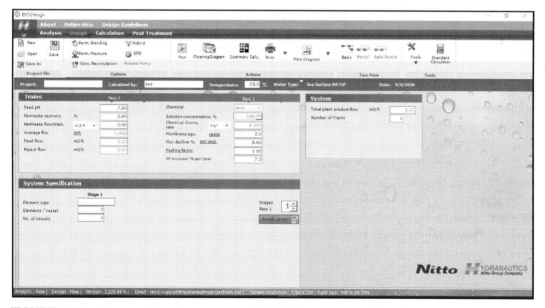

FIGURE 8.26 Primary design input screen of IMSDesign program of Hydranautics. Source: *Courtesy: Hydranautics—A Nitto Group Company (Personal Communication with Hydranautics' team. Retrieved from https://membranes.com/solutions/software/).*

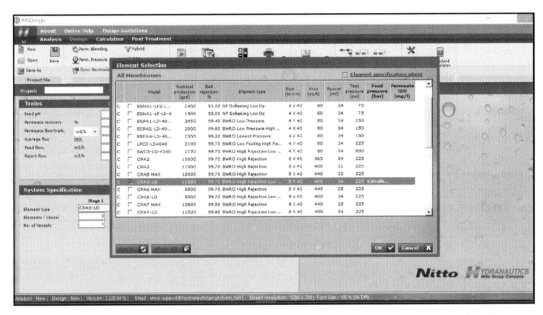

FIGURE 8.27   Membrane element selection in the design input screen of IMSDesign program of Hydranautics. Source: *Courtesy: Hydranautics—A Nitto Group Company (Personal Communication with Hydranautics' team. Retrieved from https://membranes.com/solutions/software/).*

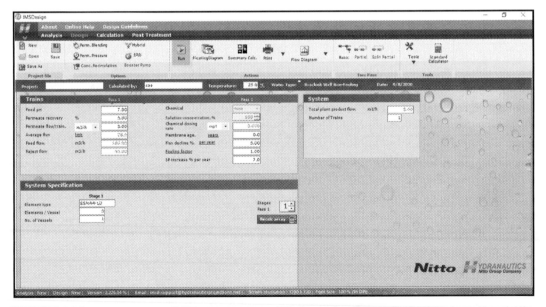

FIGURE 8.28   Program run in IMSDesign program of Hydranautics. Source: *Courtesy: Hydranautics—A Nitto Group Company (Personal Communication with Hydranautics' team. Retrieved from https://membranes.com/solutions/software/).*

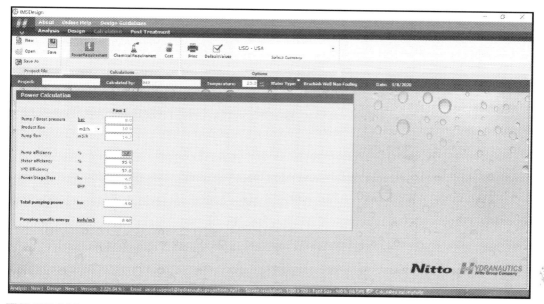

FIGURE 8.29 Calculation screen of IMSDesign program of Hydranautics. Source: *Courtesy: Hydranautics—A Nitto Group Company (Personal Communication with Hydranautics' team. Retrieved from https://membranes.com/solutions/software/).*

8.3.2.3 Calculation

The third step in the RO design process is the calculation. In this stage, the program computes and shows the values for chemical requirement, power requirement, and cost. After inputting the design parameters, IMSDesign performs the computation. The program computes the quality of the product as well as decides whether the design limits have been surpassed. Fig. 8.29 is the calculation screen of IMSDesign program of Hydranautics.

8.3.2.4 Post-treatment

The fourth step in the RO design process is the post-treatment. During the post-treatment of the sample, the user will be able to define the options for permeate treatment and the subsequent changes in ionic concentrations and pH. It has separate sections for cations, anions, and saturations. Fig. 8.30 is the post-treatment screen of Hydranautics IMSDesign program.

8.3.2.5 Analysis save

This is the last stage in the RO design process, and the user can save the analysis here (Fig. 8.31).

Several research groups have carried out case studies for RO system designing using the IMSDesign software for water treatment (Aghababaei, 2017; Mehta & Patel, 2014). Aghababaei (2017) developed a design method based on a simulation technique for optimizing RO desalination systems. The design was made with the use of Hydranautics design software, version 2011. The study confirmed that Hydranautics design software gave accurate design with the least possible error.

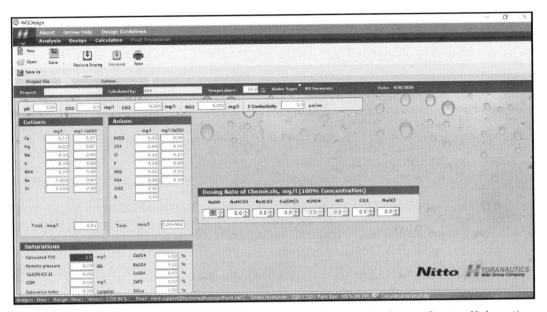

FIGURE 8.30   Post-treatment screen of IMSDesign program of Hydranautics. Source: *Courtesy: Hydranautics— A Nitto Group Company (Personal Communication with Hydranautics' team. Retrieved from https://membranes.com/solutions/software/).*

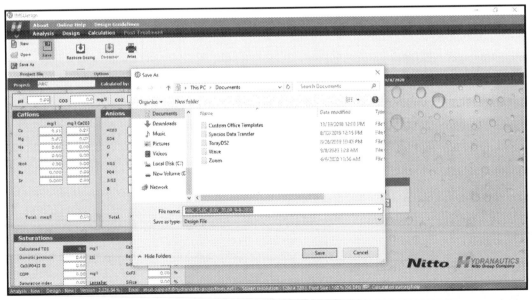

FIGURE 8.31   Analysis save of IMSDesign program of Hydranautics. Source: *Courtesy: Hydranautics—A Nitto Group Company (Personal Communication with Hydranautics' team. Retrieved from https://membranes.com/solutions/software/).*

8.4 TORAYDS2 2.2.00.189 (*Toray Industries, Inc., 2020*)

The design software TorayDS2 2.2.00.189 is the latest software version from TORAY Industries. The software can be downloaded using the below link:

https://ap3.toray.co.jp/toraywater/

The latest version TORAYDS2 2.2.00.189 (released on October 8, 2020) has the following features as described by the manufacturer

- DS2 is truly multilingual and permits input and output in several different languages including Russian, Korean, Arabic, Japanese, Chinese, and European languages.
- The design software DS2 has an option for Pitzer-based osmotic pressure computations as a development tool.
- DS2 permits straight export of every calculation run to Excel along with automatic variations in age and temperature exported to Excel.
- This design software uses the SQL database technology to store and organize projects. The project databases could be effectively shared over a network.
- For the majority single-stage projects and many two-stage projects, the feature "autocalculation" can be used. As the user makes changes, the program recalculates automatically.
- The design software DS2 is able to import the older version DS1 project files of the user.
- "Teach Mode" in DS1 is presently substituted by several "Templates" and "Guide Me Mode" in DS2 for a brief learning curve andfast generation of results required.
- DS2 allows more options in the system design (bypass, ERD, partial permeate, etc.)

The design software TORAY DS2 has five tabs namely Start-Main, Project, Feed Data, RO Design, and Report.

8.4.1 How to start a project

8.4.1.1 *Guide Me Mode*

"Guide Me Mode" function is presented in the latest DS2 version for providing a simple guide on how to make a simple (one-pass) RO projection. The "Guide Me Mode" function could be turned on/off from the "Start" tab (on the right side of the page), or through "Project Configuration" menu. The function would guide the user with the "To Do List," which will illustrate to the user what to be done in each phase, "Navigation Arrow," which would show the user where the action must be done, "Colored Input Field," which would illustrate the user which field is required or filled, and "Auto Layout Function," which will help the user automatically design the basic vessel configuration with the input data. The user can fine-tune the configuration to match the preference after finishing the basic design employing the guide me mode function. Fig. 8.32 is the Guide Me Mode in the DS2 design software program.

8.4.1.2 *Basic templates*

The DS2 program provides various basic templates that could be employed as the basic design for the comfort of the user. The basic templates available would be different based on the selected feed water type. The user would be able to scale-up or scale-down depending on the selected template accordingly to the required system design size. Fig. 8.33 is the basic templates in the DS2 design software program.

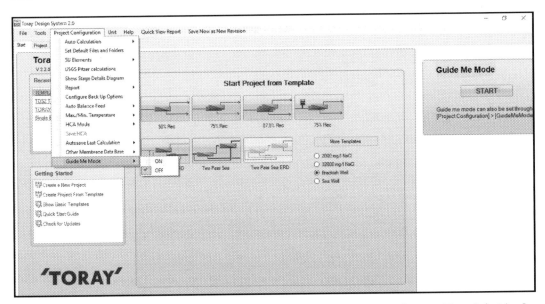

FIGURE 8.32   The Guide Me Mode in DS2 Design software program. Source: *Courtesy: Toray Industries, Inc. (DS2 Guideline. (2020). Toray Industries, Inc. Retrieved from https://ap3.toray.co.jp/toraywater/).*

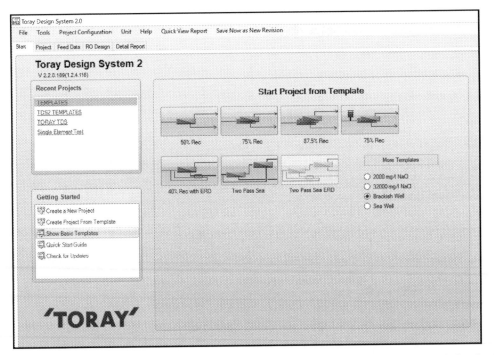

FIGURE 8.33   The Basic Templates in DS2 Design software program. Source: *Courtesy: Toray Industries, Inc. (DS2 Guideline. (2020). Toray Industries, Inc. Retrieved from https://ap3.toray.co.jp/toraywater/).*

8.4.1.3 Start a new project

The user also can create and design a new project from scratch without any template, by following the below instructions.

- For starting a new project, the user can click "Create a New Project" in the "Start" tab or move straight to the "Project" tab.
- Click "Start New Project" for clearing the existing background information. Fig. 8.34 is the tab for creating a new project in the DS2 Design software program.
  - ⊛ The user should not forget to save the current running project before clicking.
  - ⊛ The user can input the project name, case description and revision description. Fig. 8.35 is the project data screen in DS2 Design software program
    - — It is suggested that the description must be stated very clear, as it will be referred to in the final report.
    - — The user can go back to the "Project" tab after he finishes all the design processes for editing the present information and for saving the current system design.
  - ⊛ After entering all the necessary project information input, the user can click "Save" for saving the project in the software. Subsequently, click "Open" for moving to the "Feed Data" tab and start to design the reverse osmosis system.

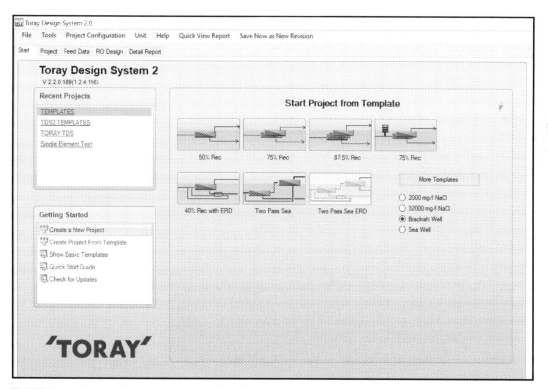

FIGURE 8.34 Creating a new project in the DS2 Design software program. Source: *Courtesy: Toray Industries, Inc. (DS2 Guideline. (2020). Toray Industries, Inc. Retrieved from https://ap3.toray.co.jp/toraywater/).*

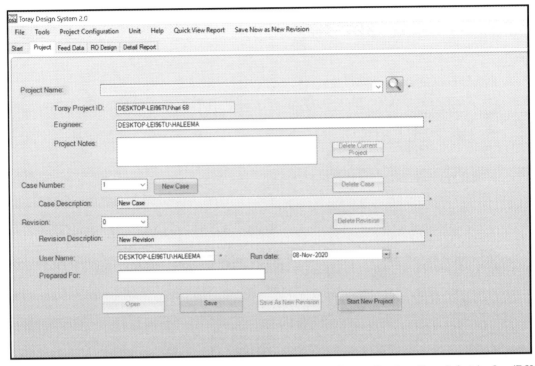

FIGURE 8.35   Project data screen in DS2 Design software program. Source: *Courtesy: Toray Industries, Inc. (DS2 Guideline. (2020). Toray Industries, Inc. Retrieved from https://ap3.toray.co.jp/toraywater/).*

8.4.1.4 Designing a reverse osmosis system

It is always simpler to imagine a design process using a case study. As an example, let us design a brackish water RO system with 65.0% recovery rate that generates 270 m³/day of water (feed: sodium chloride 2000 mg/L, brackish surface, pH 7, temperature: 25°C). The user can start with the feed water data input.

8.4.1.4.1 Feed data tab

In this section, the user should enter the detailed information of feed water quality input, which will be used for the designed system.

Selection of water type   DS2 software will work with a selected guideline value based on the feed water type the user wants to treat in the RO system. It is extremely important to match this water type selection with the actual feed water condition for a very robust design of a system.

For selecting the water type, the user can do it from the pull-down option in the "Stream Information" box. In this case study, the user can select "Brackish Surface" (Fig. 8.36).

Feed water quality data input   The user should enter detailed information on the ions present in the feed water (Fig. 8.37), along with its pH and temperature. Each ion concentration could be mentioned as mg/L, mEq/L, or ppm $CaCO_3$.

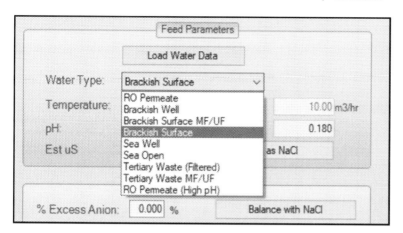

FIGURE 8.36 Feed parameters screen in DS2 Design software program. *Source: Courtesy: Toray Industries, Inc. (DS2 Guideline. (2020). Toray Industries, Inc. Retrieved from https://ap3.toray.co.jp/toraywater/).*

Cations

Current Stream : 1

Ions	mg/l	mEq/L	ppm CaCO3
Ca	0.01	0.0025	0.12
Mg			
Na	786.75	34.2217	1712.62
K			
Ba			
Sr			
NH4			
Fe			
Totals	**786.8000**	**34.2242**	**1712.7498**

Anions

Ions	mg/l	mEq/L	ppm CaCO3
HCO3	0.10	0.0016	0.08
Cl	1213.25	34.2200	1712.54
SO4			
NO3			
F			
Br			
B			
SiO2			
PO4			

FIGURE 8.37 Ions input screen in DS2 Design software program. *Source: Courtesy: Toray Industries, Inc. (DS2 Guideline. (2020). Toray Industries, Inc. Retrieved from https://ap3.toray.co.jp/toraywater/).*

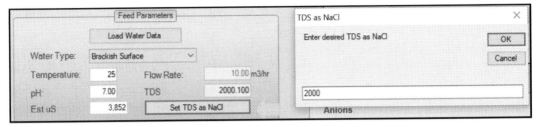

FIGURE 8.38 pH, temperature, and total dissolved solids (TDS) input screens in DS2 Design software program. Source: *Courtesy: Toray Industries, Inc. (DS2 Guideline. (2020). Toray Industries, Inc. Retrieved from https://ap3. toray.co.jp/toraywater/).*

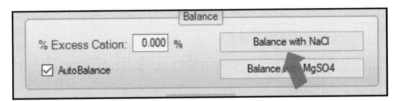

FIGURE 8.39 Balance box in DS2 Design software program. Source: *Courtesy: Toray Industries, Inc. (DS2 Guideline. (2020). Toray Industries, Inc. Retrieved from https://ap3.toray.co.jp/toraywater/).*

If the user does not know the ion composition in the feed water, however, knows the total dissolved solids, then the user could assume all the total dissolved solids as sodium chloride (NaCl_ using "Set TDS as NaCl" button inside the "Feed Parameters" box) (Fig. 8.38). The user could also set the feed water pH and the temperature here. In this case study, the user should enter the feed as 2000 mg/L total dissolved solids as sodium chloride, pH 7, and temperature of 25°C.

If the anions and cations entered are not balanced, then the user can balance them with NaCl or MgSO$_4$ using the button in the "Balance" box (Fig. 8.39).

After completing all the required feed water data input, the user can move to the "RO Design" tab.

8.4.1.4.2 Selecting unit

The user needs to select the units that he wants to be used in the project. The user can also check the currently used units on the top left of the "RO Design" page. For changing the units, click "Unit" on the menu-bar, and select the units the user wants to use in the pop-up window (Fig. 8.40).

8.4.1.4.3 Filling in flow diagram

For starting the designing process, the user needs to decide the main goal of the system related to the water flow, which is two out of the three data of feed flow, RO recovery, and product flow in the flow diagram (the other information will be automatically calculated). In this case study, we already know the recovery rate (65%) and the product flow (270 m^3/d). This information can be used for filling the flow diagram (Fig. 8.41).

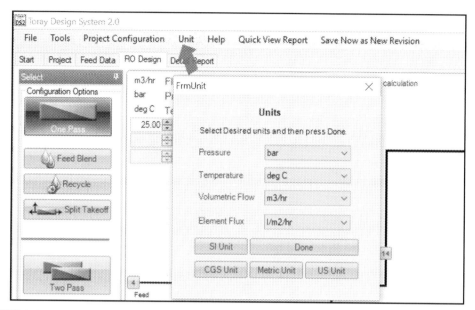

FIGURE 8.40 Selection of units in DS2 Design software program. Source: *Courtesy: Toray Industries, Inc. (DS2 Guideline. (2020). Toray Industries, Inc. Retrieved from https://ap3.toray.co.jp/toraywater/).*

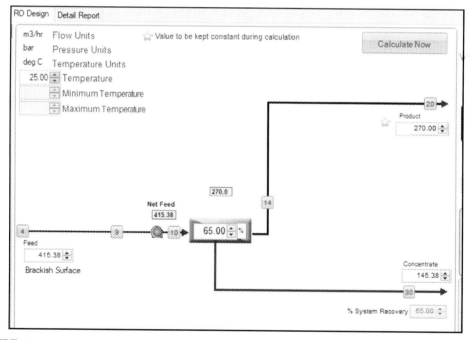

FIGURE 8.41 Filling-in the flow diagram in the DS2 Design software program. Source: *Courtesy: Toray Industries, Inc. (DS2 Guideline. (2020). Toray Industries, Inc. Retrieved from https://ap3.toray.co.jp/toraywater/).*

If the user wants to do some modifications in the basic flow data, then the user can click the star icon to choose which parameter he wants to set as constant, then set the parameter the user wants to modify.

8.4.1.4.4 System flux and required total element number calculation

The guideline in this software will help the user to design a robust RO system in most of feed water types. The user can find it in the "Help" tab > "Design Guidelines" (Fig. 8.42). Moreover, it will provide the user a guide on the recommended operating flux for the RO system, which is considered to be a very important value for the RO design.

For a feed from a brackish surface with conventional pretreatment, the recommended flux is between 18 and 23 L/m^2 per hour. Hence. The user can take 20 L/m^2 per hour for this case study. And the effective membrane area of the standard 400 ft^2 8-inch RO element is around 37.3 m^2/element. With these two data, the user can determine the total number of required 8-inch RO elements.

$$\frac{270\,(m^3/d) \times 1000\,(L/m^2)}{24\left(\frac{h}{d}\right) \times 20\left(\frac{L}{m2h}\right) \times 37.3(m^2)} = 15.08 \sim 15\text{pcs}$$

From this calculation, roughly 15 pcs of RO elements will be needed for the system.

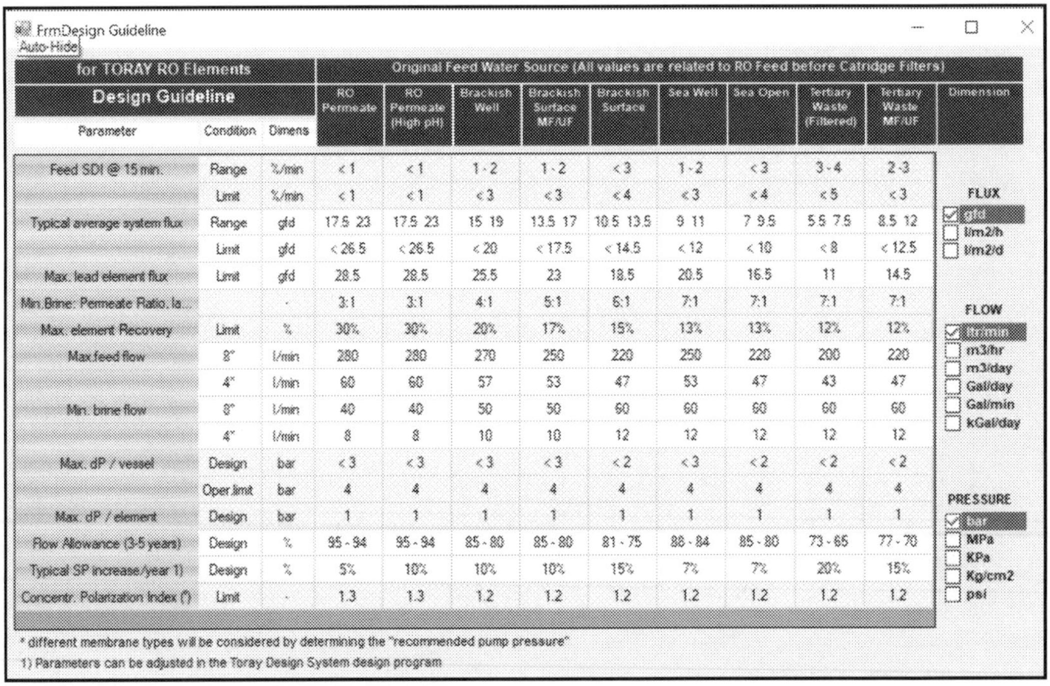

for TORAY RO Elements			Original Feed Water Source (All values are related to RO Feed before Catridge Filters)									
Design Guideline			RO Permeate	RO Permeate (High pH)	Brackish Well	Brackish Surface MF/UF	Brackish Surface	Sea Well	Sea Open	Tertiary Waste (Filtered)	Tertiary Waste MF/UF	Dimension
Parameter	Condition	Dimens										
Feed SDI @ 15 min.	Range	%/min	< 1	< 1	1 - 2	1 - 2	< 3	1 - 2	< 3	3 - 4	2 - 3	
	Limit	%/min	< 1	< 1	< 3	< 3	< 4	< 3	< 4	< 5	< 3	FLUX
Typical average system flux	Range	gfd	17.5 23	17.5 23	15 19	13.5 17	10.5 13.5	9 11	7 9.5	5.5 7.5	8.5 12	☑ gfd
	Limit	gfd	< 26.5	< 26.5	< 20	< 17.5	< 14.5	< 12	< 10	< 8	< 12.5	☐ l/m2/h
Max. lead element flux	Limit	gfd	28.5	28.5	25.5	23	18.5	20.5	16.5	11	14.5	☐ l/m2/d
Min. Brine: Permeate Ratio. la..		·	3:1	3:1	4:1	5:1	6:1	7:1	7:1	7:1	7:1	
Max. element Recovery	Limit	%	30%	30%	20%	17%	15%	13%	13%	12%	12%	FLOW
Max.feed flow	8"	l/min	280	280	270	250	220	250	220	200	220	☑ lt/min
	4"	l/min	60	60	57	53	47	53	47	43	47	☐ m3/hr
Min. brine flow	8"	l/min	40	40	50	50	60	60	60	60	60	☐ m3/day
	4"	l/min	8	8	10	10	12	12	12	12	12	☐ Gal/day ☐ Gal/min
Max. dP / vessel	Design	bar	< 3	< 3	< 3	< 3	< 2	< 3	< 2	< 2	< 2	☐ kGal/day
	Oper.limit	bar	4	4	4	4	4	4	4	4	4	
Max. dP / element	Design	bar	1	1	1	1	1	1	1	1	1	PRESSURE
Flow Allowance (3-5 years)	Design	%	95 - 94	95 - 94	85 - 80	85 - 80	81 - 75	88 - 84	85 - 80	73 - 65	77 - 70	☑ bar
Typical SP increase/year 1)	Design	%	5%	10%	10%	10%	15%	7%	7%	20%	15%	☐ MPa ☐ KPa
Concentr. Polarization Index (*)	Limit	-	1.3	1.3	1.2	1.2	1.2	1.2	1.2	1.2	1.2	☐ Kg/cm2 ☐ psi

* different membrane types will be considered by determining the "recommended pump pressure"
1) Parameters can be adjusted in the Toray Design System design program

FIGURE 8.42 Design guideline in DS2 Design software program. Source: *Courtesy: Toray Industries, Inc. (DS2 Guideline. (2020). Toray Industries, Inc. Retrieved from https://ap3.toray.co.jp/toraywater/).*

8.4.1.4.5 Deciding stage and vessel configuration

Derived through the values in the guideline, the stage configuration in a system mainly depends on the system recovery rate. In general, it will follow the following "rule of thumb."

If the recovery rate is below 50%, then the number of stage required will be one.

If the recovery rate is between 50% and 75%, then the number of stages required will be two.

If the recovery rate is above 75%, then the number of stages required will be three.

For seawater RO unit, because of the limitation from the osmotic pressure of brine, in the majority cases, the recovery rate would be less than 50.0%, so the one-stage design is the most frequently used.

And the element quantity in each stage is generally put as the half amount of the previous stage. Hence, the proportion of the element quantity in stage 1, stage 2, and stage 3 will be roughly 4:2:1.

Thus, in this case, with the recovery of 65.0%, the system will have two stages. For the first stage, the number of RO element will be 10 and for the second stage, will need five RO elements.

Subsequently, the user should decide the length of the PV employed in the system. Generally, it would be in the range of 4–7 elements per vessel. Let us consider the number of elements per vessel to be 5. Hence, with a simple calculation, the first stage will have two vessels, and the second stage will have one vessel. On the "RO Design" tab, the user can input this information in the upper right part of the page (Fig. 8.43).

The user should not forget to set the number of "elements in vessel" for each stage, because it will not be set automatically if the user changes the number in other stages.

8.4.1.4.6 Selecting element type

DS2 can help the user to sort out the most suitable element model by using the "Element Selection Criteria" function (Fig. 8.44). For activating this function, the user should click the "Element Selection Criteria" button on the right side of the page. It will show the user a pop-up that allows the user to select the criteria he or she wants to select.

In this case study, Standard Brackish Water RO 8 inch with the standard 400 ft^2 membrane area will be required. The user should check the (Globally Available), (8 inch/400 ft^2) and (BWRO Standard pressure) boxes.

After sorting, the user would find the TM720D-400 as the model that is the best fit with the criteria. The user can apply this model by choosing from the drop-down list (Fig. 8.45) or by dragging and dropping after sorting the suitable element by clicking the "Element Selection Criteria" button.

The user can also modify the age of each element or even put multiple models of elements inside a single vessel. However, in this case study, for making it simple, we can

FIGURE 8.43 Deciding stage and vessel configuration (Reverse Osmosis Design) in DS2 Design software program. Source: *Courtesy: Toray Industries, Inc. (DS2 Guideline. (2020). Toray Industries, Inc. Retrieved from https://ap3. toray.co.jp/toraywater/).*

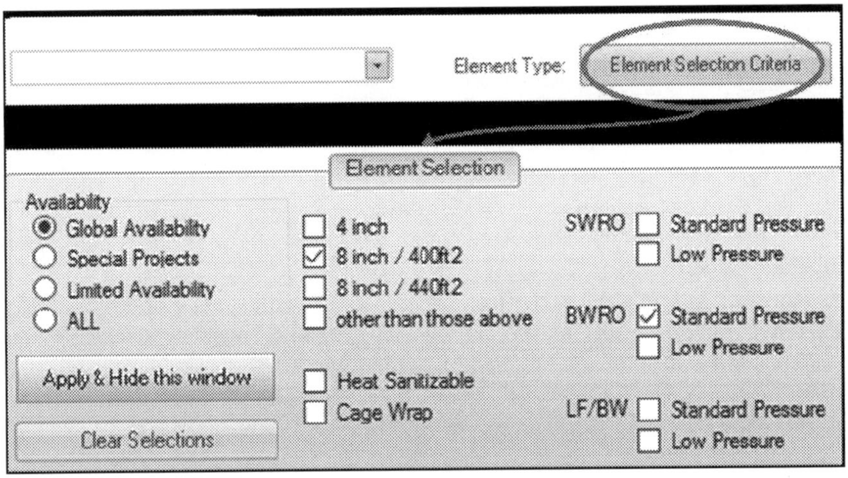

FIGURE 8.44 Element Selection Criteria in DS2 Design software program. Source: *Courtesy: Toray Industries, Inc. (DS2 Guideline. (2020). Toray Industries, Inc. Retrieved from https://ap3.toray.co.jp/toraywater/).*

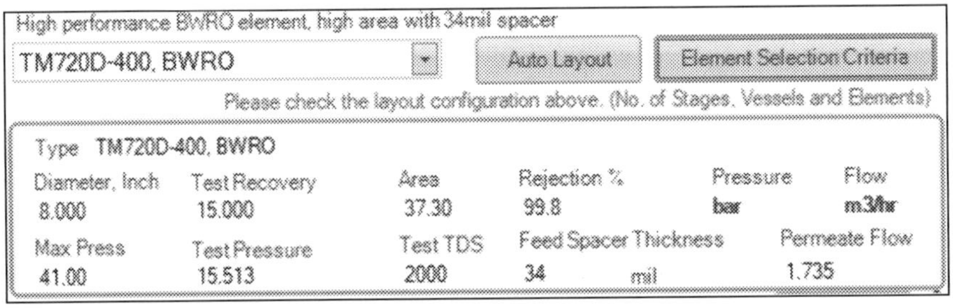

FIGURE 8.45 Drop-down Element Selection in DS2 Design software program. Source: *Courtesy: Toray Industries, Inc. (DS2 Guideline. (2020). Toray Industries, Inc. Retrieved from https://ap3.toray.co.jp/toraywater/).*

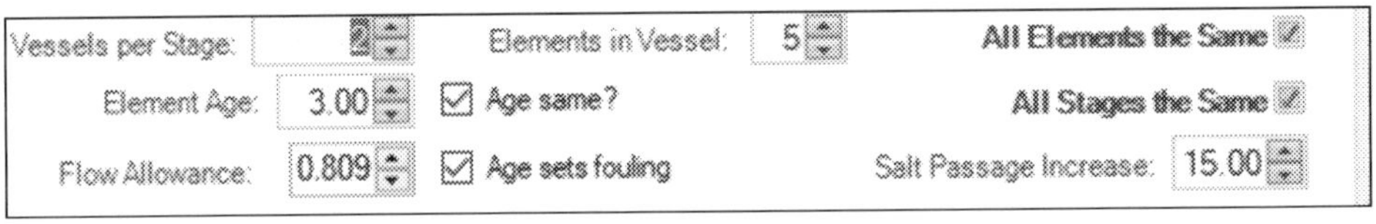

FIGURE 8.46 Elements and stages input data in DS2 Design software program. Source: *Courtesy: Toray Industries, Inc. (DS2 Guideline. (2020). Toray Industries, Inc. Retrieved from https://ap3.toray.co.jp/toraywater/).*

make all elements the same model and age by checking the "Age same?" box, "All Elements the Same" box, and "All Stage the Same" box on the upper right side of the design page (Fig. 8.46). Subsequently, click "Apply & Hide this window" in the "Element Selection" box.

8.4.1.4.7 Executing calculation

With this, the user has completed the designing. Then, he can click the "Calculate Now" button for executing the calculation. For seeing the result quickly, the user can employ the result table on the page bottom. First, set which part of the system the user wants to know in detail. The user may choose feed water after the pump (Icon No. 10), Permeate water (No. 14), and Brine (No. 30). Subsequently, drag each of the icons to the column in the bottom of the page (Fig. 8.47).

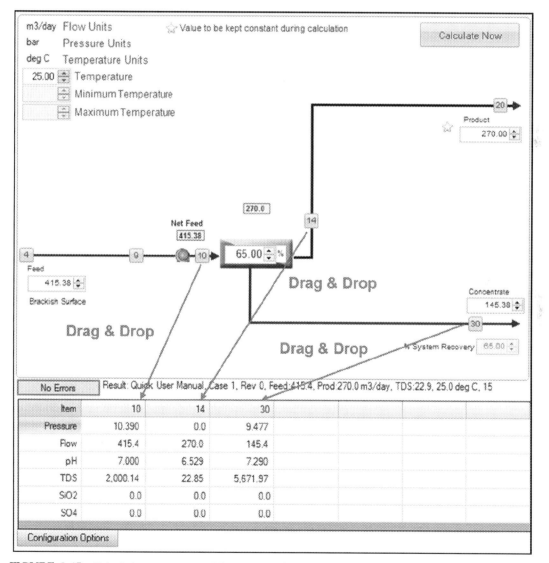

Item	10	14	30			
Pressure	10.390	0.0	9.477			
Flow	415.4	270.0	145.4			
pH	7.000	6.529	7.290			
TDS	2,000.14	22.85	5,671.97			
SiO2	0.0	0.0	0.0			
SO4	0.0	0.0	0.0			

FIGURE 8.47 Calculation execution in DS2 Design software program. Source: *Courtesy: Toray Industries, Inc. (DS2 Guideline. (2020). Toray Industries, Inc. Retrieved from https://ap3.toray.co.jp/toraywater/).*

8.4.1.5 Saving project and issuing a report

The user can save the project by going back to the "Project" tab and then click the save button at the page bottom. The user can also save it as a new revision for a case, or even make the present design as a new case. The user should describe each case and revision clearly as it will be stated on the issued report.

DS2 also permits the user to issue a report for the projection. There will be three versions of the report that the user can employ as per his or her requirement:

Quick View Report: This report will show the user the most basic data of the projection result in maximum of four pages. This will be shown as a new window once the user clicks "Quick View Report" in the menu bar. Also, he or she can print or save it as a PDF. This would be a very convenient tool for doing a quick check on the overall result of the current projection.

Short and Long Reports: Instead of "Quick View Report," the user can also select to issue the short or the long report. These two versions will provide the user with more detailed information about the projection results. The main difference between these two is "Long Report" will provide the user with information of each membrane condition in the designed system, whereas the "Short Report" will not.

8.5 LewaPlus (LANXESS)

LewaPlus allows for the planning of RO, UF, and ion exchange (IX) systems under a variety of system configurations. In the water treatment industry, this design software offers not only the possibility to treat one stream after the other (one dimensional) but design complex treatment systems (two dimensional) with different technologies.

The latest LewaPlus software can be downloaded using the below link:

https://lpt.lanxess.com/lewaplus-software/

The main features of LewaPlus are as follows:

- Not cloud-based that makes the software also accessible offline and warranties data security.
- Portable software version with entire functionality available that does not require any Windows administrative rights for installation.
- A single water analysis data entry screen, with RO scaling calculations, suitable for both ion exchange and RO applications.
- Clear interface and reliable and rapid results of the calculation.
- Design flexibility: Parameters could be widely modified and values outside the recommended borders are highlighted.
- Comprehensive output of RO system design parameters and effluent (permeate) quality in an easy to manipulate the printed output.
- Product scout and cross-reference tool to allow a proper selection of the right resin or membrane type.
- Direct access to technical documentation.
- Regular updates for keeping the software up-to-date and further enhancing its functionality with new features.

- Data exchange to technical LANXESS experts for optimization and a final check of your calculation.
- Design water treatment systems just by dragging and dropping the selected technologies in the requested position.
- Available in 11 different languages.

RO processes and functionalities

The RO module calculates the design and performance of a reverse osmosis plant for brackish and seawater applications. It helps for simulating the influence on performance and quality at different parameters such as salinity, temperature, and salt passage.

- Immediate calculation of system performance, including feed pressure calculations and product water quality on a single screen in the RO module.
- Advanced RO system configuration options with automatic calculation of a recommended array for optimization of system sizing.
- Support for hybrid PV configurations for achieving reduced feed pressure or to equalize permeate flux distribution along the membrane unit.
- Support for split-partial two-pass designs and various concentrate recirculation options.
- RO normalization tool for data comparison under standard conditions.
- Power consumption module to calculate system power costs for a variety of energy recovery device configurations.
- Post-treatment module for chemical addition, pH adjustment, and degasification of the final product water.
- Adjustable capital and operating cost module using historical industry data as default values. The model follows RO industry conventions for assignment of water costs in project bid situations.

The RO module is a tool to calculate the design and performance of a reverse osmosis plant. It helps to simulate the influence on performance and quality at different parameters like temperature, salt concentration, and salt passage.

The design process consists of the following steps:

1. Water analysis
2. Design
    a. Hybrid design
    b. Split partial two-pass design
    c. How to handle calculation failures
3. Power and energy calculation
4. Post-treatment
5. Cost calculation
6. Report generation

8.5.1 Water analysis

Primarily, the designer needs to enter the project data as shown in Fig. 8.48. Then, click on the Water Analysis icon, as shown in Fig. 8.49. On the water analysis page, in the upper

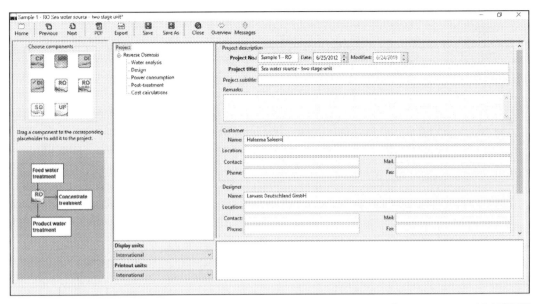

FIGURE 8.48   Project information input screen of LewaPlus reverse osmosis design program of LANXESS. Source: *Courtesy: LANXESS (Personal Communication with LANXESS team. https://lpt.lanxess.com/lewaplus-software/).*

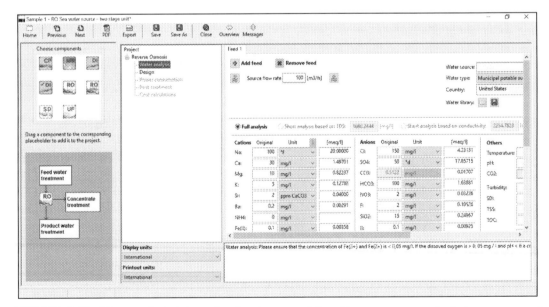

FIGURE 8.49   Water analysis information input screen of LewaPlus reverse osmosis design program of LANXESS. Source: *Courtesy: LANXESS (Personal Communication with LANXESS team. https://lpt.lanxess.com/lewa-plus-software/).*

part of the screen, there are buttons for managing the water analysis on the left side and entry fields for a general description of the water analysis on the right side.

The buttons for adding and removing a feed allow the designer to blend up to three different feed water analyses to define the raw water analysis to be used for the RO plant design:

8.5.2 Design

The design screen (Fig. 8.50) provides several options to design a reverse osmosis membrane system. On the left side, the various design options can be selected; on the right side, a diagram of the current configuration is presented. Additionally, there are options to fine-tune the configuration. Moreover, the water composition of the permeate, the concentrate, and a list with details for each membrane element are presented in different tabs.

8.5.3 Power consumption

The power and energy screen shown at the top of the screen tab (Fig. 8.51) for each pass in the designed system displays the pass details. For each pass, the characteristic parameters of the required high-pressure pump (HPP) are shown. If an additional booster pump or an additional energy recovery device was selected during the design process, the power and energy consumption data for these devices are shown as well. The bottom of the Power and Energy screen provides an important summary of the Energy consumption of the designed RO system, including a summary of total motor power (in kW), and specific power consumption (in kWh/m^3 of product water).

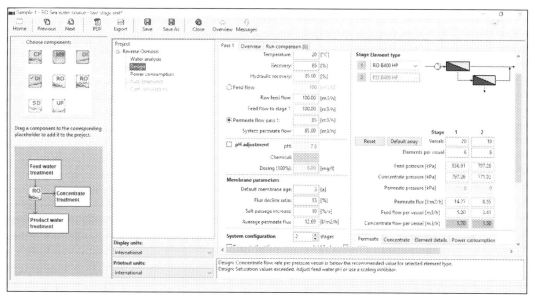

FIGURE 8.50 Design information input screen of LewaPlus reverse osmosis design program of LANXESS. Source: *Courtesy: LANXESS (Personal Communication with LANXESS team. https://lpt.lanxess.com/lewaplus-software/).*

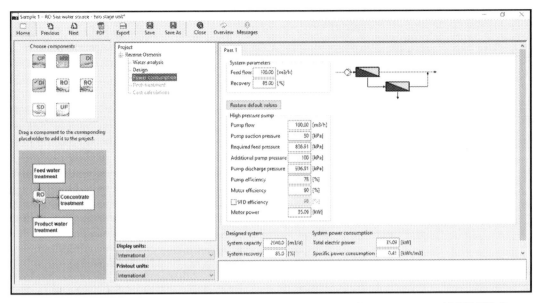

FIGURE 8.51   Power consumption screen of LewaPlus Reverse osmosis Design program of LANXESS. *Source: Courtesy: LANXESS (Personal Communication with LANXESS team. https://lpt.lanxess.com/lewaplus-software/).*

8.5.4 Post-treatment

The post-treatment screen has three parts (Figs. 8.52 and 8.53). At the topmost section of the screen, the water composition before the currently selected treatment step is displayed. At the bottom of the screen, the water composition after the final treatment step is shown. The controls for selecting treatments and configuring individual treatment steps are located in the middle of the screen. The middle section is divided into three areas: on the left side, a list of available treatments is shown. In the middle, the sequence of currently applied treatment steps is shown. Lastly, on the right side, the configuration options for the current (highlighted) applied treatment step are shown.

8.5.5 Cost calculations

The capital and operating costs module calculates total water costs using an innovative cost model that allows assignment of cost structures specific for local conditions, or recent (5-year historical data) industry default values. The costs can be adjusted to reflect the planned cost of capital and local costs for all RO system cost inputs. The program also permits for change of percent distribution of major components of the desalination system cost. The system cost calculated by this program includes the components starting with chemical dosing systems and cartridge filters, through pumping unit, RO membrane unit, and post-treatment unit.

The cost calculation screen has three parts as shown in Figs. 8.54 and 8.55. For the whole cost screen, it can be specified whether it should be included in the printout. At the topmost

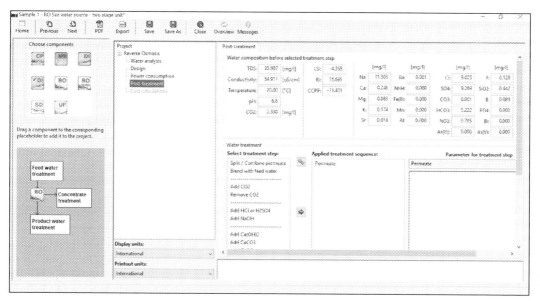

FIGURE 8.52 Post-treatment screen (top portion) of LewaPlus Reverse Osmosis Design program of LANXESS. Source: *Courtesy: LANXESS (Personal Communication with LANXESS team. https://lpt.lanxess.com/lewa-plus-software/).*

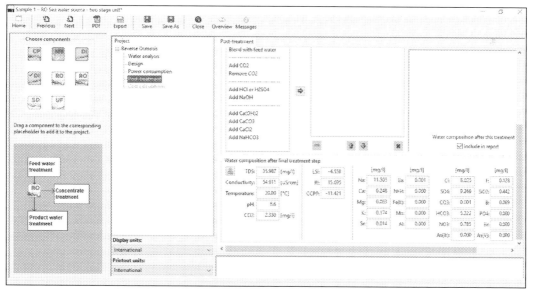

FIGURE 8.53 Post-treatment screen (bottom portion) of LewaPlus Reverse Osmosis Design program of LANXESS. Source: *Courtesy: LANXESS (Personal Communication with LANXESS team. https://lpt.lanxess.com/lewa-plus-software/).*

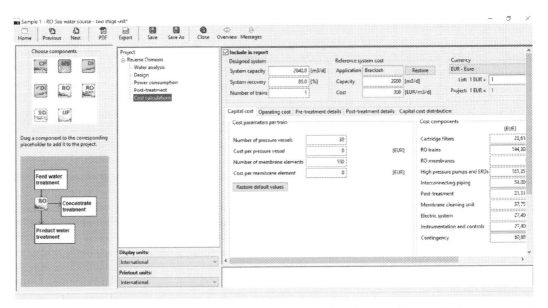

FIGURE 8.54 Cost calculation (top section) screen of LewaPlus Reverse Osmosis Design program of LANXESS. Source: *Courtesy: LANXESS (Personal Communication with LANXESS team. https://lpt.lanxess.com/lewaplus-software/).*

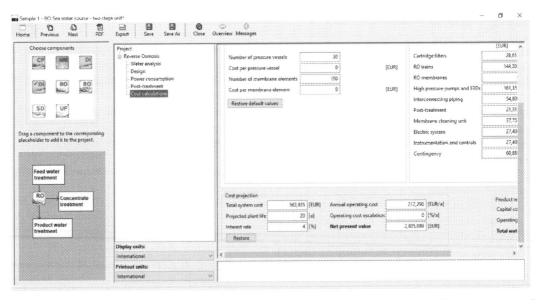

FIGURE 8.55 Cost calculation (bottom section) screen of LewaPlus Reverse Osmosis Design program of LANXESS. Source: *Courtesy: LANXESS (Personal Communication with LANXESS team. https://lpt.lanxess.com/lewaplus-software/).*

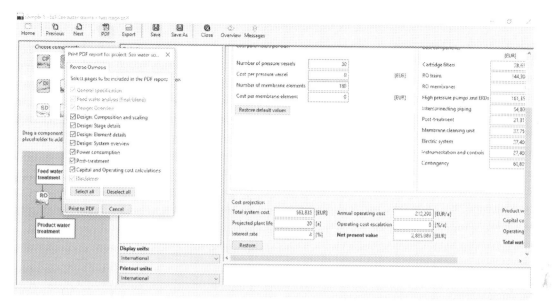

FIGURE 8.56 Report generation screen of LewaPlus Reverse Osmosis Design program of LANXESS. Source: *Courtesy: LANXESS (Personal Communication with LANXESS team. https://lpt.lanxess.com/lewaplus-software/).*

section of the screen, parameters for the designed RO system and the reference RO system cost are presented. The cost of the reference system is based on the water source and system capacity. Additionally, the designer can select the currency to be used for the cost calculations, and a methodology for adjusting and updating the currency are located on the right of the screen. In the middle of the screen, the user has several tabs detailing the system costs. Each tab opens up a specific set of cost parameters that are either user-defined, or default values within the software (that the designer can override). The costs for the major cost blocks (plus contingency) are calculated based on user inputs and cost distribution default values within the software. At the bottom of the screen, a cost projection of the designed RO system is shown. The total system capital cost and annual costs are calculated. The designer can input specific values that influence total costs, including the projected plant life, interest rate (cost of capital), and operating cost escalation.

8.5.6 Report generation

After entering all the above details, it can be saved and the report can be generated, as shown in Fig. 8.56.

8.6 Winflows 4.01 (Suez Water Technologies & Solution)

Winflows 4.01 is the software tool, from SUEZ Water Technologies & Solutions, to design RO applications. This program is designed to allow quick, easy, and intuitive design of RO systems. This tool was developed in Visual Studio in .Net platform. It can be used for designing an integrated desalination system, by making use of the following add-on features along with the RO system

- Chemical dosing—feed and product
- Argo Analyzer tool (in feed and concentrate streams) for antiscalant selection
- Gas stripping—feed and product
- Cartridge filters
- Electrodeionization (EDI) for permeate stream
- Energy recovery devices (ERD)—work and pressure exchangers, turbochargers, and pelton wheel
- Feed bypass and concentrate recycle options
- Split permeate option
- Mixing of multiple feed water analyses

The latest Winflows software can be downloaded using the below link:
https://www.suezwatertechnologies.com/resources/winflows
After successfully installing Winflows on the system, the Winflows 4.01 icon is automatically created on the desktop. Winflows session can be started either by double-clicking the icon or through program files or by double-clicking the existing Winflows *.win file. At any given time, three different sessions of Winflows could be opened simultaneously.

Data could be entered via the Design Assistant or from the Flowsheet or from any of the Menu commands.

In the below section, the design of a reverse osmosis system is illustrated using the Design Assistant to get a first-pass design and then the addition of a few refinements for achieving the targeted water quality. The following criteria describe the example system:

The information in Table 8.2 will enable the development of a first-pass design within a couple of minutes using the Design Assistant.

8.6.1 Starting a project using Design Assistant

The Design Assistant, accessible from the main menu bar, steps the user through the steps involved in the design of a reverse osmosis system in a logical and orderly manner. Two approaches are available from the Design Assistant.

- Winflows Step Design—manual entry of data.
- Auto Design—Winflows fills in default data. Defaults can be overridden by the user. The auto design will only function for single-stage designs.

The Winflows Step Design and Auto Design functions will carry the user through each of the steps necessary to complete a design. "Next" moves the user to the next step, "previous" or "back" goes to the prior step. "Exit" quits the Design Assistant but any changes made in previous dialogs are saved.

- Project Information (Optional): Enter any project information the user wants to save here.
- Feed Analysis or Specifying the Feed (Required): However, pressing "Next" with no data present will allow the Design Assistant to set the feed water analysis to 100 ppm NaCl.
- Flowsheet Configuration (Optional): The default is a single stage or the current configuration set by the user.

TABLE 8.2 Feed stream composition.

Compound or element	mg/L
Calcium	83
Magnesium	25
Sodium	106
Potassium	8
Ammonia	2
Barium	0.05
Strontium	0.1
Fluoride	1
Nitrate	5
Bromide	0
Phosphate	1
Boron	0
Silica	5
Bicarbonate	149.39
Carbonate	0.11
Carbon dioxide	18.82
Total dissolved solids	724.49
Gal/min feed	60
Permeate pH	7
Number of passes	1
Feed bypass	Yes
Feed predosing	Yes
After feed dosing	Yes
Product dosing	Yes
Argo analyzer	Yes
Feed stripping	Yes
Product stripping	Yes
High-pressure pump	Yes
Electrodeionization	Yes
Energy recovery devices	Yes

Courtesy: SUEZ Water Technologies & Solutions (Commercial Engineering Lead, Water Technologies & Solutions. (2020). Zsirai, T. SUEZ Water Technologies & Solutions. Retrieved from https://www.suezwatertechnologies.com/resources/winflows).

- Chemical Dosing (Optional): The default is no chemical treatment.
- Chemical Stripping (Optional): The default is no chemical stripping.
- Flow Rates or Specifying Flow Rates (Required): No defaults are initially set; however, data set from the flowsheet will appear in this dialog as defaults.
- Pump (Required): The user should specify an inlet pressure in the pump to calculate the efficiency and power consumed.
- Array Specification (Required): The user should specify an array layout and at least one element/housing and one housing per bank. The Auto Design function will calculate a default array for single-stage systems.
- Pressing the finish button on the Array Specification Dialog activates the calculation algorithm and presents the user with the results.

As shown in Fig. 8.57, click on the File > Design Assistant > Auto Design option, then the project information input tab will appear.

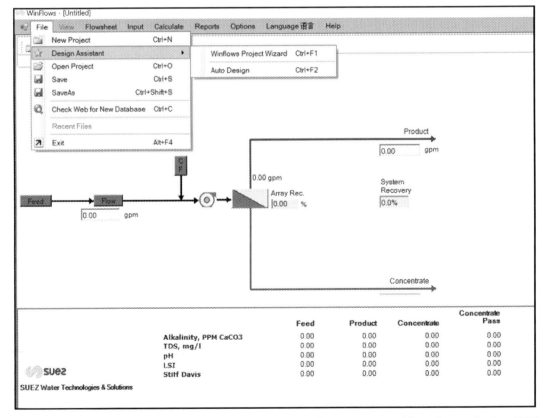

FIGURE 8.57 Starting a project in Winflows Reverse osmosis Design program of Suez. Source: *Courtesy: SUEZ Water Technologies & Solutions (Personal Communication with SUEZ Water Technologies & Solutions team. Retrieved from https://www.suezwatertechnologies.com/resources/winflows).*

8.6.2 Entering the project information

This dialog box allows the user to enter the project information. This information is used for identifying reports generated by Winflows. This dialog box can be opened by selecting the Project Info command on the Input menu. As shown in Fig. 8.58, the project information can be entered in the tab. Then click "NEXT."

8.6.3 Flowsheet configuration

The next step in the design assistant (expert mode) is to specify the features that will control the design: the characteristics or volume of the permeate stream, dosing chemical, stripping, feed bypass, recycles, EDI, and ERDs. As shown in Fig. 8.59, the flowsheet configuration information can be entered. Then click "OK."

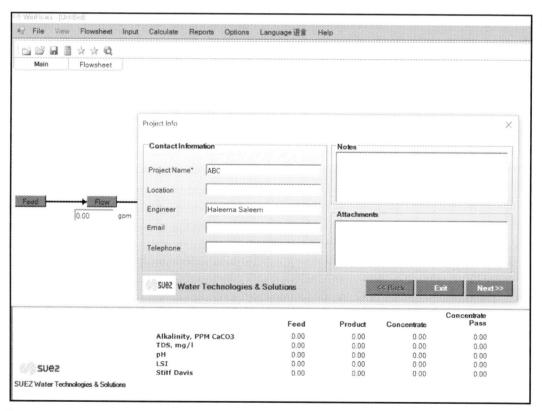

FIGURE 8.58 Entering project information in Winflows Reverse osmosis Design program of Suez. *Source: Courtesy: SUEZ Water Technologies & Solutions (Personal Communication with SUEZ Water Technologies & Solutions team. Retrieved from https://www.suezwatertechnologies.com/resources/winflows).*

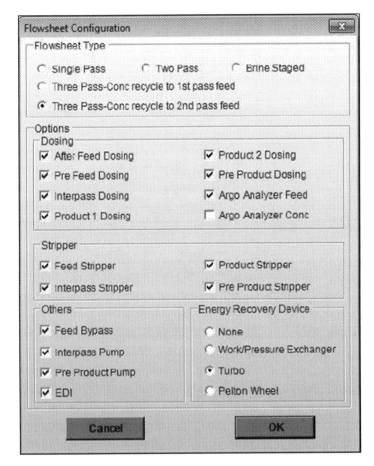

FIGURE 8.59 Entering the flowsheet configuration information in Winflows Reverse osmosis Design program of Suez. Source: *Courtesy: SUEZ Water Technologies & Solutions (Personal Communication with SUEZ Water Technologies & Solutions team. Retrieved from https://www.suezwater-technologies.com/resources/winflows).*

8.6.4 Feed water specification

The next step in the design assistant, expert mode is to provide information to describe the chemical composition of the Feed Stream. This page provides information to describe the chemical composition of the Feed Stream. As shown in Fig. 8.60, the feed water analysis data can be entered in the tab. On this screen, the users have the option of entering the concentration of selected elements or compounds or they can select a typical feed stream from the drop-down list labeled Water Analysis. In this example, a custom composition is entered. To ensure ionic balance, the program will adjust the level of either sodium or chloride. Then click "OK."

8.6.5 Flow rate specification

On this screen, the user can enter the RO recovery, flow rate, dosing, and pH. The flow rate screen is shown in Fig. 8.61.

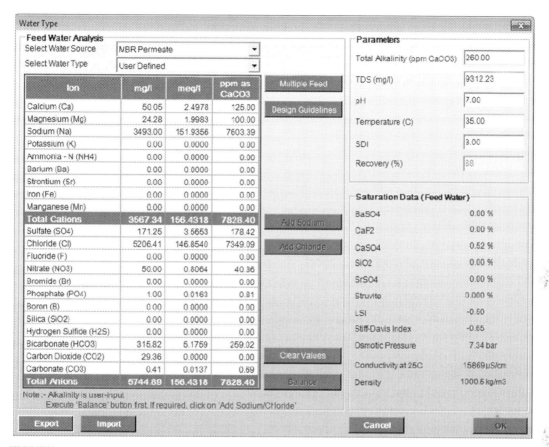

FIGURE 8.60   Feed water analysis page in Winflows Reverse Osmosis Design program of Suez. Source: *Courtesy: SUEZ Water Technologies & Solutions (Personal Communication with SUEZ Water Technologies & Solutions team. Retrieved from https://www.suezwatertechnologies.com/resources/winflows).*

8.6.6 Array specification

The final step in the design assistant is to specify the number of banks, number of housings, and type and number of RO elements in each housing. The design assistant does all of this for the user upon entry to the window. Fig. 8.62 is the array specification tab in Winflows RO Design program of Suez. The design assistant configures the first array only. In this example, the second-pass array was made identical to the first-pass array. The user has the option of modifying the configuration of either array to optimize permeate quality and to minimize system cost. When the user presses Finish, the design is calculated and the results are compiled for review.

8.6.7 Explorer

This window (Fig. 8.63) automatically opens after the program has calculated the results of the selected design. The system explorer presents the user with a summary of the

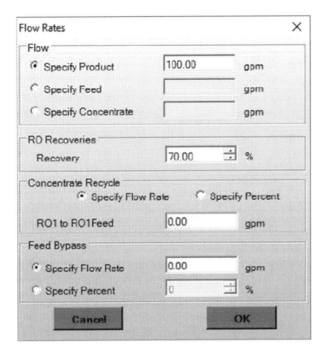

FIGURE 8.61 Flow rate specification tab in Winflows Reverse Osmosis Design program of Suez. Source: *Courtesy: SUEZ Water Technologies & Solutions (Personal Communication with SUEZ Water Technologies & Solutions team. Retrieved from https://www. suezwatertechnologies.com/resources/winflows).*

results and alerts the user to any errors or potential problems with the design. To view the errors, the user can press the F11 key or select Errors from the View menu. The error analysis will give helpful suggestions on how to resolve the problem.

8.6.8 Report

This window is opened by selecting Generate reports on the Reports menu. In the Report Options section, the user can select whichever sections that need to be present in the report. The report contains the following sections on separate sheets.

- Input Data Summary: User can view all the input data are entered.
- Results Summary: User can view the summary of the calculation results.
- Process Data Sheet: User can view all the calculated output on the process side, like flow, Pump, stripper, dosing, and Argo Analyzer.
- Streams Info: User can view all the streams data.
- Element Detail Data: User can view the array data in detail.
- EDI: User can view the EDI input and output data.
- ERD: User can view the ERD's (work exchanger/pressure exchanger, Turbo charger, and Pelton Wheel) output data.
- Capex/Opex: User can view the capital and operational costs in the project.

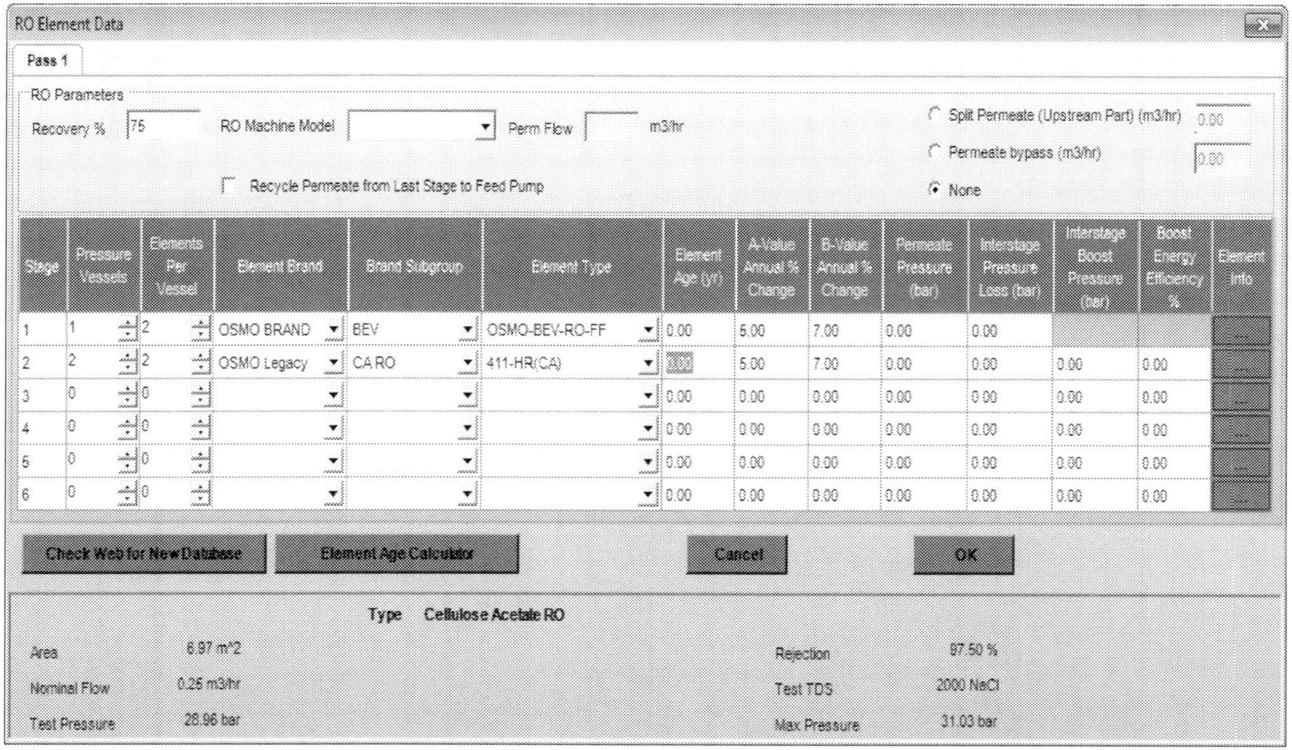

FIGURE 8.62 Array specification tab in Winflows Reverse Osmosis Design program of Suez. Source: Courtesy: SUEZ Water Technologies & Solutions (Personal Communication with SUEZ Water Technologies & Solutions team. Retrieved from https://www.suezwatertechnologies.com/resources/winflows).

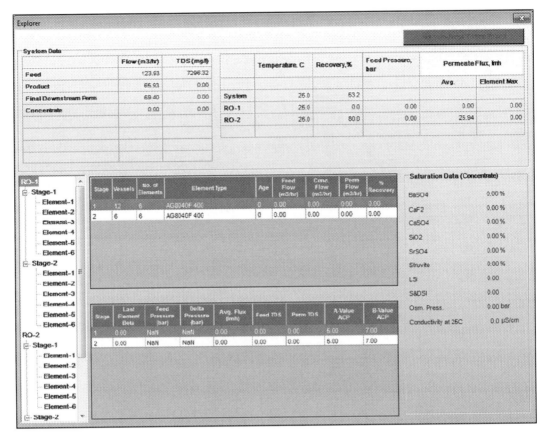

FIGURE 8.63 Explorer tab in Winflows Reverse Osmosis Design program of Suez. Source: *Courtesy: SUEZ Water Technologies & Solutions (Personal Communication with SUEZ Water Technologies & Solutions team. Retrieved from https://www.suezwatertechnologies.com/resources/winflows).*

8.7 ROAM Ver. 2.0 (MICRODYN-NADIR)

RO Application for MICRODYN-NADIR (ROAM) is the new design software for estimating system designs for MICRODYN RO products.

The latest ROAM Ver 2.0 software can be downloaded using the below link:

https://www.microdyn-nadir.com/software-downloads/

Main features:

- Projection software used to estimate system designs using MICRODYN RO and NF products.
  - Models element performance in a one- or two-pass system
  - Estimates the required feed pressures
  - Estimates the permeate quality coming out of the RO/NF system
- Software launched in November 2019.

- The modern interface is simple and user-friendly.
- The program provides easy-to-read reports.
- The program works on operating systems using Windows 7 or later.

Step-by-step instruction for RO system designing using ROAM

8.7.1 ROAM home screen

- Upon opening the program after installation, you will see the program's home screen (Fig. 8.64).
- The picture to the right is what the program's home screen looks like. Users have three options here:
  - Create a new projection.
  - Open an existing projection saved in the user's library.
  - Exit the program.

8.7.2 Raw water analysis

- User inputs the feed water quality (Fig. 8.65).
  - Feed temperature and pH
  - The concentration of individual cations and anions
- Once the user has filled out all of the necessary information, they can click on the "Calculate" button.

FIGURE 8.64 Home screen of ROAM. Source: *Courtesy: MICRODYN-NADIR GmbH (Personal Communication with MICRODYN-NADIR GmbH team. Retrieved from https://www.microdyn-nadir.com/software-downloads/ MICRODYN-NADIR, Goleta, CA, USA).*

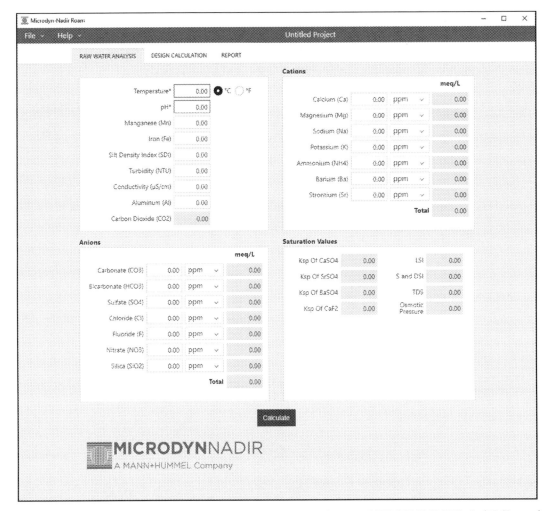

FIGURE 8.65 Raw water analysis Screen of ROAM. Source: *Courtesy: MICRODYN-NADIR GmbH (Personal Communication with MICRODYN-NADIR GmbH team. Retrieved from https://www.microdyn-nadir.com/software-down-loads/MICRODYN-NADIR, Goleta, CA, USA).*

  — If needed, the program will automatically add sodium/chloride to ensure feed water is neutral in charge.
- Users can save this feed water quality in their library for later access.
- The user then proceeds to the "Design Calculation" tab of the program.

8.7.3 Design calculation

- User inputs the system design (Fig. 8.66):
  ∘ Select one or two pass.

* Choose imperial or metric units.
* Enter system specifics (product flow, adjusted pH, recirculate flow, etc.).
* Enter system array (up to four stages).
* Choose between various 4-inch and 8-inch MICRODYN RO & NF products.
* Program will provide preliminary flux, flow, pressure, beta, and concentration calculations when the "Calculate" button is selected.
• Users can save this system design in their library for later access by going to File > Save or Save As.

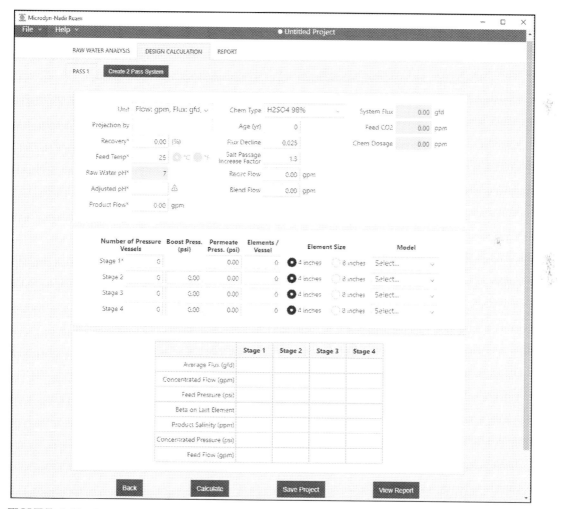

FIGURE 8.66 Design calculation screen of ROAM. Source: *Courtesy: MICRODYN-NADIR GmbH (Personal Communication with MICRODYN-NADIR GmbH team. Retrieved from https://www.microdyn-nadir.com/software-downloads/MICRODYN-NADIR, Goleta, CA, USA).*

8.7.4 Report

- User can view the full projection report (Fig. 8.67) within the program by clicking on "View Report" on the Design Calculation page.
- User make any changes in the previous "Raw Water Analysis" and "Design Calculation" tabs prior to downloading a complete report.
- User can download this easy-to-read report as a PDF by clicking on the icon on the upper, right-hand corner (as depicted by the arrow) of the report.
- Report includes the following, as shown in Figs. 8.68—8.70:
  - Estimated adjusted feed water, concentrate, and permeate qualities
  - Estimated feed and concentrate pressures
  - Summarizes system design
  - Includes individual element information
  - Water saturation indexes
  - Any design warnings and recommendations

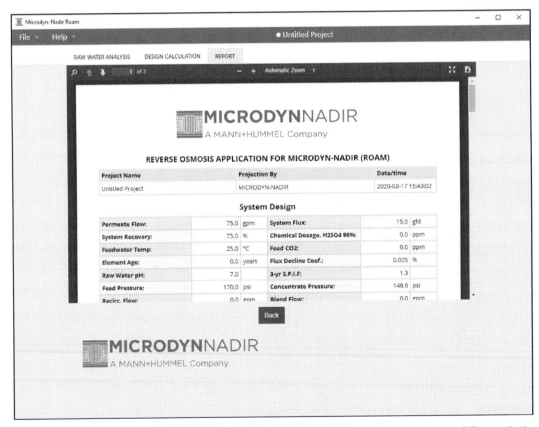

FIGURE 8.67 Report screen of ROAM. Source: *Courtesy: MICRODYN-NADIR GmbH (Personal Communication with MICRODYN-NADIR GmbH team. Retrieved from https://www.microdyn-nadir.com/software-downloads/ MICRODYN-NADIR, Goleta, CA, USA).*

REVERSE OSMOSIS APPLICATION FOR MICRODYN-NADIR (ROAM)

Project Name	Projection By	Date/time
Untitled Project	MICRODYN-NADIR	2020-08-17 15:43:02

System Design

Permeate Flow:	75.0	gpm	System Flux:	15.0	gfd
System Recovery:	75.0	%	Chemical Dosage, H2SO4 98%:	0.0	ppm
Feedwater Temp:	25.0	°C	Feed CO2:	0.0	ppm
Element Age:	0.0	years	Flux Decline Coef.:	0.025	%
Raw Water pH:	7.0		3-yr S.P.I.F:	1.3	
Feed Pressure:	170.0	psi	Concentrate Pressure:	149.6	psi
Recirc. Flow:	0.0	gpm	Blend Flow:	0.0	gpm

Water Quality

	Raw Water		Adjusted Feed Water		Concentrate		Permeate	
Ion	mg/l	CaCO3	mg/l	CaCO3	mg/l	CaCO3	mg/l	CaCO3
Calcium (Ca)	0.00	0.00	0.00	0.00	0.00	0.00	0.00	0.00
Magnesium (Mg)	0.00	0.00	0.00	0.00	0.00	0.00	0.00	0.00
Sodium (Na)	790.00	1717.39	790.00	1717.39	3154.61	6857.84	14.45	31.41
Potassium (K)	0.00	0.00	0.00	0.00	0.00	0.00	0.00	0.00
Ammonium (NH4)	0.00	0.00	0.00	0.00	0.00	0.00	0.00	0.00
Barium (Ba)	0.00	0.00	0.00	0.00	0.00	0.00	0.00	0.00
Strontium (Sr)	0.00	0.00	0.00	0.00	0.00	0.00	0.00	0.00
Carbonate (CO3)	0.00	0.00	0.00	0.00	0.00	0.00	0.00	0.00
Bicarbonate (HCO3)	0.00	0.00	0.00	0.00	0.00	0.00	0.00	0.00
Sulfate (SO4)	0.00	0.00	0.00	0.00	0.00	0.00	0.00	0.00
Chloride (Cl)	1219.88	1718.14	1219.88	1718.14	4871.31	6861.00	22.27	31.37
Fluoride (F)	0.00	0.00	0.00	0.00	0.00	0.00	0.00	0.00
Nitrate (NO3)	0.00	0.00	0.00	0.00	0.00	0.00	0.00	0.00
Silica (SiO2)	0.00	0.00	0.00	0.00	0.00	0.00	0.00	0.00
TDS	2009.88	3435.53	2009.88	3435.53	8025.92	13718.84	36.72	62.78
pH	7.00		7.00		7.00		7.00	

FIGURE 8.68 System design and water quality details in the ROAM report. Source: *Courtesy: MICRODYN-NADIR GmbH (Personal Communication with MICRODYN-NADIR GmbH team. Retrieved from https://www.microdyn-nadir.com/software-downloads/MICRODYN-NADIR, Goleta, CA, USA).*

Stage Information

Stage	Array	No of Elements	Element	Feed Flow (gpm)	Conc. Flow (gpm)	Pressure Feed (psi)	Pressure Conc. (psi)	Pressure Boost (psi)	Pressure Permeate (psi)	Beta	Avg. Flux (gfd)
1	2x6	12	8040-BW-400	100.0	44.4	170.0	150.4	-	0.00	1.16	16.7
2	1x6	6	8040-BW-400	44.4	25.0	170.4	149.5	20.00	0.00	1.08	11.7

Individual Elements

Stage	Element No.	Permeate Flow (gpm)	TDS	Beta	Flux (gfd)	Net Driving Pressure (psi)
1	1	5.38	15.40	1.11	19.36	143.1
1	2	5.09	18.27	1.11	18.32	135.5
1	3	4.81	21.90	1.12	17.31	128.0
1	4	4.52	26.59	1.13	16.28	120.4
1	5	4.22	32.88	1.14	15.19	112.3
1	6	3.89	41.67	1.16	13.99	103.4
2	1	4.16	44.34	1.09	14.99	110.9
2	2	3.80	53.70	1.09	13.68	101.2
2	3	3.44	65.54	1.09	12.38	91.5
2	4	3.07	80.99	1.09	11.05	81.8
2	5	2.70	101.51	1.09	9.71	71.8
2	6	2.33	128.88	1.08	8.38	62.0

FIGURE 8.69 Stage Information and Individual Elements details in the ROAM report. Source: *Courtesy: MICRODYN-NADIR GmbH (Personal Communication with MICRODYN-NADIR GmbH team. Retrieved from https:// www.microdyn-nadir.com/software-downloads/MICRODYN-NADIR, Goleta, CA, USA).*

Water Saturation Indexes

	Raw Water	Feed Water	Concentrate
Langelier Saturation Index	0.00	0.00	0.00
S and DSI Index	0.00	0.00	0.00
CaSO4/Ksp*100%	0.00	0.00	0.00
SrSO4/Ksp*100%	0.00	0.00	0.00
BaSO4/Ksp*100%	0.00	0.00	0.00
CaF2/Ksp*100%	0.00	0.00	0.00
SiO2 Saturation	0.00	0.00	0.00
Osmotic Pressure, psi	23.00	23.00	92.38

*** Calculations are accurate within +/- 10% ***

*** This projection is not to be used for warranty purposes ***

Contact

Europe	Americas	Asia
Germany: +49 611 962 6001	USA: +1 805 964 8003	China: +86 592 677 5500
Italy: +39 0721 1796201	Brazil: +55 11 3378 7500	Singapore: +65 6457 7533
info@microdyn-nadir.com	info@microdyn-nadir.com	infochina@microdyn-nadir.cn

MICRODYNNADIR
A MANN+HUMMEL Company

FIGURE 8.70 Water Saturation Indexes details in the ROAM report. Source: *Courtesy: MICRODYN-NADIR GmbH (Personal Communication with MICRODYN-NADIR GmbH team. Retrieved from https://www.microdyn-nadir. com/software-downloads/MICRODYN-NADIR, Goleta, CA, USA).*

8.8 CSMPRO 6.0 (Toray Advanced Materials Korea)

CSMPRO 6.0 is the latest version of CSM RO projection software. Microsoft's.NET technology has been used to develop program tools to make complicated computations easy for users. In CSMPRO 6.0, the salt passage rate prediction was improved under low- and high-concentration conditions, and the salt passage rate of some ions was improved to be accurately predicted based on a laboratory test and field experience (Design Software).

Features of CSMPRO6.0

— Improves accuracy of membrane performance prediction
— Improves accuracy of salt passage rate prediction of low/high concentrations
— Improves accuracy of salt passage rate prediction according to pH
— Differential pressure by the thickness of the feed spacer is reflected

The latest CSMPRO 6.0 software can be downloaded using the below link:
http://www.csmfilter.com/csm/03result/Software.asp

A pilot-scale study was carried out by Park et al. Park and Kwon (2018) to analyze the long-term stability of low-pressure RO membrane operation. The average flux variation of the individual RO elements in a vessel and the economic feasibility of low-pressure RO were also evaluated through a simulation study using a reverse osmosis system design software CSMPro 5.0. The simulation study predicted that the average operating pressure difference ratio (%) between two RO membranes decreased from 24.4% to 17.8% and a substantial quantity of low-pressure RO elements (83.3%) must be replaced to meet the designated water criteria only after 2 years' operation.

8.9 IPSE (Integrated Process Simulation Environment) software—(SimTech Simulation Technology)

IPSEpro, from SimTech Simulation Technology, is a software system comprising a set of software modules for creating process models for an extensive range of applications and for utilizing these models all the way through the lifecycle of process plants. Ready-to-use solutions are accessible for a wide range of applications such as desalination, thermal power, and geothermal energy. IPSEpro covers the complete lifecycle of a process plant from conceptual design to plant operation (*SimTech Simulation technology*. 2020).

The latest IPSE software can be downloaded using the below link:
http://simtechnology.com/CMS/index.php/ipsepro

8.10 Case study

In order to effectively optimize the design of RO seawater desalination system, it is important to consider the system operation not just at the design condition but also at variable conditions differing from the design conditions. As an example, a membrane system should be designed in such a way that it should be able to generate the necessary water quantity with the worst combination of cold, high-salt concentration water and fouled/aged membranes.

The study by Verhuelsdonk et al. (2010) discusses the IPSE software, which allows all of the appropriate process units to be united into a single simulation environment, thus permitting simple as well as rapid simulation. The IPSE software package allows the designer to define simulation equations for the various process modules, and for connecting those modules with a graphical interface. This software is considered to be open-ended, in such a way that the designer can define any equations behind any specific graphical symbol. In this study, the system performance at different operating conditions was analyzed, and also the effect of overflush on system performance was examined. Table 8.3 presents the different parameters used in this study.

Two distinct seawater conditions were simulated, demonstrating the extremes of the anticipated operating range. These include cold (20°C temperature), high salt concentration (41,891 ppm) conditions and warm (35°C temperature), low salt concentration (37,903 ppm) conditions. Both of these conditions were simulated with 5-year-old membranes and with new membranes, due to the fact that ageing influences the membrane salt passage and permeability.

The system is assessed at each condition with either a variable speed HPP or a flow control valve. In each case, the designer defines the number of PVs, the feed water quality, flow rate, the product water flow rate and pressure, number of PX220 units, and the overflush ratio. With the variable speed pump, the control valve position is set to 100%. With the control valve, the pump speed is fixed at 60 Hz, and the design software automatically computes the control valve position, the energy recovery device feed pressure, the membrane feed pressure, the brine reject pressure, the booster pump speed, and the quality of product water. Furthermore, the specific energy demand, considering the booster pump and the feed pump, is computed. Table 8.4 presents the performance depending on salinity and feed water temperature. It can be noted that in the worst case with high salinity and low temperature, the power demand with a variable frequency drive is greater than with a control valve. This effect is due to the 97% variable frequency drive efficiency. From Table 8.4, the user could rapidly appreciate the needed operating range of the valves and pumps.

This design software permits the user to analyze system optimization potentials, which could be extremely hard to examine with conventional simulation tools. For evaluating the impact of

TABLE 8.3 Seawater reverse osmosis train details.

Parameters	Values
Overall recovery	40%
Production of product water	12,000 m^3/d
Type of membrane element	Hydranautics SWC4 +
Number of elements per vessel	7
Seawater feed pressure	2.0 bar
Number of pressure vessels	154

Reproduced from Verhuelsdonk, M., Attenborough, T., Lex, O., & Altmann, T. (2010). Design and optimization of seawater reverse osmosis desalination plants using special simulation software. Desalination, 250(2), 729–733.

TABLE 8.4 Performance depending on salinity and feed water temperature.

Temperature	°C	20		35	
Feed TDS	ppm	41,891		37,903	
Membrane age	years	0	5	0	5
Valve position (open)	%	31.67	51.38	28.38	30.13
HPP power (valve)	kWh	1160	1161	1153	1153
HPP power (variable-frequency drive)	kWh	1068	1186	945	1012
Membrane pressure	bar	60.64	67.08	53.58	57.32
HP brine pressure	bar	59.35	65.77	52.32	56.03
LP brine pressure	bar	1.23	1.23	1.23	1.23
Booster pump speed	rpm	808.8	812.3	806.1	810.4
Booster pump power	kWh	62.01	62.62	61.04	61.82
Permeate TDS	mg/L	188.6	278.8	271.4	399.0
Spec. energy (HPP valve)	kWh/m^3	2.444	2.447	2.427	2.430
Spec. energy (HPP VFD)	kWh/m^3	2.261	2.497	2.011	2.147

Reproduced from Verhuelsdonk, M., Attenborough, T., Lex, O., & Altmann, T. (2010). Design and optimization of seawater reverse osmosis desalination plants using special simulation software. Desalination, 250(2), 729–733.

TABLE 8.5 Overflush effect on performance.

Overflush (%)		5.0	0.0	5.0
Mixing (%)		7.1	6.0	4.8
Flow (m^3/d)	ERI high-pressure feed	18,610	17,733	16,848
	Reverse osmosis feed	30,784	29,901	29,017
Salinity (mg/kg)	Reverse osmosis feed	41,939	41,766	41,579
	Permeate	186.8	188.9	9191.2
Recovery (%)	Recovery (%)	38.98	40.13	41.35

Reproduced from Verhuelsdonk, M., Attenborough, T., Lex, O., & Altmann, T. (2010). Design and optimization of seawater reverse osmosis desalination plants using special simulation software. Desalination, 250(2), 729–733.

variable overflush, the net seawater feed flow and the permeate flow are fixed, so that there is no effect on the pretreatment system size. As the overflush is increased, only less feed water will be available to feed the membranes, and it is hence essential to increase the membrane recovery (computed by the design software). Consequently, there is an increase in membrane feed pressure, which is in contrast to the decrease in membrane feed pressure arising from the poor mixing, and reduced membrane feed salt concentration.

Table 8.5 presents the overflush impact on flow and salt concentration with 20°C, high salt concentration, and new membranes. It was noted that the overflush has almost no effect on the product water salinity. With no overflush, the recovery was noted to be 40.13%. In the event that the overflush is increased to 5.0%, the feed flow at the energy recovery device decreases to about 16,848 m^3/d, and to obtain the required product flow of 12,000 m^3/day, the membrane recovery should be increased to 41.35%. The optimization is carried out both with a variable speed high pressure pump and with a control valve. Considering the specific energy demand with variable frequency drive-controlled HPP, it was noted that using a variable frequency drive, there is a minimal energy requirement at almost 3.0% overflush. This is due to that fact that the booster pump requires less power with more overflush due to the low flow rate, while the HPP requires more power. With increasing overflush, the membrane recovery as well as the feed pressure increases. Considering the specific energy demand with valve-controlled HPP, it was noted that there is no optimum overflush, but the increasing overflush decreases energy usage.

Thus, this case study demonstrated that the integrated process model could be employed for different user-defined analyses. Calculations could be performed extremely fast, even for cases that would need an iterative method if conventional approaches were used.

Reference

Agarwal, M., Singh, K., Dohare, R. K., & Upadhyaya, S. (2016). Process control and optimization of wastewater treatment plants using simulation softwares: A review. *International Journal of Advanced Technology and Engineering Exploration, 3*(22), 145.

Aghababaei, N. (2017). Reverse osmosis design with IMS Design software to produce drinking water in Bandar Abbas, Iran. *Journal of Applied Research in Water and Wastewater, 4*(1), 314–318.

Altaee, A. (2013). Theoretical study on feed water designs to reverse osmosis pressure vessel. *Desalination, 326,* 1–9.

APAC Communications, DuPont Water Solutions. (2020). Wang, M.C. personal communication. Retrieved from https://www.dupont.com/water/resources/design-software.html.

Commercial Engineering Lead, Water Technologies & Solutions. (2020). Zsirai, T. personal communication. SUEZ Water Technologies & Solutions. Retrieved from https://www.suezwatertechnologies.com/resources/winflows.

CSMPRO 6.0. (2020). Design Software. Retrieved from http://www.csmfilter.com/csm/03result/Software.asp.

DuPont Water Solution's. (2020). FilmTec™ RO technical manual; Version 3; Form No. 45-D01504-en, Rev. 3; April 2020.

Gaublomme, D., Strubbe, L., Vanoppen, M., Torfs, E., Mortier, S., Cornelissen, E., ... Nopens, I. (2020). A generic reverse osmosis model for full-scale operation. *Desalination, 490,* 114509.

Haryati, S., Hamzah, A. B., Goh, P. S., Abdullah, M. S., Ismail, A. F., & Bustan, M. D. (2017). Process intensification of seawater reverse osmosis through enhanced train capacity and module size–Simulation on Lanzarote IV SWRO plant. *Desalination, 408,* 92–101.

Hydranautics—A Nitto Group Company. (2020). Shah, M.J. *Global Marketing & Product Manager,* personal communication. Retrieved from https://membranes.com/solutions/software/.

Jiang, A., Ding, Q., Wang, J., Jiangzhou, S., Cheng, W., & Xing, C. (2014). Mathematical modeling and simulation of SWRO process based on simultaneous method. *Journal of Applied Mathematics, 2014,* 908569.

LANXESS Deutschland GmbH, Business Unit Liquid Purification Technologies. (2020). Lipnizki, J. *Head of Technical Marketing Membrane,* personal communication. https://lpt.lanxess.com/lewaplus-software/.

Mehta, K. P., & Patel, A. S. (2014). Reverse osmosis design with hydranautics design software for industrial waste water reuse. *Journal of Environmental Research and Development, 9*(2), 421–430.

Membrane Projection Software. (2020). Filtration + Separation, 57(2), 8. doi:10.1016/s0015–1882(20)30056–2.

MICRODYN-NADIR GmbH. (2020). Parsley, H. *personal communication, September 22, 2020.* Retrieved from https://www.microdyn-nadir.com/software-downloads/MICRODYN-NADIR, Goleta, CA, USA.

Park, H. G., & Kwon, Y. N. (2018). Long-term stability of low-pressure reverse osmosis (RO) membrane operation—A pilot scale study. *Water, 10*(2), 93.

RO Design and Design Software. (2015). *Reverse osmosis: Industrial processes and applications* (2nd (Ed.), pp. 255–282). Jane Kucera. Available from http://doi.org/10.1002/9781119145776.ch10.

SimTech Simulation technology. (2020). IPSEpro. Retrieved from http://www.simtechnology.com/CMS/index.php/ipsepro. (Accessed 17 September 2020).

Talaeipour, M., Nouri, J., Hassani, A. H., & Mahvi, A. H. (2017). An investigation of desalination by nanofiltration, reverse osmosis and integrated (hybrid NF/RO) membranes employed in brackish water treatment. *Journal of Environmental Health Science and Engineering, 15*(1), 18.

Toray Industries, Inc. (2020). DS2 Guideline. Retrieved from https://ap3.toray.co.jp/toraywater/.

Verhuelsdonk, M., Attenborough, T., Lex, O., & Altmann, T. (2010). Design and optimization of seawater reverse osmosis desalination plants using special simulation software. *Desalination, 250*(2), 729–733.

CHAPTER

9

Performance Monitoring, Process Parameters, and Their Control

9.1 System operation

For the effective operation and maintenance of the system, it should have a properly planned pretreatment and an initial plant start-up procedure. The scaling and fouling of membrane can decrease the product water flow rate and salt rejection, and these membrane issues are caused mainly by the poor pretreatment and inappropriate operation (Jiang et al., 2017; Saleem & Zaidi, 2020c). Mechanical damage along with chemical damage of the reverse osmosis system comprising the membranes might also occur from wrong operations. The different reverse osmosis system components are discussed in Chapter 2 in detail.

9.1.1 Initial start-up

Before beginning the procedure for system start-up, the pretreatment verification, membrane elements loading, instrument calibration, and other system inspection must be completed.

9.1.1.1 Equipment

The initial start-up of the system is usually carried out immediately after the loading of membrane elements. For a start-up, some additional equipment are suggested, and this must also be part of the equipment at the site. These include (i) silt density index (SDI) measuring device, (ii) thermometer, (iii) chemicals for sanitization, cleaning, as well as preservation, (iv) safety glasses when working with chemicals, (v) spare elements, (vi) pH meter, (vii) scale to weigh one element, (viii) bottles for water samples, (ix) conductivity meter, (x) single-element test stand, and (xi) analysis equipment for color, total organic carbon (TOC), redox potential, free chlorine, silica, iron, sulfate, chloride, alkalinity, calcium, and total hardness.

9.1.1.2 Prestart-up examination and commissioning audit

After loading the membrane elements into the pressure vessels (PVs) and before the membrane unit start-up, the operator has to ensure that the complete pretreatment unit is operating as per the specifications. If the pretreatment involved changing of the chemical characteristics of the source water, then a complete examination of the water reaching the membrane unit should be performed. Further, the silt density index and turbidity should be found, and the absence of chlorine must be confirmed. The source water intake should be stable in relation to (i) pH, (ii) temperature, (iii) turbidity, (iv) SDI, (v) flow, (vi) bacteria, and (vii) conductivity.

9.1.1.3 Sequence of start-up

Proper start-up of RO water treatment units is important for preparing the membranes for effective service and for preventing the damage of the membrane because of extreme pressure/flow or hydraulic shock. The correct start-up sequence also aids to make sure that the system operational parameters conform to design conditions such that the desired water quality, as well as production targets, could be accomplished. Initial system performance measurement is considered to be a critical aspect of the start-up procedure. Documented results of this assessment can be used as a benchmark, against which the ongoing system operational performance could be measured.

DuPont Water Solutions recommends the below start-up sequence of reverse osmosis system (DuPont Water Solution, 2020):

1. Prior to the initialization of the start-up sequence, carefully rinse the pretreatment section for flushing out debris and other contaminants without letting the feed reach the membrane element.
2. Check the entire valves for ensuring that the settings are accurate. The feed pressure control, as well as concentrate control valves, must be entirely open.
3. Use low-pressure water at a lower flow rate for flushing the air out of the membrane elements and PVs. The flushing must be carried out at 30−60 psi gauge pressure. The complete product water and reject flows must be directed to a proper waste collection drain in the course of flushing.
4. At the time of flushing process, the operator has to check entire pipe connections and valves for any leaks. The connections should be tightened where essential.
5. Subsequent to the system flushing for a minimum of 30 minutes, the feed pressure control valve should be closed.
6. The operator should make sure that the reject control valve is open.
7. Gradually crack open the feed pressure control valve (feed pressure must be below 60 psi).
8. The high-pressure pump can be started now.
9. Gradually open the feed pressure control valve to increase the feed pressure and feed flow rate to the membrane elements till the design reject flow is attained. The feed pressure increase to the elements must be lesser than 10 psi/sec for achieving an easy start.
10. Gently close the reject control valve untill the ratio of product water flow to reject flow approaches, however do not go beyond the design ratio (recovery). The system

pressure must be continuously checked for ensuring that it does not exceed the design limit.

11. Steps 9 and 10 must be repeated till the design permeate and reject flows have been attained.
12. The system recovery can be calculated and compared to the design value of the system.
13. The addition of pretreatment chemicals should be checked. Then, calculate the feed water pH.
14. The Stiff & Davis Stability Index (S&DSI) or the Langelier Saturation Index (LSI) of the concentrate should be checked by measuring conductivity, pH, alkalinity, and calcium hardness levels and then essential calculations should be carried out.
15. The operator should permit the system to operate for 1 hour.
16. Then the initial readings of all operational parameters should be taken.
17. The permeate conductivity from each PV should be checked for verifying that all vessels conform to the expected performance.
18. After one to two days of operation the recorded plant operational data such as feed pressure, temperature, differential pressure, conductivity, recovery, and flows readings must be reviewed. Simultaneously, draw samples of feed water, product water, and reject water for composition analysis.
19. The system performance should be compared to the design values.
20. Check the right operation of mechanical and instrumental safety equipment.
21. Then, switch the product water flow from the drain to the standard service position.
22. System can be locked into automatic operation.
23. Utilize the primary system performance data obtained in Steps 16 through 18 as a reference for assessing forthcoming system performance. The system performance should be monitored frequently during the initial week of operation for checking its correct performance in the course of this significant early stage.

9.1.1.4 Start-up performance and stabilization of membrane

The start-up performance of a reverse osmosis membrane unit and the time needed for attaining the stabilized performance will depend on the membrane's prior storage conditions. If properly stored, the dry membranes and the wet preserved membranes attain similar stabilized performance after a few hours or some days of operation. The wet membrane flow performance is usually stable from the very beginning, whereas the dry membranes will start at a little higher flow.

9.1.1.5 Special systems—double-pass reverse osmosis

During the operation of a double-pass system, the first-pass system should have been operational for at least one day before the product water of the first pass is fed to the second-pass membrane. Else, a permanent flux decline of the second pass might develop. The pH of feed water to the two passes must be adjusted for the best results in salt rejection. A final permeate conductivity of less than 1 $\mu S/cm$ will be attained normally from brackish water (BW) with two-pass brackish water reverse osmosis (BWRO) membrane systems.

9.1.2 Start-up of operation

As soon as the membrane system is started, preferably it must be kept operating at constant conditions. In practice, membrane facilities must be shut down and restarted repeatedly. These start/stop cycles will cause pressure change along with flow change, resulting in membrane element mechanical stress. Hence, the start/stop occurrence must be minimalized, and the normal operating start-up sequence must be as smooth as possible. Ideally, the same sequence is suggested for the initial start-up. Most significant is a gradual feed pressure rise, specifically for seawater reverse osmosis (SWRO) facilities. The usual start-up sequence can be normally automated by the utilization of programmable controllers as well as remotely operated valves. The instrument calibration, the working of safety devices and alarms, prevention of corrosion, and a leak-free operation must be tested and maintained regularly.

9.1.3 Shutdown of reverse osmosis system

A reverse osmosis system is designed to function on a continuous basis. On the other hand, in practice, the RO membrane systems will be started up and shut down quite frequently. During the shutdown phase of the membrane system, the system should be flushed, preferably with product water or otherwise with superior quality feed water, for removing the higher salt concentration from the PVs till the conductivity of reject matches the conductivity of feed water. Flushing can be performed at lower pressure (approximately 40 psi). A higher feed flow rate is at times advantageous for a cleaning effect; nevertheless, the maximal pressure drop per membrane element and per multielement PV should not be exceeded. At the time of low-pressure flushing, the PVs of the final stage of a reject-staged system are generally exposed to the highest feed flow rate and consequently, they demonstrate the greatest pressure drop.

The water utilized for flushing must not contain pretreatment chemicals; particularly scale inhibitors should not be present. Hence, any injection of chemicals, if used, must be stopped prior to flushing. After the system flushing, the feed valves must be closed fully. If the reject line ends into a drain below the level of the PVs, then an air break must be used in the reject line at a location higher than the highest PV.

When the high-pressure pump (HPP) is turned off, and the feed/reject side had not been rinsed with product water, a temporary product water reverse flow would happen due to natural osmosis phenomena. This reverse flow is at times known as product water draw-back or suck-back. This product water draw-back alone or along with a feed-side flush might contribute to an advantageous cleaning effect. In the event that the product water line is pressurized at the time of operation and the system will shut down, then the membrane may be exposed to a static permeate backpressure. For avoiding this backpressure membrane damage, the static permeate backpressure should not go beyond 5 psi anytime.

If the system has to be shut down for more than two days, the operator should ensure that:

- The RO system is sufficiently protected against microbiological growth. Hence, consistent flushing should be done every 24 hours.
- The membrane elements must not dry out. Dry membrane elements would lead to irreversible flux loss.
- Whenever applicable, the RO system must be shielded against temperature extremes.

The RO membrane facility could be stopped for 24 hours with no preservations and precautions for microbial fouling. If feed water for flushing every 24 hours is not accessible, chemical preservation is essential for lengthier stops than 48 hours.

9.2 Instrumentation and controls

Instrumentation and controls could be as complex as a supervisory control and data acquisition (SCADA) system or as simple as a manual control having automated shutdown features for pumps as well as membrane safety.

9.2.1 Supervisory control and data acquisition system

SCADA is a system of software and hardware elements that permits the industry to:

- Control industrial processes at remote locations or locally
- Monitor, collect, and process real-time information
- Directly interact with devices like motors, pumps, valves, sensors, and more through human−machine interface (HMI) software
- Record events into a log file.

The SCADA system of a plant consist of an information network and a control network. At the very least, the information network comprises servers/workstations, personal computers, hubs, switches, printers, etc. situated at the central control room. SCADA systems are very important for industries as they help to maintain system efficiency. The control network comprises remote input/output (RIO) panels, programmable logic controllers (PLCs), serial data cables or fiber optic cables, and execution components related to important processes and equipment like variable frequency drives (VFDs) and motorized valves.

Programmable logic controllers: Completely automated desalination facility consists of PLC control panels for main plant components provided with a redundant programmable logic controller as well as an operator interface.

Human−machine interface: HMI comprises incident recording, data presentation, trend functions, alarm functions, graphic control screens, etc. for monitoring and controlling the whole plant.

Local control panels: For large-sized and medium-sized plants, local input/output control panels can be used. These panels are arranged next to all main facilities, particularly in remote sites, which are quite far from the main control panel.

9.2.2 Plant monitoring and control system

The control system of a plant will have manual control provision and automated control provision. At the very least, manual/automatic controls for the below systems will be usually provided:

- The whole plant: normal start-up, emergency shutdown, and normal shutdown
- Pump station for the intake of source water
- Pretreatment process
- HPP for membrane feed
- Reverse osmosis membrane racks
- System for the recovery of energy
- Systems for clean-in-place (CIP) and flush
- System for the spent cleaning solution
- Transfer pumps for product water
- Degasifier and scrubber system
- Chemical feed system
- Reject management system.

Instrumentation: The basic instrumentation needed for monitoring and controlling any reverse osmosis system comprises control valves and devices for the measurement of temperatures, pH, conductivity, pressure, flow, and liquid levels. The location of the instrument is chosen in a manner that the instruments are subjected to substantial vibrations and excessive turbulence. Instruments must be available without any difficulty because they need regular calibrations as well as repairs.

1. Rotameters: Rotameters are simple industrial flowmeters and appropriate just for small package facilities. These are suitable and affordable equipment for indication of chemical feed flows of small volume. Even though rotameters could be provided with a flow signal transmitter, this type of configuration is very rare. The precision of the rotameter is extremely sensitive to the measured liquid density and viscosity and to the particulate content in the measured stream.
2. Magnetic flowmeters: These are well-known flow measurement devices used in larger capacity membrane facilities. These flowmeters devices are employed in the majority water streams available in reverse osmosis facilities, with the probable exception of certain product water streams. Lower conductivity water will not be providing a correct flow reading, and hence these flowmeters should be wisely chosen. Vortex shedding meters could be employed for lower conductivity applications.
3. Conductivity analyzers: These types of analyzers can be used to monitor the quality of the raw water as well as the product water. The conductivity sensors are inline sensor units having a local indication and a transmitter for accurate continuous remote monitoring, indicating, and recording. Even though conductivity meters are typically set up online, the valved sample points to measure salinity/conductivity employing portable device must also be supplied at main positions, like the feed to the reverse osmosis membrane system, the source water intake pump station, the reject discharge, and the permeate lines from the different reverse osmosis trains.

4. Electronic pressure transmitters: These are microprocessor-controlled pressure measurement devices and could give reading precision of 0.1% of their span, which will be very crucial while measuring differential pressures in certain locations at a reverse osmosis facility.
5. Liquid-level sensors: A liquid-level sensor is an instrument to measure the liquid height and converting it to an electrical signal, which is sent to other instrumentation for displaying, monitoring, or controlling the liquid level. The operational parameters of these type of sensors are similar to those used in standard water treatment facilities. Typically, these sensors are embedded in corrosion-resistant housing. The liquid-level sensors are equipped with automated air temperature as well as density compensation.
6. pH and temperature analyzers: In reverse osmosis desalination plants, online electronic temperature and pH analyzers and transmitters are commonly used. The online temperature analyzers should be fixed on the feed line to the reverse osmosis unit, if the temperature of the water is anticipated to change remarkably. A pH analyzer is usually set-up on the permeate line.

In a study performed by Feliu-Batlle et al. (2017), a fractional-order robust controller for controlling a reverse osmosis desalination plant in the south of Spain was proposed. The plant dynamics were experimentally identified. A multivariable mathematical model was obtained, which demonstrated large plant parameter variations. This suggested control system consisted of a diagonal compensator of the less variable dynamics, a decoupler, and a fractional-order proportional integral controller in each of the two loops. The fractional orders of these last controllers have been optimized for improving the phase margin robustness.

The study by Sobana and Panda (2014) implemented model-based controls for the simulation of the RO desalination process for both regulatory and servo issues. Pressure from HPP and recycle ratio are manipulated inputs to the reverse osmosis process, whereas product water flow rate, pH, and concentration are regarded as output. Control configuration for this 2×3 process is selected employing regular interaction analysis approaches. Two control strategies, namely, multivariable proportional-integral controller with decoupler under the decentralized scheme and multiloop internal model control—proportional-integral controller under the centralized scheme, were adopted. The non-square model for synthesizing the model-based control purpose is formulated using the mass balance continuity equation and thus model parameters are assessed. In the two cases, satisfactory results were obtained by reducing the fluctuations in the controlled parameters in the closed-loop control systems. The two controller performances were compared using the integral of absolute error performance index criteria. Results with model predictive control are promising for its application in the desalination process.

9.3 System performance calculation

9.3.1 Calculation of membrane system performance—manual method

The most commonly used method for projecting the reverse osmosis system performance is to calculate product water flow in accordance with the net driving pressure

(NDP) model as well as base calculations of product water salt concentration on salinity gradient between permeate and feed as a driving force of the salt transportation. Reference conditions will be the nominal element performance, defined by membrane manufactures, as tested at standard testing conditions. For basic system configuration, the single point calculations could be carried out manually as presented below. On the other hand, repeated calculations needed for process design optimization are performed employing the computer programs obtained from all main manufacturers of the membrane.

Following are the steps for manual calculation process (Wilf & Awerbuch, 2007):

1. Select membrane element type and the system average permeate flux (APF) value in accordance with the feed water source type.
2. The specific permeability of the selected membrane element is calculated employing the nominal test conditions as well as the nominal element performance.
3. The required net driving pressure can be calculated using the values of system average permeate flux and specific permeability.

   Specific flux, or specific permeability, describes the membrane material with respect to the rate of water flux driven by the applied NDP gradient.

$$\text{Specific permeability} = \frac{\text{Average permeate flux}}{\text{Net driving pressure}} \qquad (9.1)$$

4. Select the system recovery ratio on the basis of the type of feed water, project specifications, or composition of feed water, and calculate the average feed water salinity.

$$\text{Average feed water salinity} = 0.5 \times C_f\left(1 + \frac{1}{(1-r)}\right) \qquad (9.2)$$

   in which, C_f is feed concentration and r is recovery rate in decimal fraction.

5. The corresponding average feed osmotic pressure can be calculated.

$$\text{Osmotic pressure}, P_{osm} = R(T + 273)\sum (m_i) \qquad (9.3)$$

   where P_{osm} is osmotic pressure in bar, T is temperature, R is universal gas constant, and $\sum (m_i)$ is total of molar concentration of all the components present in a solution.

6. As regards the system array, permeate backpressure, and pressure drop per stage, assumptions can be made.
7. Subsequently, the required feed pressure can be calculated.

$$\text{Net Driving Pressure} = P_f - P_{os} - P_p - 0.5P_d\,(+P_{osp}) \qquad (9.4)$$

   where P_f is feed pressure, P_p is permeate pressure, P_{os} is average feed osmotic pressure, P_{osp} is permeate osmotic pressure, and P_d is pressure drop across reverse osmosis elements.

8. Estimate the permeate salinity on the basis of nominal element salt passage (SP), average system permeate flux, average feed salinity, and element permeate flux at nominal testing conditions.

The below example demonstrates the system performance calculation (Wilf & Awerbuch, 2007).

Example 9.1

Consider a brackish water dual-stage reverse osmosis system. The feed salinity is noted to be 2,500 ppm total dissolved solids, and the recovery rate is 85%. The feed water temperature is 25°C and the average flux rate is 27.20 L/m².h (16 GFD). Calculate the specific element performance, system performance, and permeate salinity.

Specific element performance calculation

The element type used is ESPA2. This high-permeability membrane has a membrane area of 36.80 m². The performance of the membrane element can be evaluated by calculating the specific permeability. The specific permeability of a membrane element is calculated employing the nominal test conditions and the nominal element performance.

Element type—36.80 m² membrane area, brackish

Nominal element performance is 34.07 m³/day at 10.30 bar pressure

Rejection of salt 99.60% at 38.60 L/m².h (22.70 GFD) flux rate

Nominal test condition—1,500 ppm sodium chloride feed salinity and 15.0% recovery rate

During the nominal test,

$$\text{Average feed water salinity} = 0.5 \times 1,500 \left(1 + \frac{1}{(1 - 0.15)} \right) = 1,632 \text{ ppm sodium chloride}$$

Avg. feed osmotic pressure = 1.25 bar = 18.10 psi
Nominal net driving pressure = 10.30−1.25 = 9.05 bar = 131.20 psi

$$\text{Specific permeability} = \frac{\text{Average permeate flux}}{\text{Net driving pressure}} = \frac{38.6}{9.05} = 4.26 \text{ L/m}^2.\text{h.bar} = 0.17 \text{ GFD/psi}$$

System performance calculation

$$\text{System net driving pressure needed} = \frac{\text{Average permeate flux}}{\text{Specific permeability}} = \frac{27.20}{4.26} = 6.40 \text{ bar (93.0 psi)}$$

Friction pressure drop per stage will be 2.0 bar (29.0 psi), and hence for the total system, it will be 4.0 bar (58.0 psi).

Permeate backpressure is 0.50 bar (7.20 psi).

Feed salinity is 2,500 mg/L total dissolved solids and osmotic pressure is 1.90 bar (28.0 psi).

$$\text{Avg. feed osmotic pressure} = 1.90 \times 0.5 \left(1 + \frac{1}{(1 - 0.85)} \right) = 7.30 \text{ bar (105.80 psi)}$$

System feed pressure required: 6.40 + 7.30 + 4.00 + 0.50 = 18.20 bar (264.0 psi)

Product water salinity

$$\text{Avg. feed salinity} = 0.5 \times C_f \left(1 + \frac{1}{(1 - r)} \right) = 0.50 \times 2,500 \left(1 + \frac{1}{(1 - 0.85)} \right) = 9,583 \text{ ppm}$$

$$\text{Salt passage} = \frac{\text{Salt concentration in the permeate}}{\text{Mean salt concentration in the feed stream}} \times 100\%$$

$$\text{Permeate salinity} = 9,583 \times \left(1 - \frac{99.60}{100}\right)\left(\frac{38.60}{27.20}\right) = 54 \text{ ppm}$$

Permeate salinity is a function of system permeate flux rate and average feed salinity.

9.3.2 Reverse osmosis performance calculations utilizing computer programs

In order to correctly design the reverse osmosis system, the projected reverse osmosis unit performance is calculated employing the computer programs designed by membrane element manufacturers.

Using the below steps, the calculation process can be performed:

1. The composition of feed water is entered.
2. Feed water pH is adjusted, if needed for preventing the calcium carbonate ($CaCO_3$) scaling.
3. Enter the feed water temperature, system recovery rate, and permeate flow rate.
4. Select the membrane elements type as well as array.
5. Membrane array to be adjusted for getting average product water flux value within the limits suggested for a specific water source.
6. Computations of permeate composition, feed pressure, and concentrate scaling indices performed by computer.
7. Display of system parameters and computation results.

The computer program will perform computations the same way as it was illustrated in Example 9.1. The calculations begin with the nominal performance of elements chosen for determining the salt transport and the specific permeability. On the other hand, the computations are performed in an element-by-element approach. Computer programs for reverse osmosis system performance calculations can be downloaded for free from membrane manufacturer websites. The various membrane projection software programs and the related calculations are explained in Chapter 8.

9.4 Process parameters and equipment performance monitoring in reverse osmosis system

The process parameters and equipment performance monitoring are needed for satisfying the below objectives (Kucera, 2019):

- Maintains production of the design water quality and quantity.
- Stops system from operation at conditions that may lead to equipment damage or personnel injury.
- Retains equipment operation within the design process limits.
- Keep up the required sequence and timing of equipment operation.

- Stores and processes the operating information, generating the reports, displaying the data.
- Facilitates estimate of maintenance requirement.
- Facilitates controlled intervention in system operation.

The following process parameters are monitored in reverse osmosis facilities:

- Turbidity of source water
- Temperature of source water
- Conductivity of source water
- Free (combined) chlorine of source water
- Flow of source water
- Raw water pump suction and discharged pressure
- Pretreatment chemical dosing rates
- Filtration unit head loss
- Pressure drop of cartridge filters
- Filter effluent silt density index and turbidity
- HPP suction and discharged pressure
- pH of feed water
- Feed water pressure
- Total chlorine of feed water
- Permeate pressure of reverse osmosis unit
- Permeate flow of reverse osmosis unit
- Permeate temperature of reverse osmosis unit
- Permeate conductivity of reverse osmosis unit
- Permeate pH of reverse osmosis unit
- Concentrate pressure of reverse osmosis unit
- Concentrate flow of reverse osmosis unit
- Post-treatment chemical dosing rate
- Turbidity of product water
- Free (combined) chlorine of product water
- Reverse osmosis permeate storage tank level

The monitoring activities carried out for protecting the RO plant equipment involves monitoring the operating parameters of the main equipment. This comprises setting alarms and shutoff switches for indicating off-limit conditions of the succeeding parameters:

- Turbidity of water
- pH of water
- Temperature of water
- Flow of treatment chemicals
- Water storage tank level
- Chemical storage tank level
- Pump discharged pressure
- Pump suction pressure
- Pressure drop in cartridge filters
- Free (combined) chlorine concentration
- Feed pressure

- Pressure of permeate
- Temperature of permeate
- Conductivity of permeate
- pH of permeate
- Concentrate pressure
- Concentrate flow
- Temperature of electric motors
- Pressure drop in reverse osmosis system

9.5 Reverse osmosis automation system

The level of instrumentation and control of a plant is very important in a plant operation (Jassim et al., 2003). With the availability of more instruments, the control becomes easier.

Instruments can measure appropriate values used for evaluating the performance (inputs) of a plant, based on which valves and pumps (outputs) can be controlled either automatically by a programmable logic controller (PLC) or manually (Lenntech). There are three types of automation systems in a reverse osmosis plant, namely, basis, standard, and advanced type of instrumentation and control.

Basic instrumentation and control: This includes the indicating instruments. Pressure and flows are assessed using a basic local indicator that does not have any power requirement. Only the conductivity meter provides an analog signal due to the fact that it is the major performance criterion of the facility. The HPP is not equipped with a VFD, and hence pressure and flows must be adjusted manually if the plant is subject to temperature deviations. These types of plants will be controlled by a ready-made microcontroller having a very limited amount of outputs.

Standard instrumentation and control: This includes a permeate flow transmitter, oxidation—reduction potential, and/or chlorine sensors to ensure that the dechlorination and/or chlorination stages are efficient. pH and temperature measurements can be added for monitoring the product water quality. The high-pressure pump can be provided with a VFD if the temperature deviations are significant; however, the feed or concentrate valve remains manual.

Standard instrumentation and control: This type of instrumentation and control is recommended for industries with a Distributed Control System. The entire instruments will be analog and monitored using a data logger or a programmable logic controller. This type of plant is controlled using a touch panel through a programmable logic controller with analog and digital inputs and outputs. Hence, the HPP is usually frequency controlled and the reject valve proportionally regulated.

9.5.1 Main reverse osmosis process parameters and their control

The reverse osmosis membrane performance is influenced by the following parameters (Lior et al., 2012):

- Salinity of feed
- Temperature of feed

- Recovery rate of the system
- Design permeate flux
- Surface condition of the membrane
- Energy recovery units and HPP equipment

Conversely, in practical terms, the prospect of changing the above-listed parameters for optimizing the existing system performance is restricted by the design and configuration of the system.

9.5.1.1 Salinity of feed

In most of the seawater sites, feed TDS varies within a narrow range not going beyond 5.0%–10.0%. Reverse osmosis systems are designed for providing the rated output at maximal feed TDS. In the event of low salinity, the system will be able to operate at a low feed pressure or a high recovery rate. At certain sites, very vast salinity variations might be experienced, mostly due to rainwater influx, and such conditions will be addressed at the design stage for providing adequate flexibility of pumping–power recovery device. Hybrid membrane systems can also be used in some exceptional situations. The salinity control of feed water could be accomplished to a certain degree in BW systems getting feed water from multiple wells. For seawater systems, the feed water salt concentration fluctuations are mostly weather-associated phenomena.

9.5.1.2 Temperature of feed

The feed water temperature fluctuation is usually related to the natural weather conditions. If the reverse osmosis system is positioned near the sea then high-temperature seawater will be available, which will enhance the feed water temperature. The feed water temperature fluctuations can be used to a certain extent for reducing the power requirements by regulating the recovery rate and feed pressure. As an example, high temperature of feed water can result in reduction in the feed pressure or increase in recovery rate while keeping constant feed pressure. This will be possible only if the product water salt concentration at higher temperature is within the design limits. The feed water temperature fluctuations are related to natural phenomena. In locations where high-temperature feed water is accessible, certain control of feed water temperature might be controlled by mixing water from cold source and hot source.

9.5.1.3 Recovery rate of system

Reverse osmosis systems can be designed for operating in the limited range of two fundamental operational parameters, namely product water flux and recovery rate. In most of the commercial systems, the pretreatment and pumping equipment available controls the maximum feed flow rate. Hence, in order to get constant feed flow, the recovery rate reduction will lead to low product water flow. At constant feed water temperature conditions, the increase in recovery needs increased feed pressure for maintaining the same product water output. Hence, the motor of the booster pump or HPP must be provided with a VFD to enable the feed pressure to increase. For an enhancement in the temperature of feed water under 25°C, it will be conceivable to enhance the rate of recovery parallel to the membrane permeability rise. On the basis of feed water salt concentration, at temperatures greater than 25°C, generally the osmotic pressure increase effects will be the same or greater than those because of the water permeability rise with temperature.

The rate of recovery for an existing system might be regulated by:

1. Adjustment of the throttling position of the valve on the reject line.
2. Adjustment of the electric motor rpm of the pump employing the variable frequency drive and altering the throttling position of the reject valve, in parallel.
3. Permitting variations in the temperature of feed water.

9.5.1.4 Design permeate flux

The flux rate of product water generally depends on the net driving pressure. A flux rate increase will need operation at a high recovery rate, which will need an extra feed pressure increase. A different approach that will lead to an increased rate of flux is the decrease in membrane area in operation. Adjusting the area of the operating membrane is not a real-world approach in the case of commercial systems, except for systems having high variations in feed water salinity.

The product water flux rate for an existing system could be varied by

1. Adjustment of feed pressure by varying the electric motor rpm of the pump using VFD and by regulating the throttling position of the reject valve, in parallel.
2. Permitting variations in the temperature of feed water.

9.5.1.5 Surface conditions of membrane

The membrane age increase will practically lead to increased salt passage (SP) and reduced water permeability. A parallel membrane fouling process has the same effect on the performance of membrane. The membrane compaction effect is considered to be non-reversible. The fouling effect could be reversed to a certain extent by the proper cleaning of the membrane. This will contribute a different aspect to the optimization process. As an example, the capability of increasing the rate of recovery is limited to a narrow range with old membranes than with new membranes, because of high SP.

9.5.1.6 Energy recovery units and high-pressure pump equipment

Presently, large commercial facilities universally use centrifugal pumps to pump feed water to the RO element. For seawater plants, the latest pumping configuration comprises pressure centers. In this type of configuration, a limited number of high-capacity HPPs will be connected to a common feed manifold, connected to a number of reverse osmosis trains. In a similar way, the reject manifolds from the entire reverse osmosis trains will be inter-linked to the energy recovery devices (ERDs). Usage of high-capacity pumps offers greater pump performance relative to that of a conventional configuration, that is, a small pump and ERD devoted to a single train. Due to the higher effectiveness of high-capacity pumps and new ERDs, there is very less energy penalty when certain reverse osmosis trains are off-line because of maintenance tasks and when the reverse osmosis unit functions at a low recovery rate (Wilf & Bartels, 2005).

In the case of brackish water reverse osmosis system, the ERDs, which are increasingly popular, are the turbochargers. It has two pump impellers that are attached coaxially. These two impellers possess blades arranged in opposite directions. Hence, the impeller that gets the reject stream operates as a driving force for the second impeller that offers boost to the feed water. Typically, turbochargers are employed for increasing interstage pressure in two-stage BW systems. The pressure boost offered by the turbocharger will be as per the pressure and flow of the reject stream. Also, the turbochargers are relatively inexpensive devices and are extremely compact. The turbocharger operation will not need any particular control equipment or operator interference for adjusting its operation, after the primary setting.

9.5.2 Control system specific requirement comparison in seawater reverse osmosis and brackish water reverse osmosis plants

Desalination operations in SWRO and BWRO facilities are almost the same in the majority cases. Their main differences are associated with operating pressure, feed salinity, system configuration, and recovery rate. Feed temperature and feed salinity in BWRO units, that typically process well water, are comparatively constant. The stability of feed temperature and salinity leads to operation at a narrow feed pressure range.

BWRO plants have a multistage configuration, and the number of desalination stages is normally two for BW units. The potential for operation optimization in BW systems is restricted to adjusting the product water output as per the drinking water demand fluctuations. Larger units will be composed of several wells and reverse osmosis units. The schedule of operation will depend on the scheduled sequence of fine-tuning operation of water sources and reverse osmosis units as per the requirement. Improved quality water sources and low-energy utilizing unit will be employed as a base load, and the less-efficient unit will be set online subsequently for satisfying the enhancing demand. The operation scheme can be adjusted periodically as per the performance evaluation of the unit, conducted offline.

The feed pressure in the SWRO system can vary over a wide range. With certain exemptions, seawater demonstrates reasonable salinity variations and comparatively wide variations in feed water temperature. The typical variation in the salinity of seawater is in the range of almost 5%. This is a salinity variation of approximately 2,000 mg/L total dissolved solids, which is the same as an average osmotic pressure variation of almost 0.2 MPa. The feed pressure should also vary approximately by the same amount for compensating for a decrease in NDP. In certain sites, the salinity of seawater is influenced by river discharge or low salinity overflow at the time of rains. Here, the salinity variations might be highly significant, sometimes in the range of almost 50.0%. The seawater temperature fluctuations are very common. If we compare the winter conditions of a power plant operational at low capacity and summer conditions of a high capacity operational power plant, then the temperature variation can extent to 20°C. This difference only can vary the specific water permeability of the reverse osmosis membrane by almost 40.0−60.0%. Moreover, there is an effect of membrane fouling and compaction that can lead to an additional permeability variation till 20.0%. A different parameter that could influence the operation of the system is SP. The SP is usually influenced by the temperature and the condition of the membrane surface. Increased SP might need the increased operation of second-pass reverse osmosis.

9.6 Reverse osmosis system performance normalization

Reverse osmosis system performance is considered to be the collective performance of specific membrane elements (Ghobeity & Mitsos, 2010). All membrane elements in a PV, arranged serially, work at dissimilar values of feed pressure and feed salinity. The performance and operation conditions will be considerably different from the corresponding nominal values. Moreover, system performances are extremely influenced by a variation in the operational parameters: feed pressure, recovery rate, temperature, and feed salinity.

Normalizing the reverse osmosis data permits the user to compare the reverse osmosis membrane performance to a set standard that does not rely on the varying operating conditions. The generic approach of reverse osmosis performance normalization is described in the American Society for Testing and Materials (ASTM) procedure (Standard practice for standardizing reverse osmosis performance data, 2010). Data normalization with respect to the preliminary start-up system performance is beneficial to demonstrate any performance variations at the time of the operations. Complete records are needed in the event of any system performance warranty claim inclusive of elements.

In real commercial reverse osmosis systems, the following normalization methods are implemented:

1. Normalization to the reference or preliminary operational condition of the reverse osmosis facility.
2. Normalization to the nominal element testing condition.
3. Salt transport value and water transport value calculations for the operational RO elements.

During the normalization calculation, the flows, salinities, and pressure information are first reduced to the average values. The values will be considered to be representative for a membrane element situated someplace in the middle of the system, on the feed-reject cross-section line. The averages can be computed on the basis of feed-reject values. Subsequently, in reference to this data, the salt and water permeability can be determined. In the above-mentioned normalization method #1, every set of the system performance information is being recalculated to the preliminary operational conditions: net driving pressure, average feed salt concentration, and temperature.

All the previously mentioned performance normalization methods would contribute the right demonstration of the performance trend of the membrane unit. The benefit of the first approach is that, apart from the normalized permeate flow and SP, it also delivers the pressure drop trend. Pressure drop is considered to be a significant pointer of an initial stage of fouling, which leads to element feed channel blockage. In the second normalization method, the system performance is determined and presented as an average element performance, it would achieve, if it is tested at the nominal testing condition. The third normalization method will be the same as the first method. In the third normalization method, the performance calculation of the reverse osmosis system is reduced to an average element performance. Subsequently, on the basis of this information, the water permeability and salt permeability are found out.

Normalization of permeate flow and SP is adapted from the salt transport relations. As per the relations, the SP will be a function of salinity gradient and product water quantity accessible for dilution. Hence, SP at specified operating conditions $SP_{(1)}$ is associated with various operational conditions appropriately to the correspondent average permeate flux (APF) rate:

$$SP_{(2)} = \frac{SP_{(1)}APF_{(2)}}{APF_{(1)}} \tag{9.5}$$

Considering two specific operating conditions, #1 and #2, permeate flow Q_p at condition #1 can be associated with operational condition #2 appropriately to correspondent temperature correction factors (TCF) and NDP.

$$Q_{p(2)} = Q_{p(1)} \times \frac{NDP_{(1)}}{NDP_{(2)}} \times \frac{TCF_{(1)}}{TCF_{(2)}} \tag{9.6}$$

9.7 Reverse osmosis system normalization software programs

9.7.1 RODataXL (Hydranautics)

RODataXL is a powerful reverse osmosis normalization program from Hydranautics, which assists the user in interpreting the information obtained from a water treatment facility using Hydranautics reverse osmosis membranes. This program generates operational graphs, normalization charts that make efficiency, reliability, and performance goals more feasible by permitting the user to address issues before they develop problems that considerably influence the operations.

"Normalization" computer programs, like RODataXL, graphically denote normalized permeate flow, feed-to-reject pressure drop, and percent salt rejection. These normalized parameters are calculated by comparing the operation on a specific day to the first day of operation. Modifications are made for changes in main operating variables like pressures, recovery, feed total dissolved solids (TDS), and temperature. In this manner, the performance decline, which is not related to operating parameters, could be identified and treated. The RODataXL normalization program is in compliance with American Society for Testing and Materials (ASTM) Standard D 4516–85 "Standard Practice for Standardizing Reverse Osmosis Performance Data."

9.7.1.1 New project entry

When the user opens the RODataXL Microsoft Excel file for the very first time, he or she would be able to do the following:

Select Display Units to be Metric or American engineering units as displayed in Fig. 9.1. The Display Units cannot be modified later on in the program; hence, the user has to save a copy of the original file.

As shown in Fig. 9.2, select the language, select the number of trains or stages in the system, select date format (e.g., 10/27/20 or 27/10/20), and then click "Apply."

Parameter	American Units	Metric Units
Pressure	psi	bar
Temperature	°F	°C
Salinity	ppm	µS/cm
Flow	gpm	m³/hr
Element Permeate Flow	gpd	m³/day

FIGURE 9.1   Selection of display units. *Source: Courtesy: Hydranautics—a Nitto Group Company (Personal communication with Hydranautics team. https://membranes.com/solutions/software/).*

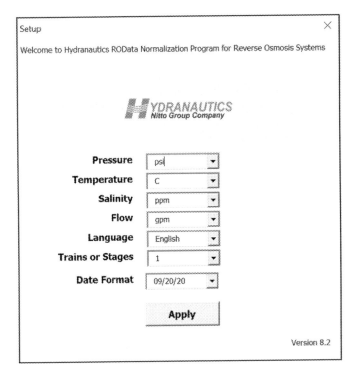

FIGURE 9.2 Selection of basic details. *Source: Courtesy: Hydranautics—a Nitto Group Company (Personal communication with Hydranautics team. https://membranes.com/solutions/software/).*

9.7.1.2 Reference data

This brings up the System Reference Data entry screen for the new project, as presented in Fig. 9.3. Only one set of System Reference Data is permitted for each Train or Stage of a Project.

Train or stage (XX): The user has to identify the train or stage using numbers. The user can have any number of Trains or Stages he or she wants for a Project, with each Train or Stage possessing its own System Reference Data.

Date (XX/XX/XX): The user has to enter the date by manual entry or by using the drop-down calendar.

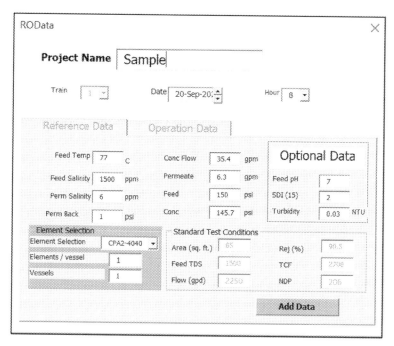

FIGURE 9.3   Selection of reference data. *Source: Courtesy: Hydranautics—a Nitto Group Company (Personal communication with Hydranautics team. https://membranes.com/solutions/software/).*

Hour (XX): The user should enter the hour the data point was logged using the military time (24-hour clock) of 0–23 hours.

Feed Temp (XXX.X): The temperature of the reverse osmosis feed has to be entered. The temperature range permitted is 1–50°C or 33–122°F.

Feed salinity (XXXXX.X): Enter the salinity of the reverse osmosis feed as it enters the first stage of reverse osmosis. The feed salinity is a measurement of the total dissolved solids content and is specified in ppm (American) or μS/cm (metric). MicroSiemens/cm (μS/cm) is also termed as micromhos/cm (μmhos/cm) in some parts of the world. PPM (parts per million) is also termed as mg/L (milligrams per liter).

Permeate salinity (XXXXX.X): The salinity of the reverse osmosis permeate should be entered here. The permeate salinity is a measurement of reverse osmosis permeate quality and is reported as ppm or mg/L or μS/cm. This value should be a lower value as compared to the feed salinity.

Permeate backpressure (XXXX.X): The user should enter the permeate backpressure as it leaves the reverse osmosis unit. American units are psi (pounds per square inch) and metric units are bar.

Concentrate flow/reject flow /brine flow (XXXX.X): The concentrate flow of reverse osmosis should be entered here. Metric units are m^3/h (cubic meters per hour) and American units are GPM (gallons per minute).

Permeate flow (XXXX.X): The permeate flow of reverse osmosis should be entered here. Permeate flow is also termed as product flow. Metric units are m^3/h and American units are GPM.

Feed pressure (XXXX.X): The user has to enter the feed pressure as it enters the first stage of the reverse osmosis system. Metric units are bar and American units are psi.

Concentrate pressure (XXXX.X): Enter the concentrate pressure as it leaves the final stage of the reverse osmosis unit. Metric units are bar and American units are psi.

Optional data: Silt density index (SDI), feed pH, and turbidity are optional information.

Element type: Select the model of the reverse osmosis element from the drop-down screen. The standard test conditions screen will automatically be filled in when an element is chosen. These data for the listed elements cannot be modified.

Number of elements/vessel (XXX): The user should enter the number of reverse osmosis elements per pressure vessel. Normally the maximum number is eight elements per vessel.

Number of pressure vessels (XXX): The user should enter the total number of reverse osmosis pressure vessels for the complete train or stage. The program computes the system flux by multiplying the number of elements/vessel by the number of pressure vessels by the available membrane area of each element.

Add: When clicking on the Add button, the System Reference Data is recorded and the file is created for the new Train or Stage of this Project. The user could proceed to enter System Operational Data once the System Reference Data has been entered.

Continue to add data by clicking the Data Entry button in the top left corner of the spreadsheet.

9.7.1.3 Normalization calculations

Units for Calculation: When only a single unit is specified for a parameter, for example, 0°C, then the subsequent calculations need only that unit to be used for achieving the correct value. When two units are specified, for example, GPM or m^3/h, then metric or American values can be used for the succeeding calculations; however, the user should stay with one unit, metric or American values for achieving the correct values.

The value of the RODataXL normalization program becomes apparent after a review of the calculations carried out. On the other hand, the user should remember that the true value of normalized data not only lies in one set of data, but in the trended values of the normalized data over time.

Equations used in the program will be listed in the spreadsheet under the "Help-File" tab. These equations match the code, which can be viewed by entering visual basic and going to the main module "ModMain" and going to the subroutine "Formulas."

9.7.1.4 Graphs

The user could select to view a number of graphs of the system operating data and normalized data. The graphs will be automatically generated and the data can be plotted relative to time. If there is more than one set of data for a specific day, the data set with the lowest hour value will be plotted. The graphs plot a straight line between data points.

The following graphs will be generated after the data input:

Permeate salinity versus time: It plots the actual permeate salinity in ppm or $\mu S/cm$.

Feed pressure versus time: It plots the actual reverse osmosis feed pressure in psi or bar.

Permeate flow versus time: It plots the actual permeate flow in m^3/h or GPM.

Salt passage versus time: It plots the actual percent salt passage for the complete reverse osmosis system. The calculated value is the actual permeate salinity divided by the "average" salinity of the reverse osmosis feed and reject.

Normalized permeate flow versus time: It plots the normalized permeate flow in GPM or m^3/h, relative to the System Reference Data at start-up.

Normalized salt passage versus time: It plots the normalized percent salt passage of the system relative to the System Reference Data at start-up.

Normalized delta-P versus time: It plots the normalized feed-to-concentrate pressure drop in psi or bar, relative to the System Reference Data at start-up. The normalized delta-P value reflects adjustments to pressure drop because of changing feed and concentrate flows.

Salt transport coefficient (STC) versus time: The significance of this number is that it measures the effectiveness of the membrane in how fast it permits the passage of salts. The value is reported as m/sec (meters per second).

Water transport coefficient (WTC) versus time: The significance of this number is that it measures the effectiveness of the membrane in how fast it permits the passage of water. The value is reported as m/sec.kPa (meters per second per kilopascal).

9.7.1.5 Analysis

The value of a normalization program is the capability to recalculate the rapid system performance at some point in time and compare its performance to the Reference Data operating conditions.

The following variations in operating parameters will reduce the actual permeate flow of a system:

- A reduction in feed water temperature with no variation in feed pump pressure.
- A reduction in reverse osmosis feed pressure by throttling down the feed valve.
- An increase in permeate backpressure with no variation in feed pump pressure.
- An increase in the feed total dissolved solids (or conductivity) since this increases the osmotic pressure that has to be overcome to permeate water through the membrane.
- An increase in the system recovery rate. This increases the average feed/concentrate total dissolved solids, which then increases the osmotic pressure.
- Membrane surface fouling.
- Fouling of the feed spacer that leads to an increase of feed-to-concentrate pressure drop (delta-P), which starves the back-end of the system of net driving pressure (NDP) to produce product water.

The below changes in operating parameters will lead to actual lower quality product water, as indicated by an increase in permeate total dissolved solids as ppm or conductivity:

- An increase in feed water temperature with the system adjusted to maintain the same product water flow.
- A reduction in the system permeate flow, which decreases the water flux and causes less-permeate water to dilute the amount of salts that have passed through the membrane.
- An increase in the feed total dissolved solids (or conductivity) since the reverse osmosis will always reject a set percentage of the salts.

- An increase in the system recovery rate increases the average feed/concentrate total dissolved solids of the system.
- Membrane surface fouling.
- Membrane surface damage that permits more salt passage.

9.7.2 Toray Trak (TORAY; TorayTrak Guideline)

The normalization software of Toray Membrane, known as TorayTrak, allows Toray reverse osmosis element users to efficiently evaluate the current membrane element performance in the reverse osmosis systems.

9.7.2.1 *Toray Trak features list*

9.7.2.1.1 New Features Version 3.1.5 (October, 2019)

This software is provided as Macro-Free Microsoft Excel programs with five versions for dealing with various process system designs and diverse operational data collection points. The process schemes available are:

- Single-stage system: TorayTrak_OneStage_PTotal.xlsx
- Split permeate single-stage system: TorayTrak_OneStage_Split.xlsx
- Two-stage system with first and second stage product water flow rate monitoring: TorayTrak_TwoStage_PF1_PF2.xlsx
- Two-stage system with the first stage and overall product water flow rate monitoring: TorayTrak_TwoStage_PTotal_PF1.xlsx
- Two-stage system with the second stage and overall product water flow rate monitoring: TorayTrak_TwoStage_PTotal_PF2.xlsx

9.7.2.1.2 Provisional version (93.1.5) (October, 2019)

This "xls" version is tentatively provided to users who could not correctly save the results of calculations using version 3.1.5. Toray assures that the users will be able to obtain the same calculation results with version (3.1.5) (Operation, Maintenance, and Handling Manual for membrane elements):

- 1−93.1.5. One stage system: TorayTrak_OneStage_PTotal_v93.1.5.xls
- 2−93.1.5. Split permeate one-stage system: TorayTrak_OneStage_Split_v93.1.5.xls
- 3−93.1.5. Two-stage system with first and second stage permeate flow rate monitoring: TorayTrak_TwoStage_PF1_PF2_v93.1.5.xls
- 4−93.1.5. Two-stage system with first stage and overall permeate flow rate monitoring: TorayTrak_TwoStage_PTotal_PF1_v93.1.5.xls
- 5−93.1.5. Two-stage system with second stage and overall permeate flow rate monitoring: TorayTrak_TwoStage_PTotal_PF2_v93.1.5.xls

The accuracy of normalization, particularly regarding the temperature effect on the operation performance of reverse osmosis membranes, has been increased by modifying the temperature correction. This normalization software of Toray Membrane can model and normalize performance for Toray reverse osmosis elements, which are registered in TorayDS2 projection software.

The below is a general overview of Toray Trak for a single-stage system as an example (TorayTrak Guideline).

9.7.2.2 Configuration

Toray Trak—Configuration Page is shown in Fig. 9.4.

1. Click on the "Configuration" Tab. All required system information is in green fill.
2. From the drop-down list, enter the Toray membrane model numbers, the number of pressure vessels, and the number of elements/pressure vessel.
3. Subsequently, choose the desired engineering units. These units should remain constant for all data inputs in the workbook.

9.7.2.3 Input data

Toray Trak—Input Data Page is shown in Fig. 9.5.

1. Click on the "Data" tab, where performance data of membrane can be entered.
2. B row 6, enter membrane performance data from columns A to K. All columns titled in yellow-colored field after the Date entry must contain data out to column K.
3. Baseline data to establish recommended cleaning lines are generated by taking the average of the data entered in rows 7−10.

9.7.2.4 Trend graph

The trend graphs of the normalized membrane performance data by the overall system are automatically presented in the following three tabs "NormDP," "NormSP," and "NormPerm."

Toray Trak—NormPerm graph is shown in Fig. 9.6.
Toray Trak—NormSP graph is shown in Fig. 9.7.
Toray Trak—NormDP graph is shown in Fig. 9.8.

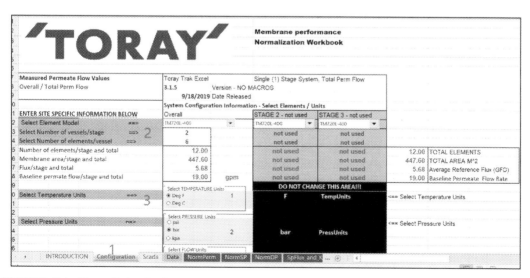

FIGURE 9.4 Toray Trak—Configuration page. *Source: Courtesy: Toray Industries, Inc. Ref (TorayTrak Guideline. Toray Industries, Inc, Website: https://ap3.toray.co.jp/toraywater/).*

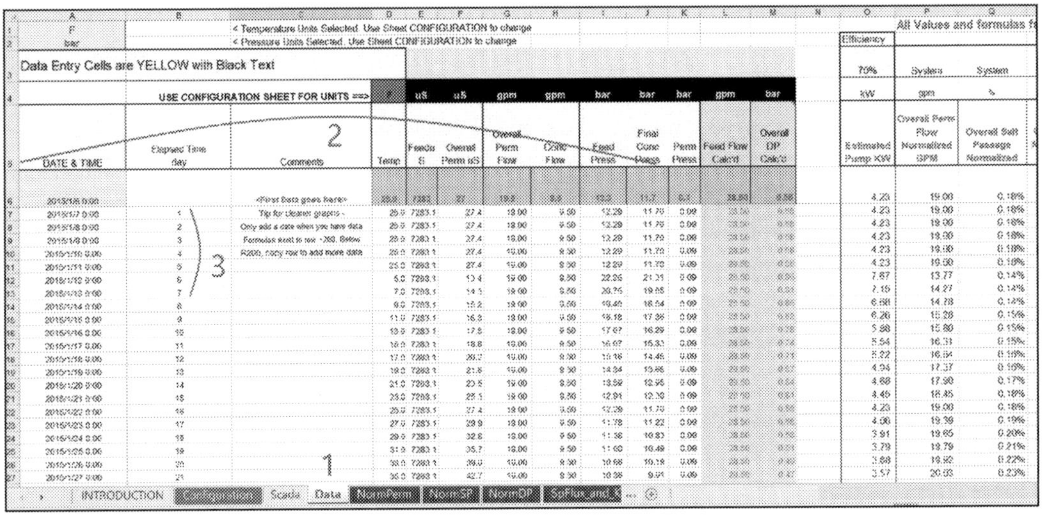

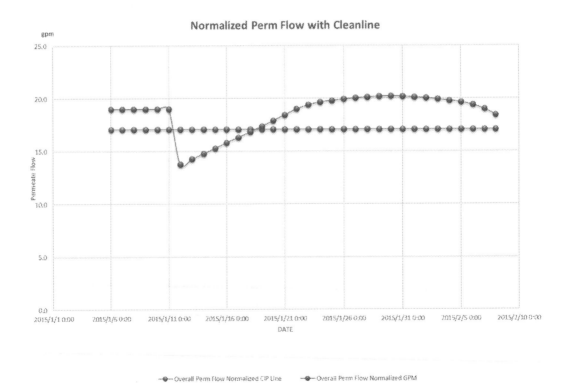

FIGURE 9.5 Toray Trak—Input data page. *Source: Courtesy: Toray Industries, Inc. (TorayTrak Guideline. Toray Industries, Inc, Website: https://ap3.toray.co.jp/toraywater/).*

Normalized Perm Flow with Cleanline

FIGURE 9.6 Toray Trak—NormPerm graph page. *Source: Courtesy: Toray Industries, Inc. (TorayTrak Guideline. Toray Industries, Inc, Website: https://ap3.toray.co.jp/toraywater/).*

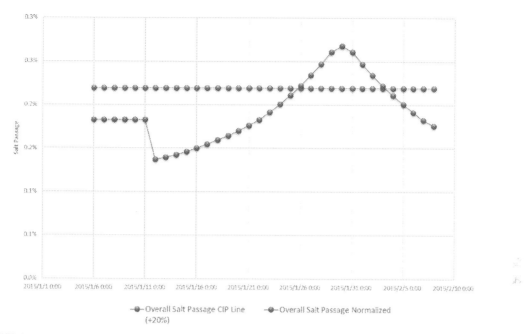

FIGURE 9.7 Toray Trak—NormSP graph page. *Source: Courtesy: Toray Industries, Inc. (TorayTrak Guideline. Toray Industries, Inc, Website: https://ap3.toray.co.jp/toraywater/).*

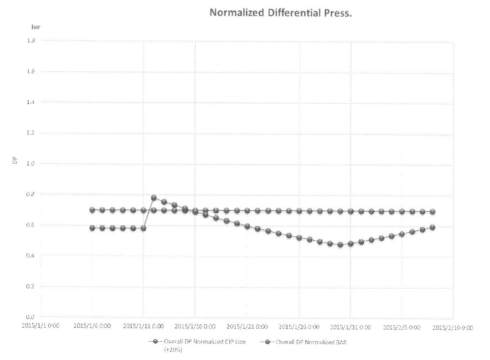

FIGURE 9.8 Toray Trak—NormDP graph page. *Source: Courtesy: Toray Industries, Inc. (TorayTrak Guideline. Toray Industries, Inc, Website: https://ap3.toray.co.jp/toraywater/).*

9.8 Fouling process of membrane elements

Membrane fouling affects the performance of membrane and can happen with any membrane types and feed water types. Fouling can even occur in secondary reverse osmosis units that process the reverse osmosis permeate. Fouling effect can be reasonable in its early stages; however, it can speed up fast and lead to a substantial performance decline in case it is not addressed in time. The different types of fouling and its prevention techniques are discussed in Chapter 5. Researchers are continually carrying out studies for improving the membrane performance and for reducing the RO membrane fouling (Emadzadeh et al., 2015; Matin, Khan et al., 2011; Matin et al., 2011; Misdan et al., 2012; Ozaydin-Ince et al., 2013; Saleem & Zaidi, 2020a,b; Saleem et al., 2020; Shenvi et al., 2015; Zaidi et al., 2019)

Correction of fouling conditions is a three-step process, as stated below:

1. Early identification of fouling phenomena
2. Fouling condition recognition and their mitigation
3. Membrane performance correction

The very efficient method for early identification of fouling process is by means of regular assessment of normalized performance trends: salt transport or SP, water permeability or product flow, and pressure drop. Variations in normalized performance value, exceeding the preliminary decrease of salt transport and water transport, will be an indication of fouling phenomena. The typical process design presumptions are that because of fouling, the water permeability would drop and the SP would rise by 5.0%–10.0% each year. The decline in the performance at a greater rate could only be allowed if some extra safety margin has been applied at the time of process design else the performance of the system would be away from the project specifications (Wilf, 2015).

The proper identification of fouling conditions is considered to be a complicated stage-by-stage process. It will begin with the assessment of the updated quality and composition of feed water. It is normal that the process design of a new reverse osmosis facility will be based on estimated source water constitution, as the feed water sources are typically developed at the time of construction of the system. The real feed water must be examined with regard to the concentration of scaling components and reverse osmosis feed water quality indicators like silt density index, total organic carbon, and turbidity. In case that the raw water has increased fouling potential, then beneficial data on the nature of foulants could be acquired by examining the deposits on the SDI filter pad employing energy dispersive X-ray (EDX) and scanning electron microscopy analyses. These analyses might give some indication about the pretreatment efficiency and assist to determine the kind of foulants that are reaching the reverse osmosis elements along with the feed water. Schneider et al. (2005) carried out an illustration of a complete investigation of a fouled element. For clarifying the membrane fouling development process, the entire fouled membranes along a feed channel of a double-stage commercial reverse osmosis system (six elements in each stage) for wastewater reclamation were autopsied and studied by Xin et al. (2020).

The reverse osmosis system fouling process is typically recognized through the performance normalization results. In the case of BW multistage systems, more information about the fouling process is obtained through the performance normalization of individual membrane stages, instead of total unit performance normalization. A higher performance decline rate in the first stage is an indication of the arrival of foulants with the feed water. The characteristic fouling process noted in the first stage is the increase in pressure drop because of feed channel blockage by biological growth or colloids. Also, a different process noted specially in the first stage at the start of the fouling phenomena is the drop in flux as a result of the organic adsorption. If the membrane elements in the final stage are severely affected, then it is probable that fouling phenomena is due to the higher content of fouling components because of a higher recovery rate. The more frequent fouling process, seen in the final phase, is the development of inorganic scaling, either sulfate or carbonate. In some cases, a mixed layer consisting both of organic foulant and inorganic foulant is formed on the surface of the membrane. As soon as the existence of fouling phenomena in the reverse osmosis system has been confirmed by means of performance normalization, the succeeding step will be to detach the membrane elements for assessment. For impactful assessment, a lead element and a tail element are needed. In some cases, a complete load of a single PV is detached for examination. The initial step will be to inspect the appearance of the elements and examine their weight. The presence of slime on the outside surface of the element demonstrates the biological fouling. Particle accumulation on the inlet element area shows insufficient filtration system operation. The reddish-brown deposit typically shows residue of iron flocculant from the pretreatment unit. The next step in the examination procedure will be the testing of the performance of elements at the nominal testing conditions and make a comparison of the results with ex-factory testing results.

The last stage in the examination process of elements is the assessment of the efficiency of cleaning operations. On the basis of the results from the examination of elements, the cleaning operations will be chosen and verified primarily on the single element. Membrane elements for performing the preliminary cleaning examinations must be chosen from locations in a system that are severely damaged by fouling. The efficiency of elements must be verified before the cleaning procedure and also subsequent to the completion of the cleaning operation. In case that several cleaning solutions are being assessed, it is suggested for testing the element performance after the application of every single cleaning solution, at least at the time of the preliminary cleaning efforts. The cleaning operations that have been noted to be the most efficient in single element cleaning tests are used for restoring the efficiency of membrane elements in the industrial reverse osmosis train.

9.9 Restoration of performance

9.9.1 Chemical cleaning

Membrane element cleaning in reverse osmosis train is carried out with the cleaning unit (Kucera, 2019). It comprises cartridge filter, heater, cleaning tank, connecting piping, and recirculation pump. Large cleaning units will have separate tanks to dissolve and mix the cleaning solutions. The construction material of the cleaning unit must be chosen such that it can withstand high- and low-pH cleaning solutions (pH 2–11) at temperatures up to 50°C. The cleaning tank size and the cleaning pump capacity depend on the number of PVs that would be cleaned at a single occasion (Wilf, 2015). At the time of cleaning operation, the cleaning solution flow rate per vessel must be 7.0–9.0 m^3/h (almost 30.0 to 40.0 GPM). The cleaning tank must hold an adequate volume of cleaning solution for providing a minimum 5 minutes of pump capacity. As an example, if 48 PVs have to be cleaned at a single time, then the cleaning tank operating volume must be almost 19.0 m^3 (5,000 Gallons). For larger reverse osmosis systems, the cleaning unit connecting piping will be permanently connected to the entire trains. Valves or detachable piping segments can be used for connecting/disconnecting a particular train or train segment to the cleaning system.

The sequence of cleaning operation involves:

1. Reverse osmosis train flushing using product water
2. Connection of train or train segment to the cleaning system
3. Preparation of cleaning solution in the cleaning system
4. Recirculation of cleaning solution for about one to four hours through the reverse osmosis train
5. Cleaning solution flushing
6. Repetition of Steps 2–5 with subsequent cleaning formulation or reconnecting cleaned train to HPP and restoration of regular operations

As the membrane cleaning will be more effective at higher temperatures, cleaning must be carried out at cleaning solution temperatures in the range of 35–40°C. These solutions could be obtained from specific suppliers or generic cleaning formulation could be used. From all the main membrane manufacturers, the composition of generic cleaning formulation could be acquired.

The frequently used low-pH generic cleaning formulation is a 2% citric acid solution with a pH of the solution around 2.5. This cleaning solution of citric acid is very efficient in the removal of metal hydroxide deposits and dissolving of carbonate scaling. In the case that fouling deposit consists of mostly metal hydroxides or calcium carbonate, then a temporary process using feed water acidified to lower pH (about 4.5–5.0) with mineral acid (sulfuric acid or hydrochloric acid) might be adequate for restoring the performance of the membrane. Cleaning, by operation at lower feed pH, will be conceivable only if discharge of lower pH reject is permitted by regional laws at a specific location.

The high-pH generic cleaning formulation comprises solution of sodium hydroxide along with ethylenediaminetetraacetic acid or sodium dodecylbenzenesulfonate. These cleaning solutions will have a pH of about 10–11 and will be very effective in removing the organic matter deposits from the surface of the membrane. The surfactants or EDTA are vital components of a high-pH cleaning solution, and their presence results in increased separation of surface deposits that consist of calcium ions embedded in the organic fouling layer. In most instances, the fouling layer will be of a blended nature, consisting of both organic and inorganic matters. The effective cleaning process is the application of low-pH cleaning succeeded by the high-pH formulation application.

Reverse osmosis units operating on well water feed rarely require the cleaning of the membrane. The frequency of cleaning will be typically lesser than one cleaning per two to three years of operation. Reverse osmosis units treating the surface water feed will need membrane cleaning very frequently. In the case of properly designed and operated SWRO systems, the cleaning process can be carried out at one- to two-year interval. For seawater systems with poor quality feed water, the frequency of cleaning required can be greater. For estimating the operational expenses, the cleaning process budget can be typically calculated on the basis of two cleaning cycles per year. If regular cleanings are needed, then this can be considered as a sign of insufficient pretreatment operation (Pervov, 1991). Table 9.1 lists the recommended cleaning chemicals by DuPont Water Solution. Table 9.2 lists the organic fouling cleaning solutions and Table 9.3 lists the biofouling cleaning solutions, as recommended by DuPont Water Solution.

The temperatures and pH listed in Table 9.1 are applicable for BW30LE, BW30, XLE, LE, TW30HP, TW30, SW30, SW30XLE, SW30HR LE, and SW30HR membrane elements. BW30LE, low-energy brackish water RO element; BW30, brackish water RO element; XLE, extra-low-energy element; LE, low-energy element; TW30HP, tap water high-performance

TABLE 9.1 Simple cleaning solutions, as recommended by DuPont Water Solution.

Cleaner→	0.1% (W) NaOH and 1.0% (W) Na$_4$EDTA, pH 12, 35°C max.	0.1% (W) NaOH and 0.025% (W) Na-DSS, pH 12, 35°C max.	0.2% (W) HCl, 25°C and pH 1–2	1.0% (W) Na$_2$S$_2$O$_4$, 25°C and pH 5	0.5% (W) H$_3$PO$_4$, 25°C and pH 1–2	1.0% (W) NH$_2$SO$_3$H, 25°C and pH 3–4
Foulant ↓						
Organic	Alternate	Preferred				
Biofilms	Alternate	Preferred				
Silica	Alternate	Preferred				
Inorganic colloids (silt)		Preferred		Preferred	Alternate	Alternate
Metal oxides (e.g., iron)				Preferred	Alternate	Alternate
Sulfate scales (CaSO$_4$, BaSO$_4$)	OK					
Inorganic salts (e.g., CaCO$_3$)			Preferred	Alternate	Alternate	

Courtesy: DuPont Water Solution. Adapted from DuPont Water Solution's FilmTec™ RO technical manual; Version 3; Form No. 45-D01504-en, Rev. 3; April 2020, https://www.dupont.com/content/dam/dupont/amer/us/en/water-solutions/public/documents/en/45-D01504-en.pdf.

TABLE 9.2 Organic fouling cleaning solutions, as recommended by DuPont Water Solution.

Cleaning solutions	Solution
Preferred	0.1 wt.% NaOH
	0.25 wt.% Na-DDS
	pH 12, 30°C maximum, followed by 0.2% HCl
	pH 2, 45°C maximum
Preferred	0.1 wt.% NaOH
	pH 12, 30°C maximum, followed by 0.2% HCl
	pH 2, 45°C maximum
Alternate	0.1 wt.% NaOH
	1.0 wt.% Na₄EDTA
	pH 12, 30°C maximum, followed by 0.2% HCl
	pH 2, 45°C maximum

NaOH, sodium hydroxide; *HCl*, hydrochloric acid; *Na-DSS*, sodium laurel sulfate; *Na₄EDTA*, tetra-sodium salt of ethylene diamine tetraacetic acid.
Courtesy: DuPont Water Solution. Adapted from DuPont Water Solution's FilmTec™ RO technical manual; Version 3; Form No. 45-D01504-en, Rev. 3; April 2020, https://www.dupont.com/content/dam/dupont/amer/us/en/water-solutions/public/documents/en/45-D01504-en.pdf.

TABLE 9.3 Biofouling cleaning solutions, as recommended by DuPont Water Solution.

Cleaning solutions	Solution
Preferred	0.1 wt.% NaOH
	0.025 wt.% Na-DDS
	pH 13, 35°C maximum
Preferred	0.1 wt.% NaOH
	pH 13, 35°C maximum
Alternate	0.1 wt.% NaOH
	1.0 wt.% Na₄EDTA
	pH 13, 35°C maximum

NaOH, sodium hydroxide; *Na-DSS*, sodium laurel sulfate; *Na₄EDTA*, tetra-sodium salt of ethylene diamine tetraacetic acid.
Courtesy: DuPont Water Solution. Adapted from DuPont Water Solution's FilmTec™ RO technical manual; Version 3; Form No. 45-D01504-en, Rev. 3; April 2020, https://www.dupont.com/content/dam/dupont/amer/us/en/water-solutions/public/documents/en/45-D01504-en.pdf.

element; TW30, tap water element; SW30, seawater element; SW30XLE, seawater extra-low-energy element; SW30HR LE, seawater high-rejection low-energy element; SW30HR, seawater high-rejection element; W, weight percent of active ingredient; BaSO₄, barium sulfate; $CaSO_4$, calcium sulfate; $CaCO_3$, calcium carbonate; Na_4EDTA, tetra-sodium salt of ethylene diamine tetraacetic acid; NaOH, sodium hydroxide; HCI, hydrochloric acid; Na-DSS, sodium laurel sulfate; NH_2SO_3H, sulfamic acid; H_3PO_4, phosphoric acid; $Na_2S_2O_4$, sodium hydrosulfite.

In a study conducted by Madaeni and Samieirad (2010), FT-30, a hydrophilic polyamide (PA) membrane, was used for industrial wastewater treatment. This membrane was fouled with inorganic and organic matters for about 540 minutes and subsequently cleaned using various chemical solutions for dissolving the deposits from the surface of the membrane. The cleaning solutions used were different chemicals inclusive of acids (sulfuric acid, nitric acid, hydrochloric acid), base (sodium hydroxide), surfactant (sodium dodecyl sulfate), complexing agent (ethylene diamine tetraacetic acid), and their combination. For achieving superior cleaning efficiency, the effects of physical factors (time, temperature, and velocity) were examined. Flux recovery and resistance removal were employed for the cleaning efficiency demonstration. The results confirmed that the acids were not effective in flux recovery; on the other hand, the two-stage caustic and detergent cleaning including sodium hydroxide-sodium dodecyl sulfate succeeded compared to acid-provided efficient recovery.

The effectiveness of different chemical cleaning approaches for a reverse osmosis membrane fouled by a wastewater effluent was studied by Ang et al. (2011). The utilization of single and paired chemicals and chemical combinations from four different categories of cleaning solutions including metal chelating agents, alkaline solutions, salt solutions, and surfactants was investigated. The results confirmed that by properly pairing chemical agents that have balancing cleaning mechanisms, a superior cleaning efficiency could be obtained. Particularly, sodium hydroxide demonstrates an excellent capacity for enhancing the overall cleaning efficiency when mixed with other chemical agents, probably because of its capability for loosening the fouling layer. The conclusions obtained have substantial implications to mitigate the fouling in reverse osmosis membranes for wastewater reuse applications. Practical cleaning approaches, where two or more chemical agents are used in combination or in sequence, could be formulated for achieving optimum performance recovery, whereas decreasing the overall chemical usage.

In most cases, a cleaning process is able to restore certain lost permeability and decrease pressure drop. On some rare occasions, salt rejection is increased. Typically, it can be the same or could even drop subsequent to cleaning. In the event that the cleaning operation does not cause adequate performance improvement, then the replacement of element is the only useful solution for the improvement in performance. Typically, a significant fraction of elements in the system must be substituted for achieving evident performance enhancement. If elements with the lowest efficiency could be recognized early in the reverse osmosis system, then the number of elements that need replacement could be reduced.

9.9.2 Direct osmosis cleaning

Direct osmosis is a process commonly used in reverse osmosis systems treating feed water with high salinity. During the shutdown of the reverse osmosis system, there will be a reverse flow of product water through the membrane back to the feed section. This can be used in a novel cleaning approach termed as Direct Osmosis-High salinity membrane performance restoration (Sagiv et al., 2008). This method involves the introduction of a high salinity solution to the HPP suction, when the reverse osmosis unit is in operation. This injection will last for a number of seconds. The high pulse will flow through the membrane elements from the feed to the reject. In the course of high salinity wave flow through the feed channels of reverse osmosis membrane elements, the flow through the membrane will be reversed. This reverse product water flow will lift the foulants from the surface of the membrane. The continued feed-reject flow wipes the lifted foulant particles from the elements, out of the reverse osmosis membrane unit. This approach of restoration of membrane efficiency is specifically effective in removing the colloidal deposits and controlling the membrane biofouling.

9.10 Case studies

9.10.1 Case study 1: Performance evaluation and design of reverse osmosis desalination plant in Egypt

A study was conducted to design and evaluate the performance of a reverse osmosis plant for treating ground water. The main objective of the study by Abdel-Fatah et al. (2016) was to design a reverse osmosis desalination plant for treating ground water to be used for agricultural, industrial, or domestic applications. The team confirmed that the study can be appropriate anywhere close to a water source and can resolve the issue of water in the deserts of Egypt. The study proposed a unique reverse osmosis design for providing 3,600 m^3/day on three years' projection on the basis of feed water with a 50% design product water recovery rate. This system will be producing product water with almost 100 mg/L total dissolved solids while operating on well-feed water with almost 25−30°C temperature and total dissolved solids of 10,000 mg/L.

The reverse osmosis units are designed to conservative standards for adaptability during feed water quality fluctuations. The proposed design included a feed water flush cycle for minimizing the membrane fouling and piping corrosion at the time of system shutdown. The system also included all suitable controls and instrumentation for performing an automatic operation. The main equipment used include chemical feed systems, cartridge filters, reverse osmosis ump, RO membranes, product water flush/cleaning system, and system instrument, control and power panel. The cleaning system included 1,000-gallon polyethylene tank, centrifugal pump, polyvinyl chloride butterfly/ball valves, temperature gauge, pressure gauge, tank-level switch, cleaning hoses, and polyvinyl chloride piping. The system will be flushing the brackish water and reject stream from the reverse osmosis unit at every shutdown, thus minimalizing the corrosion on the system's metallic components and extending the reverse osmosis membrane life. As regards the instrumentation and control, a central reverse osmosis control panel is provided. This panel housed the entire system-related controls and instrumentation. The controls are programmable logic controller-based. For the entire alarm function, indicating lights are provided. The main power is 380 V, 3 Phase, 50 HZ, whereas the control power is 220 V, 1 Phase, 50 HZ.

By carrying out multiple variable optimizations, it was noted that the best recovery that contributed the maximum revenue using the minimum cost is 50%, the optimal number of elements is 132 arranged in 22 PV, and 6 elements were chosen per PV for operating at the safer side. Rather than employing a conventional RO plant, an energy recovery device was used that saved about $30,556.28/year. Moreover, a degasifier was also employed for reducing the carbon dioxide content and thereby reducing the amount of sodium hydroxide used in permeate post-treatment. Multiple variable optimizations were performed on different varieties of membrane systems such as the number of stages, flow factor, and recovery.

9.10.2 Case study 2: Fouling behavior of marine organic matter (MOM) in reverse osmosis membranes of a real-scale SWRO desalination plant, South Korea

The effects of MOM characteristics on the fouling layer composition of seven parallelly configured RO membrane modules were recognized to contribute useful information about the fouling behavior of marine organic matter in a real-scale SWRO desalination plant, South Korea (Lee et al., 2020). Therefore, the physicochemical properties (i.e., functional group compositions, fluorescence spectral characteristics, and molecular weight distribution) of marine organic matter in the feed water and treated water and the extracted foulants from the reverse osmosis membranes used were characterized by different analytical techniques for identifying the interactions between marine organic matter and the membrane surfaces closely associated with the fouling formation in the reverse osmosis membranes. This desalination plant consisted of a dissolved air flotation with ball filters (dissolved air flotation) process, ultrafiltration, and reverse osmosis membranes intermittently operated for almost four years (Fig. 9.9). Table 9.4 shows the reverse osmosis membrane physicochemical properties.

Four types of water samples, inclusive of feed water (seawater from the offshore near the SWRO facility), the effluent from the dissolved air floatation process, the ultrafiltration membrane permeate, and the reverse osmosis membrane permeate, were obtained from the desalination facility for examining the inorganic and organic constituent removal and the changes of marine organic matter properties in the feed water and treated water by each process of the real-scale desalination facility. The foulant samples have been obtained from two types of the fouled reverse osmosis membranes, collected from the first RO and seventh RO modules in the same PV employing three types of cleaning agents, that is, acid solution (0.1 N hydrochloric acid), base solution (0.1 N sodium hyroxide), and deionized water. Three coupons of the fouled reverse osmosis membranes were soaked in 500 mL of each cleaning agent for 6 hours with adequate stirring for extracting the foulants, based on their physicochemical properties. The pH of the extracted reverse osmosis foulant samples was

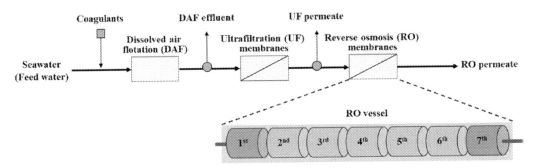

FIGURE 9.9 The diagrammatic representation of the real-scale seawater desalination plant, South Korea. *Source: Reproduced from Lee, Y.G., Kim, S., Shin, J., Rho, H., Lee, Y., Kim, Y.M., & Chon, K. (2020). Fouling behavior of marine organic matter in reverse osmosis membranes of a real-scale seawater desalination plant in South Korea. Desalination, 485, 114305.*

TABLE 9.4 Reverse osmosis membrane physicochemical properties.

Type	Reverse osmosis membrane
Manufacturer	Toray Chemical, Korea
Membrane code	RE 16040-SHF
Materials	PA thin-film composite
Roughness (nm)	86.1 ($\pm$ 13.7)
Zeta potential (mV)	-24.0 ($\pm$ 1.3)
Contact angle ($^\circ$)	73.6 ($\pm$ 7.6)
Nominal pore size (μm)	< 0.005

Reproduced from Lee, Y.G., Kim, S., Shin, J., Rho, H., Lee, Y., Kim, Y.M., & Chon, K. (2020). Fouling behavior of marine organic matter in reverse osmosis membranes of a real-scale seawater desalination plant in South Korea. Desalination, 485, 114305.

adjusted utilizing hydrochloric acid and sodium hyroxide solutions to a pH range of 5.5–6.5, and subsequently pretreated using filters and stored at 4°C before the analyses.

From the results, it was noted that the irreversible membrane fouling comprising both hydrophobic and hydrophilic marine organic matter fractions was accountable for the development of the fouling layers on the surface of the reverse osmosis membrane. Further, the complexes with multivalent metal ions (i.e., aluminum, calcium, copper, iron, and magnesium) and hydrophobic marine organic matter fractions played an important role in the formation of the irreversible RO membrane fouling in South Korea. The fouled reverse osmosis membrane showed the lower contact angle and the negative surface zeta potential, relative to the clean reverse osmosis membrane (Table 9.5) as hydrophilic marine organic matter fractions intensely contributed to the fouling formation in the reverse

TABLE 9.5 The surface features of the fouled and cleaned reverse osmosis membranes.

		Contact angle (°)	Surface zeta potential at pH 7 (mV)
First RO membrane	Fouled RO membrane (first)	41.0 (±2.1)	− 3.1 (±1.2)
	Cleaned membrane (acid)	39.1 (±1.5)	− 16.5 (±1.3)
	Cleaned membrane (base)	28.3 (±3.7)	− 19.1 (±2.2)
	Cleaned membrane (deionized water)	40.4 (±4.3)	− 17.5 (±2.3)
Seventh RO membrane	Fouled RO membrane (7th)	47.2 (±3.1)	− 5.9 (±0.2)
	Cleaned membrane (acid)	33.5 (±4.0)	− 11.2 (±1.4)
	Cleaned membrane (base)	24.3 (±3.7)	− 28.6 (±3.3)
	Cleaned membrane (deionized water)	42.7 (±4.9)	− 18.3 (±2.6)

Reproduced from Lee, Y.G., Kim, S., Shin, J., Rho, H., Lee, Y., Kim, Y.M., & Chon, K. (2020). Fouling behavior of marine organic matter in reverse osmosis membranes of a real-scale seawater desalination plant in South Korea. Desalination, 485, 114305.

osmosis membranes. Subsequent to the cleaning process with deionized water, acid and base solutions, the surface zeta potential of the fouled reverse osmosis membrane was efficiently recovered; however, its contact angle was considerably reduced. This was more noticeable for the cleaned first RO membranes compared with the cleaned seventh RO membranes (Table 9.5). These results supported the assumption that a higher deposition potential of hydrophilic marine organic matter fractions (i.e., polysaccharide-like and protein-like substances) than hydrophobic marine organic matter fractions onto the reverse osmosis membrane surfaces might result in substantial differences in the fouling layer composition of the first RO and first RO modules in the real-scale seawater SWRO desalination facility.

References

Abdel-Fatah, M. A., El-Gendi, A., & Ashour, F. (2016). Performance evaluation and design of RO desalination plant: Case study. *Journal of Geoscience and Environment Protection, 4*(2), 53−63.

Ang, W. S., Yip, N. Y., Tiraferri, A., & Elimelech, M. (2011). Chemical cleaning of RO membranes fouled by wastewater effluent: Achieving higher efficiency with dual-step cleaning. *Journal of Membrane Science, 382*(1−2), 100−106.

DuPont Water Solution's FilmTec™ RO technical manual; Version 3; Form No. 45-D01504-en, Rev. 3; April 2020, https://www.dupont.com/content/dam/dupont/amer/us/en/water-solutions/public/documents/en/45-D01504-en.pdf.

Emadzadeh, D., Lau, W. J., Rahbari-Sisakht, M., Daneshfar, A., Ghanbari, M., Mayahi, A. A., ... Ismail, A. F. (2015). A novel thin film nanocomposite reverse osmosis membrane with superior anti-organic fouling affinity for water desalination. *Desalination, 368*, 106−113.

Feliu-Batlle, V., Rivas-Perez, R., & Linares-Saez, A. (2017). Fractional order robust control of a reverse osmosis seawater desalination plant. *IFAC-PapersOnLine, 50*(1), 14545−14550.

Ghobeity, A., & Mitsos, A. (2010). Optimal time-dependent operation of seawater reverse osmosis. *Desalination, 263*(1−3), 76−88.

Jassim, A. A. A., Abdulrazaq, N., & Kahdim, A. (2003). Automation of reverse osmosis plants. *Seventh International Water Technology Conference Egypt, 187*–194.

Jiang, S., Li, Y., & Ladewig, B. P. (2017). A review of reverse osmosis membrane fouling and control strategies. *Science of the Total Environment, 595,* 567–583.

Kucera, J. (2019). *Desalination: Water from water.* John Wiley & Sons.

Lee, Y. G., Kim, S., Shin, J., Rho, H., Lee, Y., Kim, Y. M., & Chon, K. (2020). Fouling behavior of marine organic matter in reverse osmosis membranes of a real-scale seawater desalination plant in South Korea. *Desalination, 485,* 114305.

Lior, N., El-Nashar, A., & Sommariva, C. (2012). Advanced instrumentation, measurement, control, and automation (IMCA) in multistage flash (MSF) and reverse-osmosis (RO) water desalination. *Advances in water desalination,* Chapter 6, 494. John Wiley & Sons, Ltd.

Madaeni, S. S., & Samieirad, S. (2010). Chemical cleaning of reverse osmosis membrane fouled by wastewater. *Desalination, 257*(1–3), 80–86.

Matin, A., Khan, Z., Zaidi, S. M. J., & Boyce, M. C. (2011). Biofouling in reverse osmosis membranes for seawater desalination: Phenomena and prevention. *Desalination, 281,* 1–16.

Matin, A., Ozaydin-Ince, G., Khan, Z., Zaidi, S. M. J., Gleason, K., & Eggenspiler, D. (2011). Random copolymer films as potential antifouling coatings for reverse osmosis membranes. *Desalination and Water Treatment, 34*(1–3), 100–105.

Misdan, N., Lau, W. J., & Ismail, A. F. (2012). Seawater Reverse Osmosis (SWRO) desalination by thin-film composite membrane—Current development, challenges and future prospects. *Desalination, 287,* 228–237.

Operation, Maintenance, and Handling Manual for membrane elements. https://www.toraywater.com/knowledge/pdf/HandlingManual.pdf.

Ozaydin-Ince, G., Matin, A., Khan, Z., Zaidi, S. J., & Gleason, K. K. (2013). Surface modification of reverse osmosis desalination membranes by thin-film coatings deposited by initiated chemical vapor deposition. *Thin Solid Films, 539,* 181–187.

Pervov, A. G. (1991). Scale formation prognosis and cleaning procedure schedules in reverse osmosis systems operation. *Desalination, 83*(1–3), 77–118.

Reverse Osmosis Desalination Instrumentation and control, Lenntech. https://www.lenntech.com/processes/desalination/instrumentation/general/instrumentation-control.htm.

Sagiv, A., Avraham, N., Dosoretz, C., & Semiat, R. (2008). Osmotic backwash mechanism of reverse osmosis membranes. *Journal of Membrane Science, 322,* 225–233.

Saleem, H., Trabzon, L., Kilic, A., & Zaidi, S. J. (2020). Recent advances in nanofibrous membranes: Production and applications in water treatment and desalination. *Desalination, 478,* 114178.

Saleem, H., & Zaidi, S. J. (2020a). Developments in the application of nanomaterials for water treatment and their impact on the environment. *Nanomaterials, 10*(9), 1764.

Saleem, H., & Zaidi, S.J. (2020b). Innovative nanostructured membranes for reverse osmosis water desalination. Available from https://doi.org/10.29117/quarfe.2020.0023.

Saleem, H., & Zaidi, S. J. (2020c). Nanoparticles in reverse osmosis membranes for desalination: A state of the art review. *Desalination, 475,* 114171.

Schneider, R., Ferreira, L., Binder, P., & Ramos, J. (2005). Analysis of foulant layer in all elements of an RO train. *Journal of Membrane Science, 261,* 152–162.

Hydranautics—A Nitto Group Company (2020). Shah, J. Global Marketing & Product Manager, personal communication. Website: https://membranes.com/solutions/software/.

Shenvi, S. S., Isloor, A. M., & Ismail, A. F. (2015). A review on RO membrane technology: Developments and challenges. *Desalination, 368,* 10–26.

Sobana, S., & Panda, R. C. (2014). Modeling and control of reverse osmosis desalination process using centralized and decentralized techniques. *Desalination, 344,* 243–251.

Standard practice for standardizing reverse osmosis performance data, ASTM D 4516–00 (2010).

TorayTrak Guideline. Toray Industries, Inc, Website: https://ap3.toray.co.jp/toraywater/.

Wilf, M. (2015). Membrane-based desalination processes: Challenges and solutions. In *Mineral scales and deposits* (pp. 477–497). Elsevier.

Wilf, M., & Awerbuch, L. (2007). *The guidebook to membrane desalination technology: reverse osmosis, nanofiltration and hybrid systems: process, design, applications and economics.* Balaban Desalination Publications.

Wilf, M., & Bartels, C. (2005). Optimization of seawater RO systems design. *Desalination, 173*(1), 1–12.

Xin, T., Yong, C., Yun-Hong, W., Yuan, B., Tong, Y., Xue-Hao, Z., & Yin-Hu, W. (2020). Fouling properties of reverse osmosis membranes along the feed channel in an industrial-scale system for wastewater reclamation. *Science of the Total Environment, 713*, 136673.

Zaidi, S. J., Fadhillah, F., Saleem, H., Hawari, A., & Benamor, A. (2019). Organically modified nanoclay filled thin-film nanocomposite membranes for reverse osmosis application. *Materials, 12*(22), 3803.

Reverse Osmosis Membrane Performance Degradation

10.1 RO data normalization

For identifying intrinsic variations in the performance of membranes such as salt passage or permeability at the initial stages of membrane deterioration process, the system operational data should be recorded at least once a day and then normalized performance can be determined. The standard method of reverse osmosis (RO) performance normalization is explained in the American Society for Testing and Materials (ASTM) procedure (Standard practice for standardizing reverse osmosis performance data, 2010). In Chapter 8, Reverse Osmosis Design Software Programs, the different methods for RO system performance normalization and software programs (RODataXL and Toray Trak) have been discussed.

The normalization of RO data enables the user for comparing the performance of a reverse osmosis membrane to an established standard, which does not rely upon the changing operational conditions (Park & Kwon, 2018). The normalized data would evaluate the actual condition of a reverse osmosis membrane and demonstrate the accurate performance and the health of a reverse osmosis membrane. This normalized data is subsequently compared to the baseline (the data obtained when the membrane was new, cleaned, or replaced) (Reverse Osmosis Normalization).

The below raw data must be collected for determining the health of the RO membrane.

1. Feed pressure
2. Feed temperature
3. Feed conductivity
4. Permeate pressure
5. Permeate flow
6. Permeate conductivity
7. Concentrate pressure
8. Concentrate flow

Data that is not normalized could be deceptive, as there are several variables that could lead to changes that might seem to be issues, while as a matter of fact they are not. The feed water temperature is considered to be the clearest condition disturbing the RO system performance. The general rule is to evaluate a product water flow variation of 1.5%/°F. As an illustration, if a reverse osmosis unit produced 50 gallons per minute (GPM) of product water when the feed water temperature was 60°F and subsequently when the temperature of feed water decreased to 5°F, then the unit generated almost 46.0 GPM. This 4.0 GPM decrease in permeate is completely normal considering the decrease in the temperature.

10.2 Interpretation of data

An operator of a reverse osmosis system will be basically concerned about the two outcomes: the product water quantity and quality (Wang et al., 2017). As stated previously, the two factors could be influenced by numerous variables like system recovery, feed water pressure, and feed water quality variations.

According to the membrane manufacturers, the cleaning process must be started when any of the subsequent conditions occur, even though the primary flux is never restored (Singh, 2005).

1. Product water flow has decreased to 10.0%−15.0% lower than the rated flow at normal pressure.
2. Product water quality has decreased 10.0%−15.0%, and salt passage has increased 10.0%−15.0%.
3. There is an increase of 10.0%−15.0% feed pressure for maintaining the rated product water flow.
4. The differential pressure (DP) along the length of a reverse osmosis module has noticeable increase (greater than 15.0%).
5. Applied pressure has increased by almost 10.0%−15.0%.

The below relations can be used while analyzing the data and monitoring the performance of a membrane system (Singh, 2005):

1. Percent rejection (or percent retention): This is considered to be one of the most important indicators of the performance of the membrane, and it can be calculated as follows:

$$\text{Percent rejection} = \frac{\left\{ \frac{[(\text{Feed} + \text{Reject})]}{2} \text{solute conc.} - \text{Permeate solute conc.} \right\}}{\frac{[(\text{Feed} + \text{Reject})]}{2} \text{solute conc}} \times 100 \qquad (10.1)$$

2. Percent recovery (or percent yield): This is also an important indicator of membrane performance, and it could be determined as follows:

$$\text{Percent recovery} = \frac{\text{Permeate flow rate}}{\text{Feed flow rate}} \times 100 \qquad (10.2)$$

3. Concentration factor: This is also a beneficial parameter, however, not frequently used.

$$\text{Concentration factor} = \frac{1}{(1 - \text{Percent recovery})} = \frac{\text{Reject concentration}}{\text{Feed concentration}} \qquad (10.3)$$

4. Temperature correction factor (T_{CF}): Viscosity is considered to be a function of temperature, and therefore a correction factor should be applied for normalizing the flux and product water flow data.

$$T_{CF} = (1.024) \exp(t - 25) \qquad (10.4)$$

where t is the measured fluid temperature in °C.

5. Normalized productivity (N_P): The entire field data should be normalized and replotted for providing a reliable and precise basis to compare the product water flow rate over time.

$$N_P = \text{Measured flow} \times T_{CF} \times \left(\frac{\text{Design feed pressure}}{\text{Measured feed pressure}} \right) \qquad (10.5)$$

10.3 Indicators of membrane performance degradation

There are three major calculated values, which will help to provide a clear indication of the actual performance of the membrane and to accurately troubleshoot potential problems of RO system involving the quality and quantity of product water being produced by the RO system (Zhao & Taylor, 2005). By collecting operational data, and proper normalization of the collected data and subsequently trending the normalized data over time and making the comparison of the values to a baseline (start-up values when the RO membranes are new, replaced, or cleaned) the user can properly address any issues before any permanent damage to the RO membranes occurs. Fouling has always been an inevitable problem and the most important issue for membrane process engineers and researchers (Mahmoudi et al., 2019; Parlar et al., 2018). Membrane fouling leads to high operation cost because of low permeability or high transmembrane pressure drop (Cho et al., 2000; Kim et al., 2009). Scientists are constantly conducting studies to improve membrane performance and for reducing membrane fouling (Matin et al., 2011; Saleem & Zaidi, 2020a, 2020b, 2020c; Saleem et al., 2020; Zaidi, Fadhillah et al., 2019). The group of Zaidi et al. (Matin et al., 2011) reviewed the membrane fouling types and the different fouling control strategies, with a focus on the latest developments. Numerous in situ monitoring techniques including optical and non-optical probes have been developed to better understand as well as control the membrane fouling process (Li et al., 2017). An accelerated ageing technique has been reported by Antony et al. (2016), in which the membranes were subjected to cyclic fouling, cleaning, and hypochlorite exposure, and the performance was benchmarked against passively and industrially aged samples.

The three calculated parameters for performance monitoring are normalized pressure differential (NPD). Normalized salt rejection (NSR) and normalized permeate flow (NPF).

Any problem in the RO system performance generally means that at least one of the following trends observed (DuPont Water Solution, 2020; Kenji & Shingo, 2004):

1. Normalized salt passage increase: In a reverse osmosis system, this is normally related to an increase in the conductivity of product water.
2. NPF rate decrease: In reality, this will be usually observed as an increase in feed pressure for maintaining the permeate output.
3. Pressure drop (ΔP) increase: At constant flow rate the difference between feed pressure and reject pressure turns out to be higher.

From the symptoms, their type of occurrence and location, the reasons for the issue could generally be found out. In the subsequent section, the main issues in a reverse osmosis process are discussed thoroughly.

10.3.1 Normalized permeate flow

NPF is a function of the average applied transmembrane pressure, the feed osmotic pressure, the permeate osmotic pressure, and temperature (Kucera, 2015). NPF measures the amount of permeate water produced by the RO unit. Fig. 10.1 presents the NPF (GPM) versus the operating time. If the NPF reduces 10.0%−15.0% lower than the baseline value, then this points out the membrane scaling or fouling, and the RO membranes must be cleaned properly. If the NPF increases, then this suggests that there is a damage to the RO membrane. The damage could be due to a mechanical issue (e.g., an O-ring failure) or a chemical attack (e.g., an attack from an oxidizer such as chlorine) on the membrane.

The different factors that lead to a decrease or an increase in the NPF are discussed in the following section.

10.3.1.1 Decrease in normalized permeate flow

The scaling and fouling of the membrane can both result in a decrease in the NPF. Moreover, the membrane compaction would also lead to a reduced product water flow (Abbas & Al-Bastaki, 2001; Al-Bastaki & Abbas, 2004; Pais & Ferreira, 2007).

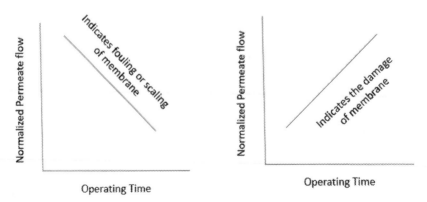

FIGURE 10.1   Normalized permeate flow versus the operating time.

If the RO system experiences a decrease in NPF and the issue could be localized, then the main rule is:

1. Initial stage issue: Particulate matter deposition; initial biofouling
2. Final stage issue: Scaling
3. Issue in the whole stages: Advanced fouling

A decrease in the NPF performance might be combined with a decreased, increased or normal salt passage. On the basis of these various combinations, conclusions regarding the causes might be obtained.

10.3.1.1.1 Reduced flow along with normal salt passage

Reduced flow of the permeate along with normal salt passage might have the subsequent causes:

Biofouling and natural organic matter   Membrane biofouling is confirmed by the below-listed variations in the operational parameters, mostly at the system front end:

1. Product water flow decreases while operated at consistent feed pressure as well as recovery.
2. Recovery will reduce while operated at consistent feed pressure, if biofouling is developed to large biomasses.
3. Feed pressure must be higher if the product water flow is to be maintained at consistent recovery. However, the feed pressure increase will be destined to fail if carried out for an extended time, as it may increase the fouling, turning it extremely hard to clean afterward.
4. Differential pressure rises severely when the biofouling is substantial or when coupled with the colloidal fouling. As the ΔP across the pressure vessel (PV) could be a sensible fouling indication, provisions to set up differential pressure monitoring instruments be incorporated for every stage in a reverse osmosis unit.
5. The salt passage continues to be normal or even low at the start, increasing as soon as the fouling turns out to be high.
6. The increased number of microbes in water samples taken from the permeate, reject, or feed streams confirm the start or the existence of biological fouling.
7. Biofilms feel slippery while touching and will usually have an unpleasant smell
8. A rapid biofouling test is the burn test: In this test, a biofilm sample will be collected using a spatula and incinerated over a lighter flame. The burnt biofilm smell will be the same as a burnt hair smell.

The reasons for biofouling are generally the consolidation of bioactive feed water and an inadequate pretreatment process.

The remedial actions include the following:

1. Cleaning and sanitizing the whole system, comprising the pretreatment unit along with the elements. A partial cleaning and disinfection lead to a quick recontamination.
2. High pH soak as well as rinse

3. Proper setting up or optimization of the pretreatment unit as per the source water fouling possibility
4. Setting up of fouling resistant (FR) elements.

The different techniques for biological fouling control are discussed in Chapter 5, Pretreatment: Fouling and Scaling Control, in detail.

Aged preservation solution If the preservation solution (bisulfite solution) is extremely warm, extremely old, or oxidized by oxygen, then the elements or RO systems preserved in the solution could also turn out to be biologically fouled. In general, an alkaline cleaning helps for restoring the product water flow.

Incomplete wetting RO elements that are allowed to dry out, might cause a decreased permeate flow, due to the fact that the fine pores in the polysulfone (PSF) layer have not been wetted.

10.3.1.1.2 Low flow along with low salt passage

Organic fouling The organic matters present in the feed water could be adsorbed on the surface of membrane and can cause a decrease in flux, particularly in the initial period. In various instances, the adsorption layer will act as an extra barrier for the dissolved salts, leading to a low passage of salt. Organic matters having a higher molecular mass and with cationic or hydrophobic groups could develop a similar effect. Examples include cationic polyelectrolytes or oil traces, which are occasionally employed in the pretreatment process. Organics will be extremely hard to separate from the surface of the membrane.

Organic fouling can be identified by performing the following analysis/test:

1. Analysis of deposits from cartridge filter and silt density index filter pads.
2. Analysis of the incoming water for grease and oil, along with organic contaminants generally.
3. Checking of the pretreatment coagulants and filter aids, particularly cationic polyelectrolytes.
4. Checking of the surfactants and cleaning detergents.

The following corrective measures for the prevention of organic fouling should be considered:

1. Cleaning of organics: Certain organics could be cleaned effectively, while some cannot be cleaned.
2. Accurate pretreatment: Use minimum dosage of coagulants; proper monitoring of the feed water variations for avoiding the overdose.
3. Modification of pretreatment, that is, oil/water separators.

Membrane compaction and intrusion The membrane compaction and intrusion are commonly related to the lower product water flow along with increased salt rejection. The compaction of the membrane is due to the applied pressure as well as temperature causing compression to the membrane, which might lead to a decrease in flux and the passage of salt. Intrusion is the membrane plastic deformation when pressed upon the product water channel spacer with extreme force and/or temperature. The permeate spacer pattern will be imprinted on the membrane. The intrusion will be usually related to lower flow. Practically, intrusion and compaction might happen simultaneously and will be extremely hard to differentiate from each other.

Even though the membrane demonstrates slight intrusion and compaction when performed correctly; however, considerable intrusion and compaction may happen under the below conditions:

1. Increased temperature
2. Increased feed pressure
3. Water hammer: This could happen when the high-pressure pump (HPP) is started with air in the unit.

New elements should be added to the system or damaged elements should be replaced for compensating for the decrease in flux. In case that the unused elements are fixed along with the used membrane element, then the unused elements must be loaded into the system tail position for protecting them from extremely higher flux operation. Unused elements must be distributed uniformly into parallel locations. The vessels loaded completely with unused elements should not be fixed in parallel with other vessels consisting of entirely used elements. This arrangement will lead to a variable flow distribution and recovery of the distinct vessels. As an illustration, in case that six elements of a 4(6): 2(6) system (in the first stage, four pressure vessels with six membrane elements per vessel and in the second stage, two pressure vessels with six membrane elements per vessel) must be replaced, then the new elements must be placed in positions four, five, and six of each of the two PVs of the second stage. Similarly, if six elements should be included, then the elements must be placed in positions five and six of the three PVs of the second stage of an enlarged 4(6): 3(6) unit (in the first stage, four pressure vessels with six membrane elements per vessel and in the second stage, three pressure vessels with six membrane elements per vessel). If this arrangement cannot be feasible for some reasons, then at least positions one and two of the 1st stage must not be loaded with new elements.

10.3.1.1.3 Low flow along with high salt passage

Low flow along with increased salt passage is considered to be the most frequently encountered condition for the failure of a plant. In the following section, the potential reasons for low flow and high salt passage are discussed.

Scaling

The scaling of RO membranes might occur once the sparingly soluble salts are concentrated inside the membrane element above their solubility limit. The characteristic scenario is a brackish water RO (BWRO) unit operating at higher recovery with no proper pretreatment process. Scaling generally begins in the final stage and subsequently progresses gradually to the upstream stages. Water consisting of higher concentrations of calcium, sulfate and/or bicarbonate can cause scaling of a membrane unit within some hours. The scaling from fluoride or from barium will be usually slow due to their lower concentration involved.

For identifying the scaling:

1. Properly check the analysis of feed water for the scaling possibility at the normal system recovery.
2. Analyze the concentration of reject for composition silicate, fluoride, sulfate, strontium, barium, calcium, pH, Stiff & Davis Saturation Index (for seawater), and Langelier Saturation Index (for brackish water). Also, calculation of the mass balance for these salts, analyzing also the feed water and the product water.
3. Examine the reject side of the system for scaling.
4. Calculate the weight of a tail element: Usually, the scaled membrane elements will be heavy.
5. Do autopsy of the tail element as well as the analysis of the membrane for scaling: Under a microscope, the crystalline structure of the deposits could be seen. The scaling type could be recognized by chemical analysis, inductively coupled plasma (ICP), or energy dispersive X-ray fluorescence (EDXRF) analysis.
6. Scaling will be hard and rough while touching, same as sand paper, and will be difficult to wipe off.

The corrective measures include the following:

1. Cleaning using an acid and/or an alkaline ethylenediaminetetraacetic acid (EDTA) solution. A proper examination of the spent solutions will aid to confirm the cleaning effect.
2. Optimize cleaning on the basis of the existing scaling salts
3. Fluoride scaling: Reduced recovery, adjustment of antiscalant type or dosage.
4. Sulfate scaling: Decreased recovery, adjustment of antiscalant type or dosage.
5. Carbonate scaling: Reduced pH, adjustment of antiscalant dosage.

The different scale control techniques are discussed in detail in Chapter 5, Pretreatment: Fouling and Scaling Control.

Metal-oxide fouling

This type of fouling occurs mostly in the initial stage. The issue could be very easily localized when product water flow meters are fixed in each array individually.

Common sources for the metal oxide fouling are:

1. Presence of aluminum or iron in feed water
2. Hydrogen sulfide (H_2S) with air in feed water leads to elemental sulfur and/or metal sulfides
3. Corrosion of vessels, piping, or components upstream of elements.

For identifying the fouling due to metal-oxide, the following should be carried out:

1. Proper analysis of feed water for aluminium and iron
2. Examination of the system components for any corrosion evidence

Iron fouling could be easily recognized from the appearance of the element. Fig. 10.2 shows the image of an iron fouled feed side of the membrane element.

The below corrective measures for reducing the metal oxide fouling should be followed:

1. Adjustment, correction, and/or modification of the pretreatment
2. Membrane elements cleaning as applicable
3. Retrofitting of piping or system components using suitable materials

FIGURE 10.2 Image of an iron fouled feed side of the membrane element. *Courtesy: DuPont Water Solutions;* Source: *Reproduced from DuPont Water Solution's FilmTect RO technical manual. (2020). Version 3; Form No. 45-D01504-en, Rev. 3; April 2020, https://www. dupont.com/content/dam/dupont/amer/us/en/water-solutions/ public/documents/en/45-D01504-en.pdf (Accessed October 8, 2020).*

Colloidal fouling

Colloidal fouling is considered as the most severe fouling issue occurring in membrane processes (Abdullah, et al., 2018). This type of fouling of RO elements can severely affect the membrane performance by decreasing productivity and in certain cases salt rejection (Ning, et al., 2005; Tang, et al., 2011). An initial sign for this type of fouling is normally an increase in pressure differential across the system. The colloidal particles are regarded as abundant in the natural waters. Colloids cover a wide-ranging size range, from several nanometers to a few micrometers (Ismail, et al., 2019). Examples of aquatic colloids include suspended matter, manganese oxide, aluminium oxide, iron oxide, colloidal silica, clay minerals, and calcium carbonate precipitates.

For identifying the colloidal fouling:

1. Review the recorded feed water silt density indexes. There might be issues sometimes because of the irregular excursions or pretreatment upsets.
2. Properly analyze the residue from silt density index filter pad.
3. Properly analyze the accumulations on the prefilter cartridge.
4. Inspect and analyze the deposit on feed scroll end of first-stage lead membrane element.

10.3.1.2 Normalized permeate flow increase

An increase in NPF usually results from a leak, either because of a crack in the RO membrane itself or due to some issues with the membrane module hardware, or due to oxidizer (chlorine) exposure.

10.3.2 Normalized salt rejection

Normalized salt rejection (NSR) shows how good the RO membrane is in rejecting the salts (contaminants) and consequently, the NSR will be associated with the product water quality. If the NSR decreases, then the amount of salts going through the RO membrane will be increasing (low-quality product water). A decrease in NSR point toward the scaling or fouling or degradation of the RO membrane.

A good performing RO membrane must provide 97.0%–99.0% salt rejection, and a membrane will be considered to be unsatisfactory when the salt rejection of the membrane decreases to 90.0% or even lower. During a normal RO membrane operation, a steady decline in NSR can be seen during continuous usage. RO membranes generally last for quite a few years, that is, 3–5 years (with proper pretreatment it can last for 5–7 years, depending on the water quality), before they need replacement and a steady reduction in NSR is a standard sign of a membrane getting aged. An appropriate cleaning procedure for RO membranes can help to improve the NSR.

The NSR could be beneficial in recognizing the biofouling problems. When the biological fouling is an issue, often the NSR will essentially increase and the NPF will decline. This is due to the fact that the biofoulant will truly seal small imperfections in the RO membrane thus increasing the salt rejection. After some time, the biofoulant layer will age and starts to expire, several species like carbon dioxide, organic acids, methane will be diffused through the membrane, and eventually influencing the quality of the product water (lesser rejection of salt leading to a low NSR).

10.3.2.1 High salt passage

10.3.2.1.1 High salt passage along with high permeate flow

Leak

Serious mechanical damages of the membrane elements or of the product water tubings could permit the feed or reject to enter into the product water, particularly while operating at higher pressures. The vacuum test would demonstrate a distinctive positive response. The vacuum test details will be discussed in Chapter 11, Reverse Osmosis System Troubleshooting.

Membrane oxidation

An increased salt passage combined with an increased NPF is typically because of oxidation damage. Whenever ozone, bromine, free chlorine, or other oxidizing chemicals will be present in the source water, then the front-end membrane element will be normally more affected relative to others. In general, the membrane attack will be favored by a neutral to alkaline pH. Moreover, the oxidation damage might occur by disinfection using an oxidizing agent, when the oxidation is accelerated by the existence of some metals, or when temperature and pH limits are not considered. In this case, uniform damage is possible.

When tested with the vacuum decay test, a FilmTec membrane element having an oxidation damaged membrane will remain mechanically undamaged (DuPont Water Solution, 2020). The chemical membrane damage could be made noticeable with the help of a dye test on membrane coupons or on the element. To confirm the oxidation damage, the autopsy of one element and examination of the membrane could be performed. All damaged elements should be replaced. No corrective action will be possible.

10.3.2.1.2 High salt passage along with normal permeate flow

Permeate backpressure

The membrane may tear as soon as the permeate pressure goes beyond the reject pressure by greater than 5.0 psi at a specific time. This type of damage could be recognized by probing and by performing the leak test and confirmed by a visual inspection at the time of autopsy.

During the unrolling of a leaf of a backpressure damaged element, the exterior membrane will normally show creases parallel to the product water tubing, typically by the exterior glue line. The membranes will delaminate and develop blisters against the feed spacer (as shown in Fig. 10.3). Cracking of membrane will happen mainly in the edges between the outer glue line, the feed-side glue line, and the reject-side glue line.

Abrasion of membrane surface

Crystalline or sharp-edged metal particles present in the feed water might enter into the feed channel and scratch the surface of the membrane. It will lead to an increase in the passage of salt from the lead membrane elements. The source water should be properly checked for such particles. Appropriate microscopic examination of the surface of the membrane would also confirm the damage. Membranes that are damaged should be replaced. The prefiltration should be confirmed to deal with these types of issues. It has to be ensured that no particles are discharged from the pump and the high-pressure pipe, and the pipe should be flushed out properly before the start-up procedure.

Telescoping

Membrane elements are mechanically damaged by an impact termed as telescoping, in which the exterior membrane layers of the element will unravel and extend downstream over the remaining layers. An ordinary telescoping will not essentially cause membrane damage; however, in extremely serious instances, the glue line and/or the membrane could be cracked. Telescoping will be due to the extreme pressure drop from feed to reject. It should be ensured that a thrust ring is employed with 8-inch elements for supporting the outer diameter of the membrane elements. Telescoping damage could be recognized by a leak test and by probing. It is recommended to replace the damaged membrane element and take corrective action against the causes.

O-ring leakage

O-ring leakage could be identified by a probing technique. Proper inspection should be carried out for the O-rings of couplers, end plugs, and adapters for accurate installation. Replacement of damaged and old O-rings should be done. O-rings might leak when exposed to some chemicals, or due to mechanical stress, for example, element motion arising from water hammer. Appropriate element shimming in a PV is important for

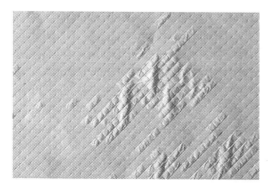

FIGURE 10.3 Membrane with permeate backpressure damage. *Courtesy: DuPont Water Solutions;* Source: *Reproduced from DuPont Water Solution's FilmTec™ RO technical manual. (2020). Version 3; Form No. 45-D01504-en, Rev. 3; April 2020, https://www.dupont.com/content/dam/dupont/amer/us/en/water-solutions/public/documents/en/45-D01504-en.pdf (Accessed October 8, 2020).*

minimizing the wear to the seals. At times, O-rings have basically not been fixed, or they are inappropriately fixed or moved out of their correct position at the time of element loading.

10.3.3 High pressure drop

High differential pressure (DP), also termed as ΔP or delta P or pressure drop from feed to reject, is the difference between the feed and concentrate pressure at the time of water flow through one or more RO membrane elements (LANXESS). It is considered to be a system operational issue due to the fact that the system flux profile will be disturbed in a manner that the lead element has to operate at extremely higher flux whereas the tail element will work at an extremely lower flux. The feed pressure increase signifies that the energy consumption will also be increased. A high pressure drop leads to an increased force in flow direction on the element feed side. The stress acting on the final element in the PV will be the maximum, and it has to tolerate the total forces produced by the ΔP of the upstream element. According to Dupont Water Solutions, the upper limit of the DP per multielement PV will be 50.0 psi, and per single fiber-glassed element it will be 15.0 psi. When the limits are crossed, just for a shorter period, the element may turn out to be telescoped and mechanically damaged (DuPont Water Solution, 2020).

An increase in the DP at constant flow rate is typically because of the presence of scale, foulant, or debris within the element flow channel. It generally happens along with a declining product water flow. An increased pressure drop occurs when the recommended maximum feed flow rates are crossed. It could also occur when the feed pressure accumulates very fast at the time of start-up. The effect would be substantially increased in the presence of a foulant, particularly biofilm results in a high pressure drop. Biofilm effects on RO membrane flux have been comprehensively explained and evaluated in recent literature by Gutman et al. (2012). Fig. 10.4 confirms that the high pressure drop because of biofouling has pushed out the feed spacer (DuPont Water Solution, 2020). In a study by Vrouwenvelder et al. (2009), pressure drop development and biofilm accumulation in membrane fouling simulators have been studied without permeate production as a

FIGURE 10.4 High ΔP because of biofouling has pushed out the feed spacer. *Courtesy: DuPont Water Solutions; Source: Reproduced from DuPont Water Solution's FilmTec™ RO technical manual. (2020). Version 3; Form No. 45-D01504-en, Rev. 3; April 2020, https://www. dupont.com/content/dam/dupont/amer/us/en/water-solutions/ public/documents/en/45-D01504-en.pdf [Accessed on October 8, 2020].*

function of the process parameters such as substrate load, substrate concentration, flow direction, and linear flow velocity. Assessment of the biofouling studies with a membrane fouling simulator resulted in the following conclusions: (2) the pressure drop increase was associated with the accumulated biomass amount as well as linear flow velocity; (2) biomass accumulation was associated with the substrate load (substrate concentration and linear flow velocity), and (3) a flow direction variation in the PVs decreased the pressure drop increase.

Water hammer commonly occurs when the system gets started-up prior to the complete flush out of the air. This can happen at the preliminary start-ups or at the operation start-up, when the system is permitted to drain. It has to be made sure that the PVs are not under vacuum when the facility is shut down, or else air could reach the system. While starting-up a partly drained RO unit, the pump might perform as if it had no or little backpressure. It would be sucking water at high velocity, consequently hammering the membrane element. Moreover, the HPP could be harmed by cavitation.

The feed-to-reject DP is considered to be a measure of the hydraulic flow resistance of water through the system. This differential pressure will be extremely dependent on the water temperature and on the flow rates through the element flow channel. Hence, it is recommended that the product water and reject flow rates should be maintained constant for detecting and monitoring any element plugging that causes an increase in DP. The information on the extent as well as the location of the DP increase will be a beneficial tool for identifying the cause of the problem. Hence, it is advantageous and helpful to monitor the DP across each array and the overall feed-to-reject DP.

Some of the frequent causes and techniques for the prevention of high DP are examined in the following section.

10.3.3.1 Brine seal problems

The damage of the brine seal can lead to a sudden rise in DP. Brine seals could be damaged or "turned over" at the time of setting up or because of hydraulic surges. "Surge pressure," "hydraulic transient," or, in water applications, "water hammer" is a kind of hydraulic transient that refers to sudden pressure changes in a pipe system that can have devastating consequences, like ruptured valves and collapsing pipes. This can lead to a certain amount of feed water bypass around the element and low flow and velocity through the element, accordingly going beyond the maximal element recovery limit. As soon as this happens, the element will be more prone to scaling as well as fouling. When a fouled membrane element in one of the multielement PVs turns out to be more plugged, there will be a higher probability for the downstream elements to be fouled because of inadequate reject flow rates within that vessel.

10.3.3.2 Precipitated antiscalant

When the polymeric organic antiscalants contact with residual cationic polymeric floc-culants, or with multivalent cations such as aluminium, they would develop gummy pre-cipitants that could deeply foul the lead elements. In this case, cleaning would be extremely hard, and frequent application of an alkaline ethylenediaminetetraacetic acid solution might help.

10.3.3.3 Biological fouling

Biological fouling normally leads to a significant rise in the DP at the lead end of the RO system. Biofilm will be quite thick and gelatinous, consequently forming a high flow resistance.

10.3.3.4 Scaling

Scaling can lead to an increase in the tail-end differential pressure. It has to be ensured that scaling control is appropriately taken into consideration (please refer to Chapter 5: Pretreatment: Fouling and Scaling Control), and proper cleaning of the membranes should be carried out using suitable chemicals. Moreover, it has to be made sure that the designed system recovery would not be crossed.

10.3.3.5 Deterioration of pump impeller

The majority of the multistage centrifugal pumps use a minimum of one plastic impeller. As soon as a pump starts giving problems like misalignment of the pump shaft , the impellers will deteriorate and throw off small plastic shaving. This shaving could move in and physically plug the lead-end RO element. The discharge pressure of the RO pump should be monitored before any control valve as part of a periodic maintenance program to check if the pump is retaining the output pressure. If it is not maintaining its output pressure, then it might be deteriorating.

10.3.3.6 Breakthrough of pretreatment media filter

At times, various finer media from carbon, multimedia, sand, weak acid cation exchange resin, or diatomaceous earth pretreatment filter might breakthrough into RO feed water. The use of particulate detection methods is an important technique for monitoring the filtration efficiency and detection of the filtration breakthrough. Out of the different techniques available, the techniques that focus on employing light scatter from particles that might pass through a damaged filter include turbidity and particle counting. Both these could be used as online (process) monitoring techniques.

10.3.3.7 Bypass in cartridge filter

Cartridge filter will help to shield the RO unit from outsized fragments that could physically block the flow channel in the lead-end element. This type of blocking could occur when the cartridge filter is loosely fitted in its housing and connected without employing the interconnector. Occasionally the cartridge filters would be deteriorated when in operation because of the presence of incompatible materials or due to hydraulic shock. Cellulosic filters must not be included because they might deteriorate as well as plug the elements.

10.4 Calculations

10.4.1 Normalized permeate flow calculation

$$\text{NPF} = \text{Permeate flow} \times \left(\frac{\text{Baseline avg. NDP}}{\text{Avg. NDP}}\right) \times \left(\frac{\text{Baseline } T_{CF}}{T_{CF}}\right) \qquad (10.6)$$

where Avg. NDP is average net driving pressure, T_{CF} is temperature correction factor.

$$\text{Concentrate factor} = \frac{\text{Permeate flow} + \text{Concentrate flow}}{\text{Concentrate flow}} = \frac{\text{Reject concentration}}{\text{Feed concentration}} \qquad (10.7)$$

$$\text{Concentrate TDS} = \text{Feed TDS} \times \text{Concentrate factor} \qquad (10.8)$$

$$\text{Avg. NDP} = \left(\left(\frac{\text{Feed pressure} + \text{Reject pressure}}{2}\right) - \left(\frac{\text{Feed TDS} + \text{Reject TDS}}{200}\right)\right)$$
$$- \text{Permeate pressure} \qquad (10.9)$$

In RO applications, the below equation is employed for calculating the temperature correction factor, applied for the water permeability calculation.

$$T_{CF} = \frac{1}{\exp}\left(C \times \left(\left(\frac{1}{298}\right) - \left(\frac{1}{273 + \text{Temp}(°C)}\right)\right)\right) \qquad (10.10)$$

where C is a constant, characteristic of membrane barrier material. In the case of a polyamide membrane, the value of C will be 2500–3000.

In a work by Kim et al. (2016), the team introduced the corrected normalized permeate flux using the fitted osmotic pressure and temperature correction factor equations for minimizing the fluctuation at the time of a non-fouling situation.

Calculation—Example 10.1

A reverse osmosis system operates at a temperature range of 15°C–29°C. At a feed water temperature of 29°C, the NDP1 needed for the design capacity is 7.5 bar (108.9 psi). What net driving pressure is needed at 15°C for maintaining the design permeate capacity? The value of constant C (Eq. 10.10) for the membrane used is 2700.

For 29°C the $T_{CF1} = \frac{1}{\exp}\left(2700 \times \left(\left(\frac{1}{298}\right) - \left(\frac{1}{273 + 29}\right)\right)\right) = 0.88$

For 15°C, the $T_{CF2} = \frac{1}{\exp}\left(2700 \times \left(\left(\frac{1}{298}\right) - \left(\frac{1}{273 + 15}\right)\right)\right) = 1.369$

NDP2 = NDP1 × TFC2/TFC1 = 7.5 bar × 1.369/0.88

NDP2 = 11.66 bar = 169.11 psi

10.4.2 Normalized salt rejection calculation

$$\text{Normalized salt rejection (NSR)} = 100 - \left(\left(\text{Salt passage} \times \left(\frac{\text{Permeate flow}}{\text{Baseline permeate flow}}\right)\right.\right.$$
$$\left.\left. \times \text{TCF}\right) \times 100\right) \qquad (10.11)$$

where permeate TDS = permeate conductivity $\times$ 0.67.

A conversion factor is required for converting the total dissolved solids to conductivity. This conversion factor will be dependent on the types of salts and minerals dissolved in the water. The conversion factor could be found in published tables. In case the actual conversion factor is not found, then 0.67 is commonly employed as an approximate conversion factor. The value will be fairly accurate for the majority of natural waters.

$$\text{Salt rejection} = 1 - \left(\frac{\text{Pemeate TDS}}{\text{Feed TDS}}\right) \tag{10.12}$$

$$\text{Salt passage} = 1 - \text{Salt rejection} = 100\% \times \left(\frac{\text{Pemeate salt concentration}}{\text{Feed stream mean salt concentration}}\right)$$

$$T_{CF} = \frac{1}{\exp}\left(C \times \left(\left(\frac{1}{298}\right) - \left(\frac{1}{273 + \text{Temp}(^\circ C)}\right)\right)\right)$$

where C is a constant, characteristic of membrane barrier material. In the case of a polyamide membrane, the value of C will be 2500–3000.

10.4.3 Normalized pressure differential

The pressure differential (PD) can help to identify if the membranes are dirty. The normalized pressure differential would account for changes in temperature and flow. An increase in the normalized pressure differential would be an initial warning of fouling and/or scaling. For preventing complicated problems, the membrane must be cleaned if the normalized pressure differential becomes 15.0% or higher than the baseline.

$$\text{NPD} = \text{Pressure drop} \times \frac{\text{Baseline average flow}}{\text{Average flow}} \tag{10.13}$$

where

$$\text{Pressure drop} = \text{Feed pressure} - \text{Concentrate pressure}$$

$$\text{Average flow} = \left(\frac{\text{Pemeate flow} + \text{Concentrate flow}}{2}\right)$$

Baseline/standard average flow is the standard flow data collected at the initial 5–7 days, and this will be the basis for which all other data can be compared.

10.5 Case studies

10.5.1 Case study 1: Performance analysis of BWRO desalination plant

The performance analysis of a multistage multipass medium-sized spiral wound BWRO desalination plant (1200 m^3/day) of Arab Potash Company (APC) in Jordan was evaluated by the modeling and simulation by Al-Obaidi et al. (2018). To achieve the goal, a mathematical model for the spiral wound membrane RO process based on the principles of the solution diffusion model was developed.

The desalination plant feed water (pH in the range of 7.45–7.59) is fed from groundwater salt wells and subsequently pumped into collection tanks at the Arab Potash Company, and then directly pumped to the water treatment facility. The desalination plant consists of various process unit operations and devices including pretreatment process, high-pressure pumping, RO membrane module, post-treatment, and finally the brine disposal. The pretreatment system of the RO plant is the media-filter stage that consists of three multimedia filters comprising silica, anthracite, and garnet for removing the colloidal and suspended particles by media filtration while the water flows through a filter media bed. The second unit of BWRO plant consists of two HPPs and one stand-by pump at both first and second passes, respectively of RO plant to drive the feed water into the RO membrane. The third unit of BWRO is the RO module. Fig. 10.5 is the BWRO plant flowsheet that consists of 20 PVs and 120 membranes. The BWRO plant configuration comprises both product water and retentate reprocessing. The first pass consists of two parallel stages of six PVs with configuration (4:2) while the second pass consists of two PVs with configuration (2:1:1).

In this study, several new expressions were suggested for salt rejection and recovery rate. In addition to the utilization of specific correlations for estimating the influence of

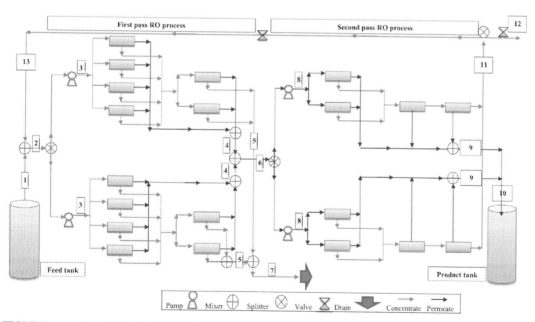

FIGURE 10.5 Schematic diagram of brackish water reverse osmosis desalination plant of Arab Potash Company. Source: *Reproduced from Al-Obaidi, M. A., Alsarayreh, A. A., Al-Hroub, A. M., Alsadaie, S., & Mujtaba, I. M. (2018). Performance analysis of a medium-sized industrial reverse osmosis brackish water desalination plant. Desalination, 443, 272–284.*

operating temperature on the model transport parameters. The model deals with the effect of operational conditions on the solution physical properties by considering the concentration polarization and the variable mass transfer coefficient. The model prediction was compared against actual operational data and showed reasonable agreement with small errors and hence applied for analyzing the impact of the operating conditions in a sensitivity simulation research work. The simulation was performed at a specific range of operating conditions within a 20% increment from the base case of actual plant information. This work evaluated the effect of numerous operation parameters including operating pressure, temperature, feed water flow rate, and feed salinity in the performance of the BWRO plant. This will in turn help to understand the effect of operation parameters on the efficiency of the facility and explains the range of the performance of the plant.

The sensitivity investigation of 20% variation of operation parameters illustrates the below:

1. Insignificant impact on entire plant salt rejection for all the tested operating parameters. Especially, the plant salt rejection is somewhat reduced with the salinity of feed water and operational temperature. Conversely, it is little increased with operating pressure and feed water flow rate.
2. The whole plant recovery is slightly increased with the salinity of feed water, considerably decreased with the flow rate of the feed water, remarkably increased with temperature and operating pressure.
3. Flow rate of feed water and operating pressure positively affect the permeate salinity whereas the operating temperature and feed water salinity adversely affect.

The mathematical modeling of the Arab Potash Company BWRO facility demonstrated in this study is considered to be a beneficial tool for investigating the plant performance for a longer-time operation because of the expectation of performance drop caused by scaling and fouling issues. Similarly, it will expedite the plant optimization studies needed for identifying the most efficient method of energy saving. Remarkably, this model developed could also be used for estimating the performance of RO seawater desalination plants of any size.

10.5.2 Case study 2: Performance assessment of a reverse osmosis plant

Performance analysis of a reverse osmosis plant has been experimentally presented by Idrees (2020). Several parameters that could influence the performance indirectly or directly were assessed. It was confirmed that the feed water properties have a significant impact on the permeate water properties, whereas the feed pressure has a substantial impact on the rejection percentage, salt passage, and recovery. Also, it was noted that when the temperature is increased, the salt passage also increased comparatively in product water. Table 10.1 presents the water parameters and operation conditions, as per the lab analysis. Fecal coliforms and total coliforms were not found in permeate water. The flow rate was maintained accurately to the designed membrane capacity. Different treatment methods coupled to pre-, intra-, and post-treatment methods were examined as well as selected as per the properties of feed water. It was noted that feed water pH increased

TABLE 10.1 Water parameters and operation conditions (as per the lab analysis).

Parameters	Unit	Feed water	After dosing	Permeate water	Reject water
Total dissolved solids	mg/L	738	722	16.75	1482
Hardness	mg/L	215.60	246.40	12.32	451.73
Chlorides	mg/L	38.28	40.20	7.65	82.31
pH	–	7.64	7.35	6.56	7.48
Sulfates	mg/L	289	284	8.0	510
Carbonates	mg/L	Below detection limit	Below detection limit	Below detection limit	Below detection limit
Color	mg/L Pt Co	Below detection limit	Below detection limit	Below detection limit	Below detection limit
Odor	–	Odorless	Odorless	Odorless	Odorless
Fecal coliforms	–	8	14	0	0
Total coliforms	–	11	32	0	0
Flow rate	m³/h	6.9	6.9	4.02	2.88
Pressure	psi	20	95	90	90

Reproduced from Idrees, M. F. (2020). Performance analysis and treatment technologies of reverse osmosis plant—A case study. Case Studies in Chemical and Environmental Engineering, 100007.

to a range of 3.0–10.0 resulted in an increase in metal ion rejection, salt rejection, and permeate quality. It was also confirmed that the permeate pH increased progressively by the increase in the feed water pH. Moreover, the plant economics was assessed and through the analysis of various recovery options, the reusing of rejection by mixing it with the raw water was tested, and thus the plant operated at zero discharge.

10.5.3 Case study 3: Autopsy of a fouled RO membrane element used in a brackish water treatment plant

The autopsy results of a spiral wound RO membrane after almost 1 year of service in a water treatment facility were studied by Tran et al. (2007). Even though the fouling of membrane is conventionally assessed by the flux reduction with time, this technique is insufficient for the characterization of fouling progress in a reverse osmosis process. It was demonstrated that when the product water flux is prominently affected, then the membrane will be seriously fouled in such a way that the restoration of its original permeability might be very difficult. Fouled membrane autopsies can be performed to understand better the physicochemical processes controlling the fouling. Different chemical and structural analyses are employed in this study, which includes X-ray diffraction (XRD), Fourier transform infrared spectroscopy (FTIR), Gas chromatography/mass spectrometry (GC—Ms), and inductively coupled plasma mass spectrometry (ICP-Ms). Both the cross section and top section of the fouled membrane were examined for providing additional understanding about the fouling layer development.

Optical images of the surface of the membrane after and before the feed spacer removal are presented in Fig. 10.6A and B, respectively. It could be noted that areas below or in the vicinity of the spacer strands have been concealed by brownish stains, while the level of staining in areas situated far away has been normally less severe and varied substantially. The assessment of areas adjacent to the strands at higher magnifications confirmed the random existence of microbes, as presented in Fig. 10.6B inset. The results from inductively coupled plasma atomic emission spectroscopy analysis are illustrated in Table 10.2. It was noted that the main elements detected are calcium (2760 mg/L), aluminium (2570 mg/L), and phosphorus (1225 mg/L). Moreover, small amounts of sulfur (865 mg/L), iron (590 mg/L), potassium (110 mg/L), magnesium (320 mg/L), silicon (410 mg/L), and sodium (190 mg/L) were also detected. A comparatively higher concentration of chlorine was found (1430 mg/L). The scanning electron microscopy/energy dispersive X-ray spectroscopy examination of the membrane cross-section provides additional information about the fouling layer development. Micrographs of a thin fouling layer at various magnifications and associated energy dispersive X-ray spectroscopy analysis are presented in Fig. 10.7.

The results obtained from various techniques have been consistent and complement each other. Following are the conclusions obtained from this study:

1. The degree of fouling was noted to be uneven across the surface of the membrane with areas under or in the vicinity of the feed spacer strands being influenced the most. The fouling in areas situated far away from the strands have been normally less serious,

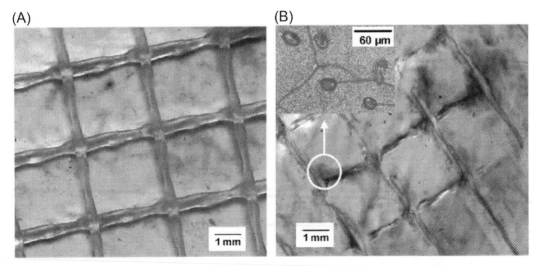

FIGURE 10.6 Optical images of the surface of membrane before (A) and after (B) the feed spacer was separated. *Source: Reproduced from Tran, T., Bolto, B., Gray, S., Hoang, M., & Ostarcevic, E. (2007). An autopsy study of a fouled reverse osmosis membrane element used in a brackish water treatment plant.* Water Research, 41(17), 3915–3923.

TABLE 10.2 Results of inductively coupled plasma atomic emission spectroscopy analyses of deposits scraped from the surface of the fouled membrane.

Elements	Concentration
Silver	60
Aluminium	2570
Barium	14
Calcium	2760
Chlorine	1430
Chromium	24
Copper	20
Iron	590
Potassium	110
Magnesium	320
Sodium	190
Nickel	22
Phosphorus	1225
Sulfur	865
Silicon	410
Strontium	23
Titanium	5.4
Zinc	35
Zirconium	19

Reproduced from Tran, T., Bolto, B., Gray, S., Hoang, M., & Ostarcevic, E. (2007). An autopsy study of a fouled reverse osmosis membrane element used in a brackish water treatment plant. Water Research, 41(17), 3915–3923.

however, varied substantially. The results highlighted the significance of local variations in the hydrodynamic conditions in the characterization of RO fouling.

2. The main inorganic elements in the fouling layer were aluminium, calcium, and phosphorus. The usage of phosphonate-based antiscalant and aluminium-sulfate coagulant could result in increased levels of phosphorus and aluminium. Small amounts of iron, sulfur, silicon, magnesium, potassium, and sodium were also available. Also, the fouling layer consisted of polysaccharides, proteins, and aromatic and aliphatic compounds derived from humic substances.

3. Fouling has been developed through various stages, as confirmed in the differences in the structure as well as the composition of the fouling layer on the basis of its thickness. Particularly, it comprise a primary thin fouling layer of an amorphous matrix with incorporated particulate matters. The amorphous matrix consisted of organic–aluminium–phosphorus complexes and the particulate

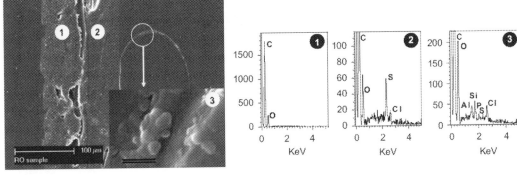

FIGURE 10.7 Scanning electron microscopy micrographs demonstrating a thin fouling layer at various magnifications as well as the associated energy-dispersive X-ray spectroscopy analysis. Bar (inset) = 500 nm. Source: *Reproduced from Tran, T., Bolto, B., Gray, S., Hoang, M., & Ostarcevic, E. (2007). An autopsy study of a fouled reverse osmosis membrane element used in a brackish water treatment plant. Water Research, 41(17), 3915–3923.*

matter was commonly aluminium silicates. Consequently, as the fouling layer attained almost 5−7 μm thickness, a secondary amorphous material started to deposit on the topmost area of the prevailing fouling layer. This secondary amorphous material had no particulate matter or any inorganic elements within it, however acted as a substrate upon which aluminium silicate crystals develop exclusively in the absence of other foulants inclusive of natural organic matter.

10.5.4 Case study 4: Performance evaluation of a reverse osmosis desalination plant

The performance assessment of a reverse osmosis desalination plant—Wadi Ma'in, Zara, and Mujib Plant, Jordan—was reported by Mohsen and Gammoh (2010). This desalination plant uses conventional pretreatments (such as coagulation, sedimentation, and filtration) and RO membranes for desalination. The plant produces 55 million m³/year desalinated water of 1500−2000 mg/L salinity This desalination plant obtains brackish water from three distinct sources: Wadi Mujib, Wadi Ma'in, and Zara Springs.

The design basis for the RO system is given below:

1. Maximum total dissolved solids of 1980 mg/L
2. Operating temperature 20°C−40°C
3. pH range maintained between 6.6 and 6.8 upstream of the RO system
4. Silica range of 18.55 mg/L (average) to 26.26 mg/L (worse-case)
5. Aluminum concentration of less than 0.02 mg/L after pretreatment
6. Iron concentration of less than 0.02 mg/L after pretreatment

Plant operation data for the RO unit from August 2006 through April 2008 has been considered for evaluating its performance. An assessment of the RO unit operation confirmed that there are two different operating trends for RO skids 61−68. In the course of the first 100−150 days of operation for each skid, the RO system performance (that is, pressure drop, product water production, system recovery, etc.) was less stable than standard operating procedure. After almost 150 days of operation, the RO unit performance has been noted to be more stable.

The daily average water production is noted to be 100,000 m^3 with a total dissolved solid of about 185 ppm. The feed water rate noted is 5100 m^3/h with a value of 1460 ppm total dissolved solids, while the designed values are 6000 m^3/h feed water input rate and 1500 ppm feed water total dissolved solids. The combination of low differential pressure across the RO system, low RO recovery, and high production of product water in the course of the first 150 days of operation might suggest that the operating conditions at this period are more appropriate for the RO system design than at a higher recovery. An increased permeate production was noted at the lower recoveries (between 75.0% and 85.0%) than when operated higher than 85.0%

It has been observed from this study that in the first 150 operating days, the overall RO unit recovery at this operation time frame has been constantly less than 85.0%, the minimum target value for the project. NPF per RO skid averaged almost 675 m^3/h. The NPF during this time frame was considered to be extremely erratic, however, was at its maximum over the whole 2-year operation period. During this period, the normalized pressure drop across the RO system was the lowest, and the salt rejection of the system was the highest, even though there was some variation. The controlled setpoints were more consistent and stable after the first 150 days of operation. The RO system was operated at a recovery steadily greater than 85.0%. The NPF reduced to almost 10.0% from the first 150 days of operation and stabilized to 600 m^3/h on average. The feed pressure for stages "one" and "three" increased by almost 2.0−3.0 bar over the following 300 days of operation. The normalized pressure drop across the RO unit slowly increased by almost 50.0%−100.0%. The RO membrane salt rejection was decreased during this period.

10.5.5 Case study 5: RO membrane autopsy and organic matter investigation

In a study by Jeong et al. (2016), the performances of pretreatment processes (inclusive of coagulation, dual-media filtration, polishing with cartridge filter together with antiscalants) used at Perth seawater desalination facility in Western Australia were characterized with respect to biological and organic fouling parameters. These evaluations were performed using assimilable organic carbon (AOC), three-dimensional fluorescence excitation emission matrix (3D-FEEM), and liquid chromatography with organic carbon detector (LC-OCD). With a capacity of 144 million liters per day, this facility is capable to produce up to 17% of Perth's drinking water requirement. Fig. 10.8 illustrates the process flow diagram at Perth Seawater Desalination Plant.

The RO membrane used in Perth Seawater Desalination Plant is SW30HR LE-400 (FILMTEC). The RO membrane was operated for 8 years. The RO membrane modules were taken from three different locations from one of the racks in the first pass RO (Rack 7): RO-A (7 Top) (fifteenth element of row A-positioned at the rack end), RO-B (3 Top) (twelfth element of rack A-located at the rack middle) and RO-C (3 Bot) (fifth element of

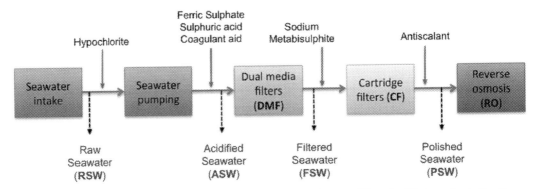

FIGURE 10.8 Process flow diagram at Perth Seawater Desalination Plant in Western Australia. Source: *Reproduced from Jeong, S., Naidu, G., Vollprecht, R., Leiknes, T., & Vigneswaran, S. (2016). In-depth analyses of organic matters in a full-scale seawater desalination plant and an autopsy of reverse osmosis membrane. Separation and Purification Technology, 162, 171–179.*

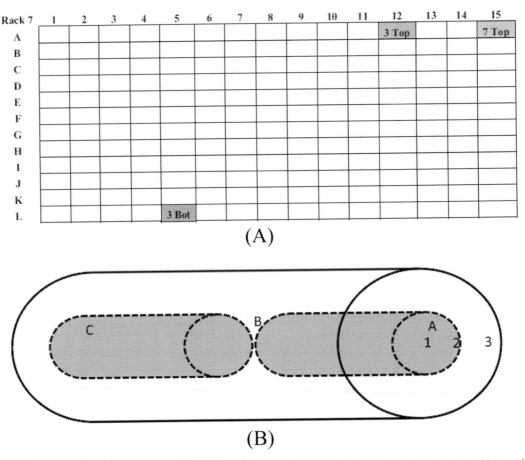

FIGURE 10.9 (A) Location of reverse osmosis membranes in rack 7 at Perth Seawater Desalination Plant and (B) sample locations of fouled reverse osmosis membranes used for autopsy in the study (A-1 to C-3). Source: *Reproduced from Jeong, S., Naidu, G., Vollprecht, R., Leiknes, T., & Vigneswaran, S. (2016). In-depth analyses of organic matters in a full-scale seawater desalination plant and an autopsy of reverse osmosis membrane. Separation and Purification Technology, 162, 171–179.*

row L which is positioned at the rack bottom), as shown in Fig. 10.9. The RO recovery was 45% and the average conductivity of rack 7 was 523 $\mu S/cm$. The RO cleaning followed the procedures stated here: (1) a low pressure flush using product water for removing the service concentrate as well as foulants, (2) cleaning solution prepared has been fed to the 1st stage and an optional soak and recirculation sequence were then employed, and (3) a low-pressure cleaning rinse using product water has been carried out for removing the chemical traces from the RO skid and the cleaning skid. Used RO membranes acquired from Perth Seawater Desalination Plant have been cut into several small segments. Three layers (from 3 to 1 toward the product water line) at three distinct locations (along the length) have been taken randomly for autopsy.

The results attained from this study are presented below:

1. Dissolved organic carbon in raw seawater is mostly comprised of humic substances and low molecular weight neutrals. The main difference in the raw seawater in the hot weather from the cold weather was the increased concentration of low molecular weight neutrals. This can be associated with the seasonal degree of algal and biological activities and the photodegradation in the sea.
2. Pretreatment processes decreased the particulate fouling possibility with a minor decrease in high molecular weight organic matter like biopolymers and humic substances. With the pretreatments at the plant, the biofouling potential was not reduced.
3. Biopolymers and humic substances have been the foremost organic foulants on the cartridge filter. Fouled cartridge filter released organic foulants, and this leads to an increase in organic fouling on the RO membrane. The cartridge filter replacement could be assessed by the organic content in produced water from the cartridge filter.
4. The maximum fouling on RO occurred in the front location of the RO rack. Inorganic scalant existing on the fouled RO membrane will be developed from the feed water, materials present in the plant, and the chemicals used in the pretreatment processes. Humic substances and low molecular weight neutrals were noted to be the main organic foulants on the fouled RO membrane. In short, it is important to properly control these organic and inorganic foulants before the RO process to effectively limiting the membrane fouling.

10.5.6 Case study 6: Fouling investigation of a full-scale seawater RO desalination facility on the Red Sea: membrane autopsy and pretreatment efficiency

In a study by Fortunato et al. (2020), the membrane autopsy was carried out on a large-scale seawater RO desalination facility situated on the Red Sea coast. This desalination plant is located 100 km north of Jeddah, Saudi Arabia. Fig. 10.10 demonstrates the process flow diagram of this seawater desalination facility. This facility had a capacity of 40,000 m^3/day. Seven membrane elements were set up in each PV (8 inches, Toray Industries). This desalination facility uses a two-pass configuration for meeting the drinking water boron level regulations in Saudi Arabia. This facility employed four seawater RO trains and four BWRO trains. Each seawater RO skid train has a single stage, with 140

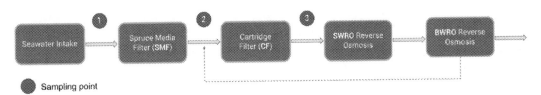

Sampling point

FIGURE 10.10 Process flow diagram of the desalination facility in Saudi Arabia on the Red Sea coast. Sampling points for assessing the pretreatment effectiveness at the time of the process. Source: *Reproduced from Fortunato, L., Alshahri, A. H., Farinha, A. S., Zakzouk, I., Jeong, S., & Leiknes, T. (2020). Fouling investigation of a full-scale seawater reverse osmosis desalination (SWRO) plant on the Red Sea: Membrane autopsy and pretreatment efficiency. Desalination, 114536.*

TABLE 10.3   Characteristics of the Red Sea water obtained from the seawater intake pump.

Parameters	Unit	Mean value
DOC	mg/L	1.33 (± 0.10)
Alkalinity	mg/L as $CaCO_3$	120.0 (± 2.0)
TDS	g/L	42.2 (± 0.2)
Turbidity	NTU	3.6 (± 0.2)
Conductivity	mS/cm	64.1 (± 0.2)
pH	–	7.9 (± 0.1)
Temperature	°C	22.0 (± 2.0)

Source: *Reproduced from Fortunato, L., Alshahri, A. H., Farinha, A. S., Zakzouk, I., Jeong, S., & Leiknes, T. (2020). Fouling investigation of a full-scale seawater reverse osmosis desalination (SWRO) plant on the Red Sea: Membrane autopsy and pretreatment efficiency. Desalination, 114536.*

PVs and 980 membrane elements. Each BWRO skid train has a dual-stage design, with 34 PVs and 238 membrane elements.

Various techniques have been used for characterizing the fate and the nature of the foulants in the process, including scanning electron microscopy-energy dispersive X-ray spectroscopy (SEM-EDS), inductively coupled plasma mass spectrometry (ICP-Ms), liquid chromatography-organic carbon detection (LC-OCD), total suspended solids, and adenosine triphosphate analyses. The pretreatment efficiency in reducing the fouling possibility was evaluated by examining the seawater after the intake feed pump, examining subsequent to the spruce media filter, and examining after the cartridge filter. The Red Sea water characteristics fed into the facility are presented in Table 10.3. Membrane autopsy has been carried out on the module operated for long term. The autopsies were carried out on three membrane modules taken from various positions inside the same PV (1st (lead; L), 4th (middle; M), and 7th (tail; T)).

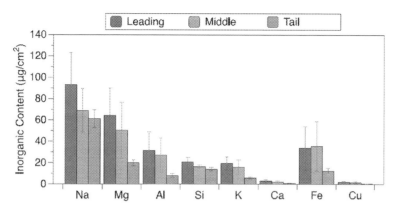

FIGURE 10.11 Inorganic composition of the foulants accumulated on the reverse osmosis membrane, as confirmed by inductively coupled plasma mass spectrometry. Source: *Reproduced from Fortunato, L., Alshahri, A. H., Farinha, A. S., Zakzouk, I., Jeong, S., & Leiknes, T. (2020). Fouling investigation of a full-scale seawater reverse osmosis desalination (SWRO) plant on the Red Sea: Membrane autopsy and pretreatment efficiency. Desalination, 114536.*

The results of the aforementioned study could be summarized as follows:

1. The dissolved organic carbon of the seawater reaching the facility essentially consisted of low molecular weight neutral organics. The pretreatment process showed some organic matter removal.
2. The organic fraction distribution noted in the fouling layer depend on the location of the module in the PV. The biopolymer content reduced from the leading to the tail module ($L = 29\%$, $M = 18\%$, and $T = 3\%$), whereas the low molecular weight neutrals demonstrated an opposite trend ($L = 45\%$, $M = 50\%$, and $T = 82\%$).
3. The scanning electron microscopy-energy dispersive X-ray spectroscopy and inductively coupled plasma mass spectrometry analyses (Fig. 10.11) of the fouling observed on the membrane and cartridge filter confirmed the existence of inorganic deposits (sediments) primarily composed of iron, aluminum, as well as magnesium silicate.
4. The inorganic sediments reached the facility from the shoreline seawater intake and collected on cartridge filter and the membrane module.
5. An increase in total suspended solids, and adenosine triphosphate was observed in the seawater collected after the cartridge filter, indicating the inappropriate cartridge filter replacement time.

10.5.7 Case study 7: Performance analysis of a reverse osmosis plant in Qatar

A performance study was carried out in a Qatar-based Umm Al Houl RO desalination plant established in 2015 (Acciona Agua). The plant produces 284,000 m³ water per day. The NPF, normalized salt passage, and normalized differential pressure data were collected over a 3-year period (2017−19). This operational data was monitored and the results are presented in Figs. 10.12−10.14. From Fig. 10.12, a gradual decreasing trend of NPF with operating time can be observed. In 2017 beginning, the NPF value was an average 1079 m³/hour, whereas in 2019 it was an average 932 m³/hour. The overall decrease in

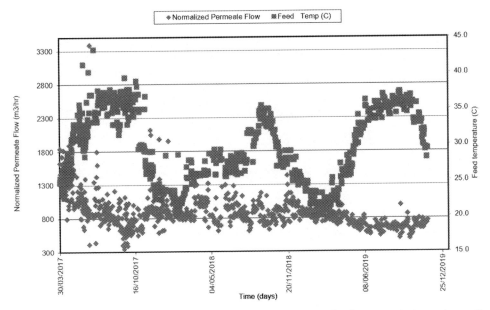

FIGURE 10.12 Normalized permeate flow over a 3-year period. *Courtesy: Acciona Agua. Personal communication with Acciona Agua team. Website: https://www.acciona.com/projects/middle-east/.*

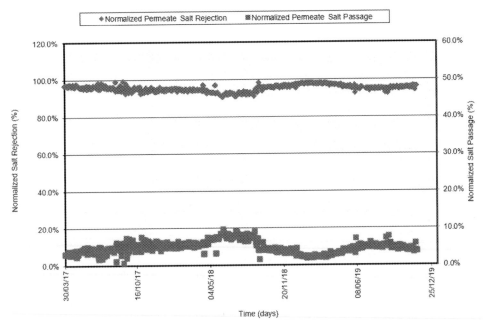

FIGURE 10.13 Normalized permeate salt rejection over a 3-year period. *Courtesy: Acciona Agua. Personal communication with Acciona Agua team. Website: https://www.acciona.com/projects/middle-east/.*

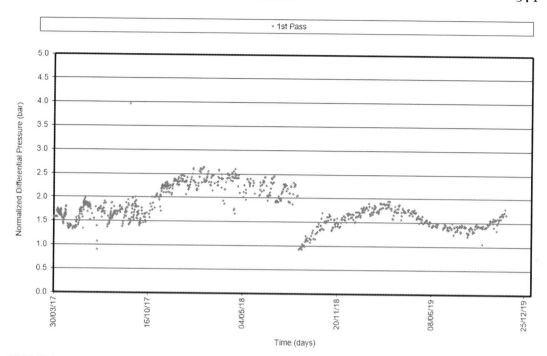

FIGURE 10.14 Normalized differential pressure over a 3-year period. *Courtesy: Acciona Agua.* Source: *Personal communication with Acciona Agua team. Website: https://www.acciona.com/projects/middle-east/.*

NPF is 5% per year. This is still better than the typical 7% flow loss per year caused by SWRO membranes age. The NSR ranged from 96.9% to 95.7%, from years 2017 to 2019 (Fig. 10.13). The data from Fig. 10.13 confirmed that the normalized differential pressure of the 1st pass increased with the operation time from 1.59 bar (2017) to 2.32 bar (2018), then dropped back to 0.93 bar after chemical cleaning.

References

Abbas, A., & Al-Bastaki, N. (2001). Performance decline inbrackish water filmTec spiral wound RO mem-branes. *Desalination, 136*, 281–286.

Abdullah, N., Rahman, M. A., Dzarfan Othman, M. H., Jaafar, J., & Ismail, A. F. (2018). Membranes and mem-brane processes. *Current Trends and Future Developments on (Bio-) Membranes*, 45–70. Available from https:// doi.org/10.1016/b978-0-12-813549-5.00002-5.

Al-Bastaki, N., & Abbas, A. (2004). Long-term performanceof an industrial water desalination plant. *Chemical Engineering Science, 43*, 555–558.

Al-Obaidi, M. A., Alsarayreh, A. A., Al-Hroub, A. M., Alsadaie, S., & Mujtaba, I. M. (2018). Performance analysis of a medium-sized industrial reverse osmosis brackish water desalination plant. *Desalination, 443*, 272–284.

Antony, A., Branch, A., Leslie, G., & Le-Clech, P. (2016). Impact of membrane ageing on reverse osmosis perfor-mance – Implications on validation protocol. *Journal of Membrane Science, 520*, 37–44.

Cho, J., Amy, G., & Pellegrino, J. (2000). Membrane filtration of natural organic matter: Factors and mechanisms affecting rejection and flux decline with charged ultrafiltation (UF) membrane. *Journal of Membrane Science, 164*, 89–110.

DuPont Water Solution's FilmTec™ RO technical manual. (2020). Version 3; Form No. 45-D01504-en, Rev. 3; April 2020, https://www.dupont.com/content/dam/dupont/amer/us/en/water-solutions/public/documents/en/45-D01504-en.pdf (Accessed October 8, 2020).

Fortunato, L., Alshahri, A. H., Farinha, A. S., Zakzouk, I., Jeong, S., & Leiknes, T. (2020). Fouling investigation of a full-scale seawater reverse osmosis desalination (SWRO) plant on the Red Sea: Membrane autopsy and pretreatment efficiency. *Desalination, 496*, 114536.

Gago, G. H. (2020). O&M Desalination Middle East Director, Acciona Agua, personal communication. Website: https://www.acciona.com/projects/middle-east/.

Gutman, J., Fox, S., & Gilron, J. (2012). Interactions between biofilms and NF/RO flux and their implications for control—A review of recent developments. *Journal of Membrane Science, 421*, 1–7.

Idrees, M. F. (2020). Performance analysis and treatment technologies of reverse osmosis plant—A case study. *Case Studies in Chemical and Environmental Engineering, 2*, 100007.

Ismail, A. F., Khulbe, K. C., & Matsuura, T. (2019). RO membrane fouling. *Reverse Osmosis, 189*–220. Available from https://doi.org/10.1016/b978-0-12-811468-1.00008-6.

Jeong, S., Naidu, G., Vollprecht, R., Leiknes, T., & Vigneswaran, S. (2016). In-depth analyses of organic matters in a full-scale seawater desalination plant and an autopsy of reverse osmosis membrane. *Separation and Purification Technology, 162*, 171–179.

Kenji, O. E., & Shingo, O. (2004). "eCUBE aqua" Application portfolio for reverse osmosis membrane diagnosis.

Kim, M., Park, B., Lee, Y. J., Lim, J. L., Lee, S., & Kim, S. (2016). Corrected normalized permeate flux for a statistics-based fouling detection method in seawater reverse osmosis process. *Desalination and Water Treatment, 57*(51), 24574–24582.

Kim, S., Lee, S., Kim, C.-H., & Cho, J. (2009). A new membrane performance index using flow-field flow fractionation (fl-FFF). *Desalination, 247*(1–3), 169–179. Available from https://doi.org/10.1016/j.desal.2008.12.022.

Kucera, J. (2015). *Performance degradation. Reverse osmosis: Industrial processes and applications* (2nd ed, pp. 305–311). Scrivener Publishing LLC, Published 2015 by John Wiley & Sons, Inc, Reverse Osmosis. doi: 10.1002/9781119145776.ch12.

Li, X., Mo, Y., Li, J., Guo, W., & Ngo, H. H. (2017). In-situ monitoring techniques for membrane fouling and local filtration characteristics in hollow fiber membrane processes: A critical review. *Journal of Membrane Science, 528*, 187–200.

Mahmoudi, E., Ng, L. Y., Ang, W. L., Chung, Y. T., Rohani, R., & Mohammad, A. W. (2019). Enhancing morphology and separation performance of polyamide 6, 6 membranes by minimal incorporation of silver decorated graphene oxide nanoparticles. *Scientific Reports, 9*(1), 1–16.

Matin, A., Khan, Z., Zaidi, S. M. J., & Boyce, M. C. (2011). Biofouling in reverse osmosis membranes for seawater desalination: Phenomena and prevention. *Desalination, 281*, 1–16.

Mohsen, M. S., & Gammoh, S. (2010). Performance evaluation of reverse osmosis desalination plant: A case study of Wadi Ma'in, Zara and Mujib plant. *Desalination and Water Treatment, 14*(1–3), 265–272.

Ning, R. Y., Troyer, T. L., & Tominello, R. S. (2005). Chemical control of colloidal fouling of reverse osmosis systems. *Desalination, 172*(1), 1–6.

Pais, J. A. G. C. R., & Ferreira, L. M. G. A. (2007). Performance study of an industrial RO plant for seawater desalination. *Desalination, 208*(1–3), 269–276. Available from https://doi.org/10.1016/j.desal.2006.06.017.

Park, H. G., & Kwon, Y. N. (2018). Long-term stability of low-pressure reverse osmosis (RO) membrane operation—A pilot scale study. *Water, 10*(2), 93.

Parlar, I., Hacıfazlıoğlu, M., Kabay, N., Pek, T. Ö., & Yüksel, M. (2018). Performance comparison of reverse osmosis (RO) with integrated nanofiltration (NF) and reverse osmosis process for desalination of MBR effluent. *Journal of Water Process Engineering, 29*, 100640. Available from https://doi.org/10.1016/j.jwpe.2018.06.002.

Principles of Reverse Osmosis Membrane Separation, Lewabrane, LANXESS. https://lpt.lanxess.com/uploads/tx_lxsmatrix/01_lewabrane_manual_ro_theory.pdf (Accessed October 19, 2020).

Reverse Osmosis Normalization. https://puretecwater.com/reverse-osmosis/reverse-osmosis-normalization (Accessed October 12, 2020).

Saleem, H., Trabzon, L., Kilic, A., & Zaidi, S. J. (2020). Recent advances in nanofibrous membranes: Production and applications in water treatment and desalination. *Desalination, 478*, 114178.

Saleem, H., & Zaidi, S. J. (2020a). Developments in the application of nanomaterials for water treatment and their impact on the environment. *Nanomaterials, 10*(9), 1764.

Saleem, H., & Zaidi, S. J. (2020b). Innovative nanostructured membranes for reverse osmosis water desalination. Qatar University Annual Research Forum and Exhibition (QUARFE 2020), Doha, 2020, https://doi.org/10.29117/quarfe.2020.0023.

Saleem, H., & Zaidi, S. J. (2020c). Nanoparticles in reverse osmosis membranes for desalination: A state of the art review. *Desalination, 475,* 114171.

Singh, R. (2005). *Water and membrane treatment. Hybrid membrane systems for water purification* (pp. 57−130)). Elsevier. Available from http://doi.org/10.1016/b978-185617442-8/50003-8.

Standard practice for standardizing reverse osmosis performance data, ASTM D 4516−00 (2010).

Tang, C. Y., Chong, T. H., & Fane, A. G. (2011). Colloidal interactions and fouling of NF and RO membranes: A review. *Advances in Colloid and Interface Science, 164*(1−2), 126−143.

Tran, T., Bolto, B., Gray, S., Hoang, M., & Ostarcevic, E. (2007). An autopsy study of a fouled reverse osmosis membrane element used in a brackish water treatment plant. *Water Research, 41*(17), 3915−3923.

Vrouwenvelder, J. S., Hinrichs, C. W. G. J., Van der Meer, W. G. J., Van Loosdrecht, M. C. M., & Kruithof, J. C. (2009). Pressure drop increase by biofilm accumulation in spiral wound RO and NF membrane systems: Role of substrate concentration, flow velocity, substrate load and flow direction. *Biofouling, 25*(6), 543−555.

Wang, X. N., Liu, Y., Pan, X. H., Han, J. X., & Hao, J. (2017). Parameters for seawater reverse osmosis product water: A review. *Exposure and Health, 9*(3), 157−168.

Zaidi, S. J., Fadhillah, F., Saleem, H., Hawari, A., & Benamor, A. (2019). Organically modified nanoclay filled thin-film nanocomposite membranes for reverse osmosis application. *Materials, 12*(22), 3803.

Zhao, Y., & Taylor, J. S. (2005). Assessment of ASTM D 4516 for evaluation of reverse osmosis membrane performance. *Desalination, 180*(1−3), 231−244. Available from https://doi.org/10.1016/j.desal.2004.11.089.

Reverse Osmosis System Troubleshooting

11.1 Troubleshooting of reverse osmosis system

There are several factors associated with reverse osmosis (RO) system fouling and some operational warnings, when monitored and documented, provide the operators with important information on the proper troubleshooting of the problem. The main goal of troubleshooting is to recognize the irregularities in the membrane system that leads to performance degradation, and to explore the modes of membrane system malfunctions, for ultimately restoring the original performance of the membrane (Kucera, 2007). The capability to accurately diagnose and correct the detected problems in the RO system is very important for successfully maintaining the unit online. In general, there are two categories of malfunctioning that a reverse osmosis system could suffer, that is, the acute type and the chronic type. Acute abnormalities happen when there is a transitory variation in the quality of feed water or any pretreatment system issue. These acute abnormalities must be managed rapidly before they get an opportunity to cause scaling, fouling, or degradation of the RO membranes. On the other hand, the chronic abnormalities can take a longer time for manifesting themselves and can bring about membrane scaling, degradation, or fouling before the operator could decide a definite action. Scientists are continuously carrying out studies for improving membrane performance and for reducing membrane fouling (Saleem & Zaidi, 2020a, 2020b, 2020c; Saleem et al., 2020; Zaidi et al., 2019). In the majority cases, there will be more than one reason for the drop in the membrane performance.

The performance of a reverse osmosis system is determined by the quantity and quality of permeate that is produced. Therefore, permeate flow and salt rejections are considered to be significant parameters for the proper evaluation of a reverse osmosis system. In some cases, the variations in the differential pressure (ΔP) of the element can be employed as an additional parameter for the assessment of RO system performance. A performance issue might happen when these factors are outside the protected value range. Even though the system performance might still be satisfactory, any unpredicted variations in these parameters must be strictly investigated and corrective actions must be taken instantly for preventing further damage to the system. Several factors influence the permeate flow, salt rejection, and element pressure drop. These are discussed in Chapter 10, Reverse Osmosis Membrane Performance Degradation, in detail. The performance variations unexplained by operation conditions involve troubleshooting.

11.2 Systematic fault detection and its solution

Periodic maintenance and proper record keeping are a critical part for efficiently operating a reverse osmosis membrane unit. If a system shows increased salt passage, then the operation data should be carefully reviewed before the fault could be identified accurately. Operation data like feed water analysis, feed pressure, total flow (permeate and feed), interstage pressures, pretreatment operations, percent recovery, chlorine concentration (or other oxidizers), alone or in combination, can have an effect on the performance of the RO membrane system leading to an increased passage of salt. For a proper evaluation of a reverse osmosis system, a single data point would not provide adequate information.

11.2.1 Startup checklist

Before the actual start of a reverse osmosis membrane system, the operator needs to review the prestartup checklist received from the system manufacturer. In case that the checklist is unavailable, then the operator has to get it from the system manufacturer. This prestartup checklist would help the system operator to confirm that the construction materials are right, gauges and valves are in place, fittings are tight and intended instrumentation is set up.

The data collection at the time of the preliminary startup of the RO membrane system is extremely significant. The information obtained through the early five to seven days would be considered as the foundation for which all other data can be related. The important information to be collected includes analysis of raw water, proper analysis of feed water including turbidity, TDS, silt density index (SDI), pH, and temperature. Moreover, the information on permeate flow rate, concentrate flow rate, concentrate TDS, permeate TDS, and conductivity should be obtained. As soon as this baseline is set, following the normalized data would be more significant.

11.2.2 Keeping of records

A proper maintenance program starts with appropriate record keeping. An operational logbook or electronic database for recording the operating pressures, chemical residuals, chemical additions, pretreatment process conditions, temperature, salt rejection, flow rates, and recovery rates would help the operator to accurately track the changes in the RO membrane system performance and operations. In general, the system that is used for tracking the RO membrane system performance would have trending capability, permitting the operator to graphically denote the performance over a period of time.

The data should be then normalized to a standard set of conditions. For distinguishing between real performance changes and normal operating changes, the system salt rejection and the permeate flow rate should be compared to a baseline performance or a reference based on the design. The start-up data is extremely beneficial in such situations. Without this preliminary data, the baseline to which the normalized data is compared would be restricted to the designed performance data, and not the real performance data. Membrane manufacturers will provide the data normalization software programs that could be set up on any computer system. The different RO system performance normalization methods and software programs (RODataXL and Toray Trak) are discussed in Chapter 9, Performance Monitoring, Process Parameters, and Their Control.

11.2.3 How trouble can be avoided?

The most appropriate way to prevent the RO system trouble is to avoid it at the initial stage. Some tips for RO design are stated below (Hydranautics):

1. Design the RO system with access to a thorough water analysis. In case of some seasonal variations (mostly in surface sources) or changing sources (mostly in municipal sources), the operator should try to obtain all the possible recent analyses.
2. Performing SDI test. It is a test used for characterizing the fouling probability of a feed stream. SDI is carried out on the basis of determining the rate of plugging a 45 μm filter employing a constant 30 psig feed pressure for a definite time period. SDI15 is the SDI test that will run for 15 minutes.
3. Perform proper pretreatment. The operator should ensure that the RO system design has suitable pretreatment processes.
4. Careful designing of the RO system flux rate, particularly if there is a fouling possibility. A decreased permeate water flow rate for a specified membrane area decreases the convective foulant deposition at the surface of the membrane. The surface water flux must range from 8.0 to 14.0 gallons per square foot of membrane area per day (GFD) and for well sources, it can be from 14.0 to 18.0 GFD.
5. Careful designing of the RO recovery rate. A conservative feed water percent recovery decreases the foulant concentration. When a reverse osmosis system has a 50% recovery, then the salt level in the reject would be two times the salt level in the feed water flow. When the recovery increases, the scaling possibility will also increase.
6. Maximizing the cross-flow velocity (CFV) in the membrane element. A conservative design will maximize the feed stream CFV and reject stream CFV. An increased CFV will decrease the salt concentration and foulant concentration at the surface of the membrane by improving their diffusion back into bulk feed stream above the surface of the membrane.
7. Selecting the appropriate membrane for the application. In certain cases, a low-fouling composite or a neutrally charged cellulose acetate blend RO membrane element will be a better choice as compared to a negatively charged composite polyamide (PA) RO membrane element for wastewater or surface water sources.

11.2.4 Identification of a problem

Variations in system operation parameters will definitely have an impact on the system performance. For example, an increase in the feed TDS will increase the feed pressure necessities by almost one psi for every 100 mg/L total dissolved solid increase because of higher osmotic pressure requirement. This would also increase the product water conductivity as the RO unit will continuously reject a fixed percent of the salts. A 10°F increment in the temperature of feed water would reduce the pressure requirement of feed pump by 15.0%. An increment in the system percent recovery will increase the concentrate TDS which will in turn increase the conductivity of product water. The reject water concentration will be 2 times greater at 50.0% recovery, 4 times greater at 75.0% recovery, and 10 times greater at 90.0% recovery. Lastly, a decrease in the product water flow would lead to higher conductivity, if the same recovery is retained due to the fact that the salt passage through the membrane will not be dependent on the water passage through the membrane, which leads to less product water for diluting the salts that have passed through.

It is highly recommended to normalize the logged operation data for determining if there is an issue in the RO system. The normalization software programs graphically represent the normalized salt passage (NSP), normalized permeate flow (NPF), and feed-to-concentrate ΔP. The normalized parameters can be determined by the comparison of operation data on a specific day to the initial day operation. Adjustments can be made for variations in main operation variables like pressure, recovery, feed TDS, and temperature. In this manner, the performance decline not related to operational parameters could be recognized and treated.

11.2.5 Primary indicators of performance issues

Having correct data about the RO membrane system and the understanding for interpreting that data accurately is the basis for effective troubleshooting of the system issues. There are three main indicators of performance issues that are mentioned below.

The primary indicators are:

1. NPF loss
2. Normalized salt rejection (NSR) loss
3. System differential pressure increase

The below details will help for properly detecting the problem and its causes.

11.2.5.1 Loss of permeate flow

The NPF loss is generally indicated by the equivalent increase in feed pressure needed to generate the required quantity of water. The permeate flow loss issue could be combined with variations in the salt passage, which are considered to be the secondary indicators of a performance issue. Fig. 11.1 demonstrates the low permeate flow with secondary indicators.

The below details can help to identify the problem due to fouling. Some of the foulants affect the front end of the system, whereas others affect the back end of the system. The RO troubleshooting table (given in Section 11.2.6) can be used for determining the nature of the foulant (Hydranautics).

1. Was the RO system shutdown properly? In certain cases, the concentrate water from the service operation must be rinsed out of the system at the time of shutdown. Else,

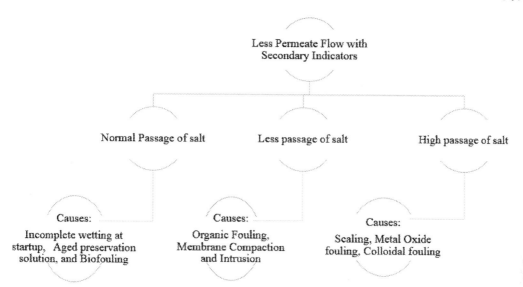

FIGURE 11.1   Less permeate flow with secondary indicators.

the inorganic foulants could precipitate onto the membrane surface. RO permeate is considered to be the best flush water source.

2. Was the RO system stored properly? An inappropriately stored system (particularly under warm conditions) could create a severe biofilm issue.

3. Was the target pH or scale inhibitor concentration achieved? In case the operator is acidifying for lowering the feed pH or adding scale inhibitors for the calcium carbonate scaling control, he has to confirm if target pH or scale inhibitor concentration is achieved. Else, the operator should do an acid cleaning.

4. Has the ΔP between the feed line and concentrate line increased higher than 15.0%? The ΔP increase confirms that fouling of the feed path and a flow restriction over the surface of the membrane is happening. The ΔP monitoring across stages will offer the operator the benefit of examining if the fouling is restricted to a specific stage, which could assist in identifying the possible foulants.

5. For seawater system, is the flushing carried out using the permeate water at shutdown? This flushing can help to remove the higher concentration of ions that might precipitate out of the solution. At the very least, feed water could be used, but it is suggested to use product water for the flushing purpose.

6. Is fouling occurring in the cartridge filters? Proper inspection of the RO feed cartridge filter should be carried out for foulants, and this is comparatively easy.

11.2.5.2 Loss of salt rejection or an increase in normalized salt passage

An increase in NSP as confirmed by a specific salt or conductivity measurement is considered to be an operational problem indicator. This NSP increase can be combined with secondary indicators like the variations in NPF. Fig. 11.2 demonstrates the increased salt passage with secondary indicators. The causes may range from the leakage of O-rings on

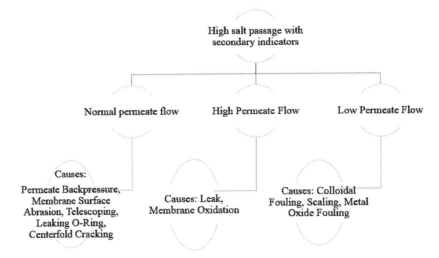

FIGURE 11.2   Increased salt passage with secondary indicators.

the interconnectors to the surface abrasions to membrane layer oxidation to scale. Certain issues can be fixed, whereas some other issues are permanent and cannot be fixed. In the long run, all issues can be avoided.

As an example, if the interconnector O-rings are not frequently lubricated as well as placed appropriately within the permeate tube, then it might cause the feed water to pass by the O-ring and leak into the product water. This will lead to an increased conductivity for the product water with negligible effect on the product water flow rate. Similarly, abrasions on the RO membrane surface (possibly due to sand or other abrasive materials present in the feed water) could lead to reduced salt rejection with the least effect on the product water flow rate.

In case that an O-ring leakage is suspected, then the operator should probe the RO system and generate permeate conductivity profiles for every pressure vessel (PV). The resultant conductivity profiles will support to the identification of the problematic section.

The PA-based RO membranes will be subjected to irreversible damage by oxidants like peroxide, bromine, chlorine, etc. As soon as the membrane surface undergoes oxidation, the rejection of salt will decrease and the product water flow rate will increase. Sometimes, at the initial stages of oxidation, the rejection of salt might slightly increase, before decreasing. It is very important to make sure that the entire oxidizing chemicals in the feed water are fully neutralized before reaching the RO system.

The salt rejection loss demonstrates itself as an increased product water conductivity. This might be because of an O-ring leakage, membrane surface fouling, or degradation. The below questions can help the operator to identify the source of the issues (Hydranautics). The operator should confirm that the product water conductivity has not increased higher than 15.0%.

1. Do all the PVs in a stage have almost similar product water conductivity? The operator needs to measure the product water quality by stage and by PV, as far as possible. The

PV with a considerably increased permeate conductivity will possibly have a defective O-ring, a disconnect, or a deteriorated RO membrane.

2. Was the composite membrane exposed to any strong oxidant like chlorine? This kind of exposure can damage the membrane.
3. Was the cellulose acetate membrane exposed to some pH extremes? This kind of exposure may damage the membrane. Possible causes of pH extremes are defective metering pump, dry acid tank, flushing/storage in non-acidified water, etc.
4. Are the instruments precise? The operator should confirm that the entire instruments are correctly calibrated.
5. Do the elements look damaged or discolored? The operator should properly inspect the RO elements for physical damage or for foulants.
6. Are the actual conductivity as well as the temperature of the feed water same as the design criteria? In the case that actual feed water has greater TDS or is warmer as compared to the design criteria, then this could lead to inconsistency. The detailed water analyses of the RO feed, permeate, and concentrate should be obtained. Then, analyses results must be compared to the design projection of the RO element fabricator.
7. Will the permeate pressure goes beyond the feed pressure at certain times? If the product water is pumped to a raised location, and there is no check valve on the product water lines, at shutdown, the permeate pressure could go beyond the feed pressure. This could lead to the expansion and rupture of the membrane envelopes.
8. Are the O-rings in perfect condition? The O-rings could flatten or crack with age. This would lead to the development of leaks. The periodic O-ring replacement is an economical cost-effective and good preventive maintenance step. Otherwise, PVs can be probed for finding defective O-rings.

11.2.5.3 Increase in system differential pressure

The difference between the feed pressure and the reject pressure is termed as the system differential pressure (Park & Kwon, 2018). Normally, this will be a value that is used to gauge the system performance deviations, same as salt passage and permeate flow.

Differential pressure could be calculated across a system, in a single stage, or even a single PV. Based on the location of the abnormal differential pressure, there could be several distinctive causes and secondary indicators are possible.

If the operator thinks there is an issue,

1. Once the operator has dismissed any mechanical failures as the RO problem source, then he or she should determine the suspected foulants and carry out cleaning or a sequence of cleanings.
2. The cleaning solution must be obtained and examined for the foulants removed, pH variation, or color variation. The cleaning efficiency could be confirmed by positioning the RO unit back into service.
3. If the operator is not sure about the foulants or the cleaning solutions to be used and the procedures, then he or she can contact the companies who are specialized in the proprietary cleaning chemical supply and off-site assessments of RO elements.
4. If all the other assessments fail in finding the causes of fouling, then a destructive autopsy could be carried out. The RO element can be cut open and unrolled with

analytical investigations run on the membrane and the foulants for determining the issue.

11.2.6 RO troubleshooting grid

Variation in the salt passage, permeate flow, and the ΔP are indications that could be related to definite causes in several cases. A general idea of symptoms, their probable causes, as well as corrective measures, as recommended by DuPont Water Solution, are provided in the troubleshooting grid (Table 11.1).

TABLE 11.1   Symptoms, their probable causes, and corrective measures.

Permeate flow	Salt passage	Differential pressure	Direct cause	Indirect cause	Corrective measure
	Increasing (main indicator)	Not changing	Oxidation damage	Free chlorine, ozone, KMnO$_4$	Replacement of membrane element
	Increasing (main indicator)	Not changing	Membrane leak	Permeate backpressure; abrasion	Replacement of membrane element, cartridge filtration improvement
	Increasing (main indicator)	Not changing	O-ring leakage	Incorrect installation	Replacement of O-rings
	Increasing (main indicator)	Not changing	Product tube leaking	Damaged at the time of element loading	Replacement of element
Decreasing (main indicator)			Scaling	Inadequate scaling control	Cleaning, scaling control
Decreasing (main indicator)			Colloidal fouling	Inadequate pretreatment	Cleaning, improve pretreatment
Decreasing	Not changing	Increasing (main indicator)	Biofouling	Contaminated source water, inadequate pretreatment	Cleaning, disinfection, improve pretreatment
Decreasing (main indicator)	Not changing	Not changing	Organic fouling	Oil; cationic polyelectrolyte water hammer	Cleaning, pretreatment improvement
Decreasing (main indicator)	Decreasing	Not changing	Compaction	Water hammer	Replacement or addition of elements

DuPont Water Solution's FilmTec™ RO technical manual (2020) Version 3; Form No. 45-D01504-en, Rev. 3; April 2020, https://www.dupont.com/content/dam/dupont/amer/us/en/water-solutions/public/documents/en/45-D01504-en.pdf [Accessed November 5, 2020].

11.2.7 Cleaning and maintenance

The guidelines for RO membrane system cleaning vary slightly from manufacturer to manufacturer; however, the standard rule will be as following: The system should be cleaned: (i) if the NSP is increased by 5.0%, (ii) if the NPF is reduced by 10.0%, or (iii) if the normalized differential pressure is increased by 15.0%.

Normalized values should be used for this interpretation and are considered to be the standard data for which all other data must be compared. Once the problem area is recognized, based on the warnings, the operator should decide on how the membrane system can be cleaned and decide how it can be prevented from happening again. In most cases, system performance could be recovered by employing hydrochloric acid for scaling and sodium hydroxide for organic fouling. For keeping the biological fouling to a minimal level, a weekly maintenance dosage of a compatible biocide can be applied. The standard cleaning chemicals are alkaline cleaners and acid cleaners. The alkaline cleaners can be employed for removing the organic fouling including biological fouling while the acid cleaners can be employed for removing the inorganic precipitates and iron. In general, sulfuric acid is not suggested for cleaning due to its possibility of calcium sulfate precipitation in the system. The RO permeate is regarded as the ideal water quality for the cleaning of the system; however, the pretreated raw water can also be used. Several raw waters are very much buffered and might lead to increased acid as well as caustic usage for achieving the required pH of 2 and 12, respectively. Ibrahim et al., (2020) studied the fouling behavior and the cleaning behavior of bis(triethoxysilyl)ethane-derived organosilica membranes, which is considered as an advanced type of RO membranes that offer high thermal stability and chlorine tolerance. With hydrothermally stable bis(triethoxysilyl)ethane membranes, the chemical-free cleaning was confirmed.

Even though the membrane manufacturers have their cleaning guidelines for flow rates, temperature, and pH, the maximal cleaning temperature will be normally 50°C. Certain membrane manufacturers permit for highly aggressive cleaning conditions and have set greater than standard pH and temperature limits for cleaning purposes. Moreover, there will be cleaning flow rates recommended for different element sizes that would simplify the cleaning procedure. For several 8-inch elements, a low-flow cleaning is recommended to be carried out initially.

During the selection of a cleaning solution, numerous factors should be considered, such as the membrane type, membrane element compatibility, and the foulants to be removed (SUEZ Water Technologies & Solutions). The cleaning solution should fall within the pH ranges listed for the specific element. Moreover, the cleaning solution should not consist of some chemical substances that are not compatible with the element, like some surfactants and, in certain cases, oxidizing agents like chlorine. The entire foulants should be removed by means of a clean-in-place (CIP) process before the occurrence of irreversible membrane damage. On the other hand, it will be very easier for removing the foulants during the start of the fouling process, as compared to a cleaning when a thick layer of fouling is developed. Therefore, a CIP must be carried out when there are considerable evidences about the start of the fouling process.

11.2.7.1 Suggested cleaning equipment

It is recommended to have a cleaning solution-mixing tank with a cover and a temperature gauge. Suitable valving, cartridge filter, recirculation pump, pressure gauges, pH monitor, sample ports, pH monitor, and flow meters are also suggested. While choosing a cleaning system equipment, the construction material of the system's components must be physically and chemically compatible with the cleaners and temperatures to be used. A cartridge filter on the cleaning solution return-to-tank or feed line to the cross-flow filtration machine will be removing the particles dislodged from the membrane elements (SUEZ Water Technologies & Solutions).

11.2.7.2 Amount of cleaning solution required

For determining the amount of cleaning solution needed, the operator should estimate the hold-up volume of the cleaning loop piping and the membrane element housings. Subsequently, add adequate water to the CIP tank for preventing it from emptying while filling the system. During the start of the cleaning cycle, the process water in the system must be discharged to drain as it will be displaced by the cleaning solution. For estimating the recirculation dimension of CIP tank, calculate the system hold-up volume and subsequently multiply it by two. For the hold-up volume in the membrane element housing, use the below estimation given the housings are filled with the maximal number of membrane elements.

1. Twenty liters for all 8-inch elements (five gallons/element)
2. Four liters for all 4-inch elements (one gallons/element)

Calculation 11.1

For cleaning 10 eight-inch-diameter PVs with six membrane elements per PV, the below calculation can be applied:

To find the volume in vessels,

$$V_{vessel} = \pi r^2 l \tag{11.1}$$

in which, V_{vessel} = volume in vessel, r = radius, and l = length

Here, $r = 4$ inches and $l = 20$ ft.

$$V_{vessel} = 3.14 \times (4 \text{ inch})^2 \times 20 \text{ ft.} \times \frac{0.0069 \text{ ft.}^2}{\text{inch}^2} \times \frac{7.48 \text{ gal}}{\text{ft.}^3} = 52 \frac{\text{gal}}{\text{vessel}}$$

Therefore, the volume in 10 vessels,

$$V_{10 \text{ vessels}} = 52 \times 10 = 520 \frac{\text{gal}}{\text{vessel}} = 1.97 \text{ m}^3$$

To find the volume in the pipe, assuming total length 50 ft. and 4-inch SCH 80 pipe
The internal diameter of the pipe is 3.82 inches, and thus radius = 1.91 inches
Here, $r = 1.91$ inches and $l = 50$ ft.

$$V_{\text{pipe}} = 3.14 \times (1.91 \text{ inch})^2 \times 50 \text{ ft.} \times \frac{0.0069 \text{ ft.}^2}{\text{inch}^2} \times \frac{7.48 \text{ gal}}{\text{ft.}^3} = 30 \text{ gal}$$

Therefore, the total volume in 10 vessels and pipe,

$$V_{10 \text{ vessels+pipe}} = 520 + 30 = 550 \text{ gal} = 2.08 \text{ m}^3$$

Hence, the cleaning tank must be almost 550 gal (2.08 m^3).

11.2.7.3 Clean-in-place protocol

In the majority of events, clean using a lower pH cleaner initially, except when sulfate scaling, silica scaling, or grease/oil fouling is suspected. Colloidal fouling could be covered by slow forming scaling. It should be separated by low-pH cleaners initially to uncover the silt and hence making it accessible to be removed using high-pH cleaners.

The below mentioned general cleaning procedure is recommended by Suez Water Technologies & Solutions (SUEZ Water Technologies & Solutions):

1. Properly inspect the cartridge filters, cleaning tank, and hoses. If necessary, clean the tank and flush the hoses. New cartridge filters should be installed. A 5 μm or tighter rating filter can be used on the cleaning unit.
2. Fill the cleaning tank with RO product water or deionized water. Turn on the agitator or tank recirculation pump.
3. The cleaner can be slowly added to the cleaning tank and allowed for proper mixing.
4. The temperature of the solution should be checked. If the solution temperature is lower than the suggested level, then adjust the heating control for providing optimum temperature. If a heater is unavailable, then recirculate the cleaning solution by employing the high-pressure pump of the membrane system. This might help to attain higher temperature.
5. The solution pH must be checked. If pH is very low, then increasing the pH using sodium hydroxide, or other chemicals as suggested by the membrane manufacturer. If pH is very high, adjust using hydrochloric acid.
6. Circulate the solution through one stage at a time in the feed flow direction for 10−30 minutes. For ensuring that this maximal flow is not crossing the limits, it is highly recommended not to go beyond 0.7 bar of ΔP per membrane element and 3 bars per PV.
7. Pressure must be lower enough such that minimum permeate is produced at the time of cleaning, but always less than 4.2 bars; 2.5−4 bars for the RO membranes, and 1.5−2.5 bars for the other membrane types (microfiltration, ultrafiltration, and nanofiltration). Increased pressure would result in higher permeation and holds foulant to the membrane surface.
8. The membranes are soaked for about 25 minutes. This will increase the cleaner effectiveness.
9. In case that the 1st-stage cleaning solution turns out to be discolored or turbid, then drain the tank and prepare a new cleaning solution before continuing. In the event that the solution temperature or pH reaches out of the suggested limit, then a fresh

solution must be prepared. In any case, a fresh cleaning solution must be prepared for every stage.

10. Rinsing can be done using RO permeate prior to returning the system back to service.
11. While returning the system to service, divert the product water to drain till any residual cleaning solution has been rinsed from the system.

In case that a second cleaning is required, always rinse the system till getting a neutral pH in both concentrate and permeate. The above-stated procedure can be repeated.

11.2.8 Cleanup and disposal

When handling the acids and bases, proper precautions should be taken for ensuring the safety of the operator and others around him or her. Make sure that the base and acid storage areas are separate and the spilled areas are cleaned. While handling the chemicals, always wear suitable personal protective safety equipment. Li et al. (2016) studied the cleaning effects of oxalic acid under ultrasound on the used RO membranes using an online cleaning and monitoring system. Fig. 11.3 is the diagrammatic representation of an ultrasonic-chemical cleaning system for the used RO membrane cleaning. The results obtained from this study could be employed as a reference basis to design a large-scale chemical and ultrasonic cleaning system of the used RO membrane.

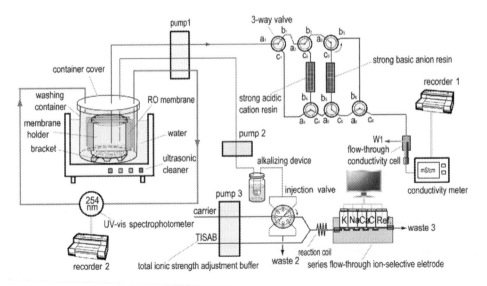

FIGURE 11.3 An ultrasonic-chemical cleaning system for cleaning the used reverse osmosis membranes. Source: *Reproduced from Li, Y. S., Shi, L. C., Gao, X. F., & Huang, J. G. (2016). Cleaning effects of oxalic acid under ultrasound to the used reverse osmosis membranes with an online cleaning and monitoring system. Desalination, 390, 62–71.*

11.3 Investigative strategies

Different investigative strategies that can be used for RO system troubleshooting. These investigative techniques are the following:

1. Mechanical assessment
2. Common performance problems
3. System design and performance projections
4. Assessment of data
5. Testing of water sample
6. Testing of membrane integrity
7. Profiling a reverse osmosis array
8. Probing
9. Replacement of O-rings
10. Shimming
11. Replacement of RO membrane elements
12. Membrane autopsies

Generally, more than one of these strategies are required for getting a total picture of system operation and to figure out what may have caused the performance loss. In a study carried out by Tow and Rencken (2016), a fouling visualization apparatus was designed for elucidating the centimeter-scale mechanisms of organic fouling and cleaning in RO. Alginate can be employed as a model organic foulant and dyed with methylene blue, which is confirmed not to affect fouling or cleaning and to adequately highlight the gel for visualization at lower salinity (upto 1% sodium chloride). The cleaning videos confirmed that foulant cake swelling and wrinkling could facilitate gel detachment and removal. Despite their efficiency in slowing fouling, spacers could hinder the removal of detached foulant pieces by blocking their path.

Sim et al. (2018) examined various approaches for monitoring RO membrane fouling, either by the characterization of the feed fouling potential or by the detection of the membrane fouling condition. The different techniques critically reviewed in this paper could be incorporated into the RO process for providing data on the nature of the foulants deposited on the RO membrane.

In the following section, the different investigative strategies for RO system troubleshooting are discussed in detail.

11.3.1 Mechanical assessment

Probably, the first line of the examination must be a mechanical assessment of the system for eliminating causes associated with valves or instrumentation. Regular examinations include an instrument examination; they must be tested and calibrated. Also, the filters must be inspected for channeling. Valves also must be examined for determining whether they are operating properly. Softeners and filters have to be inspected to decide if the flux or specific flow rate is within the design guideline. When the mechanical problems are either eliminated or recognized as the reason for reduced performance, the RO system performance ought to be assessed to confirm if some other factors are causing the reduced performance.

11.3.2 Common performance problems

There are some common performance problems that need to be examined while RO system troubleshooting. These issues are the following:

1. Visual investigation: Basic visual examinations of pretreatment along with RO systems can demonstrate the potential for scaling or fouling.
   * Inspect tanks and piping for biogrowth.
   * The feed side of a PV must be opened and assessed for fouling. Biofouling will have an odor and will feel slippery.
   * The concentrate side of a PV should be opened and examined for scaling. If scaling is present, then the surface will be rough.
2. Application of chemical: Dosages and controls need to be checked to guarantee appropriate chemical feeding. This includes caustic/acid, chlorine or other oxidizers, antiscalants, and coagulants in pretreatment.
3. Frequency of cartridge filter replacement: Higher replacement rates (at 2-week regular intervals or lower) can represent a fouling issue. Lower replacement rates (at a one-month regular interval or more) can result in microbial fouling because the microorganisms grow in the "old" cartridges.
4. Frequency of membrane cleaning: The RO systems that require more than four cleanings each year might have a membrane fouling or/and scaling issue.

11.3.3 System design and performance projections

11.3.3.1 System design

The RO system design greatly affects the potential for scaling or fouling of the membranes. The recovery, water flux, concentrate flow, and feed water flow affect the membrane's ability to scale and foul. The flow rates influence the concentration polarization boundary layer in which scaling and fouling happen. Recovery and flux will affect the contaminant concentration within the boundary layer. The membrane scaling as well as fouling could be limited if proper design guidelines are followed. Moreover, if the design guidelines are not properly followed, then membrane scaling and fouling are more prone to occur. If a reverse osmosis system gives indications of performance problems, the system design must be investigated. In case that the fluxes and flow rates do not agree with the design guidelines, then the impact of the variations having on the membrane performance must be carefully assessed.

11.3.3.2 Performance projection

As discussed in Chapter 9, Performance Monitoring, Process Parameters, and Their Control, the performance projection software programs for the RO system design can be used to assess the RO system performance. During practical application, the performance may fluctuate within $\pm 15.0\%$ from nominal values (given in the projection) (DuPont

Water Solution, 2020). The rejection of salt might also change for specific membrane module, however, would not be lower than the minimum specified rejection of salt. Accordingly, the actual performance might not match with performance projections; however, must be close for systems larger than around 125 gallons per minute (GPM) (for a module diameter of about 8 inch). As an example, the actual permeate TDS must not be higher than almost 1.5 times the concentration projected. The product water flow must vary by not more than almost ±5.0%. Should the actual performance shifts significantly from the projected one, then the membrane scaling, fouling, or degradation might happen. Also, it has to be noted that there is a time period after the new membrane installation, when the membrane performance will not be stable. At this timeframe, which could last as long as about 14 days of consistent operation, the salt passage and permeability of the membrane both drop. This drop in performance is because of compaction and will be worsened for seawater and wastewater applications. Other reasons behind this decrease are not very clear but may include the hydration degree of the membrane upon start-up. The initial permeability as well as salt passage can drop up to 10% at this time period of destabilized performance (Wilf et al., 2007). Moreover, the permeate backpressure must be considered when assessing the performance of a reverse osmosis system. Otherwise while running projections, the actual performance would demonstrate an increased feed pressure relative to the projected. Consequently, whenever the observed operating pressure is higher than the projected, the projection should be examined to decide whether the permeate pressure was considered. In the event that the permeate pressure was included in the development of projection, then scaling or fouling of the membrane can clarify the dissimilarity between the predicted and the actual operating pressures. Also, the ΔP should be evaluated. The majority of the projection programs consider a piping loss of around 5.0 psi per stage in addition to the ΔP through the membrane module in the projection program. If the actual ΔP surpasses the predicted ΔP, then there can be two clarifications: membrane scaling or fouling or pressure sensor position causing deceptive readings. The restrictions in the concentrate and feed headers can prompt a higher value than the pressure drop predicted. The location of the pressure sensor must be near to the PV to prevent these restrictions and away from increased-disturbance areas, for example, valves.

11.3.4 Assessment of data

Data, especially normalized data, should be assessed for determining the nature of the decline in the performance of the membrane. Differential pressure, NSR, and NPF must be assessed to determine the performance trends. Also, normalized data must be used to decide when the membrane must be cleaned. This data will help to decide whether the membranes have been cleaned timely, or if probable scaling or/and fouling can end up permanently due to the fact that the membranes were not cleaned timely. Further, the efficiency of cleaning could be assessed by following the normalized data trend. The performance after cleaning should return to starting conditions, and also the efficiency must stay at such a level for a prolonged time period. Also, the normalized performance could be used to check the improvements in pretreatment, for the system which could be effectively cleaned.

11.3.5 Testing of water sample

The testing of water sample cannot be underestimated in regards to the significance it has to the efficiency of the RO unit. The source of water plays a major role in deciding the extent of membrane scaling, fouling, and degradation that the RO unit may undergo. During the troubleshooting process, the source of water must be specified. In the case that it is municipal water, then the ultimate source should be specified. The pretreatment unit must be examined for ensuring that it meets the feed water challenges.

A comprehensive water examination is additionally necessary. Particular species could be identified that might have already scaled, fouled, or degraded the membranes. The mass balance around the membrane is a troubleshooting method that will assist in figuring out the species which have scaled or fouled the membrane. A species is chosen, for example, aluminum or iron, and subsequently, a mass balance is carried out around the RO unit. The total species in the feed side must be equal to the sum of the species in the concentrate and the product water together. It has to be taken into consideration that this is mass conservation, and not the concentration. In this way, in order to calculate the amounts, the user should take the concentration, and then it can be multiplied with the flow rate. This would generate an amount for each unit time, which would be appropriate, if the entire timeframes are identical. If a species appears short in the permeate plus concentrate versus the feed mass balance, it is possible that this species has been deposited on the membrane or somewhere else in the membrane module.

11.3.6 Testing of membrane integrity

In the case that the NPF is high or the NSR is low, the membrane integrity might be questionable. A membrane element with an increased salt passage must be first tested if leakages are present with feed water or reject water leakage into the element product waterside. Leakages might occur through surface damage of membrane by scratches or punctures or by delamination and membrane physical damage, for example, by water hammer or permeate backpressure. The vacuum decay test is considered to be a straightforward method for the integrity of a spiral wound RO membrane module. This testing is used to recognize the leakage inside the membrane modules instead of leakage because of any chemical attack. The vacuum decay test needs the separation of a specific membrane module or the whole PV. Then, a vacuum is pulled on the membrane and the decay rate in pressure is noted. A decay of higher than 100 mbars for every minute is an indication of membrane leakage. ASTM Standards D6908 (Standard Practice for Integrity Testing of Water Filtration Membrane Systems, 2007) and D3923 (Standard Practices for Detecting Leaks in Reverse Osmosis Devices, 2003) can be referred for a comprehensive review of this procedure.

Prior to testing, the membrane element should be drained from water existing in the feed channels as well as in the permeate leaves. The PV to be tested should not have water in it. The element product water tube should be evacuated and then isolated. The vacuum decay rate confirms the mechanical integrity or a membrane element leak. A mechanically undamaged membrane element and a chemically damaged membrane will still hold the vacuum; however, a membrane mechanically damaged will not. The vacuum decay testing is beneficial as a screening technique and is not considered as an examination for absolute leak confirmation. Conversely, this testing permits the identification of leaking elements or O-rings within a shorter timeframe. This test will also help to identify between the mechanical membrane damage and the chemical membrane damage.

The procedure of the vacuum decay test is as follows:

1. The element should be drained first.
2. Seal a single end of the product water tube using an appropriate leakage-tight cap.
3. The other end of the product water tube should be connected to a vacuum gauge and a valved vacuum source.
4. The membrane element must be evacuated to 100–300 mbars absolute pressure.
5. The isolation valve to be closed and the vacuum gauge reading can be noted. The rate at which the vacuum decays should be noted. A rapid decay could confirm the leakage presence.
6. Gradually release the vacuum and permit the element to reach atmospheric pressure before disconnection.
7. This testing must be repeated several times for confirming its reproducibility.

A total PV testing permits the inclusion of adapters and couplers into the leakage testing. The procedure will be similar as mentioned before, with the variation that the product water port at a single side of the PV should be closed, and the vacuum will be pulled from the product water port of the other side. Feed port as well as reject port might be open.

11.3.7 Profiling a reverse osmosis array

Profiling and probing are the two methods commonly used together for determining if the RO unit membranes are leaking or scaled and also to determine their position. These methods must be employed when the conductivity in the RO system product is high. When there is a problem in a reverse osmosis system, the capability for isolating the problem to a specific location within the system contributes valued data regarding the nature of the problem. This would determine the corrective actions like replacement of membrane element, replacement of O-ring, or cleaning. When the system is in operation, then an online investigation can be carried out. This involves examining the performance trends in NPF and NSR and the relating these trends with indications of known problems. The data for online examination is obtained from the operating data log of the plant and the basic plant design package. The completeness, accuracy, and frequency of the operating log are very important for the effective troubleshooting. Moreover, the instrumentation accuracy is also very critical. It cannot be overemphasized that prior to the troubleshooting of any plant issue, the good working order as well as proper calibration of meters and gauges should be first assured.

11.3.8 Probing

Probing involves the determination of the concentration of permeate at different points inside the PV. As soon as the profiling has isolated a salt rejection issue to a specific PV, or a set of PVs, then probing could be employed for isolating the issue further. It involves introducing flexible tubing through one of the product water connections of PV as a means of diverting the product water from a particular zone inside the membrane elements. Subsequently, this water will be tested for conductivity using a portable meter. Probing should be carried out when the RO unit is in operation. Adequate time (almost 30 seconds) must be permitted between samples for ensuring that water from the new sampling position has totally displaced the water inside the tubing.

A standard conductivity profile demonstrates a stable upsurge in the product water generated at the PV feed side toward the PV reject end. An abnormally higher deviation from this profile will indicate the source of the higher salt passage issue. Generally, the O-ring issues will be specified by a step variation in the conductivity profile at adapter/coupler sites, whereas a noticeable rise outside this region signifies a leak from a membrane element, for example, because of backpressure damage. The standard conductivity profile will also depend on the position of the probing tube entry and on the flow direction of the product water out of the probed vessel. Fig. 11.4 presents a structure with probing from the reject end of the PV with the product water flowing to the reject side also. The first sample from the feed side end of the PV signifies the product water generated at precisely that position. When the tube is slowly pulled out from the PV, the sample signifies the combined product water that is generated upstream of the sample position. The final sample signifies the product water of the complete vessel (DuPont Water Solution, 2020).

Fig. 11.4 shows how the probing procedure is applied. From the conductivity readings obtained, a graph is plotted. The lower curve in the graph represents the expected graph if none of the membrane elements or O-rings are malfunctioning. The upper curve in the graph indicates the source of the high salt passage problem.

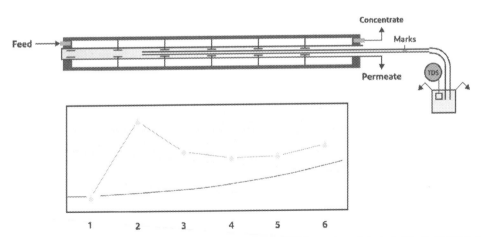

FIGURE 11.4   Location of high salt passage (probing). Source: *DuPont Water Solution's FilmTec™ RO technical manual (2020). Version 3; Form No. 45-D01504-en, Rev. 3; April 2020, https://www.dupont.com/content/dam/dupont/amer/ us/en/water-solutions/public/documents/en/45-D01504-en.pdf. Accessed November 5, 2020.*

11.3.9 Replacement of O-rings

The movements of the spiral wound membrane (SWM) elements inside the PV can frequently lead to abrasion and breaking of the O-ring that seals the interconnector to the element product water tube.

A rapid increase in permeate conductivity, not together with a clear increase in product water flow rate can point out a broken or missing O-ring. Profiling and then probing to determine if just individual PVs or membrane elements are causing the higher conductivity will confirm if the O-ring damage is the issue.

For replacing the O-ring, the RO unit must be shutdown and permitted to drain by opening the sample valves. Subsequently, the end-cap will be removed. Generally, the O-ring damage will be visible. The O-ring can be replaced by hand, wetting with some lubricant, if required.

11.3.10 Shimming

Some movements of the membrane elements within the PV housings can be very normal. This occurs because the ΔP across the membrane element could bring them to compress. The membrane fouling or higher flow rate could lead to substantial movements, mainly during the system startup. During its shutdown, the membrane elements would subsequently relax. This movement would lead to rubbing against the interconnector O-ring, specifically in the lead end membrane elements. As time passes, this will make them abrade and probably break. At the time of extreme pressure drops, the O-rings could be totally dislodged and blow out of their slots.

The possibility for this movement must be minimalized by ensuring that the membrane element fits firmly inside the PV. Any slop must be taken up with shims. Shims are parts of plastic piping with an interior diameter that just fits over the exterior of an end connector, commonly the end connector between the vessel end-cap and the lead end element. Shimming is the practice of introducing parts of material (shims) between two parts for taking up free space and thereby helping to restrict the movements. In the case of a membrane system, plastic washers can be employed as shims.

11.3.11 Replacement of reverse osmosis membrane elements

Sometimes, it is necessary to replace the RO membrane elements. This will be confirmed after the required troubleshooting and other corrective actions. While replacing, the RO system must be shutdown as well as drained. Before setting up, the new membrane element serial numbers must be documented specifying their proposed site in the system. This would be beneficial for having the proper comparison between the test data of membrane manufacturer with the performance of the system. The removal of both the vessel end-caps might be essential. Subsequently, the membrane elements could be removed in their normal flow direction, and this would prevent their brine seal from congestion against the PV. The replacement membrane elements could be inserted in the feed end of the PV and used to push the other elements through.

To aid the fitting, the U-cup brine seals and the interconnector O-rings could be carefully lubricated using glycerin. Every interconnectors must have the O-rings fitted. It has to be noted that never put the brine seals on both ends of a membrane element.

After replacing the element, any gaps must be limited with shims. The end-caps can then be fitted and the RO system can be started up. It must be filled using low-pressure water before starting the high-pressure pump. New elements must be rinsed to remove any residual preservative chemicals present. System operation information must be collected subsequent to the stabilization of the RO system performance (within 24 hours).

11.3.12 Membrane autopsies

RO membranes will undergo some performance decline with time, and this will mainly depend on the pretreatment efficiency. The performance drops because of membrane scaling or fouling, and in several cases, could be restored to a certain extent through chemical cleaning. On the other hand, the performance drop because of irreversible membrane polymer damage will be difficult to be restored.

The autopsy of the membrane is one of the most widely recognized techniques for determining the origin as well as the nature of the RO membrane fouling. This strategy consolidates consecutive completion of different standardized tests in lab conditions on membrane elements obtained from RO trains having reduced efficiency (Chesters et al., 2011). Xin et al. (2020) carried out the autopsy of the entire fouled membranes along a feed channel of a two-stage commercial-scale RO system for wastewater reclamation. The six elements in each stage have been autopsied and examined, for identifying the membrane fouling development process. The salt rejection and water flux efficiency of the fouled membranes at the tail and head were the lowest out of the 12 membrane elements, thereby confirming extremely severe fouling on these membranes as presented in Fig. 11.5. The test results confirmed that various types of fouling did not happen always at the same position even though microorganisms could produce organic matter, thus leading to organic fouling. It was observed that the seventh element has been extremely biofouled, as compared to the eighth element; however, they experienced a similar degree of organic

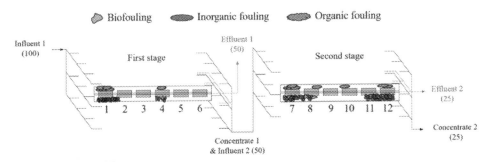

FIGURE 11.5 Fouling characteristics of the two-stage reverse osmosis unit. Source: *Reproduced from Xin, T., Yong, C., Yun-Hong, W., Yuan, B., Tong, Y., Xue-Hao, Z., ...Yin-Hu, W. (2020). Fouling properties of reverse osmosis membranes along the feed channel in an industrial-scale system for wastewater reclamation. Science of the Total Environment, 136673.*

fouling. Almost 70% of metal elements, mainly Fe, have been deposited on the first membrane element. Even though the iron content in the feed water was very less as compared to the concentration of magnesium and calcium, the iron content on the first three elements was remarkably greater than any other membrane elements, confirming that iron was very easily deposited on the RO membranes.

The physical dissection of a membrane, i.e, autopsy, followed by laboratory analyses for identification of its foulants and scalants constitute one of the most definitive methods of troubleshooting a reverse osmosis system. The autopsy should include all the procedures applied to identify the decline in RO performance and conclude with a well-written report and recommendations. The following tests are required to be carried out during a membrane autopsy:

1. Membrane external visual examination
2. Measurement of membrane weight
3. Bubble test and vacuum test of specific membrane
4. Standard membrane performance analysis
5. Cleaning analysis of membrane
6. Internal visual examination and testing
7. Fujiwara test.
8. Dye test.
9. Chemical identification of foulants/scalants
10. Weight loss on ignition testing
11. Microscopic evaluation

The first five tests are carried out with the entire membrane elements in the real conditions at which they were harvested, and are known as nondestructive testing. Subsequent to the completion of the tests from 1 to 5, the membrane elements will undergo dissection, and specimens of the membrane envelopes will be gathered to carry out the examinations from 6 to 11. Due to the fact that the performance of the tests from 6 to 11 needs the destruction of the membrane, these tests are also known as destructive testing. The cutting process of the testing membrane elements for obtaining the specimens for carrying out the destructive tests is known as "membrane autopsy" and is broadly used in reality for the complete membrane assessment.

A short explanation of all the 11 membrane autopsy tests in addition to their purpose as well as the potential result is given in the following section. Generally, the tests are performed as a minimum on a single front-membrane element, situated in the initial location close to the feed water entrance to the RO PV, and on a single tail-RO element situated at the final location in the RO PV. The front RO element will experience the greatest degree of microbial, colloidal, organic, and particulate fouling, whereas the final RO membrane element will be possibly exposed to the most serious scaling conditions that might be occurring in the PVs. Ideally, it is suggested to carry out membrane autopsy on a total set of the entire membrane elements in a minimum of a single PV inside the RO train. The membrane set chosen must be a representation of the performance difficulties of the complete RO train and of the RO desalination facility, as far as possible. If the RO unit is suffering from various issues, it is recommended to harvest a few complete sets of RO elements working at conditions similar to the corresponding RO system difficulties. The membrane autopsy can be usually carried out by university or commercial labs dedicated for such services, which are considered to have all the required test equipment, instrumentation, and experienced personnel to perform the testing. It is recommended that the RO membrane taken for an autopsy should have the original factory documents, if available, for the membrane elements.

Finally, there should be a submittal of the "final report," documenting step by step all the findings of the autopsy. The report should also include conclusions and recommendations that will ensure that the observed deficiencies that caused the decline in membrane performance do not recur.

In the following section, different tests required to be carried out during a membrane autopsy are described.

11.3.12.1 External visual examination of the membrane

Once the RO membrane elements are taken to the testing lab, these will be initially investigated by visual examination for identifying the possible physical damage of the brine seals, the core tubes, the fiberglass casting, as well as antitelescoping devices (ATDs), also for telescoping, extrusion of the brine spacers, and the antitelescoping device discoloration. Moreover, the exterior of the front-end ATDs will be assessed for

FIGURE 11.6 Protrusion of membrane spacer. Source: *Reproduced from Voutchkov, N. (2017). Diagnostics of membrane fouling and scaling. Pretreatment for reverse osmosis desalination (pp. 43–64). Elsevier.*

FIGURE 11.7 Fiberglass casting physical damage. Source: *Reproduced from Voutchkov, N. (2017). Diagnostics of membrane fouling and scaling. Pretreatment for reverse osmosis desalination (pp. 43–64). Elsevier.*

FIGURE 11.8 Membrane telescoping. Source: *Reproduced from Voutchkov, N. (2017). Diagnostics of membrane fouling and scaling.* Pretreatment for reverse osmosis desalination *(pp. 43–64). Elsevier.*

accumulation of solids, biofilm, and uncommon odors. The tail end-cap surface is generally examined for the collection of the mineral scale crystals. Protrusion of membrane spacer (Fig. 11.6), fiberglass casting physical damage (see Fig. 11.7), and membrane telescoping (Fig. 11.8) are indications of permanent damage of membrane because of the membrane element exposure to high feed pressure surge, mishandling, or extreme fouling. Brownish/reddish staining of ATDs is caused generally by overdosage of ferric coagulants. Thus, mainly, the external visual examination includes the following:

1. Examination of external construction of element for damage during operation/shipping including fiberglass, ATDs, and brine seal.
2. Examination of feed and concentrate ends before and after removal of ATD.

11.3.12.2 Measurement of membrane weight

The entire RO membranes will be assessed and their weight can be compared with that of fresh RO elements of similar sizing. The extent of the weight increase will be a demonstration of the membrane fouling level. A typical new 8-inch RO element will have a weight of approximately 36 lbs (16.4 kg). Under typical working conditions, the entire membrane elements will undergo fouling over time, and in this manner, the weight of these elements increases. Generally, an irreversible weight upsurge of the front RO membrane element more than 40 lbs (18 kg) is a sign of heavy fouling.

11.3.12.3 Bubble test and vacuum test of specific membranes

Bubble test and vacuum test can be employed for identifying if the RO membranes are damaged mechanically by severe fouling, backpressure, or might have production defects and/or membrane envelope glue line damage. These types of damages will reduce the RO membrane envelope integrity, typically bring about increased passage of salt, and eventually, result in unsuitably increased salinity of the RO permeate.

11.3.12.3.1 Bubble testing

This testing includes closing a single end of the RO element product water tube using an adapter plug, immersing the RO element into pure water, and then application of air at low pressure (3 psi/0.2 bars) to the opposite side of the product water tube for approximately one minute. If a continuous air bubble stream is discharged from the membrane element into the water, then this will be a sign of membrane integrity loss. If the membrane element integrity is not compromised, then it would cease discharging the bubbles after the first minute and would maintain the feed air pressure at an almost consistent level. This bubble testing failure of the membrane demonstrates the damage of the element, and it will be difficult to repair (Hydranautics, 2013). Normally, the bubble test is employed for identifying the membrane leaf glue damage as well as the associated rejection of salt.

11.3.12.3.2 Vacuum testing

The vacuum testing involves the application of a vacuum of 20-inch mercury (Hg) level (0.677 bar) for about 2 minutes to the open end of the product water tube of RO membranes. If the integrity of the membrane element is not compromising, then the membrane element would maintain a vacuum of about 19 bars at the completion of the analysis. During the completion of the testing, if greater than 35.0% of the vacuum is lost, then the RO membrane element will have extreme damage.

11.3.12.4 Standard membrane performance analysis

The entire RO elements will be tested at a predefined standard salinity, pH range, product water recovery, temperature, and pressure following the manufacture in order to confirm that they satisfy the performance specifications with respect to differential pressure, salt rejection, and permeate flow (this type of testing is known as factory testing). The typical testing conditions may be different for seawater RO element, brackish water (BW) RO elements, and sometimes change among the different membrane manufacturers. When a specific element is being procured, then its documents should consist of all the results of its original standard factory testing. During the standard membrane performance examination, the provided membrane elements for assessment must be retested at similar conditions, as the membranes have been tested initially by the manufacturer, and the estimations of salt passage, flow, and membrane rejection must be cross-checked. A substantial decrease in flow and increment in differential pressure (almost 8.0% per year) are generally indication of quantifiable fouling of the membrane. About a 20.0% increase in the passage of salt can be due to membrane element damage by oxidants, membrane fouling and/or mechanical integrity loss. The comparative analysis of the other performance tests discussed in this chapter can be used for identifying the most possible reasons for the performance degradation of the membranes. In the event that the entire membrane elements inside the PV have comparatively higher production loss, then this perception is an indication of the fouling development all through the length of the membrane vessel. In case that the salt passage of the entire membranes in a specific PV is increased by an almost similar degree, then it is highly probable that these membranes are exposed to oxidants. If just the specific elements have been showing the increased salt passage, then this phenomenon is possibly brought about by the loss of element mechanical integrity.

11.3.12.5 Membrane cleaning analysis

The membrane element sample will be cleaned by the application of different mixes of cleaning chemicals and durations of soak as well as flush cycles. This will help to identify the ideal mix of cleaning solutions and the sequence of application for accomplishing highest recovery of RO membrane efficiency and rejection of salt and to limit the differential pressure of membrane. On the basis of this test, the research facility finishing the membrane performance investigation prepares suggestions for economical and effective membrane cleaning.

When the product water flow rate of the element tested is very low as compared to the specified value, cleaning could be carried out. On the other hand, when the membrane is severely scaled/fouled (normally when the permeate flow is less than 50.0% of specification) or when the membrane itself is damaged, the cleaning cannot be effective.

The cleaning evaluation will include the setting up of proper cleaning procedures, their application on membrane specimens, and succeeding performance testing. The cleaning assessment might be carried out on membrane element after performance testing, or on membrane flat sheet coupons subsequent to the destructive autopsy. When the effective cleaning test has been proved, then the treatment could be applied to the entire RO unit.

11.3.12.6 Internal visual examination and testing

Internal assessment and testing can be done as a part of the membrane autopsy. The antitelescoping device (ATD) will be initially separated, and the membrane surface will be assessed for the buildup of colloidal particles, solids, telescoping, brine/feed spacer extrusion, and biofouling.

The internal visual examination mainly includes the following:

1. Examination of each individual membrane sheet for evidence of manufacturing defects, channeling, fouling, scaling, etc.
2. Examination of glue lines, and checking for any evidence of delamination.
3. Measurement of active surface area to see if it matches the surface area in the specification sheets of the membrane manufacturer.

11.3.12.7 Fujiwara test

A Fujiwara test is a qualitative test (using pyridine solution) to detect the possible oxidation of the membrane (polymer) surface by a halogen. This test is appropriate for membrane made up of PA material, which is the main material used currently for the spiral-wound RO element for the desalination process.

11.3.12.8 Dye test

The dye test with methylene blue can be carried out to detect any physical defects or deterioration of the membrane surface. Areas of damage show the dye to soak through to the permeate side of the membrane. In the event that the surface of the membrane is consistently stained with an exceptionally light blue/violet coloring, then this shows that the integrity of the membrane is safe. In case that the membrane is damaged physically, then the dye used would be absorbed on the damaged zones of the surface of the membrane and/or alongside the glue line.

11.3.12.9 Chemical identifications of foulants/scalants

The chemical and physical nature of the foulants and scalants can be studied and analyzed by Fourier transform infrared (FTIR), energy dispersive X-ray (EDX), and scanning electron microscope (SEM).

11.3.12.9.1 Fourier transform infrared spectrometer

This type of spectroscopy uses infrared radiation across a specimen and by investigating absorption as well as transmission of the radiation it produces a molecular spectrum of the specimen, and thereby the foulants/scalants can be recognized. The molecular spectrum of each compound is exclusive and the compounds on the RO membrane could be recognized by the comparison of its infrared spectrum with a range of infrared spectrums of known compounds.

11.3.12.9.2 Energy dispersive X-ray equipment

The EDX equipment can produce electronic beams that will strike the exterior of the RO membrane specimen tested which causes the X-ray discharge from the material collected on the specimen. Subsequently, graphs will be generated, which characterizes the recognized materials as peaks. Individual peaks will be a true representation of a unique chemical element.

11.3.12.9.3 Scanning electron microscope

The SEM image can be used to noticeably evaluate the state of the surface of the membrane and morphology and the topography of the fouling of the membrane. Fig. 11.9 shows a typical high-resolution SEM image of a fouled membrane.

11.3.12.10 Weight loss on ignition testing

This test will be used to determine the proportion of inorganic and organic constituents present in the foulants. In this test, the specimen of the foulant is assembled from the RO membrane surface and its weight is estimated. Subsequent to the weight estimation, the

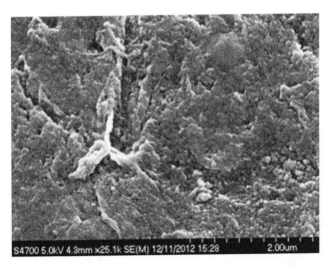

FIGURE 11.9 Typical High-resolution scanning electron microscope (HRSEM) image of fouled membrane. Source: *Reproduced from Abdullah, S. Z., Bérubé, P. R., & Horne, D. J. (2014). SEM imaging of membranes: Importance of sample preparation and imaging parameters.* Journal of Membrane Science, 463, 113–125.

specimen is initially dried at 110°C and later it is heated to 550°C. The residue weight from the specimen ignition is estimated and the change in weight reduction because of the ignition procedure is determined as the percentage of the first weight. In the case that the weight loss on ignition testing is greater than 35.0%, then this will be an indication of organic foulants. Weight loss greater than 50.0% confirms substantial organic fouling.

11.3.12.11 Microbial evaluation

The microbial testing for anaerobic bacteria and aerobic bacteria and fungi is generally performed. Specific cultures include:

1. Slime-forming bacteria culture
2. Iron-related bacteria culture
3. Sulfate reducing bacteria culture
4. General aerobic bacteria culture
5. Yeast, molds

11.4 Case studies

11.4.1 Case Study 1: Fouling characterization of reverse osmosis membranes after 11 years of operation in a brackish water desalination facility

Ruiz-García et al. (2018) characterized fouled BWRO membrane elements from a commercial-scale desalination facility by bringing together the performance data and by performing a membrane autopsy for identifying the fouling characteristics and behavior related to the feed water inorganic composition and operation conditions. According to the authors, this was the first time that membranes of a desalination facility were autopsied with a data collection along long 11 years of operation. The membrane elements examined have been taken after 11 years of operation in the facility.

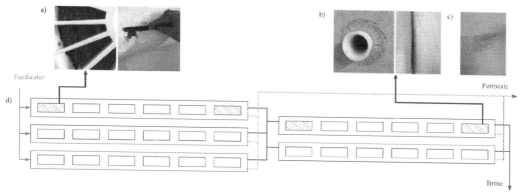

FIGURE 11.10 (A) Visual examination of the inlet of the first membrane element in the first stage, (B) visual examination of the final element in the second stage, (C) carbonate test, and (D) arrangement of the reverse osmosis system. Source: *Reproduced from Ruiz-García, A., Melián-Martel, N., & Mena, V. (2018). Fouling characterization of RO membranes after 11 years of operation in a brackish water desalination plant. Desalination, 430, 180–185.*

The RO system had five PVs in two stages. The arrangement was 3:2, that is, two stages, the first stage with three PVs and the second stage with two PVs, and the number of elements per PV was six. BW30–400 Filmtec was the membrane element used. The arrangement of brackish water reverse osmosis (BWRO) membrane elements chosen in the PV is presented in Fig. 11.10. The BWRO desalination plant had a production capacity of 15.0 m^3/day with almost 60.0% recovery. The water has been fed from groundwater well with cartridge filters and antiscalant dosing as pretreatment.

Chemical and structural characterizations were carried out by visual inspection, scanning electron microscopy with energy dispersive X-ray (SEM-EDX), and qualitative analysis test. The membrane elements have been visually examined for any physical damage and/or defects on the external surface, including the outer wrapping and the fiberglass, product water tube, brine seal, and ATDs for channeling. The fouling layer comprised of a biofilm noticed in the first membrane elements of the RO system. Biofilm containing diatoms on the entire elements were noted. The inorganic foulants mostly included aluminosilicates and calcium carbonate. After 11 years of operation, the results confirmed that under such fouling, with appropriate control and with suitable operation conditions, the elements used must not be replaced in shorter periods of operation time (Ruiz-García & Ruiz-Saavedra, 2015). Furthermore, as presented in Fig. 11.11, the NSR was relatively constant along the operation time. On the other hand, the average membrane resistance increased greater than 100.0% over the operation time, which is strongly associated with the efficiency loss in the RO system. The results confirmed that it will be feasible to preserve BWRO elements in service for up to 11 years with an appropriate conventional pretreatment. On the other hand, the membrane resistant increased more than two times over the operation period.

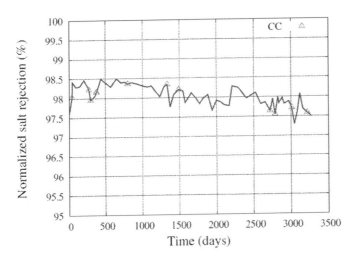

FIGURE 11.11 Normalized salt rejection. Source: *Reproduced from Ruiz-García, A., & Ruiz-Saavedra, E. (2015). 80,000 h operational experience and performance analysis of a brackish water reverse osmosis desalination plant. Assessment of membrane replacement cost. Desalination, 375, 81–88.*

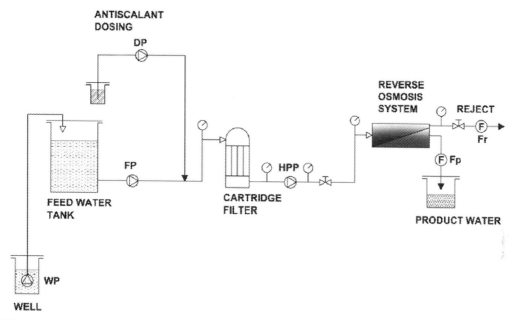

FIGURE 11.12 Desalination plant diagram. Source: *Reproduced from Ruiz-García, A., & Ruiz-Saavedra, E. (2015). 80,000 h operational experience and performance analysis of a brackish water reverse osmosis desalination plant. Assessment of membrane replacement cost. Desalination, 375, 81–88.*

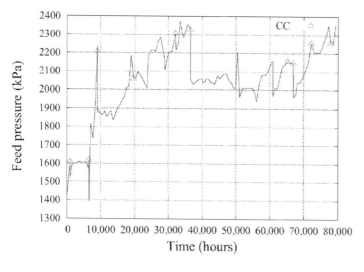

FIGURE 11.13 Feed pressure change with time. Source: *Reproduced from Ruiz-García, A., & Ruiz-Saavedra, E. (2015). 80,000 h operational experience and performance analysis of a brackish water reverse osmosis desalination plant. Assessment of membrane replacement cost. Desalination, 375, 81–88.*

11.4.2 Case Study 2: 80,000 hours of operating experience and performance examination of a brackish water reverse osmosis desalination plant

In another study carried out by Ruiz-García and Ruiz-Saavedra (2015), the team assessed the operation data accessible and energy consumption for evaluating when it will be suitable for replacing the membranes considering the long life membrane elements involved (Ruiz-García et al., 2018). The BWRO desalination facility has been operating successfully for almost 80,000 hours. The BWRO desalination facility (Fig. 11.12) is in the Gran Canaria island and it has been designed for desalinating groundwater well for irrigating crops using 5 μm cartridge filters in the pretreatment process. The RO unit was equipped with five PVs and used BW30−400 Filmtec membrane elements. The configuration was 3:2, that is, two stages, first stage with three PVs and second stage with two PVs, and the number of elements per PV was six (total 30 RO elements).

As shown in Fig. 11.13, the feed pressure was almost 1372.93 kPa during the start and increased up to 2353.60 kPa because of the performance decline, particularly compaction, scaling, and fouling over the years. A comparison between specific energy consumption increase because of the membrane performance decline, the regularity of chemical cleaning, as well as replacement of membranes with respect to costs were performed. It was confirmed that not replacing the membranes during the first 10 years of operation is considered as the most efficient choice even considering the new advanced membranes. However, the chemical cleanings have been very often and less efficient in the last 10,000 operating hours. The feed pressure has been extremely close to the high-pressure pump limit leading to the membrane replacement. Due to this, it is significant to oversize the high-pressure pump considering the feed pressure increase in a long-term operation and the membrane replacement expense.

11.4.3 Case Study 3: Biofouling in seawater reverse osmosis: the effect of module geometry and mitigation with ultrafiltration

The Pazargad seawater RO desalination facility is designed for the production of potable water as well as service water (Nejati et al., 2019). This plant is situated in Asaluyeh, on the Persian Gulf coast, in the south of Iran. The source water is obtained using two vertical pumps, over a 14 m well. Subsequent to ground filtration, the source water passes through three sand filter tanks having a coarse layer, medium layer, and fine layer. The last pretreatment step is three cartridge filter vessels each having 33 cartridge elements with a 5-μm pore size. The high-pressure pumps deliver the filtered water to the three RO trains, with each having eight PVs, each consisting of six RO membrane elements.

Weekly chemical cleaning of RO membranes in this desalination facility was a major problem as the permeate electrical conductivity surpassed the accepted limit. In the study by Nejati et al. (2019), the autopsy of the fouled RO membranes confirmed that the development of biofilm was a major reason for the system membrane fouling. A fouled SWM RO element after a prolonged operation, almost seven years, was obtained and then unrolled for identifying if there is a problem in the membrane leaves. The surface of the fouled membrane was separated into six sampling positions, such that the results can be

FIGURE 11.14 Unrolled fouled reverse osmosis membrane, in Pazargad seawater reverse osmosis plant, classified into six sampling positions. TM820H-370 (Toray) was the membrane model used. Source: *Reproduced from Nejati, S., Mirbagheri, S. A., Warsinger, D. M., & Fazeli, M. (2019). Biofouling in seawater reverse osmosis (SWRO): Impact of module geometry and mitigation with ultrafiltration. Journal of Water Process Engineering, 29, 100782.*

more explicitly interpreted. Fig. 11.14 presents the surface of the membrane which is divided into six distinct sampling locations.

Biofoulants on the RO membrane were non-uniformly distributed and deposited due to the changing nutrient concentration. A computational fluid dynamic simulation was carried out for realizing the impact of module inlet geometry on the flow inside the membrane leaves as well as on the biofouling. It has been noted that the flow resulted in larger eddies (water circular movement) with a stagnation zone that could have a smaller impact on the flow rate to the membrane module. The silicon carbide ultrafiltration membranes were examined for ensuring their efficiency in the removal of microorganisms. The test results confirmed that silicon carbide ultrafiltration membranes are efficient for removing approximately 100% of the microbes present in the feed water. Otherwise stated, the use of silicon carbide ultrafiltration membranes can be a promising and efficient technique for controlling membrane biofouling.

References

Abdullah, S. Z., Bérubé, P. R., & Horne, D. J. (2014). SEM imaging of membranes: Importance of sample preparation and imaging parameters. *Journal of Membrane Science, 463*, 113–125.

Bubble Testing of RO Membrane Elements (2013). Hydranautics Technical Service Bulletin, TSB 101.02. Retrieved from https://membranes.com/docs/tsb/TSB101.pdf.

Chesters, S., Pena, N., Gallego, S., Fazel, M., Armstrong, M., & de Vigo, F. (2011). Results from 99 seawater RO membrane autopsies. In: Proceedings of IDA World Congress, Perth, Western Australia, September 4e9, 2011.

DuPont Water Solution's FilmTec™ RO technical manual Version 3; Form No. 45-D01504-en, Rev. 3; April 2020, < https://www.dupont.com/content/dam/dupont/amer/us/en/water-solutions/public/documents/en/45-D01504-en.pdf > Accessed November 5, 2020.

Ibrahim, S. M., Nagasawa, H., Kanezashi, M., & Tsuru, T. (2020). Chemical-free cleaning of fouled reverse osmosis (RO) membranes derived from bis (triethoxysilyl) ethane (BTESE). *Journal of Membrane Science, 601*, 117919.

Kucera, J. (2007). Membranes: Troubleshooting: methods to improve system performance—Part 1. *UltraPirre Water, 24*(3).

Li, Y. S., Shi, L. C., Gao, X. F., & Huang, J. G. (2016). Cleaning effects of oxalic acid under ultrasound to the used reverse osmosis membranes with an online cleaning and monitoring system. *Desalination, 390,* 62−71.

Nejati, S., Mirbagheri, S. A., Warsinger, D. M., & Fazeli, M. (2019). Biofouling in seawater reverse osmosis (SWRO): Impact of module geometry and mitigation with ultrafiltration. *Journal of Water Process Engineering, 29,* 100782.

Park, H. G., & Kwon, Y. N. (2018). Long-term stability of low-pressure reverse osmosis (RO) membrane operation—A pilot scale study. *Water, 10*(2), 93.

Ruiz-García, A., Melián-Martel, N., & Mena, V. (2018). Fouling characterization of RO membranes after 11 years of operation in a brackish water desalination plant. *Desalination, 430,* 180−185.

Ruiz-García, A., & Ruiz-Saavedra, E. (2015). 80,000 h operational experience and performance analysis of a brackish water reverse osmosis desalination plant. Assessment of membrane replacement cost. *Desalination, 375,* 81−88.

Saleem, H., Trabzon, L., Kilic, A., & Zaidi, S. J. (2020). Recent advances in nanofibrous membranes: Production and applications in water treatment and desalination. *Desalination, 478,* 114178.

Saleem, H., & Zaidi, S. J. (2020a). Developments in the application of nanomaterials for water treatment and their impact on the environment. *Nanomaterials, 10*(9), 1764.

Saleem, H., & Zaidi, S. J. (2020b). Innovative nanostructured membranes for reverse osmosis water desalination, Qatar University Annual Research Forum and Exhibition (QUARFE 2020), Doha. https://doi.org/10.29117/quarfe.2020.0023.

Saleem, H., & Zaidi, S. J. (2020c). Nanoparticles in reverse osmosis membranes for desalination: A state of the art review. *Desalination, 475,* 114171.

Sim, L. N., Chong, T. H., Taheri, A. H., Sim, S. T. V., Lai, L., Krantz, W. B., & Fane, A. G. (2018). A review of fouling indices and monitoring techniques for reverse osmosis. *Desalination, 434,* 169−188.

Standard Practice for Integrity Testing of Water Filtration Membrane Systems (2007). ASTM D6908−03, American Society of Testing and Materials International, West Conshohocken, PA.

Standard Practices for Detecting Leaks in Reverse Osmosis Devices (2003). ASTM D3923−94(2003)e1, American Society of Testing and Materials International, West Conshohocken, PA.

Tow, E. W., & Rencken, M. M. (2016). In situ visualization of organic fouling and cleaning mechanisms in reverse osmosis and forward osmosis. *Desalination, 399,* 138−147.

Troubleshooting Your RO, Hydranautics, Retrieved from https://membranes.com/docs/trc/TROUBLES.PDF.

Voutchkov, N. (2017). Diagnostics of membrane fouling and scaling. In Anita Koch (Ed.), *Pretreatment for reverse osmosis desalination* (pp. 43−64). John Fedor, Elsevier. Available from https://doi.org/10.1016/B978-0-12-809953-7.00003-6.

Wilf, M., et al. (2007). *The guidebook to membrane desalination technology.* Italy: Balaban Desalination Publications.

Xin, T., Yong, C., Yun-Hong, W., Yuan, B., Tong, Y., Xue-Hao, Z., & Yin-Hu, W. (2020). Fouling properties of reverse osmosis membranes along the feed channel in an industrial-scale system for wastewater reclamation. *Science of the Total Environment, 713,* 136673.

Zaidi, S. J., Fadhillah, F., Saleem, H., Hawari, A., & Benamor, A. (2019). Organically modified nanoclay filled thin-film nanocomposite membranes for reverse osmosis application. *Materials, 12*(22), 3803.

Zsirai, T. Dr. (2020). Commercial engineering lead, water technologies & solutions, personal communication. SUEZ Water Technologies & Solutions, Technical bulletin, Cleaning guidelines, TB1194EN.docx Nov-16.

12

Issues Related to System Engineering and Frequently Asked Questions

12.1 Different issues related to reverse osmosis system engineering

Commonly encountered issues related to reverse osmosis (RO) system engineering include: cleaning of RO membranes on-site or off-site, RO brine disposal, membrane-based filtration and backwashing residuals, pressure-drop tradeoff, pumps, upstream priorities, membrane fouling, preventing scale formation, chlorine elimination, sodium softening and the sequencing of the softeners, whether to use sodium softeners or antiscalants, sizing of a reverse osmosis system in variable flow demand conditions and the system installation issues. Fig. 12.1 presents the different issues related to RO system engineering.

12.1.1 Cleaning of membrane: onsite versus offsite

The performance restoration by cleaning is already discussed in Chapter 9, Performance Monitoring, Process Parameters, and Their Control. This section discusses the onsite [clean-in-place (CIP)] and offsite membrane cleaning.

12.1.1.1 Onsite membrane cleaning

The benefits CIP or onsite cleaning of the membrane are the following (Kucera, 2015):

1. Cleaning can be carried out with the membrane modules onsite, and hence there will not be any requirement for a second membrane set.
2. As compared to the offsite cleaning, the cleaning process will be faster with onsite cleaning. For example, cleaning of a two-stage, 500 gallons per minute (GPM) RO skid could be carried out in just 2 days. On the other hand, the offsite cleaning could take almost a few weeks to turn back the membranes.
3. Onsite cleaning will be less expensive as compared to offsite cleaning.

The drawbacks of onsite membrane cleaning are the following:

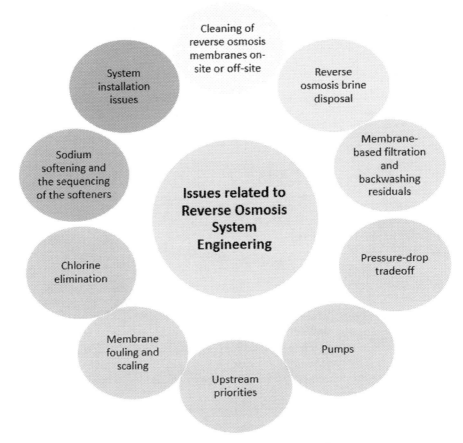

FIGURE 12.1   Different issues related to reverse osmosis system engineering.

1. Onsite cleaning will be less effective relative to offsite cleaning. Normally, the cleaning process will be limited to one cleaner per pH onsite. It is very expensive to stock the entire cleaners that may be required. Moreover, membrane manufacturers recommend adhering to their cleaning specifications.
2. The expenses for cleaning skids are not usually included in the initial capital cost.
3. Inappropriate storing and handling of the cleaning chemicals and generated wastewater by the onsite cleaning staff can be unsafe.

Improvement of CIP of a reverse osmosis desalination process using air micronano bubbles was performed by Dayarathne et al. (2017). Laboratory-scale and pilot-scale cross-flow filtration tests were carried out for evaluating the effect of air micronano bubbles on solute rejection, transmembrane pressure, and permeate flux. As this membrane CIP technique is not using any chemicals, this does not cause environmental pollution, making this a more applicable cleaning technique.

12.1.1.2 Offsite membrane cleaning

In offsite membrane cleaning, the membrane modules will be removed from the pressure vessels (PVs) and subsequently shipped offsite for a third-party cleaning. In case the RO should be in operation, the second set of membranes will be employed for replacing the membranes sent out for cleaning.

The benefits of offsite membrane cleaning are the following:

1. This type of cleaning provides professional service. The staff of the third-party cleaning company will be specially trained for the membrane cleaning purpose.
2. The offsite cleaning is normally very efficient. This type of cleaning operation involves different cleaners at their disposal for use against majority scale and foulants. If a single cleaning is not effective, then a different cleaner can be tried for improving the results.
3. The manufacturers of membranes sometimes offer special variances to offsite cleaning operations for using conditions beyond the standard membrane cleaning recommendations, as mentioned in the specifications of membranes.
4. The results of offsite cleaning are documented. Documents usually include performance testing before and after cleaning, and compare the results with specifications of that particular membrane type.

The drawbacks of offsite membrane cleaning are the following:

1. The cleaning costs of offsite cleaning will be higher than onsite cleaning.
2. A second membrane set will be required for the continuous operation of the RO system.

12.1.2 Reverse osmosis brine disposal

The desalination brine is a by-product liquid stream with higher concentrations of the majority of dissolved solids (sodium chloride) in feed, certain pretreatment additives (residual quantities of antiscalant, flocculants, and coagulants), microbial contaminants, organics, and any particulates rejected by the RO membranes (Fig. 12.2). The seawater RO unit can concentrate the salt concentration up to two times higher as compared to the feed water concentration. In a recent study, Panagopoulos et al. (2019) have evaluated present practices in the management of brine including disposal techniques and brine treatment technologies. The study also highlighted future research spaces for brine treatment technologies aiming to improve the viability and efficiency of the desalination process.

As per common RO membrane pressure resistance limitation, the RO system will be able to concentrate saline stream up to almost about 70,000 mg/L total dissolved solids (TDS) and will not be able to treat water with additional TDS. Thus, brine can be considered as saline streams of about 65,000–85,000 mg/L that cannot be treated further with seawater RO units (Brine Treatment (ZLD)).

Brine disposal can be problematic as

1. It might contain pretreatment chemicals and membrane cleaning chemicals.
2. It will affect the local aquatic life.
3. It will increase the salinity of the receiving water bodies.

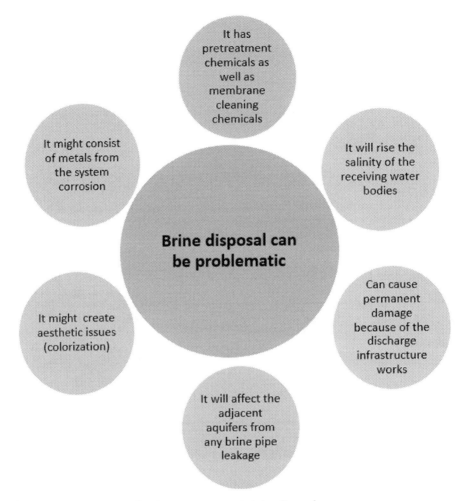

FIGURE 12.2   Different issues related to reverse osmosis brine disposal.

4. It will cause permanent damage because of the discharge infrastructure works.
5. It will impact the adjacent aquifers from any brine pipe leakage.
6. It will create esthetic issues (colorization).
7. It might consist of metals from the system corrosion (chromium, nickel, iron, copper, etc.).

The amount of brine produced will depend on the production capacity of the desalination plant along with its recovery rate (Pramanik et al., 2017). The recovery can be expressed in terms of the percent of the freshwater produced flow rate to the total feed flow rate to the system. In general, the brackish water RO will have higher recoveries from 50.0% to 90.0%, whereas the seawater RO recovery typically ranges from 30.0% to 55.0%. Higher recovery will result in a smaller volume of concentrate (increased salinity) and vice versa.

The brine volume produced in a desalination facility can be determined as

$$V_b = V_p \times \left(\frac{1-r}{r}\right) \tag{12.1}$$

where V_b = volume of brine, V_p = volume of permeate, and r = system recovery rate in percentage.

The quality of brine depends on the (1) total recovery, (2) salt rejection of desalination membrane, and (3) feed composition and its salinity.

Typically, the brackish water RO concentration factor is 4.0–10.0, whereas the seawater RO is 1.5–2.0 times. Brine TDS will depend on the feed TDS, permeate TDS, and plant recovery.

$$TDS_{brine} = TDS_{feed} \times \left(\frac{1}{1 - \text{System recovery rate}}\right) \times \left(\frac{\text{System recovery rate} \times TDS_{permeate}}{100 \times (1 - \text{Plant recovery})}\right) \tag{12.2}$$

The concentration factor can be determined as follows:

$$\text{Conc. factor} = \frac{1}{1 - r} \tag{12.3}$$

In case that the salt passage of the membrane is known, then the concentration factor could be found as follows:

$$\text{Conc. factor(\%)} = \frac{[1 - (r \times \text{Salt passage})]}{(1 - r)} \tag{12.4}$$

The concentration factor of salt is mostly limited by the increased osmotic pressure of brine. For seawater RO, this limit is approximately 65,000–85,000 mg/L. For a single-pass seawater RO system, the optimum recovery will be 40.0%–45.0% and the concentration factor varies in a range of 1.50–1.80. However, the brackish water RO plants normally have recoveries of 70.0%–90.0% and concentration factors of 4.0–10.0.

Based on the feed water quality, it is possible to predict the brine quality using the following rules:

1. The pH of brine will be greater than the feed water due to its greater alkalinity.
2. RO membranes can reject the heavy metals in a comparable ratio as magnesium and calcium.
3. The majority of organics will be rejected more than or same as 95.0% (except for the organics with lower molecular weight).
4. Ground water (brackish) RO brine might be anaerobic and consists of hydrogen sulfide.

For brine disposal, the following five conventional brine management options are available:

1. Surface water discharge
2. Evaporation ponds
3. Deep well injection
4. Land application
5. Sewer disposal

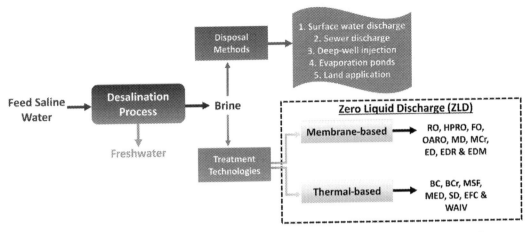

FIGURE 12.3　Different brine disposal and treatment technologies. *RO*, Reverse osmosis; *HPRO*, high-pressure reverse osmosis; *FO*, forward osmosis; *OARO*, osmotically assisted reverse osmosis; *MD*, membrane distillation; *MCr*, membrane crystallization; *ED*, electrodialysis; *EDR*, electrodialysis reversal; *EDM*, electrodialysis metathesis; *C*, brine concentrator; *BCr*, brine crystallizer; *MSF*, multistage flash distillation; *MED*, multieffect distillation; *SD*, spray dryer; *EFC*, eutectic freeze crystallization; *WAIV*, wind-aided intensified evaporation. Source: *Reproduced from Panagopoulos, A., Haralambous, K. J., & Loizidou, M. (2019). Desalination brine disposal methods and treatment technologies—A review. Science of the Total Environment, 693, 133545.*

Surface water discharge is the commonly used technique for brine disposal as this could be applied to all the desalination plant sizes. Moreover, this technique is very economical for medium to large brine flow rates. However, the brine might have an adverse impact on the marine ecosystem, and it will require difficult as well as complex permit procedures. Evaporation ponds and land application are typically applied for medium-size and small-size plants where the climate and the conditions of soil contribute to higher evaporation rates and year-round growth and harvesting of halophytic (salt-tolerant plant) vegetation. Systems for the above-mentioned two techniques will be easy to construct and operate. Conversely, these techniques have been limited to smaller brine flows, have higher footprint, and are expenses. The deep well injection technique is most appropriate for large-size and medium-size inland brackish water facilities. This technique offers moderate costs and lower energy consumption. On the other hand, this technique will be feasible only if deep confined saline aquifers are available, and also this brine management option may cause potential groundwater pollution. Sewer disposal is the commonly applied technique for the discharges of small-size desalination facilities. This technique has low operation and construction costs, is easy to implement, and has lower energy consumption. However, this brine management option may be limited to small size brine flows and can cause possible negative impacts on wastewater treatment plant operations. Fig. 12.3 presents the different brine disposal and treatment technologies.

12.1.3 Membrane-based filtration and backwashing residuals

Currently, membrane-based filtration has become very common in several applications such as pretreatment for RO systems. The membrane-based filtration processes are already discussed in Chapter 5, Pretreatment: Fouling and Scaling Control. This type of filtration can normally produce water with low suspended solid concentration, as compared to water filtered using a pressurized multimedia filter. Hence, the membrane-based filtration systems might be used as an alternate to the multimedia filtration, or probably downstream for more polishing the water as well as minimizing the RO fouling.

Microfiltration (MF) or ultrafiltration(UF) system requires frequent backwashing, will have intermittent chemical cleanings, and therefore generates residuals from both the MF and UF unit processes. Even though the backwashing frequency changes on a site-specific and a system-specific basis, this backwashing process could be usually performed in every 15–60 minutes. In the standard operation conditions, the frequency of backwash must be almost the same, thus permitting for the amount of residuals produced to be assessed accurately. In general, the residual stream generated from backwashing microfiltration or UF membrane will have a suspended solid concentration almost 10–20 times higher as compared to the feed water. Even though microfiltration or UF systems separate almost similar types of feed water components as the conventional media filter, the characteristics and volume of the residuals might be substantially different. Currently, in several applications of MF or UF for municipal water treatment, filter aids like polymers and coagulants might not be essential. In such situations, the quantity of solids separated in the backwashing process might be considerably lower as compared to a conventional filtration plant (Wang et al., 2011).

On-site treatment methods for microfiltration or UF backwash residuals will be similar to those that may be employed with conventional media filtration. The second stage of MF or UF might also be used to concentrate further the residuals and enhance the process recovery.

12.1.4 The pressure drop tradeoff

Feed channel pressure drop is considered to be an undesirable by-product of the mechanical support and mass transfer functions (Johnson & Busch, 2010). Due to the fact that several RO modules are normally used in series in a larger system, the feed-side pressure drop will definitely impact the performance of the system by decreasing the transmembrane pressure, and subsequently the production of permeate, in the downstream membrane modules. This kind of underutilization will lead to an overutilization as well as a higher fouling rate in the upstream modules. Several efforts for improving the mass transfer through optimization of the biplanar extruded net and other configurations were carried out; however, these have not led to any substantial variations to commercial spacers, which continue to be the same as those used previously. The primary reason for this is a comparatively smaller magnitude of possible advantage related to the increased mass transfer compared to that accomplished historically through continuing developments in membrane chemistry. Another reason is the mass transfer tradeoff, which links the decreased polarization to the higher pressure drop. The third reason is the reduced price of the existing spacers. The tradeoff will be movable, and spacers are proposed, which promise concurrent mass transfer along with pressure drop increase. As an example, a multilayer spacer can put an obstruction at the surface of the membrane where they could efficiently interfere with the concentration boundary layer whereas minimalizing disturbance of the bulk flow. In a study by Haidari et al. (2018), the role of the feed spacer in spiral wound membrane modules was critically analyzed. They reviewed several studies carried out in narrow spacer-filled channels for determining the impact of various geometric features of the feed spacer on hydraulic conditions. Spacer with strands of the non-circular cross-section seems to decrease pressure drop although still mixing the boundary layer. However, economical commercial-scale manufacturing approaches for such configurations have not been advanced.

12.1.5 Pumps

The two most important components in a seawater RO system are the membranes and the high-pressure feed pump (Al-Karaghouli & Kazmerski, 2012). These components are considered to be the heart of any RO system, and these need careful selection and application for effective operation.

In general, there are two types of high-pressure pumps (HPPs) used in seawater RO systems: (1) centrifugal pump and (2) positive displacement plunger pump. The plunger pump might generate higher output pressure variance (pulsation) because of its reciprocating action, which will be translated to vibration. The vibration will possibly damage the pump, other system components especially plumbing, instrumentation and the framework of the system. For minimizing the vibration damage to the system components, the pump will require a discharge pulsation dampener as well as a suction stabilizer. Another significant factor to consider is the speed of the pump in RPM. When the pump speed is slow, then only less vibration will be transferred. Mechanical design for the isolation of vibration is also an important step for minimizing the vibration damage from the pumping system.

As seawater RO pumps could produce pressures of more than 1000 psig, it is suggested that a safety switch, along with a pressure relief valve, be included in the design. Extreme damage might take place if the pump pressure goes beyond the material strengths of the RO design.

12.1.6 Upstream priorities

Larger RO systems can be designed with a proper chlorine elimination system and a scale inhibiting system with safeguards for preventing the issues with their operation and to maintain the expected life of RO membrane (Byrne, (c)). As an illustration, an oxidation–reduction potential (ORP) apparatus could constantly check the RO unit inlet water for the presence of a high concentration of any oxidants. It could be interlocked with any alarm unit for shutting down the system in case that its reading continues to be high for a longer time period. Considering the chemical injection system, the RO membrane must be shielded against both the non-existence of the injected chemicals, and also from an excess of the injected chemicals. The incorporation of an automated isolation valve on the chemical injection line positioned before their point of injection into the water line would restrict the chemical dumping when the system is shutdown.

The working of a chemical flow sensor positioned on each chemical injection line could be interlocked using a reverse osmosis system alarm and shutdown function. This could protect against defective programming that leads to the insignificant chemical being injected, and also shielding against any injection pump failures or inadequacy to refill any injection chemicals.

12.1.7 Fouling of membrane

The fouling process will not necessarily reduce the life of the RO membrane if effective cleaning is carried out. In case that the RO unit is severely fouled and the cleaning process is not very effective, then the membrane might lose performance eventually. In a recent study by Jiang et al. (2017), the membrane fouling types as well as fouling control approaches have been analyzed, with a special focus on the state-of-the-art progress. The fouling basics have been discussed previously in Chapter 5, Pretreatment: Fouling and Scaling Control, in detail including colloidal fouling, organic fouling, biofouling, and inorganic scaling. Several other studies were also carried out for minimizing the fouling in RO membranes (Saleem & Zaidi, 2020a, 2020b, 2020c; Saleem et al., 2020; Zaidi et al., 2019). Fig. 12.4 is the photographs of the fouled RO membrane with (A) coarse and (B) fine spacers.

Presently, a filter housing is commonly included on the RO system inlet that consists of a 2.5-inch-diameter cartridge filter, whose pore size will be nominally rated. The real capability to remove small-sized particles could differ significantly. Some of them just shield the RO unit against large-sized particles that could be trapped inside the membrane flow channel or cause HPP damage. This is low priced and might last for some weeks before an increased pressure drop specifying the replacement requirement. Tighter porosity filters, that could separate the majority of the arriving suspended solids, have been normally very costly and moreover need very regular replacements. Hence, the use of tighter filters will become a very economical option if the suspended solid content in the water is minimalized by upstream treatment.

In general, the suspended solids could be efficiently reduced to moderate concentration for the downstream RO unit by using a multimedia filter (Byrne, (c)). This addition in the RO water system may be adequate for preventing a higher rate of fouling that can lead to uncontrollable cleaning requirements. Multimedia filters consist of at least two different sizes or types of anthracite, crushed rock, or sand. This type of filter could be effective in

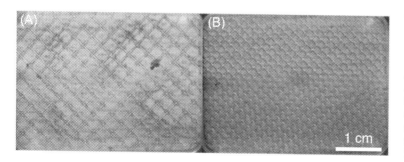

FIGURE 12.4 Photographs of the fouled reverse osmosis membrane with (A) coarse and (B) fine spacers. Source: *Reproduced from Tow, E. W., & Rencken, M. M. (2016). In situ visualization of organic fouling and cleaning mechanisms in reverse osmosis and forward osmosis. Desalination, 399, 138–147.*

removing the majority of particles that constitute the suspended solids provided that the below conditions are satisfied:

1. It will have a low collection lateral system intended for obtaining constant flow distribution across the media, while the filter operates at a lower flow velocity, and also allows the entrance of an adequate backwash flow rate for a bed expansion of 40.0%.
2. It will be sized for a 2.0 ft./s downward flow velocity.
3. The filter is backwashed beforehand its formerly removed small/fine particles are shed, which might occur before there is a substantial increase in the filter pressure drop.
4. After backwashing, the filter can be forward-rinsed at its service flow rate till the quality of effluent is adequate.

Some sources of water might contain a very high concentration of fine particles. In such events, it might be essential for sending the water through bigger reaction tanks anticipated to provide the particles additional time for coagulating into bigger particles that could be subsequently filtered more easily.

Inorganic chemical coagulants can be included in the water stream of the tank for accelerating the coagulation approach. The different chemical-based pretreatment processes are discussed in Chapter 5, Pretreatment: Fouling and Scaling Control. Coagulants are very efficient if they are initially blended thoroughly with the suspended solids. In case that the soluble metals (like manganese or iron) are present in the raw water, then a certain percentage would be oxidized by permitting the water to contact atmospheric air present in the tank, even though the percentage is normally very low. A chemical oxidizer like chlorine could be included in the water stream for oxidizing the metals into their insoluble oxides before coagulation.

In a research study carried out by Sefatjoo et al. (2020), the electrocoagulation (EC) process has been used as a designated pretreatment (Fig. 12.5) for evaluating its efficiency as well as operating expense with regard to the simultaneous turbidity and calcium removal for mitigating the scaling potential and colloidal fouling potential of RO system. The impacts of the major parameters, specifically, current density, time, initial turbidity, and initial calcium concentration on the electrocoagulation process have been evaluated and optimized by using the response surface methodology (RSM).

12.1.8 Scaling prevention

Normally, there is a minimum single salt in any naturally occurring water source that would concentrate above its solubility limit and subsequently develop scaling. Prevention of scaling is not the main issue unless the source of water has an extremely higher content of a slightly soluble salt, or unless the RO unit is being operated with an extremely higher permeate recovery (Byrne, (c)).

The development of scaling can be prevented by the injection of a chemical scale inhibitor, by water softening, or by the injection of acid into the incoming water. Typically, the most economical approach is by the use of a scale inhibitor, which normally slows down the rate of salt crystal growth when the solubility is surpassed. The injection of acid will prevent the formation of calcium carbonate scale; however, it will result in a higher CO_2

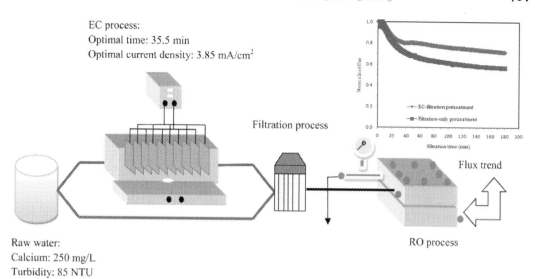

EC process:
Optimal time: 35.5 min
Optimal current density: 3.85 mA/cm²

Filtration process

Raw water:
Calcium: 250 mg/L
Turbidity: 85 NTU

RO process

FIGURE 12.5 Electrocoagulation process as a selected pretreatment for reducing fouling potential. *Source: Reproduced from Sefatjoo, P., Moghaddam, M. R. A., & Mehrabadi, A. R. (2020). Evaluating electrocoagulation pretreatment prior to reverse osmosis system for simultaneous scaling and colloidal fouling mitigation: Application of RSM in performance and cost optimization. Journal of Water Process Engineering, 35, 101201.*

concentration, which will not be removed by the RO system, and this will lead to a higher removal requirement on the downstream ion-exchange process. Moreover, the acid injection only will not provide much shield against the development of sulfate scaling or some other scaling.

The water softening has some benefits; however, it will have increased capital cost as well as operating expenses, unless there is mainly lower magnesium and calcium concentration hardness in the water. Moreover, softeners will remove other possible scale-forming ions like barium, strontium, etc., and also separates metals (aluminium, manganese, iron) that might then lead to RO system fouling. However, the softening resin will also foul with the metals, and subsequently needs intermittent chemical treatment. Scale-inhibitor providers generally use some software programs for estimating the possibility for the formation of scale. These software programs can predict the level of salts existing in the RO reject stream, and its pH, for determining the quantity of scale inhibitors required.

When a scale inhibitor is used, it is very important to properly rinse the RO system of its higher content of dissolved salts each time the RO system is shutdown. Else, the scaling particles develop and stick to the surface of the membrane at the time of shutdown. The rinsing practice must be automated and is generally carried out with lower-pressure inlet water. Lower pressure decreases the RO permeate that have a tendency for concentrating the dissolved salts. An improved rinsing can be carried out with pressurized product water if a line could be plumbed back to the RO unit from a product water storage tank system. The product water is considered to be biostatic, and its use reduces the development of biological solids inside the RO unit during shutdown.

12.1.9 Sodium softeners and its sequencing

Conventionally, sodium softeners are used as a reverse osmosis pretreatment. The sodium softeners have the ability to remove hardness as well as metals, like manganese and iron, that can cause fouling, scaling, or the degradation of RO membranes. Moreover, softeners can also help to reduce the silt density index (SDI) and suspended solids from the surface or other extremely fouling feed water before the RO process. Sodium softeners will act as an extra barrier ahead of the membrane. The drawback associated with the use of sodium softening as RO pretreatment is that the softeners should treat not just the permeate volume but also the water volume that will turn out to be the reject. Otherwise stated, the softener should be quite high for treating the complete feed water volume to the RO unit (Kucera, 2015). This would lead to two problems:

1. The softener system should be comparatively large, as the service flow rate through a softener vessel must be almost 6.0–8.0 GPM/ft². A 500 GPM RO system operating at a recovery of 75.0% will need two 120-inch-diameter softener vessels for softening the feed water and retain the preferred service flow rate whereas one unit is in regeneration.
2. Another issue is the possibility of the discharge of chloride. A single 120-inch-diameter softener vessel would produce almost 3400 gal of 10.0% brine waste just from the brining stage alone. Brining step involves the injection of brine into the resin bed. A 10.0% brine solution may consist of almost 6000 mg/L chloride, which should be diluted for the discharge.

The above-stated problems have urged various users to move the sodium softener from before the RO unit to after the unit, for polishing the RO permeate. The RO post-treatment is generally very essential due to the fact that the RO process will not reject 100.0% of the feed water hardness. Based on the permeate application, polishing using a softener for removing the hardness might be needed.

The benefits of this type of configuration are listed below:

1. The sodium softener will be treating just the RO permeate, normally almost 75.0% of the flow rate of the feed.
2. The sodium softener polishing could function at an increased service flow rate, as compared to a primary softener. Rather than being limited to 6.0–8.0 GPM/ft², polishing softeners will be able to function at 10.0–15.0 GPM/ft². The same 500 GPM RO system that needs two 120-inch-diameter softener vessels for the pretreatment will just need two 84-inch-diameter vessels for post-treating the RO permeate.
3. One 84-inch-diameter vessel will produce almost 1700 gal of 10.0% brine waste. Only 60% of chloride-free water will be required to dilute the chloride to meet the discharge limits, when compared to the sodium softener while positioned before the RO unit.

The drawback of the post-RO configuration is that the RO membranes will be now more susceptible to fouling, scaling, and degradation due to the presence of hardness and metals like manganese and iron. This issue could be solved by the addition of antiscalants for minimizing the scaling and by suitable filtration for removing manganese and iron. On the other hand, the expenses should be properly examined for determining which option (softener behind or in front of the RO unit) is more economical.

12.1.10 Reverse osmosis sizing and capacity

Appropriate sizing of a reverse osmosis unit, especially when the product water requirement is variable, could be a great task. Variable demand of product water could have actual swings in requirement usually brought on by high-level sensor as well as low-level sensor positioned in product tanks that cycle the RO feed pump ON and OFF. The perfect condition is to maintain all RO skids working uninterruptedly. The membrane idling turns the membrane more prone to scaling and fouling, particularly when a shut-down flush is not used. Moreover, frequent start-ups put the membrane through physical stress and possibly water hammer also, and soft-start motors could minimalize this.

The best method for satisfying the variable product water demand is the designing of the RO system for the average flow rate needed. As an illustration, if the product water demand is 500 GPM for 10.0% of the time and 350 GPM for 90.0% of the time, then the RO system can be designed for almost 365 GPM. A permeate tank that is capable of holding the additional 15 GPM for 90.0% of the time is needed. A reverse osmosis system online or offline depends on the level settings in the RO permeate tank. The RO feed pump trips on at the low set-point, and it trips off at the high set-point. It should be noted that the position of the level set-points must be far apart such that the RO unit stays online for the maximal time.

If a larger tank is not possible, then an alternate option will be multiple skids, a few of it might sit idle. The main point is to rotate the skids on as well as off, such that no single skid will experience the majority of the start-ups or downtime. Moreover, it is suggested that a shutdown or an offline down flush can be carried out, such that the motor can have a soft start. As an example, for the product water demand stated above, two 350 GPM skids can be used. Both skids can be online at the time of 500 GPM product water demand. A permeate tank with sufficient capacity for handling the 200 GPM excess will be needed. In case that the product water demand is constantly variable, then the finest option will be the designing of the tank for the average flow rate, as stated previously.

12.1.11 Adequate pretreatment

The most frequent reason for a total failure of a reverse osmosis system is the insufficient feed water pretreatment in the RO system (Byrne, (b)). Pretreatment is necessary because the membrane systems are susceptible to fouling (Singh, 2014). Any compromise in the pretreatment approaches, quality of equipment, and monitoring instrumentation would generally lead to operational issues in the downstream RO system. In a recent study carried out by Kavitha et al. (2019), the team makes an effort to understand the conventional process, microfiltration, UF, nanofiltration (NF), and integrated membrane systems for the seawater pretreatment, their advantages and limitations, and the effect of the pretreated water on the performance of RO membranes through the studies conducted by researchers and by considering certain case studies. Depending on the different laboratory-scale and pilot-scale studies carried out worldwide, the team noted that based on the seasonal deviation in the raw seawater feed and the necessary conditions for the RO feed, a blend of membrane and conventional pretreatment, which can operate with the renewable energy source, can contribute to an economical solution for meeting the requirement for potable water by seawater desalination process.

Filter backwash versus rinse flows: As an illustration, a commonly observed arrangement is to employ the same flow-control orifice on a multimedia filter discharge line for controlling both the flow rate of the filter backwash and the flow rate of the rinse carried out after backwashing. This will result in an approximately similar flow rate being used for the two steps. However, where a flow rate of backwash based on 12 GPM/ft^2 of cross-sectional area is suitable for attaining 40% expansion of the media granules, this same flow rate in the rinsing cycle would compact the media granules under a pressure drop of more than 10 psid (psi differential). This would be pushing the suspended particle present in the uppermost part of the media filter intensely into the media bed.

Filter flow balancing: Another frequent mistake associated with media filter will be not fixing distinct flow meters on each of the multiple filters arranged in parallel. In the absence of such flow readings, there will be no technique for knowing if the flow rates have been properly balanced amongst the filters. In case that a specific filter begins plugging up with solids, then increased flow would get diverted to the other filters. Always it should be taken into consideration that the pressure gauge on each side of the filter would not point out in case that one filter is plugging more than any other, when connected on the common outlet and inlet lines.

Observing the SDI: In case that the media filter does not have the capability of producing water that is within an SDI value of 5, then a serious mistake can occur, that is, the injection of a polymer filtration aid right beforehand the media filters. The mistake is principally deceptive due to the fact that this seems to intensely increase the filter effluent quality. However, the residual polymer breaking through the filter will not be shown in the SDI analysis or the effluent turbidity. Due to the molecular charge characteristics of the polymer, it would irreversibly bond with the RO membranes.

Suspended solids may get attached to the polymer, instead of migrating along the surface of the membrane. The fouling rate would intensify, and thus cleaning will be no longer able to reestablish the original performance. Thus, the membrane elements should require to be replaced (American Water Chemicals).

12.1.12 Proper elimination of chlorine

The polyamide-based thin-film composite (TFC) membrane normally employed in majority RO systems will not be able to handle chlorine (Byrne, (a)). Certain manufacturers of membranes have stated that their membranes are able to withstand free chlorine same as exposure of 1 mg/L over a time period of 1000 hours before a doubling of salt breakthrough will take place. However, the above-stated guideline has been generally misunderstood to mean that it will be okay to permit chlorine to infrequently come in contact with the RO membrane as an approach to decreasing the biofouling. Generally, the membrane damage will take place as soon as it has been exposed to any quantity of chlorine, and also it would be cumulative. This damage would get worsened in case that any transition metals like iron are fouled out on the surface of the membrane. Gohil and Suresh (2017) reviewed the mechanisms of TFC membrane degradation by chlorine and the techniques for its mitigation.

Sodium bisulfite is normally used for reducing the concentration of chlorine into the RO unit. However, this sodium bisulfite can also react with any dissolved oxygen present in the water. Any additional bisulfite may have a tendency for reducing the concentration of oxygen, which will increase the possibility for higher growth of biological anaerobes, the species accountable for the formation of heavy slime that could cause rapid system fouling. An indication of this will be the rotten egg smell of sulfur dioxide, as soon as the membrane vessels are opened. The sodium bisulfite optimal concentration will be challenging to maintain. The sodium bisulfite existing in chemical totes or in the injection tank may degrade with time as it undergoes a reaction with oxygen present in the atmosphere. In case that the bleach (sodium hypochlorite) is injected upstream, then the bleach concentration would also vary based on its age. Therefore, the accurate bisulfite concentration injection relative to the chlorine content might be difficult.

ORP is a comparatively economical technique to monitor the bisulfite dosage; however, this technique might not straightaway reveal the residual chlorine content. Other variables, especially, pH, could also influence its reading. When the ORP is employed for controlling the bisulfite dose on a regular operational system, then the result might be unsuccessful in case that the RO product water comes back to an upstream feed tank while the process water is not being required. At the time of minimum usage, the higher concentration of RO product water in the blended feed signifies that low alkalinity would be present. Additional bisulfite would have a higher effect on the pH of the water and leads to its drop. The reducing pH would result in an increase in the ORP reading, even though no chlorine is available.

12.1.13 Scale inhibition

The chemical scale inhibitor injection is usually the most inexpensive method for preventing the formation of scale in a reverse osmosis system (Byrne, (b)). The scale inhibitors will function by binding with the forming scaling crystals, which decreases the growth rate of the particles. The small-sized scaling particles have the tendency to remain suspended and leave the RO unit in the reject stream.

Rinsing of small-sized particle scales: A process of rinsing supersaturated salts from the RO system before the shutdown is very important for the efficiency of this mechanism. The most suitable approach is to tee in pressurized product water with an automated valve downstream of the inlet isolation valve for displacing the water in the RO unit during shutdown.

Moreover, this reduces the possibility for the development of anaerobic bacteria at the time of shutdowns by decreasing the anion concentrations in the RO unit, which are needed by the anaerobes. Moreover, this will improve the product water quality at the time of startup; hence a product water diversion system might not be required at the start-ups.

Use of blend inhibitors: The homogeneous polyacrylic acid polymers will have a greater possibility for coming out of solution because of overinjection or as a result of a reaction with aluminum or iron. In certain cases, they might just come out of solution with hardness in the event that the injected chemicals do not blend rapidly at the injection point. Blend inhibitors with two or more chemical constituents will function in an improved manner and are less likely to develop these issues.

12.1.14 Proper installation of reverse osmosis system

The RO reject stream regularly is plumbed to a discharge drain positioned under the uppermost point of the membrane PVs (Byrne, (b)). If not a vacuum-breaking valve or an automatic isolation valve is fixed on the reject line, a siphon effect (atmospheric pressure will push the liquid up and gravity will pull the liquid down) will pull on the RO system, when it is shutdown. Water will continue to flow through the line after shutdown and will pull a vacuum on the RO unit.

The vacuum would cause water to moderately drain from the RO PVs. When a reverse osmosis system drains, the entering air will transmit bacteria and fungi spores into the elements, which might lead to fouling of the membrane. At the time of RO system restart, a water hammer might take place, which could fracture the fiberglass wrap as well as the anti-telescoping device on the ends of the membrane elements. A check valve that uses a light-weighted spring might be used into the topmost of the reject discharge line for allowing air to be sucked into the line, under the vacuum condition. It must be plumbed in such a manner that, each time the RO system starts-up, it does not spit water at someone nearby. Fig. 12.6 is the inside photograph of a reverse osmosis desalination plant.

12.2 Frequently asked questions

12.2.1 General

12.2.1.1 Can a reverse osmosis system accomplish 100% separation or 100% recovery?

No, 100% recovery or 100% separation will not be possible by using the TFC membranes. With the TFC membranes, there will always be three streams, namely, feed stream,

FIGURE 12.6 Inside photograph of a reverse osmosis desalination plant. Source: *Courtesy Acciona Agua. Personal communication with Acciona Agua team. Website: https://www.acciona.com/projects/middle-east/.*

permeate/product stream, and concentrate stream. When the permeate is being generated, the feed water will become more concentrated; however, it cannot be separated 100%. Moreover, as there is a concentrate stream always, 100% recovery will not be feasible. Percent recovery is the flow rate of product divided by flow rate of feed × 100%.

12.2.1.2 What is the difference between reverse osmosis and nanofiltration processes?

NF is a membrane-based liquid separation technique that is located between UF and RO. The RO technology has the ability to remove the smallest solute molecules, even in the range of 0.0001 μm in diameter and smaller, whereas NF technology can only remove molecules in the range of 0.001 μm. NF removes the divalent ions present in water, whereas RO removes monovalent ions, which means that it desalinates the water.

In the NF process, polypiperazine amide membrane is used and this process is principally a lower-pressure form of RO, in which the permeate purity is not as important as the pharmaceutical grade water. As an illustration, the concentration of dissolved solids to be separated will be lower than what is usually encountered in seawater or brackish water. Hence, NF technology is very appropriate for the well water treatment or for the treatment of water from several surface supplies.

12.2.1.3 What is hyperfiltration?

The hyperfiltration process is the same as the RO process. RO is also called hyperfiltration. RO is a pressure-driven membrane-based separation process where a semipermeable membrane separates the solvent (usually water) from the other components present in the solution. In this process, the membrane configuration will be normally cross-flow. Here, the pore size of the membrane will be very small permitting just water and possibly very little quantity of extremely low-molecular-weight solute to pass through the membrane.

12.2.1.4 What is the relationship between feed total dissolved solids and minimum pressure?

For a reverse osmosis system to generate product water, minimal pressure must be applied to the membrane for overcoming the natural water osmotic pressure. The pressure will depend on the available ion types and their concentration in the water. The osmotic pressure will not depend on the membrane type. Approximately, every 100 mg/L of TDS contributes almost 1 psi of osmotic pressure. As an illustration, if the TDS of feed water is 2000 mg/L, then the feed water osmotic pressure will be almost 20 psi. Here, a minimum pressure of 20 psi should be applied for obtaining the permeate across the membrane. In general, the applied pressure will be at least two times the osmotic pressure for a sustainable RO system. In case that the applied pressure is not adequate for overcoming the natural osmotic pressure of water, then a pump can be used for raising the applied pressure.

12.2.1.5 What is the importance of normalization?

The efficiency of a reverse osmosis system will be dependent on various factors such as the recovery, temperature, feed pressure, and composition of feed water (DuPont Water Solution, 2020). As an example, a feed water temperature decrease of 4°C will lead to a permeate flow reduction of almost 10%. This is considered to be a usual phenomenon, and for distinguishing between such ordinary phenomena and performance variations because of fouling or other issues, the measured salt passage and product flow should be normalized. Normalization is the process of comparison of the actual performance to a specified reference performance, whereas the impacts of operation parameters are satisfactorily considered. The reference performance is usually the stabilized performance of the RO system within the initial 48–72 hours of operation.

Most membrane manufacturers and a few original equipment manufacturers have developed software programs to normalize operating data. The normalizing software programs provided by membrane manufacturers can be downloaded or requested through their websites. Normalization is important to assess the performance of a reverse osmosis system. It requires at least daily once collection of system parameters, such as pressures, flows, temperature, conductivity, etc. The data are entered into the normalization software. Normalization is not only important to determine when cleaning is needed but it also shows the behavior of the membrane system.

12.2.2 Contaminants

12.2.2.1 Will arsenic present in the feed water be removed by reverse osmosis process?

Arsenic is extensively distributed in nature in soil, water, and air. Studies have confirmed that the long-term exposure to arsenic present in drinking water could result in several health issues. The major forms of arsenic in ground and surface water include arsenite(III) and arsenate(V). The survey of arsenic content in natural waters is of significance relative to the appropriate maximum limit of 10 ppb or less for human consumption. RO is considered to be an efficient arsenic removal technology proven through bench-scale studies as well as pilot-scale studies according to a report prepared for the United States—EPA (Environmental Protection Agency, 2000).

12.2.3 Chemical injection

12.2.3.1 How an antiscalant will function in a reverse osmosis-based water treatment application?

An antiscalant addition is a pretreatment step, where it is injected into the feed water before the feed water reaches the RO membranes. The presence of antiscalant will delay the calcium or magnesium precipitation and slow down the rate of precipitation. This will prevent scale formation during the RO process. As the duration of the water in the membrane system is comparatively short at the time of the treatment, scale formation is prevented.

12.2.3.2 Under which conditions can antiscalants lead to reverse osmosis membrane fouling?

RO antiscalants can lead to membrane fouling in the RO system under the following conditions (American Water Chemicals):

Growth of fungus in the antiscalant: Certain antiscalants might contain some impurities that could offer the required nutrients as well as carbon for biological growth. In polymer-based antiscalants, it is very common to observe the growth of fungus. Reputed suppliers manufacture high-purity products and use appropriate antifungal preservatives for preventing these problems.

The dosing point very near to the acid addition point: pH will be normally adjusted using concentrated sulfuric acid. Without proper mixing, the acid will momentarily form a separate phase because of its increased density as compared to water. In case that the antiscalant is dosed instantly after the acid dosing point, it could be hydrolyzed using a strong acid.

Unsuitability with water quality: Certain antiscalants have no tolerance to aluminum and iron. Such antiscalants would develop an insoluble complex with these metals and subsequently precipitate on the surface of the RO membrane.

Increased hardness or very high dosage of antiscalant: Antiscalants carry a high negative charge relative to a typical scale forming anions like phosphate or carbonate, and hence they have a stronger interaction with magnesium and calcium. At the time of threshold inhibition, this interaction with magnesium and calcium leads to the formation of clusters and interferes with the development of a stable nucleus. On the other hand, when the concentration of hardness is very high, then the antiscalants itself could develop insoluble calcium−antiscalant salt. This will be problematic because once the active inhibitor precipitates, then all the other scales will also precipitate.

12.2.4 Pretreatment

12.2.4.1 What are the uses of granular activated carbon filters?

Carbon filters are used for reducing organics, chlorine, tannin, color, and unpleasant tastes and odors from water.

Carbon filters could remarkably reduce the below contaminants:

1. Volatile organic chemicals (VOCs): Organic chemicals that change into a vapor
2. Turbidity (cloudy water)
3. Herbicides, pesticides, and insecticides
4. Bad tastes and odors
5. Chlorine
6. Chlorine by-products like trihalomethanes

12.2.4.2 Why is chlorine added and subsequently separated before the reverse osmosis process?

Disinfectants such as chlorine are needed for minimizing the possibility of microbial fouling of RO membranes (Kucera, 2015). As soon as the RO membranes are subjected to microbial fouling, it will be extremely hard to remove the fouling. A free chlorine residual of almost 0.50−1.00 mg/L in the pretreatment unit will be required. Feed water to the RO unit should be dechlorinated before reaching the membranes due to the fact that the membranes are very sensitive to oxidizers, which could subsequently cause membrane degradation. The use of sodium bisulfite is the ideal technique for dechlorinating unless the feed water has an increased organic content; in that situation, carbon filtration at a 2 GPM/ft^3 flow rate is suggested. Sodium metabisulfite ($Na_2S_2O_5$) is normally around 33% active, and the stoichiometric dosage of $Na_2S_2O_5$ is almost 1.80 ppm per ppm-free chlorine. Hence, the stoichiometric dosage of 33% active $Na_2S_2O_5$ is 5.40 ppm. Considering safety, a factor of 1.5 is used for increasing the amount of $Na_2S_2O_5$ for ensuring the total removal of free chlorine.

12.2.5 Silt density index measurement and testing

12.2.5.1 What is silt density index?

Colloidal materials and suspended solids present in feed water are one of the main problems in the RO system. Although majority systems follow certain pretreatment including a 5-μm prefilter, these fine particles will cause fouling in the RO membrane. For having some measure of the extent of this fouling issue, a concept known as SDI is employed.

It is a test used for characterizing the fouling probability of a feed stream. SDI is carried out on the basis of determining the rate of plugging a 45-μm filter applying a constant 30 psig feed pressure for a definite time period. SDI$_{15}$ is the SDI test that will run for 15 minutes.

Generally, the spiral wound systems need an SDI value <5 and hollow-fiber systems need an SDI value <3. The majority deep well waters have an SDI of 3 and common surface water have SDI higher than 6.

12.2.5.2 After the calculation of silt density index, how to determine the requirement for reverse osmosis prefiltration?

SDI < 1, several years without colloidal fouling, no prefiltration is required.

SDI < 3, several months between cleaning, no prefiltration is required.

SDI 3–5, particular fouling likely a problem, frequent cleaning. No prefiltration is required.

SDI 5–10, unacceptable, additional pretreatment is needed. A media (sand-type) filter can be used.

SDI > 10, unacceptable, additional pretreatment is needed. A two-stage media filtration is required probably with the aid of coagulants or settling tanks.

12.2.5.3 What is the difference between silt density index and turbidity?

Turbidity is a measurement of the quantity of suspended solids in water, whereas SDI is a measurement of the fouling possibility of the suspended solids present in water. Turbidity is determined by analyzing the intensity of light scattered by the suspended particles in samples of water. It is generally expressed in terms of nephelometric turbidity units (NTU).

These two terms are not the same, and there is no direct relation between these two. However, in practical terms, the membranes demonstrate very less fouling when the turbidity of feed water is less than 1 NTU. Similarly, the membranes exhibit very less fouling at a feed SDI value of under 5.

12.2.6 Reverse osmosis membranes

12.2.6.1 Can the membranes be operated at a temperature higher than 45°C (113°F)?

TFC membrane elements are not designed for supporting the temperature greater than 45°C (113°F). However, it does not mean that the membrane elements are not able to go beyond the maximal temperature limit of 45°C. A high possibility for membrane element damage exists as the temperature exceeds 45°C, and the warranty is void. For operating or cleaning at elevated temperatures, the heat-sanitizable or high-temperature elements with various materials of construction are suggested.

12.2.6.2 What is the difference between thin-film composite and cellulose acetate membranes?

Cellulose acetate (CA) is hydrophilic in nature, and the membranes composed of CA are usually of asymmetric construction. Formerly CA membranes have been made up of cellulose triacetate, cellulose diacetate, or a combination of these materials. These CA-based membranes require a residual of chlorine for protecting these membranes from microbial attack. The CA membranes also have restricted pH operational needs. These membranes are also considered to be uncharged due to the fact that their functional groups are not polar. Because of this nonpolar nature, the CA membranes will never attract foulants to the surfaces easily. Low fouling is noted because of a smoother surface of the CA membrane.

TFC is a reverse osmosis membrane consisting of three layers joined each other. There will be two base layers (porous structure support) of the asymmetric construction and a thin third skin layer (salt rejecting layer) of PA deposited on the surface. Developed at the beginning of the 1970s, the asymmetric polyamide polymer has been used in the fabrication of the TFC spiral wound RO membrane. Polyamide polymer is the most distinctive membrane material because of its more flexible operating conditions and low-pressure requirements. Also, these PA membranes are oxidant (ozone, bromine, chloramine, chlorine, etc.) intolerant. Different types of membrane materials are discussed in Chapter 4, Transport Models, Membrane Materials and Basic Flow Patterns.

12.2.6.3 How frequent is membrane cleaning recommended?

The guideline for RO membrane system cleaning varies slightly from manufacturer to manufacturer; however, the standard rule will be as follows: The system should be cleaned: (1) if the normalized salt passage is increased by 5.0%, (2) if the normalized permeate flow is reduced by 10.0%, or (2) if the normalized differential pressure (DP) is increased by 15.0%.

For orientation, the cleaning frequency could be in the range of four/year with an SDI under 3. For an SDI of 5, the cleaning frequency can be double. Though, the frequency of cleaning will depend on the particular situation.

12.2.6.4 What is the difference between dry membrane and wet membrane?

When the thin-film membranes are fabricated, they will be normally dry. These dry membranes will have an unlimited shelf life, if they are stored appropriately. However, membranes turn out to be wet while they are flushed or tested using water. As soon as these membranes are wet, they cannot be dried again. These wet membranes should be preserved for preventing microbial growth. This can be done normally by employing a 1.0%−2.0% solution of sodium metabisulfite.

12.2.6.5 Is it possible to use various kinds of membranes in a single reverse osmosis unit?

It is not suggested to use various membranes together in a single RO unit. However, sometimes the membranes can be used together in a low-pressure system. In this kind of system, the water flux could reduce remarkably through the last several membrane modules as the osmotic pressure of the feed achieves almost the difference between the applied pressure and the pressure drop in the PV. This kind of situation is very ordinary in low-pressure municipal applications in which several systems have seven membrane modules arranged in series in a PV. In such a situation, the final two or three membranes in the last stage could be substituted using low-energy membranes. These low-pressure membranes commonly compromise salt rejection; however, higher rejection will not be very important for municipal applications in which 80−90 mg/L TDS permeate is adequate. Moreover, it has to be made sure that these low-pressure membranes must not be positioned before the standard-pressure membranes in a reverse osmosis system.

Another situation in which different membranes are used together is at the time of some emergency situations where certain membranes, but not all, are damaged in some way and should be replaced. Due to the dissimilarities among interconnections, membranes from different manufacturers are usually not mixed in a single stage of a single RO system.

12.2.6.6 Will there be a shelf life for a reverse osmosis membrane?

Shelf life will depend on the membrane conditions while they have been stored. In the case of fresh membranes, which have been stored in the unopened original bag, the shelf life will be almost 1 year (Kucera, 2015). The warranty for membranes will usually begin at the start-up of the system or 1 year after shipment, whichever approaches first. In the case of membranes that have been wet tested before the consignment from the factory must be checked once every 3 months for pH and for biological growth. When any biological growth is observed, subsequent to 6 months of storage, or if the preservative solution pH decreases lower than 3, then the membrane modules must be stored in a new preservative solution. Membranes that have not been wet tested must also be examined frequently for any biological growing. When any growth is observed, then these membranes must be soaked in a preservative solution.

The used membranes could be separated from the RO skid and stored. However, these membranes must be cleaned before storage. As soon as it is separated from the PV, all the membrane modules must be soaked in 1% solution of non-cobalt-activated sodium metabisulfite solution combined with deionized (DI) water like RO product water. In order to prevent the bacteria growth, Toray recommends the immersion of membrane elements in a 0.2−0.3 wt.% formaldehyde (HCHO) sanitization solution at pH 6−8, adjusted by sodium bicarbonate (NaHCO$_3$) (Toray, 2020). This sanitization technique is efficient and adequate for controlling the microbial activity for long-term or short-term shutdowns. However, this membrane element immersion in a HCHO sterilizing solution is not appropriate for fresh membrane elements. The membrane elements should have been in full operation at design conditions for a minimum of 72 hours before any HCHO sterilization procedure. The element exposure to HCHO before 72 hours of operation may lead to an irreversible flux loss. The modules must be soaked in a vertical position for approximately 1 hour. Subsequent to the soaking, the module must be allowed to drip and subsequently stored inside an oxygen impermeable plastic bag. It is not necessary to fill the bag using the preservative solution, due to the fact that the moisture present in the module will be satisfactory. The membrane modules could be stored for 6 months by employing this technique. The membranes that are stored must be examined once every 3 months for pH or for any microbial growth. Membranes must be cleaned in a higher pH solution before returning the membranes back to service. The modules must be always in a dark and cool location, which is out of direct exposure to sunlight and set aside from freezing.

12.2.6.7 *What is concentration polarization?*

Concentration polarization (CP) is an important factor affecting the efficiency of membrane-based separation processes (Mulder, 1991). The accumulation of solutes nearby the surface of the membrane is termed as the CP, and this will act as a barrier to the water flow across the membrane thereby decreasing the efficiency of the membrane. The water flow across the membrane will bring feed water (consisting of solute and water) to the surface of the membrane, and as the pure water passes across the membrane, the solutes will accumulate over the membrane surface. During this filtration process, the particles will deposit over the membrane and subsequently develops a cake layer. Due to the fact that the rejection mechanisms for RO are different, these solutes will remain in the solution and leads to the development of a boundary layer of greater concentration at the surface of the membrane. Consequently, the solute concentration in the feed water turns out to be polarized, with the solute concentration at the membrane surface higher as compared to the solute concentration in the bulk feed water in the feed stream. The prediction of solute CP is very important to design the RO systems, to forecast their performance, and particularly to understand the surface fouling phenomenon (Marinas & Urama, 1996).

12.2.7 Cleaning

12.2.7.1 *When to start the membrane cleaning process?*

According to the membrane manufacturers, the cleaning process must be started when any of the subsequent conditions occur, even though the primary flux is never restored (Singh, 2005).

1. Product water flow has reduced to 10.0%–15.0% lower than the rated flow at normal pressure.
2. Product water quality has decreased to 10.0%–15.0%, and salt passage has increased to 10.0%–15.0%.
3. There is an increase of 10.0%–15.0% feed pressure for maintaining the rated product water flow.
4. The DP along the length of the module a reverse osmosis unit has noticeable increased (greater than 15.0%).
5. Applied pressure has increased by almost 10.0%–15.0%.

12.2.7.2 *How to determine the amount of cleaning solution required?*

For determining the amount of cleaning solution needed, the operator should estimate the hold-up volume of the cleaning loop piping as well as the membrane element housings. Subsequently, add adequate water to the CIP tank for preventing it from emptying while filling the system. During the start of the cleaning cycle, the process water in the system must be discharged to the drain as it will be displaced by the cleaning solution. For estimating the recirculation dimension of CIP tank, calculate the system hold-up volume and subsequently multiply it by two. For the hold-up volume in the membrane element housing, use the below estimation given the housings are filled with the maximal number of membrane elements.

1. Twenty liters for all 8-inch elements (five gallons/element)
2. Four liters for all 4-inch elements (one gallons/element)

12.2.7.3 *How much time will it take for cleaning a reverse osmosis system?*

A standard two-stage RO skid can take almost 10–12 hours for cleaning, based upon the time it will take for heating up the cleaning chemical solution. In case that a prolonged soak time is needed, then it could take more time, may be up to 24 hours, including the soak period. Individual stage in a skid must be cleaned separately, irrespective of the others such that it should not contaminate any stage by scalants or foulants from another. This is the main reason for taking a longer time (up to 24 hours) for cleaning a complete RO system.

12.2.7.4 *What should be the temperature of cleaning solutions to be used for cleaning membranes?*

In general, membranes must be cleaned at a higher temperature and as well as at pH extremes as suggested by the membrane manufacturer. The different performance restoration methods such as chemical cleaning and direct osmosis cleaning have been analyzed in Chapter 9, Performance Monitoring, Process Parameters, and Their Control. Studies have confirmed that cleaning under the recommended conditions can remove more foulants and scale, as compared to the cleaning efficiency at neutral pH and ambient temperature. On the other hand, cleaning outside the suggested pH and temperature will result in membrane degradation, and this might make the membrane warranty invalid, and hence it must not be tried without any authorization from the manufacturer.

12.2.8 Commercial reverse osmosis system

12.2.8.1 *What is the difference between element recovery and system recovery?*

Element recovery is the recovery rate of a single membrane element.

$$\text{Element recovery} = \frac{\text{Permeate flow rate of single element}}{\text{Feed flow rate to the single element}} \times 100 \qquad (12.5)$$

System recovery is the cumulative recovery rate of the system.

$$\text{System recovery} = \frac{\text{Cumulative permeate flow rate of membrane elements in a system}}{\text{Feed flow rate to the system}} \times 100$$

$$(12.6)$$

As an example, if there are two parallel PVs and each PV consists of six elements, feed flow to the system is 100 GPM. As there are two PVs in parallel, the feed flow to each PV will be 50 GPM. The first membrane element in each PV will get 50 GPM of feed. In case that the first membrane element produces 5 GPM of product water and the system generates 50 GPM of product water, then, the recovery of first membrane element = 5 GPM/ 50 GPM × 100% = 10%, and the system recovery will be 50 GPM/100 GPM × 100% = 50%

12.2.8.2 What is a pressure drop and why is high-pressure drop problematic?

Pressure drop is the pressure loss from the feed end to the reject end of a membrane module or a pressure vessel. This high DP will generate an increased force, thereby pushing the element feed side in the direction of flow (CSM, Saehan Industries). The force will definitely impact on the product water tubes and the membrane element fiberglass shells in the same PV. The stress acting on the last membrane element will be the highest because it experiences the total of all the forces from the pressure drops of all the previous membrane elements.

According to the CSM RO membrane manufacturer, the higher limit of the pressure drop per multielement PV will be 60 psi (4.1 bar), and per single membrane element 20 psi (1.4 bar). As soon as the values exceed the limits, just for a shorter time period, the membrane elements will be damaged mechanically leading to telescoping and/or fiberglass shell breaking. The above-stated kind of membrane damage might not effect the performance of the membrane temporarily; however, ultimately it can lead to increased salt passage or loss in flux.

An increased pressure drop at steady flow rates might be due to the buildup of foulants, debris, and scale inside the membrane element flow channels. This will generally reduce the product water flow. A higher increase in pressure drop can occur from some operation errors like going beyond the suggested feed flow, and the quick feed pressure buildup at the time of start-up (referred to as water hammer).

The feed to concentrate pressure drop will be a measure of the hydraulic flow resistance of water through the RO system. The pressure drop will be highly dependent on the water temperature and on the flow rates through the membrane element flow channel. Hence, it is recommended that the concentrate flow rate and permeate flow rate should be maintained constant, as much as possible, for properly observing and monitoring any membrane element plugging resulting in a higher pressure drop.

The accurate information on the location and extent of the pressure drop increase contributes an important tool for identifying the problem cause. Hence, it is beneficial to monitor the pressure drop across every array and the overall feed to concentrate pressure drop

12.2.8.3 How to start-up a reverse osmosis unit?

The RO unit start-up, especially when fresh membranes are set-up, must be performed very carefully for preventing the water hammer from module crushing (Kucera, 2015). The RO system's initial start-up is discussed in detail in Chapter 9, Performance Monitoring, Process Parameters, and Their Control. Initially, prior to any start, the procedure for start-up given by the equipment vendor must be fully read as well as understood. Subsequently, for preventing any water-hammer membrane damage, the permeate valves and concentrate valves must be fully open during the start-up. The operator must not start the RO system with the closed concentrate valve and afterward opening it till the expected recovery is achieved. Moreover, the RO feed pump must be slowly started, raising the pressure at a rate not higher than 10 psi/s. In case that a variable frequency drive is used, it could be adjusted for starting up slowly. If variable frequency drive is not set up, and a centrifugal pump is used, then the concentrate valve must start open and later closed down very slowly till the expected recovery and feed pressure is achieved, ensuring that the pressure at an adequate rate. The use of an older, positive displacement pump needs a pulsation dampener and a gentle start, using the concentrate and pump recycle valves for adjusting the feed pressure and the recovery.

For preventing membrane damage, it must be adequately shimmed and the thrust ring must be fixed correctly. The thrust ring and shims would reduce or prevent the membrane module movement at the time of shutdown and start-up of the RO system.

12.2.8.4 *What are the impacts of brine seal damage?*

The brine seals could be damaged or turned over at the time of setting up or because of some hydraulic impacts (CSM, Saehan Industries). Some feed water would flow through the chasm in the damaged brine seals for bypassing around the membrane element, leading to a lower flow and velocity through the membrane element. Thus, the maximum element recovery will exceed the limit and thereby increases the fouling and scaling possibility.

When the fouled membrane element in the multielement PVs turns out to be severely plugged, then there will be a higher probability for the fouling of the downstream elements because of inadequate concentration flow rates in that PV.

12.3 Survey of the problems and challenges faced by reverse osmosis desalination/water treatment plants in Qatar

12.3.1 Reverse osmosis desalination plant in Qatar (Plant 1)

Operational issues

Q1: In your RO plant, do you experience a deviation from the required permeate quality?
A1: No, required permeate quality is fulfilled at all times
Q2: Do you experience flow rate increase or decrease while operation?
A2: No, the flow rate is stable
Q3: Do you experience salt rejection increase or decrease while operation?
A3: It depends on seawater temperature and lifetime of your RO membranes
Q4: Any leak from element parts?
A4: No.
Q5: Degradation of support membrane problem?
A5: No.
Q6: Scale formation on membranes?
A6: Scale formation is controlled by the application of an antiscalant.
Q7: Iron fouling of membranes?

A7: Small quantity of iron could reach the RO membranes from coagulant use. Autopsies of RO membranes showed a small amount of this fouling.

Q8: Biofouling issue?

A8: Yes, autopsies of RO membranes showed this fouling.

Q9: Particle fouling issue?

A9: No, autopsies of RO membranes are not showing this fouling.

Q10: Chemical fouling issue?

A10: Only in relation to the coagulant, used ferric chloride in the pretreatment.

Q11: Any membrane delamination issue?

A11: No.

Q12: What is the frequency of O-ring replacement?

A12: Along with the frequency of RO membranes replacement.

Q13: Do the elements demonstrate significantly increased conductivity with a higher flow?

A13: Conductivity increases but not significantly.

Q14: Is there a proper monitoring system?

A14: Yes, installed a full monitoring system in the LCR.

Q15: Are there any instrumentation failures?

A15: Yes, sometimes but they are not affecting our continuous RO plant water production

Membrane cleaning

Q16: What is the frequency of membrane cleaning?

A16: It depends on fouling rate, average 1 CIP/every 2 weeks

Q17: What is the cleaning procedure?

A17: Flushing with permeate water, circulation with a chemical solution, and soaking with chemical solution

Q18: Which are the cleaning solutions used?

A18: Mainly NaOH + EDTA, citric acid, and biocide used at different concentrations.

Q19: What is the average flux recovery/pressure reduction post-cleaning?

A19: NA flux recovery. The average pressure reduction is 0.2 bar

Q20: About your CIP system?

A20: It has been designed to comply with the requirements of RO cleaning according to manufacture's instructions.

Equipped with: tanks, pumps, mixers, heaters, and so on.

Membrane replacement

Q21: What is the frequency of membrane replacement?

A21: 12% per year

Q22: What is the reason for replacement (increased pressure/decreased production)?

A22: Pressure increment, permeate flow decrement, and poor permeate water quality.

Suggestions for improving the efficiency

Q23: For corrosion in intake piping materials and other equipment

A23: Select anticorrosion and suitable material for seawater, and apply preventive maintenance.

Q24: For boron and bromide level management in product water
A24: Use higher boron and bromide rejection seawater RO membranes.

12.3.2 Reverse osmosis desalination plant in Qatar (Plant 2)

Operational issues

Q1: In your RO plant, do you experience a deviation from the required permeate quality?
A1: Sometimes yes.
Q2: Do you experience flow rate increase or decrease while operation?
A2: Yes.
Q3: Do you experience salt rejection increase or decrease while operation?
A3: Yes.
Q4: Any leak from element parts?
A4: Only a few times from the connections.
Q5: Degradation of support membrane problem?
A5: No.
Q6: Scale formation on membranes?
A6: Yes.
Q7: Iron fouling of membranes?
A7: Yes.
Q8: Biofouling issue?
A8: Yes.
Q9: Particle fouling issue?
A9: Yes.
Q10: Chemical fouling issue?
A10: No.
Q11: Any membrane delamination issue?
A11: No.
Q12: What is the frequency of O-ring replacement?
A12: One to two times a year.
Q13: Do the elements demonstrate significantly increased conductivity with a higher flow?
A13: No.
Q14: Is there a proper monitoring system?
A14: In most cases, yes.
Q15: Are there any instrumentation failures?
A15: Sometimes, the pH meter

Membrane cleaning

Q16: What is the frequency of membrane cleaning?
A16: Depends. However, seawater membranes monthly.
Q17: What is the cleaning procedure?
A17: Alkaline, acid, biocide
Q18: Which are the cleaning solutions used?
A18: (1) EDTA + alkaline, (2) citric acid, (3) DBNPA

Q19: What is the average flux recovery/pressure reduction post-cleaning?
A19: Average 20%–30%
Q20: About your CIP system?
A20: Tank and low-pressure circulation pump

Membrane replacement

Q21: What is the frequency of membrane replacement?
A21: 3–5 years
Q22: What is the reason for replacement (increased pressure/decreased production)?
A22: Low production or high conductivity or increased pressure.

Suggestions for improving the efficiency

Q23: Corrosion in intake piping materials and other equipment
A23: Rare.
Q24: Boron and bromide level management in product water.
A24: No issue.

12.3.3 Reverse osmosis desalination plant in Qatar (Plant 3)

Operational issues

Q1: In your RO plant, do you experience a deviation from the required permeate quality?
A1: Permeate quality deteriorated gradually due to a decrease in salt rejection, during the long run of 7 years' operation.
Q2: Do you experience flow rate increase or decrease while operation?
A2: No abnormal variation in flow rate, other than normal flux reductions in between CIPs.
Q3: Do you experience salt rejection increase or decrease while operation?
A3: Salt rejection decreased in accordance with the age of the membrane.
Q4: Any leak from element parts?
A4: No leakage from element part.
Q5: Degradation of support membrane problem?
A5: Membrane degradation not observed.
Q6: Scale formation on membranes?
A6: Scale formation not observed.
Q7: Iron fouling of membranes?
A7: Iron fouling not observed.
Q8: Biofouling issue?
A8: We observed high weight in the lead element, and it is assumed due to biofouling.
Q9: Particle fouling issue?
A9: Particle fouling not observed.
Q10: Chemical fouling issue?
A10: Chemical fouling not observed.
Q11: Any membrane delamination issue?
A11: Membrane delamination not observed, but channeling and feed spacer dislocation noticed.
Q12: What is the frequency of O-ring replacement?

A12: O-rings for membrane interconnectors and adapters are replaced rarely, but PV side port O-ring replacement intervals are at an average of 2 years.

Q13: Do the elements demonstrate significantly increased conductivity with a higher flow?

A13: No.

Q14: Is there a proper monitoring system?

A14: All trends are monitored in SCADA.

Q15: Are there any instrumentation failures?

A15: No instrumentation failure.

Membrane cleaning

Q16: What is the frequency of membrane cleaning?

A16: Average cleaning intervals are of 3 months.

Q17: What is the cleaning procedure?

A17:

1. Flush the unit until brine conductivity equals the product water conductivity.
2. Prepare the solution and make good mixing and adjust the pH.
3. Low-flow circulation followed by soaking.
4. Check pH during the first circulation and add caustic/citric to maintain pH.
5. High-flow circulation and soaking for around five to six times for high pH and three to four times for low pH.
6. Flush out solution with permeate water until reaches the pH and conductivity.

Q18: Which are the cleaning solutions used?

A18:

- (M74) Na EDTA: 1% weight/volume.
- (M70) STPP: Until the level of foam is controllable, but less than 2% by weight/volume
- NaOH: Upto desirable pH
- Citric acid: 2% weight/volume

Q19: What is the average flux recovery/pressure reduction post-cleaning?

A19: 15%–20% average flux recovery and 1.5–2.0 bar DP reduction with constant feed pressure operation.

Q20: About your CIP system?

A20: Dedicated CIP system for both 1st and 2nd pass with associated piping, tanks, pumps, cartridge filters, and monitoring instruments like flow meters, pressure transmitters, etc.

Membrane replacement

Q21: What is the frequency of membrane replacement?

A21: No membranes replaced during the first 6 years of operation, and 15%–30% membrane replacement scheduled per year.

Q22: What is the reason for replacement (increased pressure/decreased production)?

A22: High membrane DP and decreased salt rejection.

Suggestions for improving the efficiency

Q23: Corrosion in intake piping materials and other equipment

A23: Small traces of corrosion observed in filter feed pump casing and impellers.

Q24: Boron and bromide level management in product water.

A24: 2nd-pass brackish water RO with caustic addition, succeeds to keep boron levels less than 5 ppm.

References

Al-Karaghouli, A., & Kazmerski, L. (2012). Economic and technical analysis of a reverse osmosis water desalination plant using DEEP-3. 2 software. *Journal of Environmental Science and Engineering A, 1*(3), 318–328.

American Water Chemicals. https://www.membranechemicals.com/faqs/under-which-conditions-can-antiscalants-cause-ro-membrane-fouling/ (Accessed 29 November 2020).

Brine Treatment (ZLD). https://www.lenntech.com/processes/brine-treatment-ZLD.htm (Accessed 17 November 2020).

Byrne, W. (a), *Common mistakes in design, use of reverse osmosis systems*. http://www.ethanolproducer.com/articles/5833/common-mistakes-in-design-use-of-reverse-osmosis-systems (Accessed 29 November 2020).

Byrne, W. (b), *The smart approach to reverse osmosis*. https://www.watertechonline.com/wastewater/article/15530070/the-smart-approach-to-reverse-osmosis (Accessed 29 November 2020).

Byrne, W. (c), *United States water consultant, design and care of reverse osmosis systems, Part 2: Upstream equipment*, https://www.kuritaamerica.com/the-splash/design-and-care-of-reverse-osmosis-systems-part-2-upstream-equipment (Accessed 17 November 2020).

CSM, Saehan Industries, Technical Manual. http://www.csmfilter.co.kr/searchfile/file/Tech_manual.pdf, (Accessed 29 November 2020).

Dayarathne, H. N. P., Choi, J., & Jang, A. (2017). Enhancement of cleaning-in-place (CIP) of a reverse osmosis desalination process with air micro-nano bubbles. *Desalination, 422*, 1–4.

DuPont Water Solution's FilmTec™ RO technical manual. (2020). Version 3; Form No. 45-D01504-en, Rev. 3, https://www.dupont.com/content/dam/dupont/amer/us/en/water-solutions/public/documents/en/45-D01504-en.pdf (Accessed 5 November 2020).

Personal communication with Acciona Agua team (2020). Gago, G. H. *O&M Desalination Middle East Director, Acciona Agua, personal communication*, November 8. Website: https://www.acciona.com/projects/middle-east/.

Gohil, J. M., & Suresh, A. K. (2017). Chlorine attack on reverse osmosis membranes: Mechanisms and mitigation strategies. *Journal of Membrane Science, 541*, 108–126.

Haidari, A. H., Heijman, S. G. J., & Van Der Meer, W. G. J. (2018). Optimal design of spacers in reverse osmosis. *Separation and Purification Technology, 192*, 441–456.

Jiang, S., Li, Y., & Ladewig, B. P. (2017). A review of reverse osmosis membrane fouling and control strategies. *Science of the Total Environment, 595*, 567–583.

Johnson, J., & Busch, M. (2010). Engineering aspects of reverse osmosis module design. *Desalination and Water Treatment, 15*(1–3), 236–248.

Kavitha, J., Rajalakshmi, M., Phani, A. R., & Padaki, M. (2019). Pretreatment processes for seawater reverse osmosis desalination systems—A review. *Journal of Water Process Engineering, 32*, 100926.

Kucera, J. (2015). *Reverse osmosis: Industrial processes and applications*. John Wiley & Sons.

Marinas, B. J., & Urama, R. I. (1996). Modeling concentration polarization in reverse osmosis spiral-wound elements. *Journal of Environmental Engineering—ASCE, 122*, 292.

Mulder, M. (1991). *Basic principles of membrane technology*. Dordrecht, NL:: Kluwer Academic Publishers.

Panagopoulos, A., Haralambous, K. J., & Loizidou, M. (2019). Desalination brine disposal methods and treatment technologies—A review. *Science of the Total Environment, 693*, 133545.

Pramanik, B. K., Shu, L., & Jegatheesan, V. (2017). A review of the management and treatment of brine solutions. *Environmental Science: Water Research & Technology, 3*(4), 625–658.

Saleem, H., Trabzon, L., Kilic, A., & Zaidi, S. J. (2020). Recent advances in nanofibrous membranes: Production and applications in water treatment and desalination. *Desalination, 476*, 114178.

Saleem, H., & Zaidi, S. J. (2020a). Developments in the application of nanomaterials for water treatment and their impact on the environment. *Nanomaterials, 10*(9), 1764.

Saleem, H., & Zaidi, S.J. (2020b). Innovative nanostructured membranes for reverse osmosis water desalination. Qatar University Annual Research Forum and Exhibition (QUARFE 2020), Doha, https://doi.org/10.29117/quarfe.2020.0023.

Saleem, H., & Zaidi, S. J. (2020c). Nanoparticles in reverse osmosis membranes for desalination: A state of the art review. *Desalination, 475*, 114171.

Sefatjoo, P., Moghaddam, M. R. A., & Mehrabadi, A. R. (2020). Evaluating electrocoagulation pretreatment prior to reverse osmosis system for simultaneous scaling and colloidal fouling mitigation: Application of RSM in performance and cost optimization. *Journal of Water Process Engineering, 35*, 101201.

Singh, R. (2005). Water and membrane treatment. *Hybrid Membrane Systems for Water Purification*, 57–130. Available from https://doi.org/10.1016/b978-185617442-8/50003-8.

Singh, R. (2014). *Membrane technology and engineering for water purification: Application, systems design and operation.* Butterworth-Heinemann.

Toray, Operation, Maintenance and Handling Manual. (2020). rev.: 110, Version: September 2020, https://www.toraywater.com/knowledge/pdf/HandlingManual.pdf (Accessed 29 November 2020).

Tow, E. W., & Rencken, M. M. (2016). In situ visualization of organic fouling and cleaning mechanisms in reverse osmosis and forward osmosis. *Desalination, 399*, 138–147.

United States-Environmental Ptptection Agency. (2000). *Technologies and costs for removal of arsenic from drinking water*, EPA-8 15-R-00–028, https://nepis.epa.gov/Exe/ZyPURL.cgi?Dockey = P1008RRB.TXT (Accessed 5 November 2020).

Wang, L. K., Chen, J. P., Hung, Y. T., & Shammas, N. K. (2011). *Handbook of environmental engineering: Membrane and desalination technologies* (Vol. 13). Springer.

Zaidi, S. J., Fadhillah, F., Saleem, H., Hawari, A., & Benamor, A. (2019). Organically modified nanoclay filled thin-film nanocomposite membranes for reverse osmosis application. *Materials, 12*(22), 3803.

Glossary

Anions   These are negatively charged ions generated from the disassociation of salts, bases, or acids present in aqueous solutions.

Algal bloom   It is a rapid increase in the population of algae in an aquatic system. It is also termed as marine bloom or water bloom. It can make the water cloudy or increase the water turbidity.

Algal count   It is a measure of the count of algal particles present per unit volume of the resource water. This is expressed in total count of algal cells present per milliliter of water.

Antiscalants   These are chemicals added during the pretreatment step to increase the solubility of sparingly soluble salts. In the case of reverse osmosis (RO) systems, calcium carbonate, calcium sulfate, barium sulfate, strontium sulfate, and calcium fluoride are the most frequently used antiscalant chemicals.

Antitelescoping device (ATD)   It is a rigid structure firmly connected to each end of a spiral wound RO membrane element that inhibits telescoping, unwinding, or any other undesirable motion of the membrane module.

Array   It is the physical arrangement of the pressure vessels, for example, a 6:3 array configuration is a two-stage configuration with a total of nine vessels. The first stage will have six pressure vessels and the second stage will have three pressure vessels.

Asymmetric membrane   It is a type of membrane made of the same material (polyamide or cellulose acetate) and shows an increment in porosity moving from surface to base. Also, there will be a thick porous support layer and a dense thin barrier skin on the membrane surface.

Backwash   It is an intermittent waste stream from ultrafiltration (UF) or microfiltration (MF) membrane unit. Also, it is a term for the cleaning process that normally involves intermittent reverse flow for removing the foulants collected at the surface of the membrane.

Biochemical oxygen demand (BOD)   It represents the oxygen amount used by bacteria and other microorganisms when these organisms decompose the organic matter at a particular temperature under aerobic conditions.

Biofouling   It is the type of membrane fouling due to the accumulation and the growth of microbes on the surface of the membrane or/and the adsorptive fouling of secretions from the microbes.

Brackish water   In general, brackish water is distinguished as the water of total dissolved solid levels from $1000 \, mg/L$ up until $15,000 \, mg/L$. It is not good for human consumption because of its high salinity.

Brine   It is a waste stream of the desalination process, which is a concentrated solution of higher dissolved solids. The seawater RO process can concentrate the salt concentration up to two times greater. It is also called concentrate or reject of the desalination process.

Brine seal   It is a rubber or plastic device that seals the exterior of one of the ends of a spiral wound membrane (SWM) element against the wall of the RO pressure vessel housing. This device prevents feed water from bypassing around the element and forces the feed water through the element.

Calcium bicarbonate [Ca(HCO$_3$)$_2$]   This is a salt available in the majority of natural waters. Water comprising $Ca(HCO_3)_2$ loses CO_2 while it is evaporated or concentrated by the RO process and subsequently $CaCO_3$ will get precipitated.

Cations   These are positively charged ions generated from the disassociation of salts, bases, or acids present in aqueous solutions.

Cartridge filters   These types of filters are used as extra protection for the RO membrane elements, for capturing any particles or suspended solids that may present in the feed stream. Cartridge filters could separate comparatively larger particles that could plug the RO membrane. These filters can be replaced when the suspended solids plug the filter and subsequently raise the differential pressure higher than the desired level.

Cellulose acetate (CA) membranes   Cellulose acetate is hydrophilic in nature, and the membranes composed of cellulose acetate are usually of asymmetric construction. These membranes are also considered to be uncharged due to the fact that their functional groups are not polar.

Chemical oxygen demand (COD)   This is a nonspecific test used on wastewater where a chemical oxidizing agent is reacted with certain organic matters present in water. COD is noted to be extremely accurate relative to the biochemical oxygen demand test; however, it will not measure the entire organic matters existing in the water.

Clean-in-place (CIP) unit   These are mobile skids used for cleaning membranes without removing them from the RO system. It can remove the accumulated foulants from the membranes and subsequently restore the original permeability as well as resistance.

Coagulation   It is a process where minute particles of suspended matter are united by chemical agents to form bigger-sized particles for permitting very fast settling or improved separation.

Coagulants   These are chemicals added to water to assist in flocculation. Alum is considered as the most commonly used coagulant.

Colloid   It is a mixture that has particles ranging between 1 and 1000 nm in diameter; however, they are still able to remain uniformly distributed throughout the solution.

Composite   It is a material made from two or more different materials that, when combined, are stronger relative to the separate materials by themselves.

Composite membranes   These are membranes with two or more distinct layers.

Concentrate   This is the water that is rejected during the RO process, and it consists of the majority of the dissolved solids present in the feed in a more concentrated form. Brine is another term used for concentrate.

Concentrate staging   It is a type of configuration in the spiral wound RO membrane system where the concentrate from each stage of a multistage system will become the feed for the consequent stage.

Concentration factor   It is the degree that the RO-feedwater-dissolved solids are concentrated in the reject water. The concentration factor is related to the recovery of RO system and the equation is important for the design of the system.

Concentration polarization   It is the salt concentration gradient on the high-pressure side of the surface of the membrane. The concentration of salt in this boundary layer goes beyond the bulk water concentration. This phenomenon affects the process efficiency by increasing osmotic pressure at the surface of the membrane, increasing salt passage, decreasing flux, and increasing the possibility of scale formation.

Conductivity   In aqueous solutions, conductivity is a measure of the ability of water to conduct an electric charge. The conductivity of the solution is related to the concentration of total dissolved solids.

Contact angle   It is a measure of the relative amounts of adhesive (liquid-to-solid) and cohesive (liquid-to-liquid) forces acting on a liquid and fluctuates over the range $0° \leq \theta \leq 180°$. The membrane's contact angle is the measurement of hydrophilicity of the membrane.

Cross-flow filtration   The running of the feedwater stream in parallel to the membrane surface so that it continuously removes contaminants from the membrane surface. The process is termed as cross-flow due to the fact that the direction of feed flow and the direction of filtration flow are at a 90° angle. This is an excellent method for filtering liquids with a high concentration of filterable matter.

Dead-end filtration   This is a type of filtration in which the water flow will be perpendicular to the surface of the membrane.

Demineralization   This is a type of water purification. It is a process for removing minerals from water, and generally the term is restricted to ion-exchange processes. Here, strong acid cation resin in the hydrogen form changes the dissolved salts into their corresponding acids, and strong base anion resin in the hydroxide form eliminates these acids. This process produces water with the same quality as distillation at a low cost.

Desalination This is the process of removing salt from seawater or brackish water for producing potable water, utilizing different systems. Desalination techniques are classified into thermal-based processes and membrane-based processes. All these technologies require energy for operation.

Differential pressure It is the pressure drop (ΔP) across a membrane module or unit from the feed inlet to the reject outlet. It is very important to monitor the differential pressure across each stage of the system.

Dissolved air flotation (DAF) DAF is a physiochemical treatment in which air is bubbled through the aqueous stream as microbubbles and subsequently it takes away the suspended and light organic matters to the surface, where they are skimmed off as the sludge.

Dissolved oxygen (DO) This is the quantity of oxygen that is dissolved in water at a specific time and it is expressed in terms of mg/L or ppm. DO level which is too low or too high could damage the aquatic life as well as disturb water quality.

Disinfection This is the process of destroying pathogenic microorganisms, typically by chemical agents, such as sodium hypochlorite and formaldehyde. This disinfecting process reduces the count of microorganisms without essentially destroying all those present.

Elements Elements are physical devices that house the membrane. In membrane systems, the elements are placed in series inside of a pressure vessel known as a module. Spiral wound systems possess up to six elements per pressure vessel. However, the hollow fiber RO systems typically possess just one element per pressure vessel.

Energy dispersive X-ray (EDX) spectroscopy It is an analytical technique employed for the chemical characterization or the elemental analysis of a sample. The EDS, combined with other techniques, such as Fourier transform infrared (FTIR) or scanning electron microscope (SEM), is an important diagnostic tool for analyzing the polymeric RO membrane failure and offers valued information to aid the manufacturers in designing improved membranes for RO process.

External fouling This type of fouling is caused by the buildup of deposits on the membrane surface by three different mechanisms: (1) mineral deposit accumulation (scaling); (2) development of cake of rejected solids, colloids, particulates, and other inorganic or/and organic matter; and (3) development of biofilm.

Feed channel spacers These spacers offer channels for the feed water flow and are seen in SWM elements. They are considered to be an important part of SWM modules in RO filtration. These spacers play a significant role in identifying the hydraulic conditions of the feed channel.

Feed stream This is the flow into the pretreatment stage of a reverse osmosis system. This stream will be divided to form a product/permeate stream as well as a concentrate stream.

Filtrate It is the water produced after a filtration process.

Filtration This is the process of separation of a liquid and a solid by utilizing a porous substance that can only allow the liquid to pass through.

Fourier transform infrared (FTIR) spectroscopy It is a well-known, comparatively cheap method for reviewing the structure of compounds by chemical bond vibrations. Recently, this technique has been developed for quantifying and analyzing the membrane fouling in different conditions.

Foulants These are substances that cause fouling.

Fouling This is the process of depositing solid substances on the RO membrane surface. This process could be because of the existence of biological growth, sparingly soluble salts, or suspended solids. RO membrane fouling results in a reduction in both the water quality and the amount of water produced.

Flux or water flux Flux is used to express the rate at which water permeates across a membrane and is normally expressed as volume per area per unit of time. Conventional units are liters per square meter per hour $L/(m^2 h)$ or gallons per square foot per day.

Flocculation This is the process of aggregation of destabilized particles and microflakes, and the successive development of sizeable flakes. One should add another chemical known as flocculent for facilitating the formation of flakes termed as flocs.

Hardness   It is the measure of the amount of magnesium and calcium present in water.

Hard water   It is the water that consists of an excessive number of positive ions. The water hardness is illustrated by the number of magnesium and calcium atoms existing. Soap typically dissolves poorly in hard water.

Heavy metals   These are metals that have a high elemental weight and a density of 5.0 or higher. The majority of heavy metals are considered to be harmful to humans, even in small concentrations.

High-pressure pumps   High-pressure pumps are used to pump the pretreated seawater up to the required pressure by RO membrane to achieve the desired separation of feed water into pure water and concentrated brine streams. For this application, centrifugal pumps are used, which operate in the range of 50–80 bar.

Hollow fiber (HF) element   This is one of the four potential membrane configurations (tubular, plate, and frame, and spiral wound are the others). These HF elements are made up of polyamide or CA. The pressurized feed stream will pass through the exterior of the fibers. Freshwater gets permeated and will be subsequently accumulated at the element end.

Homogenous material   It is a material of uniform composition all over, and cannot be mechanically separated into different materials.

Human machine interface   It is a panel that permits interactive monitoring as well as control of system conditions and settings.

Hydrolysis   It is a process of chemical breakdown of a membrane when it is exposed to high or low pH, temperature, and bioactivity. The hydrolysis process is generally related to CA membranes in which the acetyl groups are substituted by hydroxyl groups. This process amplifies leakage of salt (i.e., higher permeate conductivity) and a lower feed pressure necessity.

Hydrophilic   It is the water-attracting property of membrane materials.

Hydrophobic   It is the water-repelling property of membrane materials.

Internal fouling   This type of fouling results in a slow deterioration of membrane efficiency due to variations in the chemical structure of the membrane polymers and activated by chemical degradation or physical compaction.

Inorganic fouling   This type of fouling is also termed as scaling, and is normally due to the accumulation of inorganic precipitates like metal hydroxides, and scales on the surface of membrane or inside the pore structure.

Ion   It is an electrically charged particle, having a negative or a positive charge, produced by the dissociation of a salt, acid, or mineral in water.

Irreversible fouling   This type of fouling will be permanent, and cannot be removed by either chemical cleaning or by backwashing. This fouling will damage the membrane permanently, and thus the membrane has to be replaced.

Langelier saturation index (LSI)   It is considered to be a measure of calcium carbonate scaling probability and is used as a significant efficiency indicator in the RO system management. A positive value of LSI specifies that calcium carbonate can precipitate. A negative value of LSI points out the corrosive nature of water. A recommended target LSI in the RO concentrate is -0.2.

Leaf   It is a sandwich arrangement of flat-sheet, semipermeable membranes positioned back to back and separated by a fabric spacer in an SWM module.

Limiting salt   This is the salt that attains its saturated concentration initially, as the water is concentrated in a reverse osmosis system, necessary to prevent precipitation. The maximum recovery conceivable before any salt starts precipitating is termed as the allowable recovery.

Membrane   It is a material that permits the mass transfer of certain compounds whereas rejecting others, which is made to permit the separation of the feed stream and permeate stream.

Membrane configuration   It refers to the membrane geometry and its position in space with regard to the feed flow and permeate flow. The common membrane module configurations used for RO applications are spiral wound and hollow fiber.

Membrane degradation Membranes can get degraded from exposure to conditions in the system that can destroy the membrane material, such as fouling, scaling, and harsh chemicals, which influence the recovery rate and water flux.

Microfiltration (MF) MF is a pressure-driven separation process, and is considered to be extensively used in concentrating, purifying, or separating macromolecules, suspended particles, and colloids from solution. Microfiltration can also be used as a pretreatment for RO or nanofiltration (NF).

Modified fouling index (MFI) MFI is proportional to the suspended matter concentration, and is considered to be a very precise index relative to the silt density index for foreseeing the water tendency to foul RO/NF membranes.

Nanofiltration (NF) NF plays a vital role in softening the seawater. It is a membrane-separation process. An integrated membrane system of NF and UF can be used as the pretreatment for the seawater to increase the overall efficiency of the SWRO plant. Various factors, such as operating pressure, cross-flow velocity, and feed temperature, can be examined to increase the effectiveness of the process.

Nanofiltration membranes These membranes are the same as RO membranes; however, the NF membranes are not as efficient as RO in separating dissolved solids. NF membranes are generally termed as membrane softeners due to the fact that these membranes will generally reject divalent ions, the twice-positive charged hardness ions (magnesium and calcium) reasonably good, on the other hand, they demonstrate very low rejection for monovalent ions (e.g., chloride, potassium, and sodium).

Net driving pressure (NDP) NDP is considered to be the amount of the actual driving pressure accessible for forcing water across the membrane. It is the difference between osmotic pressure and feed pressure. As the value of NDP increases, the flux also increases proportionally (provided that all the remaining factors should remain constant).

Normalization It is the process of assessing the performance of a membrane system at a specific set of reference conditions, permitting the direct comparison of the daily performance of the system independent of changes to the actual operational conditions.

Normalized permeate flow (NPF) NPF measures the quantity of product water that the RO system is producing. By the normalization of the measured product flow for observed net driving pressure as well as temperature, a measure is achieved which could be used for comparing the membrane condition to the initial start-up conditions. A decrease in normalized permeate flow of about 10.0%–15.0% specifies that the cleaning of the membrane is needed.

O-rings These are used for sealing the product water tube interconnectors of side-by-side elements for preventing the intrusion of increased pressure feed stream (low quality) into the less pressure product stream (high quality). A defective O-ring would lead to a greater saline concentration of the product water in that particular area of the system.

Organic fouling This type of fouling is the collection of carbon-based material on a membrane. Natural organic matters are carbon-based compounds normally found in groundwater, surface water, and soil, due to the decomposition of plant and animal material.

Osmosis It is a natural process where water diffuses across a membrane from a low concentrated salt solution to a high concentrated salt solution.

Osmotic pressure It is the pressure that should be applied to the solution side for stopping the movement of fluid when a semipermeable membrane separates a solution from pure water. In the case of RO systems, this osmotic pressure needs to be overcome for producing product water.

Parts per million (ppm) It is considered to be a measurement of concentration, and one ppm is one-unit weight of solute per million unit weight of the solution. For the analysis of water, the ppm is the same as mg/L.

Permeability It is the capability of a medium for passing a fluid under pressure.

Permeate It is the portion of the feed water stream that passes across the membrane and is generally known as "product."

Permeate collection tube This is also termed as product collection tube, and this tube will collect the product water and directs it to a permeate water header. This tube will be in the middle of an SWM element with the "membrane-product channel spacer—membrane-feed water channel spacer" sandwich wrapped around it.

Permeate staging It is a type of configuration of RO membrane system where the permeate from each stage of a multistage system will become the feed for the consequent stage.

Physical and chemical treatment These are the processes commonly used in wastewater treatment facilities. In physical treatment, different physical methods are employed for wastewater cleaning. Processes such as sedimentation, screening, flotation, and skimming are used for removing the solids. No chemicals are involved in this physical treatment process. Chemical treatment processes include disinfection, adsorption, neutralization, ion exchange, and chemical precipitation (coagulation, flocculation).

Pilot test This is the testing of a cleanup technology under real site conditions in a lab for identifying possible difficulties prior to implementation.

Plugging It is the physical blocking of the feed water side flow passages of membranes or membrane modules.

Polymers Polymers are high molecular weight compounds derived either by the addition of several small molecules (such as polyethylene) or by the condensation of several small molecules with the removal of alcohol and water (such as nylon). The main polymers that are being utilized for preparing RO membranes are polyamides, cellulose diacetate, cellulose triacetate, cellulose acetate, etc.

Polyamide (PA) membranes Presented at the beginning of the 1970s, the asymmetric polyamide polymer has been used in the fabrication of the TFC spiral wound RO membrane. PA-based membranes are the most distinctive membrane fabrication material because of the more flexible operating conditions and low-pressure requirements.

Pore size It is the size of openings in a porous membrane, usually expressed either as an absolute (maximum) or as a nominal (average) value. Membranes applied for pressure-driven filtration processes can be differentiated based on their pore diameter.

Porosity Porosity of membrane is the void volume fraction of the membrane and as defined as the volume of the pores divided by the total membrane volume.

Pretreatment These are processes used for reducing or eliminating wastewater contaminants before going to the RO unit. Pretreatment is a requirement for any NF or RO system. Pretreatment typically involves the use of sand filters and fine prefilters. Further, chemical treatment is required in the event that biological fouling or corrosion scaling of the RO membranes is expected.

Post-treatment In a reverse osmosis system, the post-treatment term normally includes the water treatment processes which happen downstream of the RO plant system. The post-treatment frequently involves disinfection utilizing appropriate biocides, probably the inclusion of appropriate corrosion inhibitors, and adjustment of pH.

Pressure drop High differential pressure (DP), also termed as ΔP or delta P or pressure drop from feed to reject, is the difference between the feed and concentrate pressure at the time of water flow through one or more RO membrane elements.

Pressure vessels These are tubular devices that comprise the membrane elements. For HF systems, the pressure vessel is normally termed as the "permeator." In the case of SWM elements, the pressure vessels are generally termed as the pressure tubes/housings and could include up to eight membrane elements (generally it is six membrane elements).

Product channel spacer It is commonly employed for preventing the membrane from turning off on itself with the increased operating pressure. During the membrane element fabrication, these types of spacers are located between dual layers of the flat-sheet membranes. It is also known as permeate water carrier or permeate channel spacer. Product water flows in a spiral path through the permeate channel spacers into the permeate collection tubes.

Product staging It is also known as permeate staging. This is a configuration in which the permeate stream of the first group of RO pressure vessels turns out to be the feed water stream for the second group. This product staging is employed to increase the product water quality.

Product stream/permeate stream   This is the share of the feed water stream that passes across the membrane. In a reverse osmosis process, the permeate will have almost 95%–99% of the dissolved salts removed from it.

PSI (pounds per square inch)   It is a unit of pressure. In SI units, 1 psi is approximately equal to 6895 N/m^2. Different units are used to express pressure, and some of these are derived from a unit of force divided by a unit of area.

Recovery   Recovery is the percent of the feed water, which is changed into a product stream (occasionally termed as conversion). By determining the percentage recovery, one will be able to rapidly analyze whether the system is functioning outside of the proposed design.

Reject staging   This is a configuration in which the reject from one group of RO pressure vessels turns out to be the feed stream of a second pressure vessel group. This type of configuration is employed to improve water recovery.

Rejection rate   As hydraulic pressure is applied to water that is in contact with RO membranes, the water permeates across the membranes and the dissolved solids present in the water will get removed. Rejection rate is defined as the extent to which the dissolved solids are removed. The rate of rejection reduces as the feed water flows across the RO unit due to the fact that the dissolved solids in it are turning out to be increasingly concentrated.

Reverse osmosis   Reverse osmosis desalination is a physical process that works on the movement of water from a region of high solute concentration to a low solute concentration. It is the reverse of the osmosis process. This movement takes place through a special type of barrier called semipermeable membrane, which allows the flow of water and restricts the salts.

Reverse osmosis membranes   A reverse osmosis membrane is a semipermeable material that will allow the passage of water through it, whereas other substances are not allowed to pass or will pass moderately slowly. These membranes will provide an interface or barrier layer for cross-flow filtration.

Reverse osmosis membrane compaction   It is the physical membrane compression and will lead to a reduction in water flux. The compaction rate is directly proportional to the increase in pressure and temperature. This process takes place naturally over time demanding increased feed pressure.

Reverse osmosis membrane flushing   Reverse osmosis system will be normally equipped with a permanently piped membrane flushing system for automatically flushing the vessels in the RO trains on shutdown for removing residual concentrate and stop RO membranes from fouling and degradation.

Reverse osmosis skids   Reverse osmosis membrane elements are fixed in pressure vessels (PVs) which typically house six to eight elements per vessel. Multiple PVs are organized on support structures (termed as racks or skids). These skids are normally made of plastic, plastic-coated steel, or powder-coated structural steel.

Reverse osmosis trains   The combination of the RO feed pump, concentrate, feed and permeate piping, pressure vessels, couplings, valves, and other fittings (instrumentation and controls, and energy-recovery system) connected on a separate support structure (rack/skid), which could operate independently, is termed as RO train.

Salinity   It is the amount of dissolved salts that are present in water.

Salt passage   This is the amount of salt that is passed across the membrane into the product stream. The salt passage is expressed in percentage, and is considered to be a function of the concentration gradient (brine salt concentration versus the permeate salt concentration), velocity, and temperature.

Salt rejection   The term rejection is used for describing the percentage of an influent species a membrane will retain. Salt rejection is the amount of salt separated from the feed water, and is expressed in percentage. As an example, 97% salt rejection signifies that the membrane retains 97% of the influent salt. This also indicates that 3% of the influent salt will be passing through the membrane into the permeate, which is referred to as the salt passage.

Scale   This is precipitate that will be formed on the surface that is in contact with water, as the result of a chemical or physical modification. Scale formation of soluble salts is one of the main factors limiting the efficiency of RO membranes for desalination.

Scale inhibitor   It is also known as antiscalant. It is a chemical added to water for inhibiting the crystallization or precipitation of salt compounds.

Semipermeable   It is a material that will allow specific size material to pass through whereas rejecting other size material.

Silt density index (SDI)   It is a test used for characterizing the fouling probability of a feed stream. SDI is carried out on the basis of determining the rate of plugging a 45-μm filter employing a constant 30 psi feed pressure for a definite time period. SDI_{15} is the SDI test which will run for 15 min.

Solution   A solution is referred to as a condition where one or more substances are evenly and uniformly mixed or dissolved. Alternatively stated, a solution is considered to be a homogenous mixture of two or more substances. The solutions can be liquids, solids, or gases, such as seawater, potable water, or air. In this book, we are concentrating mainly on liquid solutions.

Spiral wound element   It is a type of membrane configuration that consists of "flat-sheet membrane-permeate channel spacer–flat-sheet membrane-feed channel spacer" combinations rolled up around a permeate collection tube.

Telescoping   It is the longitudinal unraveling of SWM elements, which leads to the RO membrane leaves ranging beyond the spacing material separating the leaves. The majority of manufacturers set up anti-telescoping devices on their membrane elements. Telescoping could be due to the hydraulic surges, excessive differential pressures, or by temperature extremes.

Thin film composite (TFC)   TFC is a reverse osmosis membrane consisting of and fabricated as three layers joined each other. There will be two base layers (porous structure support) of the asymmetric construction and a thin third skin layer (salt-rejecting layer) of PA deposited on the surface.

Total dissolved solids (TDS)   TDS corresponds to the entire concentration of dissolved substances existing in water. TDS comprises inorganic salts, along with a certain quantity of organic matter. The commonly found inorganic salts in water are sodium, potassium, magnesium, and calcium, which are all cations, and sulfates, chlorides, bicarbonates, nitrates, and carbonates, which are all anions.

Total organic carbon (TOC)   TOC is the carbon mass existing in a sample of water, exclusive of the carbon existing as carbonates or/and carbon dioxide. The value is obtained by catalytically oxidizing the entire dissolved carbon to carbon dioxide. The resultant carbon dioxide might be determined by infrared (IR) absorption, or it might be reduced in a furnace using H_2 to develop CH_4, which could be found using flame ionization detectors.

Total suspended solids (TSS)   TSS are the solids present in water which could be trapped using a filter. For measuring the total suspended solids, the water sample can be filtered by means of a preweighed filter. The residue deposited on the filter is subsequently oven-dried at a temperature of $103°C–105°C$ until the filter weight no longer changes. The increase in the filter weight will give the total suspended solids.

Transmembrane pressure (TMP)   The pressure that is required for pushing water through a membrane is termed as TMP. This TMP is the difference between the average feed pressure and the permeate pressure, or the pressure gradient of the membrane.

Turbidity   Turbidity is determined by analyzing the intensity of light scattered by the suspended particles in samples of water. It is generally expressed in terms of nephelometric turbidity units (NTU).

Ultrafiltration (UF)   UF is a membrane-based separation process used for removing colloidal, extremely fine particles and macromolecules from a water stream. The size of pores in a UF system varies from 0.001 to 0.1 μm. Also, the UF membrane systems are characterized by the molecular weight cutoff points.

Ultrapure water   This is the term used for characterizing electronic grade process water. Fundamentally, ultrapure water will not have colloids, particles, inorganic, and organic contaminants in it.

Zero liquid discharge (ZLD)   ZLD is a treatment process in which the plant releases no liquid effluent into surface waters, thereby totally eradicating the environmental pollution related to the water treatment.

Zeta potential   It is a significant and reliable indicator of the membrane surface charge, and information about it is very important for designing and operating the processes.

Index